Computational Automation for Water Security

Computational Automation for Water Security

Enhancing Water Quality Management

Edited by

Ashutosh Kumar Dubey

Department of Computer Science and Engineering, Chitkara University School of Engineering and Technology, Chitkara University, Solan, Himachal Pradesh, India

Arun Lal Srivastav

Department of Applied Sciences, Chitkara University School of Engineering and Technology, Chitkara University, Solan, Himachal Pradesh, India

Abhishek Kumar

Department of Computer Science and Engineering, UIE, Chandigarh University, Punjab, India

Fausto Pedro Garcia Marquez

University of Castilla-La Mancha, Ciudad Real, Spain

Dimitrios A. Giannakoudakis

Department of Chemistry, Maria Curie-Sklodowska University (UMCS), Lublin, Poland

ELSEVIER

Elsevier
Radarweg 29, PO Box 211, 1000 AE Amsterdam, Netherlands
125 London Wall, London EC2Y 5AS, United Kingdom
50 Hampshire Street, 5th Floor, Cambridge, MA 02139, United States

ISBN: 978-0-443-33321-7

For Information on all Elsevier publications
visit our website at https://www.elsevier.com/books-and-journals

Publisher: Peter Llewellyn
Acquisitions Editor: Maria Elekidou
Editorial Project Manager: Ellie Barnett
Production Project Manager: Kumar Anbazhagan
Cover Designer: Mark Rogers

Typeset by MPS Limited, Chennai, India

Contents

List of contributors

Mohammad Afzal Department of Management, Rajshree Institute of Management & Technology, Bareilly, Uttar Pradesh, India

Faheem Ahamad Department of Environmental Science, Keral Verma Subharti College of Science (KVSCOS), Swami Vivekanand Subharti University, Meerut, Uttar Pradesh, India

Syed Aqeel Ahmad Department of Civil Engineering, Integral University, Lucknow, Uttar Pradesh, India

Shaheen Efrah Ali Department of Management, Invertis University, Bareilly, Uttar Pradesh, India

Teresinha Andrade Nucleus of Applied Research to Sciences (NIAC), Federal Institute of Maranhao, Presidente Dutra, Brazil

Nuha Awang Plant Engineering Technology (PETech), Universiti Kuala Lumpur - Malaysian Institute of Industrial Technology (UniKL MITEC), Jln Persiaran Sinaran Ilmu, Bandar Seri Alam, Masai, Jahore Malaysia

Neera Batra Department of Computer Science & Engineering, MMEC, Maharishi Markandeshwar (Deemed to be) University, Mullana, Ambala, Haryana, India

Faizal Bux Institute for Water and Wastewater Technology, Durban University of Technology, Durban, South Africa

Aakriti Chauhan Department of Agriculture, Vivekananda Global University Jagatpura, Jaipur, Rajasthan, India

Avnish Chauhan Department of Environmental Sciences, Graphic Era Hill University, Dehradun, Uttarakhand, India

Prabhat K. Chauhan Department of Environmental Science, VBS Purvanchal University, Jaunpur, Uttar Pradesh, India

Rakhi Chauhan Geeta University, Panipat, Haryana, India

Kritika Chhabra Deloitte, Toronto, ON, Canada

Shikha Dhakad School of Studies Tourism and Travel Management, Jiwaji University, Gwalior, Madhya Pradesh, India

Sartajvir Singh Dhillon Computer Science Engineering, Chandigarh University, Mohali, Punjab, India

Nisha Gaur Gautam Buddha University, Greater Noida, Uttar Pradesh, India

Arpita Ghosh Symbiosis Centre for Management Studies, Symbiosis International (Deemed University), Pune, Maharastra, India

Sonali Goyal Department of Computer Science & Engineering, MMEC, Maharishi Markandeshwar (Deemed to be) University, Mullana, Ambala, Haryana, India

Inam Ul Haq Faculty of Engineering and Technology, SGT University, Gurugram, Haryana, India

Shama E. Haque North South University, Bashundhara, Dhaka, Bangladesh

Che Hafizan Che Hassan Plant Engineering Technology (PETech), Universiti Kuala Lumpur - Malaysian Institute of Industrial Technology (UniKL MITEC), Jln Persiaran Sinaran Ilmu, Bandar Seri Alam, Masai, Jahore Malaysia

Vijaya Ilango Department of General Sciences/Chemistry, Birla Institute of Technology & Science, Pilani, Dubai Campus, Dubai, United Arab Emirates

Mohd Anuar Ismail Plant Engineering Technology (PETech), Universiti Kuala Lumpur - Malaysian Institute of Industrial Technology (UniKL MITEC), Jln Persiaran Sinaran Ilmu, Bandar Seri Alam, Masai, Jahore Malaysia

Jazbi Izhar Applied A.I. Solutions Development, George Brown College's School of Computer Technology, Toronto, ON, Canada

Tabish Izhar Department of Civil Engineering, Integral University, Lucknow, Uttar Pradesh, India

Ahmad Firdaus Abdul Jalil Indah Water Konsortium Sdn. Bhd, Kuala Lumpur, Malaysia

Mohd Al-Fatihhi Mohd Szali Januddi Plant Engineering Technology (PETech), Universiti Kuala Lumpur - Malaysian Institute of Industrial Technology (UniKL MITEC), Jln Persiaran Sinaran Ilmu, Bandar Seri Alam, Masai, Jahore Malaysia

Azyyati Johari Plant Engineering Technology (PETech), Universiti Kuala Lumpur - Malaysian Institute of Industrial Technology (UniKL MITEC), Jln Persiaran Sinaran Ilmu, Bandar Seri Alam, Masai, Jahore Malaysia

Amandeep Kaur Department of Computer Science & Engineering, MMEC, Maharishi Markandeshwar (Deemed to be) University, Mullana, Ambala, Haryana, India

Gagandeep Kaur CGC College of Engineering, Landran, Mohali, Punjab, India

Akib Mohi ud Din Khanday Department of CS and Software Engineering, United Arab Emirates University, Al Ain, United Arab Emirates

Shubhra Khare Department of Applied Sciences and Humanities, Invertis University, Bareilly, Uttar Pradesh, India

Advait Vihan Kommula CGC College of Engineering, Landran, Mohali, Punjab, India

Arvind Kumar Institute for Water and Wastewater Technology, Durban University of Technology, Durban, South Africa

Shubham Kumar University of Minnesota, Minneapolis, MN, United States

Sheena Kumari Institute for Water and Wastewater Technology, Durban University of Technology, Durban, South Africa

Rwitabrata Mallick Department of Environmental Science, Amity School of Life Science, Amity University Madhya Pradesh, Gwalior, Madhya Pradesh, India

Mamta Department of CSE, Chandigarh University, Mohali, Punjab, India

Kartavya Mathur Gautam Buddha University, Greater Noida, Uttar Pradesh, India

Anshu Mehta Computer Science Engineering, Chandigarh University, Mohali, Punjab, India

Purnima Mehta Dr. Yashwant Singh Parmar University of Horticulture and Forestry, Solan, Himachal Pradesh, India

Ankita Mukherjee Indian Institute of Engineering Science and Technology, Shibpur, West Bengal, India

Neha Mumtaz Department of Civil Engineering, Integral University, Lucknow, Uttar Pradesh, India

Nur Naqiuddin Mdd Nordin Langsat-OSC Sdn. Bhd, PLO 209, Jalan Rasau, Kompleks, Kawasan Perindustrian Tanjung Langsat, Pasir Gudang, Jahore, Malaysia

Manish Pant Department of Chemistry- SoPS, DIT University, Dehradun Uttarakhand, India

Swapnil M. Parikh Department of Computer Science & Engineering, Parul Institute of Technology, Parul University, Vadodara, Gujarat, India

Srinibas Pattanaik Computer Science Engineering, Chandigarh University, Mohali, Punjab, India

Sandeep Poddar Lincoln University College, Wisma Lincoln, Petaling Jaya, Selangor Darul Ehsan, Malaysia

Aliff Radzuan Mohamad Radzi Universiti Kuala Lumpur Branch Campus – Malaysian Institute of Chemical and Bioengineering Technology (UniKL MICET), Lot 1988 Vendor City, Taboh Naning, Alor Gajah, Melaka, Malaysia

Syed Zoofa Rufai Faculty of Engineering and Technology, SGT University, Gurugram, Haryana, India

Sairul Izwan Safie Plant Engineering Technology (PETech), Universiti Kuala Lumpur - Malaysian Institute of Industrial Technology (UniKL MITEC), Jln Persiaran Sinaran Ilmu, Bandar Seri Alam, Masai, Jahore Malaysia

Vishal Kumar Sandhwar Department of Chemical Engineering, Parul Institute of Technology, Parul University, Vadodara, Gujarat, India

A.K. Sarkar Business & Administration Department, M. J. P. Rohilkhand University, Bareilly, Uttar Pradesh, India

Aparna Satapathy Department of Zoology, MDS University, Ajmer, Rajasthan, India

Diksha Saxena Department of Chemical Engineering, Parul Institute of Technology, Parul University, Vadodara, Gujarat, India

Richa Saxena Department of Biotechnology, Invertis University, Bareilly, Uttar Pradesh, India

Shivendu Saxena Department of Chemical Engineering, Parul Institute of Technology, Parul University, Vadodara, Gujarat, India

Anamika Sen Research Scholar, Department of Commerce, Bareilly College, Bareilly, Uttar Pradesh, India

Tarakeshwar Senapati Sidho-Kanho-Birsha University, Purulia, West Bengal, India

Hilal Ahmad Shah Faculty of Engineering and Technology, SGT University, Gurugram, Haryana, India

Paridhi Sharma Gautam Buddha University, Greater Noida, Uttar Pradesh, India

Richa Sharma Department of Microbiology, Mahatma Gandhi University of Medical Sciences & Technology, Jaipur, India

Abdul Gaffar Sheik Institute for Water and Wastewater Technology, Durban University of Technology, Durban, South Africa

Jabrinder Singh Department of Chemistry- SoPS, DIT University, Dehradun Uttarakhand, India

Nitesh Singh Department of Plant Pathology, Faculty of Agriculture Sciences, SGT University, Gurugram, Haryana, India

Farhan Sadik Snigdho North South University, Bashundhara, Dhaka, Bangladesh

Karthiyayini Sridharan RHS, RoswellNew Mexico, United States

Arun Lal Srivastav Department of Applied Sciences, Chitkara University School of Engineering and Technology, Chitkara University, Solan, Himachal Pradesh, India

Aastha Srivastava American Institute of Chemical Engineering, Den Haag, Zuid Holland, Netherlands

Nishtha Srivastava Department of Applied Sciences and Humanities, Invertis University, Bareilly, Uttar Pradesh, India

Vaishnavi Srivastava School of Life Sciences and Biotechnology, Chhatrapati Shahu Ji Maharaj University, Kanpur, Uttar Pradesh, India

B. Sundaravadivazhagan College of Computing and Information Sciences, University of Technology and Applied Sciences-Al Mussanah, Oman

Nishat Tasneem North South University, Bashundhara, Dhaka, Bangladesh

Pardeep Singh Tiwana CGC College of Engineering, Landran, Mohali, Punjab, India

Alok Tiwari Department of Chemical Engineering, Parul Institute of Technology, Parul University, Vadodara, Gujarat, India

Anirudh Tomar Department of Physics, Graphic Era Hill University, Dehradun, Uttarakhand, India

Sudhir K. Upadhyay Department of Environmental Science, VBS Purvanchal University, Jaunpur, Uttar Pradesh, India

Alessandro Vinciarelli Computer Science Intelligence Society, University of Glasgow, Glasgow, Scotland, United Kingdom

Syahidah Nurani Zulkifli School of Electrical Engineering, Faculty of Engineering, Univeristi Teknologi Malaysia, Skudai, Jahore, Malaysia

About the editors

Dr. Ashutosh Kumar Dubey is an Associate Professor in the Department of Computer Science and Engineering at Chitkara University School of Engineering and Technology, Chitkara University, Himachal Pradesh, India. He is a Postdoctoral Fellow at the Ingenium Research Group Lab, Universidad de Castilla-La Mancha, Ciudad Real, Spain. Dr. Dubey earned his BE and MTech degrees in Computer Science and Engineering from RGPV, Bhopal, Madhya Pradesh. He completed his PhD in Computer Science and Engineering at JK Lakshmipat University in Jaipur, Rajasthan, India. He is a Senior Member of both IEEE and ACM and possesses more than 17 years of teaching experience. Dr. Dubey has authored and edited 20 books and has published over 90 articles in peer-reviewed international journals and conference proceedings. He serves as an Editor, Editorial Board Member, and Reviewer for numerous peer-reviewed journals. His research interests encompass machine learning, renewable energy, health informatics, nature-inspired algorithms, cloud computing, and big data.

Dr. Arun Lal Srivastav is working as an Associate Professor at Chitkara University School of Engineering and Technology, Chitkara University, Himachal Pradesh, India. He received his PhD from the Indian Institute of Technology (BHU), Varanasi, in water treatment. He has done postdoctoral research at National Chung Hsing University, Taiwan. He is currently involved in the teaching of environmental science, environmental engineering, disaster management, and design thinking to the undergraduate engineering students. His research interests include water quality surveillance, climate change, water treatment, river ecosystem, soil health maintenance, engineering education, phytoremediation, and waste management. He has published >100 research papers in various prestigious journals (Elsevier, Springer, IWA, Taylor & Francis, etc.), including some book chapters and conference publications. He has edited many books with Elsevier, Springer, NOVA, IGI Global, and Wiley. Additionally, he is also one of the series editors of Elsevier and Nova Publisher on energy sustainability and e-waste management, respectively. He is also working on four government-sponsored projects (worth ~ 16 million INR) on phytoremediation, adsorption, capacity building, organic farming, leachate treatment, agro-waste management, etc. He received the prestigious "Teachers Associateship for Research Excellence (TARE) Fellowship" followed by a research grant from the Science and Engineering Research Board (SERB), Govt. of India, in 2022. He has also filed 25 patents on multidisciplinary topics, and 12 out of which have been granted by the Government of India. He has also been recognized among the 2% of worldwide scientists of Earth and Environmental Sciences jointly by Stanford University and Elsevier in the years 2023 and 2024.

Dr. Abhishek Kumar is an Assistant Director and Associate Professor in the Computer Science & Engineering Department at Chandigarh University, Punjab, India. He holds PhD from the University of Madras and completed postdoctoral research at Universidad de Castilla-La Mancha, Spain. With over 13 years of teaching experience and 170+ peer-reviewed publications, he has guided 30+ MTech projects and four PhD scholars. His research spans AI, renewable energy, image processing, and data mining. An award-winning researcher, he holds a patent and received the Sir CV Raman National Award (2018). He has authored or coauthored 7 internationally published books and edited 51 books with IET, Elsevier, Wiley, Springer, and De Gruyter. He serves as the Series Editor for four book series: Quantum Computing (De Gruyter), Intelligent Energy Systems (Elsevier), Mathematical Methods in the Digital Age, and AI for Environmental Sustainability (De Gruyter).

Dr. Fausto Pedro Garcia Marquez works at UCLM as a Full Professor (accredited as Full Professor from 2013), Spain; Visiting Professor at Shenyang University of Technology, China; Lecturer at the Postgraduate European Institute; and he has been a Senior Manager in Accenture (2013-2014). He obtained his European PhD with a maximum distinction. He has been distinguished with the prizes: Runner Prize (2023), Nominate Prize (2022), Gran Maestre (2022), Grand Prize (2021), Runner Prize (2020), Advancement Prize (2018), Runner (2015), Advancement (2013), and Silver (2012) by the International Society of Management Science and Engineering Management (ICMSEM), First International Business Ideas Competition 2017 Award (2017), etc. He has published more than 290 papers (200 JCR [98Q1; 52Q2; 39Q3; 9Q4], 65% ISI, 30% JCR, and 92% internationals), some recognized as: "Applied Energy" (Q1, IF 9.746, as "Best Paper 2020"), "Renewable Energy" (Q1, IF 8.001, as "Best Paper 2014"); "ICMSEM" (as "excellent"), "International Journal of Automation and Computing" and "IMechE Part F: J. of Rail and Rapid Transit" (most downloaded), etc. He is the author and editor of 60 books (Elsevier, Springer, Pearson, McGraw Hill, Intech, IGI, Marcombo, AlfaOmega, etc.), 150 international chapters, and 6 patents. He is the Editor of 5 international journals, Committee Member of more than 70 international conferences. He has been the Principal Investigator in 5 European projects, 7 National projects, and more than 150 projects for universities, companies, etc. His main interests are artificial intelligence, maintenance, management, renewable energy, transport, advanced analytics, and data science.

Dr. Dimitrios A. Giannakoudakis graduated as chemist from Aristotle University of Thessaloniki (AUTh) in Greece, where he also obtained two MSc degrees ((1) Physical &

Electrochemistry and (2) New Educational Technologies). He received his PhD from the City University of New York (CUNY), subdiscipline of Nanotechnology and Materials Chemistry, in 2017, with full scholarship from CUNY Graduate Center, Onassis Foundation, and Leventis Foundation. He continued as a Postdoctoral Researcher and Adjunct Lecturer at the City College of New York (CCNY), AUTh, and Institute of Physical Chemistry in Warsaw (IChF). Afterwards, he served at IChF as Adjunct Assistant Professor and Lecturer of Modern Topics in Physical Chemistry. During the COVID-19 pandemic, he moved to AUTh as a Research Associate. Nowadays, he serves as Assistant Professor and as PI of a SONATA grant at Faculty of Chemistry, Maria Curie-Sklodowska University (UMCS), Poland. His research focuses on photo/thermo/sono-catalysis (solely or in combination) for biomass valorization, (micro)plastics upcycling, environmental and energy applications, and on design, functionalization, and characterization of nano-engineered materials and nanoporous textiles. DG is included in the 2% of the top world most influential scientists (single years 2021, 2022, and 2023, Stanford list, Elsevier). He is elected vice-chair at the European Chemistry Society (EuChemS), Division of Physical Chemistry (2023–25). Up to day, he has coauthored >125 articles in leading journals (average IF as first or last author above 11) that are cited more than 5000 times, 1 monograph, 6 edited books, 17 book chapters, 4 invention patents, with his work to be presented for more than 100 times in conferences. He is on the editorial boards of the journals Molecular Catalysis (Elsevier), Carbon Research (Springer), and Sustainable Chemistry (MDPI). His experience in teaching is for more than 15 semesters in the United States, Poland, and Greece for graduate and undergraduate levels (for more details @ www.DaGchem.com).

Preface

Water security is a pressing global challenge, requiring innovative solutions to ensure access to clean, safe, and sustainable water resources. Computational automation, artificial intelligence (AI), and machine learning (ML) have emerged as transformative technologies, revolutionizing water quality management through real-time monitoring, predictive analytics, and intelligent interventions. These advancements offer powerful tools to tackle complex water quality issues, optimize resource utilization, and enhance sustainability.

This book presents a comprehensive exploration of the integration of AI, ML, Internet of Things (IoT), and advanced technologies in addressing water security challenges. The chapters cover a wide range of topics, from foundational concepts and emerging technologies to practical applications, case studies, and future trends.

This book comprises 22 chapters: Chapter 1 provides a comprehensive review of AI- and ML-based approaches for water quality monitoring, prediction, and analysis, offering insights into state-of-the-art methodologies and applications. Chapter 2 discusses advancements in water quality monitoring, highlighting the roles of MI and AI in improving environmental management. Chapter 3 explores the transformative impact of AI and ML on water quality management, focusing on innovations that enhance monitoring and analysis capabilities. Chapter 4 explores the integration of predictive modeling and IoT in water quality monitoring, analysis, and intervention, emphasizing their role in proactive management. Chapter 5 examines the applications of AI and ML technologies in improving water quality management, showcasing their potential to revolutionize the field. Chapter 6 highlights the use of AI and ML in automotive water quality monitoring, analysis, and management, presenting novel applications in automated systems. Chapter 7 explores the integration of nanotechnology with AI and ML, shaping the future of water quality management with cutting-edge innovations. Chapter 8 investigates the synergy between automation and IoT integration in empowering water security, providing a holistic view of connected technologies. Chapter 9 conducts a cost–benefit analysis of automated water quality management, discussing the financial and strategic advantages of adopting such systems. Chapter 10 focuses on the significance of automation in water treatment processes, demonstrating its role in enhancing efficiency and effectiveness. Chapter 11 reviews advances in water quality management technologies, offering a detailed account of progress in sustainability-focused solutions. Chapter 12 discusses future directions in water quality management, emphasizing the integration of advanced technologies and sustainable practices. Chapter 13 examines future trends and emerging technologies in water quality management, highlighting opportunities for innovation. Chapter 14 addresses complex challenges in water quality management, exploring emerging technologies and sustainable strategies for resolution.

Chapter 15 introduces computational automation in water quality management, providing a foundational understanding of its principles and applications. Chapter 16 presents case studies on successful implementations of automation for water quality assessment, showcasing real-world examples of impact. Chapter 17 focuses on enhancing water security through automation, detailing technical advancements and practical applications in water quality management. Chapter 18 discusses India's Jal Jeevan Mission (JJM) and the role of smart technologies in achieving rural water security, demonstrating the potential of digital solutions. Chapter 19 examines the core applications of computational automation for water security, offering a detailed analysis of key innovations. Chapter 20 explores the synergy between computational automation and sustainable tourism, linking water security with broader sustainability goals. Chapter 21 addresses ethical dilemmas in water automation, discussing the moral considerations surrounding technology adoption in this critical sector. Chapter 22 provides an in-depth discussion on AI and ML applications in automating water quality monitoring, analysis, and management, emphasizing their transformative potential.

Overall, this book is designed to explore computational automation and its intersection with AI, ML, and related technologies for advancing water quality management and ensuring water security. Equipped with practical insights, frameworks, applications, and challenges, readers can make informed decisions about adopting intelligent automation in water management. This book serves as a valuable resource for developing efficient frameworks, models, and strategies for implementing these cutting-edge technologies across diverse domains of water security and environmental management.

It is with great pleasure that we acknowledge the contributions and support of numerous individuals who made this book possible. We extend our heartfelt gratitude to all the authors who submitted chapters, whose expertise and thoughtful discussions have greatly enriched the content and ensured the success of this work. We hope readers derive immense value and gain future insights from the contributions made by the authors. This book is intended to inspire further research and innovation in the field of water security and quality management. We are deeply thankful to the team of Elsevier for their meticulous efforts and dedication in ensuring the timely publication of this book. We also express our sincere gratitude to our universities/institutions and colleagues for their encouragement and support throughout this endeavor. Finally, we extend our heartfelt appreciation to our families for their patience, understanding, and unwavering encouragement, which has been our constant source of strength.

Ashutosh Kumar Dubey

Arun Lal Srivastav

Abhishek Kumar

Fausto Pedro Garcia Marquez

Dimitrios A. Giannakoudakis

CHAPTER 1

Artificial intelligence and machine learning based water quality monitoring, prediction, and analysis: a comprehensive review

Amandeep Kaur[1], Sonali Goyal[1], Neera Batra[1] and Kritika Chhabra[2]

[1]Department of Computer Science & Engineering, MMEC, Maharishi Markandeshwar (Deemed to be) University, Mullana, Ambala, Haryana, India, [2]Deloitte, Toronto, ON, Canada

1.1 Introduction

Pure water is becoming a rare resource due to large population, pollution, and weather changes. Monitoring water quality (WQ) is so crucial, particularly for household water. Conventionally employed laboratory testing methods are expensive, time-consuming, and do not provide an immediate response (Jan et al., 2021).

1.1.1 Water pollution

Since water can dissolve more chemicals than any other liquid accessible, it is extremely susceptible to contamination (NRDC, 2021). When any hazardous material, such as chemicals or microbes, contaminates a body of water, such as a river or pond, water pollution results. Actually, it taints the water and lowers its quality (NRDC, 2021). Contaminants such as pesticides, fertilizers, and industrial waste can enter an aquifer and cause groundwater pollution. Typically, an aquifer is composed of broken rock, gravel, or sandstones (Zhu et al., 2022). The relevant aquifer is now dangerous for human use as a result of this act.

Notably, eliminating these pollutants from groundwater can be extremely expensive and challenging, if not impossible. It has been noted that contaminated water bodies may remain unusable for many years (Zhu et al., 2022). Groundwater has the ability to disperse pollution further from the source. It occurs as a result of its ability to infiltrate into lakes, streams, and oceans (Brands, Rajagopal, Eleswarapu, & Li, 2017; Ezugwu & Apeh, 2017; Geetha & Gouthami, 2016).

Computational Automation for Water Security.
DOI: https://doi.org/10.1016/B978-0-443-33321-7.00023-8

Surface Water Pollution: According to a recent USEPA survey, most of the rivers, streams, ponds, and lakes are unfit for swimming, fishing, or even drinking (Ezugwu & Apeh, 2017). It has been noted that one of the main contaminants from well-known sources is nutrient contamination, such as phosphates and nitrates. Although these nutrients are undoubtedly necessary for all animals and plants to thrive healthily, excessive use of them in agriculture might contaminate water supplies.

Furthermore, natural pollution can also taint water. For instance, when it flows through rocks, it becomes tainted with minerals like calcium, iron, manganese, and arsenic. Animal feces, volcanoes, storm silt, floods, and algal blooms are the main causes of natural pollution. Water contaminants, including bacteria, pesticides, fertilizers, viruses, suspended matter, turbidity, organics, synthetic chemicals, and heavy metals, can arise from all forms of pollution together (Kordbacheh & Heidari, 2023). These contaminants have the potential to change ecosystems, increase stress in aquatic life, lower oxygen and light availability, and cause eutrophication and nutrient enrichment (Babuji et al., 2023; Boyd, 2019; Sharma, Rajan, & Nayak, 2024).

1.1.2 Water quality

Human health is endangered by contaminated water (Bhateria & Jain, 2016; Hou, Yan, & Yu, 2023; Hounslow, 2018), which also harms aquatic life and vegetation. It might be carrying infections, viruses, and bacteria. Millions of people die each year from water-borne illnesses, according to the WHO. According to published research, additional pathogenic bacteria could appear and spread through water in the near future. Increased migration, climate change, agricultural intensification, and rapid population expansion could all contribute to it. The potential for these germs to become resistant to disinfectants is another reason why they might arise.

It is not necessary for water that comes from many sources to be pure and germ-free. For instance, freshwater from certain sources—like groundwater—may be suitable for use, and like other water bodies—lakes, ponds, rivers, or rain—it could not be the case. This is why it is essential to regularly check the WQ to quickly stop any harm from spreading to customers. WQ is based on its physical, chemical, and biological characteristics.

The application of AI and ML (Ahmed et al., 2019; Aldhyani et al., 2020; Hameed et al., 2017; Hoang et al., 2022) holds great promise for resolving the difficulties and complexities related to water management. AI has the potential to improve infrastructure operations, encourage sustainable behaviors, and manage WQ. AI can help manage water scarcity by evaluating enormous volumes of data and offering precise forecasts and insights. To predict instances of WQ, ML algorithms can examine past data on water availability, usage trends, and meteorological conditions. By using this information, water resources can be allocated more effectively through pre-planning and the use of water conservation measures. AI can also help with real-time water system monitoring, which helps minimize water loss by allowing early leak or anomaly detection.

1.2 Literature survey

In this work (Uddin et al., 2023c), AI approaches for predicting the water quality index (WQI) and the water quality category (WQC) have been developed. For the WQI prediction, artificial neural network models have been created, specifically the nonlinear autoregressive neural network (NARNET) and the long short-term memory (LSTM) method from deep learning. Additionally, the WQC forecasting has made use of three machine learning (ML) technologies: Naive Bayes, KNN, and SVM. The developed models have been assessed utilizing a few statistical techniques, and the dataset that was applied has seven principal components.

The aim of this research (Gaya et al., 2020) is to assess the index of WQ model's ability in precisely describing the overall quality of water from coastal areas through the application of an innovative classification scheme. Data on the state of the water from Cork Harbour was acquired by the Irish EPA, or Environmental Protection Agency, and used throughout the study. In this work, four ML algorithms—SVM, NB, RF, KNN, and XGBoost—were applied to identify which of the seven popular WQI designs could estimate WQ groups most accurately.

This study (Uddin et al., 2022) evaluates the water quality score (WQI) associated with the Palla station along the Yamuna River within India using ML methods and a multilinear regression analysis (MLR) as the usual linear model. The models were validated using significant river datasets. The reliability and efficiency of the models have been assessed using performance metrics such as the average square error (MSE), roots of mean squared error (RMSE), and determined coefficient. The ultimate outcome showed that AI models outperformed MLR models.

The Cork Harbour coastline WQIs have been found in the current research (Patel et al., 2022) through the use of an updated World Quality Index method that was just recently created. In lieu of continuously focusing on SI as well as weight figures to reduce model ambiguity, the research aims to identify which ML approach or method, if any at all, is the most persistent and consistent for forecasting WQIs at a given monitored point.

The article provides an examination of various ML techniques utilized to classify WQ, namely, Gradient Boost, AdaBoost, Random Forest, Decision Tree (DT), and Support Vector Machine, or SVM, techniques (Uddin et al., 2023a). The index for WQ dataset, which is available on Kaggle, is used to train the model.

The accuracy and dependability of WQI models are a big problem because of the sharp rise in WQI applications (Nova, 2023) around the globe and the absence of particular application guidelines.

To better understand how AI can be applied to water conservation concerns, this study (Sajib et al., 2023) looks at five main areas: WQ monitoring, drought prediction, smart water grids, early warning systems, precision agriculture, and water management and distribution.

To close this gap, our research (Uddin et al., 2023b) presents a novel strategy: the Groundwater Quality Index (GWQI) model, which was evaluated in Bangladesh's Savar subdistrict. The GWQI model incorporates 10 WQ indicators—six of which are heavy metals—that were collected from 38 sampling sites in the study area. To solve this issue, the authors (El Bilali, Taleb, & Brouziyne, 2021) created two brand-new weighted quadratic mean and unweighted-based root mean squared models.

Through electrical conductivity and temperature (*T*), respectively, as inputs, this paper (Madni et al., 2023) looks to calculate the variables of potential salinity, sodium adsorption rate, interchangeable sodium percentage, the pH level, the total solids dissolved, magnesium adsorption ratio, as well residual sodium carbonate.

The present research (Parween et al., 2022) solves the accuracy difficulty via the layered ensemble water vapor AutoML, or Machine model, and solves the missing values utilizing the KNN imputer.

Since rivers provide freshwater to all metropolitan communities, it is necessary but difficult to monitor the water in them. The goal of this work (Hmoud Al-Adhaileh & Alsaade, 2021) was to establish a quantitative method to evaluate the WQ state of an urban subtropical river in India. In April and June of 2021, water samples were taken for this study from five different locations along the main streams of the Mahananda River, which includes both urbanized and nonurbanized parts of Siliguri city.

In this study (Haghiabi, Nasrolahi, & Parsaie, 2018), the adaptive neuro-fuzzy inference system (ANFIS) technique was developed to forecast the WQI. K-nearest neighbors and feed-forward neural networks were used to classify the quality of the water.

1.3 Discussions

According to the prediction findings (Uddin et al., 2023c), the NARNET model outperformed the LSTM by a small margin in the WQI value prediction, while the SVM technique had the highest accuracy of 97% for the WQ prediction. In addition, the NARNET and LSTM models' evaluation accuracy is comparable, despite a slight difference in the model's regression parameter (RNARNET = 96.17% and RLSTM = 94.21%). Findings indicate that the models recommended are capable of accurately predicting WQI and identifying WQ based on higher resilience.

The XGBoost algorithm surpassed the KNN algorithm (Gaya et al., 2020) (100% correct and 0% incorrect) to accurately predict the WQ for seven WQI models, with 99.9% correct and 0.1% erroneous. The XGBoost classifier outperformed the others in correctly classifying WQ, as shown by its precision (0.99), accuracy (1.0), specificity (1.0), sensitivity (0.99), and F1 (0.99) score, which were obtained from the model validation results. Additionally, it was discovered that, in contrast to WQI models, unweighted root mean square and weighted quadratic mean for each class had greater F1 score, sensitivity, specificity, accuracy, and precision in prediction. The outcome of the study revealed that, in terms of accurate grouping, the WQM as well as RMS models would prove useful and reliable when assessing the overall condition of coastal waters.

Based on the data, the best version of both ANN as well as ANFIS (Uddin et al., 2022) exhibited a significant 10% increase in terms of accuracy over MLR during the evaluation phase. Due to a little gain in performance accuracy, the gap in precision between ANN and ANFIS is limited, suggesting that both ANN and ANFIS might be used as reputable models for the WQI estimate.

LR, RF, GNB, KNN, SVM, Extended Tree (ExT), XGB, and DT are the eight popular methods examined in this study (Patel et al., 2022). The dataset was split into a pair of categories: training 70% and testing 30% to formulate the prediction models. The cross-validation 10-fold strategy was used for checking the models. The RMSE values, the MSE, the case of MAE R2, as well as PREI metrics have been used in the study for evaluating the models' performance. The approaches that performed superior to the others were tree-based DT and combined tree-based XGB. The DT, the ExT, as well as GXB examples may be solid, reliable, and greatly decrease the variability of the models in forecasting WQIs, as determined by the results of the model efficiency and PREI.

The *Z*-score is used (Uddin et al., 2023a) to normalize the collected data before an algorithm is trained. The synthetic minority oversampling technique (called SMOTE) is employed to correct the disparities in the dataset. The outcomes of the trials demonstrate that Gradient Boost along with Random Forest provide a maximum precision of 81%. One of the main problems with the ML model is its lack of transparency, which prevents evaluation of the model's output. Explainable AI, which helps us identify which traits are most crucial, is used to solve this problem. In the context of this study, the characteristics' relevance is determined by the application of local interpretable model-agnostic explanations.

Significant uncertainties have been reported (Nova, 2023) to arise from WQI models at different stages, such as (1) selecting WQ indicators, (2) calculating subindices (SI), (3) weighting WQ indicators, and (4) aggregating SI to determine the overall index. This study offers a strong, statistically valid way for evaluating the uncertainties in the WQI model using the Monte Carlo simulation technique, and prediction uncertainties in the WQI models at each sample site were done using the Gaussian process regression approach.

An overview of the system's components and how each one is interdependent is given at a high level in the study (Sajib et al., 2023). The first domain, water distribution and management, focuses on using AI to analyze a variety of data sources to optimize the management of water resources. To ensure an effective water supply and reduce waste, various components are interconnected, including water demand prediction, leak detection, meteorological data analysis, satellite imagery processing, sensor networks, and water distribution systems.

One of the most important sources of drinking water in the world is groundwater. Continuous monitoring and management of groundwater quality are essential to achieving sustainable development goals. Despite its importance, there aren't any particular instruments available right now to evaluate the presence of heavy or trace metal contamination in groundwater. The study (Uddin et al., 2023b) used six well-established ML approaches to improve the accuracy of WQ evaluation. The model's performance was assessed based on variables like reliability, sensitivity, and uncertainty. The inclusion of heavy metals in the framework of the WQI model is a significant breakthrough of this work.

In marine habitats, "good" WQ is essential for both living creatures and inanimate objects. These models (El Bilali et al., 2021) assess the Bay of Bengal's WQ using seven parameters: pH, temperature, salinity, dissolved oxygen, total oxidized nitrogen, and molybdate reactive phosphorus. The WQ was ranked in between the "good" and "fair" categories by the weighted and unweighted models, with no appreciable variation. The models' estimated WQI scores differed greatly; for WQM, they ranged from 68 to 88 with an average score of 75, while for RMS, they ranged from 70 to 76 with an average score of 72.

Authors analyzed 520 examples of data related to 14 qualities of groundwater variables in the Berrechid reservoir in Morocco to build and assess the performance of SVR, RF, ANN, and AdaBoost models (Madni et al., 2023). According to the results, RF and AdaBoost models perform better than SVR and ANN models. The AdaBoost as well as RF models are shown to have been less responsive to input variables and more generalizable compared with the ANN as well as SVR models based on studies of their generality capabilities and input sensitivity. The algorithms developed might benefit farmers and those who make decisions nationwide with their irrigation water strategy when they are useful in evaluating irrigation WQ units of measurement.

Additionally, the suggested system's (Parween et al., 2022) performance is contrasted with seven ML techniques. The KNN imputer and missing value removal are the two settings in which the experiments are carried out. With the use of SHAP (SHapley Additive exPlanations), the contribution of each feature to prediction is described. The results show that the recommended stacking model outperforms the other models in the prediction of WQ, with 97% accuracy, 96% precision, 99% recall, and 98% F1 score (Parween et al., 2022).

The modified National Sanitation Foundation (NSF) WQI model was employed in this study (Hmoud Al-Adhaileh & Alsaade, 2021) to determine WQ status, and the principal components analysis method was employed to identify the key WQ indicators. The NSF-WQI was calculated using all WQ indicators, with the exception of water pH and TH. The majority of WQ indicators exceeded the surface water guideline limits set by the Indian Standards and the Bureau of Indian Standards.

The results in Haghiabi et al., (2018) showed that an ANFIS model was excellent at estimating the WQI values. The FFNN technique, on the other hand, achieved the best reliability (100%) for water. Furthermore, the FFNN model exhibited better resilience in noticing the WQC, even when the ANFIS model correctly predicted WQI. Furthermore, the FFNN model exhibited its greatest accuracy (100%) for WQC, whereas the ANFIS model proved effectiveness across the testing phase as well, with a coefficient of regression for predicting the WQI. Advanced artificial intelligence is utilized in this recommended approach to help manage water management and cleansing.

This study (Shams, Elshewey, & El-kenawy, 2024) analyzes the performance of AI methods, such as SVM, GMDH (group method of data handling), and ANN, to predict the Tireh River's (Iran) WQ. To achieve this, the most dataset-related, well-known components—SO_4, Ca, pH, Na, HCO_3, Cl, Mg, HCO3, etc.—were collected. Results show that SVM has the best performance and accuracy.

The GMDH model has the lowest accuracy among the models. Moreover, an analysis of the models' performance shows that GMDH and SVM models produced more dependable results than ANN.

This work (Jayaraman, Nagarajan, Partheeban, & Krishnamurthy, 2024) uses the grid search approach to fine-tune the parameters of four classification regression models. The KNN, DT, SVR model, and MLP regressor model are used to predict WQ, while the RF, XGBoost, GB, and AdaBoost models are utilized for classification. The accuracy, F1 score, recall, precision, and MCC of the classification models were assessed using five performance indicators.

Four evaluation metrics—MAE, MedAE, MSE, and R2—were calculated to evaluate the effectiveness of the regression models. The results indicated that the GB model using the grid search method has 99.5% prediction accuracy in terms of classification. The findings showed that the MLP regressor model with the grid search strategy produced the best results, equal to 99.8% to predict WQI (Jayaraman et al., 2024).

Comparative analysis of ML-based models for WQ prediction is shown in Table 1.1.

Fig. 1.1 shows the proportion of ML algorithms used for WQ prediction and monitoring in literature.

Table 1.1: Comparative analysis of machine learning-based model for water quality prediction.

Sr. No.	References	Year	Methods	Accuracy (%)	Advantages	Disadvantages
1	Gaya et al. (2020)	2023	SVM. NB, RF, KNN, XGBoost	100	Efficiency in classification, robustness	High memory requirement
2	Nova (2023)	2023	Monte Carlo Simulation	100	Helps in risk assessment and assist decision making	Computationally intensive
3	Uddin et al. (2023b)	2023	SVM, RF, NB, KNN	98	Accurate distinction between classes	Computational complexity
4	Parween et al. (2022)	2023	KNN	97	No training period, easy to implement	Does not work well with high dimensions
5	Jayaraman et al. (2024)	2024	RF, XGBoost, AdaBoost, KNN, DT	99.8	Robust to noise and overtraining	Unstable sand prone to underfitting
6	Zhang et al. (2024)	2024	SVM, Deep Neural Networks, and KNN	95	Memory efficient, more effective in high-dimensional space	Not suitable for large datasets
7	AlZubi (2022)	2024	XGBoost, Linear, support vector regression (SVR), and Ridge	91	Computationally efficient and handle large datasets effectively	Requires significant computational resources
8	Mohseni, Pande, Pal, and Alshehri (2024)	2024	Fusion of KNN and SVM	94	Simple to use	Slow with large datasets
9	Mohseni, Pande, Pal, and Alshehri (2024)	2024	ANN,SVM, RF, XGBOOST	91	Solve the curse of dimensionality	Longer training time

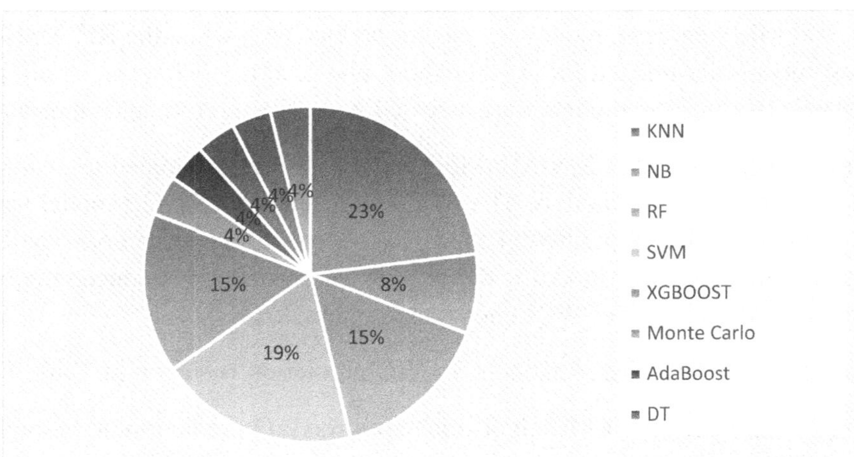

Figure 1.1
Use of machine learning algorithm in the literature.

1.4 Conclusion and future work

As a result, this study shows how well ML models—specifically, DT, Naïve Bayes, XGBoost, Gradient Boosting, Random Forest, and SVM—predict the WQI. These models' high accuracy rates highlight their potential for accurate evaluation of WQ. Future studies include more variables to have a more thorough understanding of the dynamics of WQI. This work not only demonstrates the potential of ML for predicting WQ, but it also emphasizes how to take uncertainties into account for thorough environmental assessments.

References

Ahmed, A. N., et al. (2019). Machine learning methods for better water quality prediction. *Journal of Hydrology*, *578*, 124084.

Aldhyani, T. H. H., et al. (2020). Water quality prediction using artificial intelligence algorithms. *Applied Bionics and Biomechanics*, *2020*.

AlZubi, A. A. (2022). IoT-based automated water pollution treatment using machine learning classifiers. *Environmental Technology*, *45*(12), 2299–2307. Available from https://doi.org/10.1080/09593330.2022.2034978.

Babuji, P., et al. (2023). Human health risks due to exposure to water pollution: a review. *Water*, *15*(14), 2532.

Bhateria, R., & Jain, D. (2016). Water quality assessment of lake water: a review. *Sustainable Water Resources Management*, *2*, 161–173.

Boyd, C. E. (2019). *Water quality: an introduction*. Springer Nature.

Brands, E., Rajagopal, R., Eleswarapu, U., & Li, P. (2017). *Groundwater. Internationalencyclopedia of geography: people, the earth, environment and technology*. (pp. 1–17). Oxford, UK: John Wiley & Sons, Ltd.

El Bilali, A., Taleb, A., & Brouziyne, Y. (2021). Groundwater quality forecasting using machine learning algorithms for irrigation purposes. *Agricultural Water Management*, *245*, 106625.

Ezugwu, C. N., & Apeh, S. (2017). Ground water and surface water as one resource: Connectivity and interaction. *IOSR Journal of Mechanical and Civil Engineering*, *14*, 54–59, [CrossRef].

Gaya, M. S., et al. (2020). Estimation of water quality index using artificial intelligence approaches and multi-linear regression. *IAES International Journal of Artificial Intelligence*, *8938*, *ISSN* 2252.

Geetha, S., & Gouthami, S. (2016). Internet of things enabled real time water quality monitoring system. *Smart Water*, *2*, 1–19.

Haghiabi, A. H., Nasrolahi, A. H., & Parsaie, A. (2018). Water quality prediction using machine learning methods. *Water Quality Research Journal*, *53*(1), 3–13.

Hameed, M., et al. (2017). Application of artificial intelligence (AI) techniques in water quality index prediction: a case study in tropical region, Malaysia. *Neural Computing and Applications*, *28*, 893–905.

Hmoud Al-Adhaileh, M., & Alsaade, F. W. (2021). Modelling and prediction of water quality by using artificial intelligence. *Sustainability*, *13*(8), 4259.

Hoang, T.-D., et al. (2022). Artificial intelligence in pollution control and management: status and future prospects. Artificial Intelligence and Environmental Sustainability: Challenges and Solutions in the Era of Industry 4.0 (pp. 23-43).

Hou, G., Yan, H., & Yu, Z. (2023). Application of AI identification method and technology to boron isotope geochemical process and provenance tracing of water pollution in river basins. *Sustainability*, *15*(7), 5942.

Hounslow, A. (2018). *Water quality data: analysis and interpretation*. CRC Press.

Jan, F., Min-Allah, N., Düştegör, D. (2021). IoT based smart water quality monitoring: Recent techniques, trends and challenges for domestic applications. *Water, 13*(13), 1729.

Jayaraman, P., Nagarajan, K. K., Partheeban, P., & Krishnamurthy, V. (2024). Critical review on water quality analysis using IoT and machine learning models. *International Journal of Information Management Data Insights*, *4*(1), 100210.

Kordbacheh, F., & Heidari, G. (2023). Water pollutants and approaches for their removal. *Materials Chemistry Horizons*, *2*(2), 139–153.

Madni, H. A., et al. (2023). Water-quality prediction based on H_2O AutoML and explainable AI techniques. *Water*, *15*(3), 475.

Mohseni, U., Pande, C. B., Pal, S. C., & Alshehri, F. (2024). Prediction of weighted arithmetic water quality index for urban water quality using ensemble machine learning model. *Chemosphere*, *352*, 141393.

Nova, K. (2023). AI-enabled water management systems: An analysis of system components and interdependencies for water conservation. *Eigenpub Review of Science and Technology*, *7*(1), 105–124.

Parween, S., et al. (2022). Assessment of urban river water quality using modified NSF water quality index model at Siliguri city, West Bengal, India. *Environmental and Sustainability Indicators*, *16*, 100202.

Patel, J., et al. (2022). A machine learning-based water potability prediction model by using synthetic minority oversampling technique and explainable AI. *Computational Intelligence and Neuroscience*, *2022*.

Sajib, A. M., et al. (2023). Developing a novel tool for assessing the groundwater incorporating water quality index and machine learning approach. *Groundwater for Sustainable Development*, *23*, 101049.

Shams, M. Y., Elshewey, A. M., El-kenawy, E. S. M., et al. (2024). Water quality prediction using machine learning models based on grid search method. *Multimedia Tools and Applications*, *83*, 35307–35334. Available from https://doi.org/10.1007/s11042-023-16737-4.

Sharma, K., Rajan, S., & Nayak, S. K. (2024). *Water pollution: Primary sources and associated human health hazards with special emphasis on rural areas. Water resources management for rural development* (pp. 3–14). Elsevier.

Uddin, M. G., et al. (2023a). A novel approach for estimating and predicting uncertainty in water quality index model using machine learning approaches. *Water Research*, *229*, 119422.

Uddin, M. G., et al. (2023b). Marine waters assessment using improved water quality model incorporating machine learning approaches. *Journal of Environmental Management*, *344*, 118368.

Uddin, M. G., et al. (2023c). Performance analysis of the water quality index model for predicting water state using machine learning techniques. *Process Safety and Environmental Protection*, *169*, 808–828.

Uddin, M. G., et al. (2022). Robust machine learning algorithms for predicting coastal water quality index. *Journal of Environmental Management*, *321*, 115923.

NRDC. Water Pollution Facts, Types, Causes and Effects of Water Pollution. (n.d.). https://www.nrdc.org/stories/water-pollution-everything-you-need-know Accessed 19.05.2021.

Zhang, M., Huang, Y., Xie, D., Huang, R., Zeng, G., Liu, X., . . . Lin, Z. (2024). Machine learning constructs color features to accelerate development of long-term continuous water quality monitoring. *Journal of Hazardous Materials*, *461*, 132612.

Zhu, M., et al. (2022). A review of the application of machine learning in water quality evaluation. *Eco-Environment & Health*, *1*(2), 107–116.

CHAPTER 2

Advancements in water quality monitoring: leveraging machine learning and artificial intelligence for environmental management

Gagandeep Kaur, Pardeep Singh Tiwana and Advait Vihan Kommula
CGC College of Engineering, Landran, Mohali, Punjab, India

2.1 Introduction

Water is a necessary resource that all living things share, including human beings, wildlife, and vegetation. It is also a requirement for every species. Every species has different requirements when it comes to the quality of water. Water is a crucial resource in our daily lives; thus it is critical to consistently assess its quality to ensure it remains usable. Lifecycle requires utilization of clean water to survive. Access to clean, safe drinking water is not only a fundamental human right but also a means of advancing sustainable development and well-being in general. Numerous factors, both natural and man-made, such as agriculture, industry, and urbanization, as well as processes like erosion and sedimentation, can impact the quality of water (Essamlali, Nhaila, & El Khaili, 2024; Nallakaruppan et al., 2024). Furthermore, the exacerbating impacts of population growth and climate change exacerbate the problems. One of the numerous effects of water pollution is the development of diseases like cholera, dysentery, and typhoid fever, which are extremely dangerous to human health. Waterborne infections are a major global health hazard (Maroju, Choudhari, Shaikh, Borkar, & Mendhe, 2023). The low WQ has negative consequences on more than only personal health; it also affects and impedes economic growth. Climate change has the potential to affect not only the quantity and quality of water that is accessible but also other aspects of water life. Monitoring water quality (WQM) is a crucial undertaking that requires an interdisciplinary approach. It includes understanding the variables that affect water quality (WQ), coming up with plans to enhance and manage water resources, and implementing laws and regulations to restrict human activity that could negatively affect WQ (Essamlali et al., 2024). In the past, determining the water quality index (WQI) and other aspects of water quality assessment primarily depended on formulas that were established based on a set of parameters and manual calculations.

Computational Automation for Water Security.
DOI: https://doi.org/10.1016/B978-0-443-33321-7.00009-3

Water-based agriculture, natural systems monitoring and management, and critical water- and wastewater-treatment applications have all been shown to benefit from the application of artificial intelligence (AI) techniques and machine learning (ML) models. AI/ML applications are expected to further optimize water-based applications and reduce capital costs, in addition to offering computer-assisted assistance for complex problems involving water chemistry and physical/biological processes. Smart technologies and emerging AI and ML are bridging a gap in water applications that were previously unmet by conventional methods and ways of thinking (Abbas et al., 2024; Lowe, Qin, & Mao, 2022). To stop microbiological contamination of drinking water sources or to lower contamination to levels safe for human health, a variety of barriers are employed from the source to the consumer. To promote successful development outcomes, recommendations for applying AI to water-related problems should be integrated with strategies for infrastructure development and capacity building. With such technology, communities can live healthily as long as they are able to supply the ecosystem with affordable, sustainable, and clean water. Treating and preventing the spread of water-related illnesses requires the prompt and precise detection of waterborne pathogens in drinking and recreational water sources, particularly in resource-constrained environments (Maroju et al., 2023). Water quality assessment is a basic concept of environmental management bearing profound implications for human health and well-being. Artificial intelligence (AI)-based ML algorithms have steered in a paradigm shift by enabling precise, efficient, and real-time analysis of water quality traits (Miller et al., 2024).

2.1.1 Water quality

The definition of "water quality" is frequently given in terms of the water indicators that are chemical, physical, and biological. Water quality (WQ) can be expressed through the water quality classes (WQCs) and WQI, which are two ways that different parameters can influence and act as indicators of WQ. Chemical parameters [DO, COD, BOD, total dissolved solids (TDS), nitrates, pH, etc.], physical parameters (water temperature, turbidity, electrical conductivity, solids, etc.), and biological parameters (chlorophyll-a) are some of the most measured parameters.

2.1.1.1 Water quality index

The state of the WQ is represented by the unitless numerical number known as the WQI. The result is dependent on several water parameters that represent the quality of the water. WQI is determined using the following mathematical equation:

$$\mathrm{WQI} = \frac{\Sigma q_i x w_i}{\Sigma w_i}$$

where w_i is the weight of a given parameter and q_i is the value of that parameter within the 0–100 spectrum (Cojbasic et al., 2023).

2.1.2 Benefits of advanced technologies ML and AI in water quality monitoring

The growth of AI and ML has caused big changes in many areas, including tracking WQ. These techniques provide sophisticated data analysis, identification of patterns, and development of prediction models, empowering monitoring systems to efficiently handle large volumes of data in real time. By combining these two methodologies, it becomes possible to create intelligent monitoring systems that can automatically identify abnormalities, produce notifications, and offer vital information about the current state of WQ. The utilization of hybrid advanced technologies provides a multitude of advantages in the realm of environmental monitoring and management. Here are some benefits through which ML and AI are transforming water quality monitoring (WQM):

- Data fusion and integration: Facts from many assets involving satellite imagery, Internet of Things (IoT) devices, laboratory tests, and citizen science projects can be combined by ML and AI systems. Combining these several datasets helps one to grasp WQ dynamics in a more complete sense.
- Automated sensor networks: ML algorithms can optimize the deployment and management of sensor networks for continuous monitoring of WQ parameters. These algorithms can dynamically adjust sensor placement, sampling frequencies, and data transmission protocols based on environmental conditions and monitoring objectives.
- Real-time monitoring: AI and ML-powered systems offer ongoing, immediate monitoring, allowing for swift response to fluctuations in WQ. This guarantees prompt actions to alleviate dangers and safeguard public health.
- Early warning systems: Through the utilization of historical data and predictive modeling, AI/ML algorithms have the capability to detect developing trends and patterns. This enables authorities to proactively implement preventive measures prior to a severe decline in WQ.
- Prediction modeling: Regression analysis, decision trees (DTs), neural networks, and other ML approaches can be used to create prediction models for contamination transit paths and trends in WQ. These models can forecast the influence of land-living use changes, climate variability, and anthropogenic activities on WQ, assisting in long-term planning and management strategies.
- Cost and time effective: Traditional methods for monitoring water quality require the collection of samples by hand, followed by analysis in a laboratory and interpretation of the results. This process can be both time-consuming and costly. AI and ML-powered systems streamline the process of data collection, research, and reporting, resulting in cost reduction and time efficiency.
- Data-driven decision-making: AI and ML systems offer precise and dependable data, enabling politicians, water resource managers, and researchers to make well-informed decisions on water treatment, resource allocation, and environmental conservation.
- Environment protection: AI and ML-powered systems play a crucial role in safeguarding the environment by constantly monitoring water quality. They achieve this

by identifying sources of pollution, evaluating the effects of human activities, and encouraging the adoption of sustainable behaviors.

- Improved accuracy and precision: Employing sophisticated technologies improves the level of accuracy and precision in environmental monitoring. Through the utilization of advanced algorithms and data analysis techniques, these technologies have the capability to identify intricate patterns, correlations, and abnormalities within environmental data. This enhanced level of accuracy allows for more accurate decision-making and focused responses to successfully tackle environmental concerns.
- Increased efficiency and speed: Various roles in environmental monitoring and management are simplified by advanced technologies such as AI and ML, which result in enhanced efficiency and speed. The amount of time and effort needed to develop insights and results can be reduced using automated data collection, processing, and analysis services. Advanced technologies optimize resource allocation by automating repetitive activities. This frees up valuable human resources to focus on more complicated and strategic tasks, which is possible because of the optimization of resource allocation.

2.1.3 Components of AI and Ml-powered water quality monitoring systems

Advancements of AI and ML techniques play a significant role in enhancing WQM in several ways. Some important components of ML and AI-based WQM systems are as follows:

- Sensor networks: To gather information on different aspects of WQ, AI–ML-powered monitoring systems rely on a network of sensors. To guarantee thorough coverage, these sensors are positioned strategically in water bodies, water treatment facilities, or distribution networks. Modern sensors can measure several factors at once, giving a comprehensive picture of the quality of the water (Aggarwal, Tripathi, Chottani, & Tiwana, 2021).
- Data acquisition and transmission: Information is gathered and sent to a central data management system via the sensor network. Cloud-based platforms and IoT are examples of wireless communication technologies that enable smooth data collection and transfer (Tiwana & Singh, 2023). This permits remote access to data for analysis and decision-making and guarantees real-time monitoring.
- Data processing and analysis: AI and ML algorithms are capable of processing incoming data, utilizing ML approaches, and producing actionable insights through data processing and analysis. To effectively analyze WQ, the algorithms can recognize trends, detect abnormalities, and connect different metrics. Additionally, future patterns and early warning signs of possible problems with WQ can be predicted using ML models.
- Visualization and reporting: Information on WQ is easily understood and interpreted by stakeholders because of the processing of data that is visualized through user-friendly

dashboards, reports, and interfaces. Effective communication and decision-making are made easier by visual representations, which make complex facts easier to understand.

2.2 Literature survey

One of the most valuable resources on the planet is water, which is necessary for maintaining life, sustaining ecosystems, and promoting economic activity. However, the water's quality is increasingly threatened by various factors such as industrialization, pollution, urbanization, and climate change. Ensuring sustainable water resource management, preserving ecosystems, and preserving human health all depend on timely and precise WQM. Utilizing emerging technologies such as AI and ML to enhance environmental management and WQM has gained popularity in recent years. Real-time decision-making, pattern recognition, trend prediction, and sophisticated dataset analysis are all made possible by ML and AI's potent computational tools (Fu, 2020; Kadam, Sunny, Admuthe, Bhosale, & Bhujbal, 2023). This section aims to explore the latest advancements and seeks to provide insights into the cutting-edge approaches, difficulties, and prospects in the use of ML and AI algorithms to monitor WQ.

Rajitha et al. (2024) examined the current applications of AI and ML integration in monitoring or improving WQ, which has emerged as a crucial component of environmental management. This paper offers a thorough investigation of current methods and new developments in the fields of ML and AI algorithms. WQM systems are now far more accurate and efficient because of these advancements. This research investigated the integration of developments into water treatment procedures, particularly evaluating their capability to classify and reduce pollutants more efficiently than traditional techniques, and presents an analysis of various case studies that demonstrate the application of AI-powered devices in evaluating WQ and enhancing treatment efficiency. The authors highlight notable advancements achieved through these technologies.

Nallakaruppan et al. (2024) proposed a system that seeks to automate the estimation of WQ using AI. Explainable artificial intelligence (XAI) is used by this system to provide an explanation of the important factors that affect water potability and impurity estimation. Because machine learning (ML) models are inherently black boxes that are unable to justify or explain their classifications. XAI is a white-box model with transparency and justifiability. To effectively access the drinkability of water, the proposed study makes use of a variety of ML models, like random forest (RF), DT, Gaussian naive Bayes, support vector machine (SVM), and logistic regression. With the highest accuracy and F1-score of 0.9999, along with precision and re-call values of 0.9997 and 0.998, respectively, the RF classifier is selected for investigation. As such, the authors represent a first exploration of the evaluation and management of WQ by indicators associated with their significance.

Essamlali et al. (2024) provided a thorough examination of recent advancements in WQM, with an emphasis on the use of machine learning techniques and wireless technology based on IoT. In this study, authors explored the potential applications of several IoT-based wireless technologies for WQM, including Bluetooth, cellular networks, Wi-Fi, Zigbee, low-power wide area networks, and RFID. It also investigated how to analyze and interpret the gathered data using supervised and unsupervised ML techniques. In addition to discussing recent technological advancements, this survey addressed open research topics and challenges related to applying ML and IoT for WQM.

Cojbasic et al. (2023) conducted a detailed assessment of the numerous ML methods that are employed to track and forecast river WQ. Numerous characteristics, such as dissolved oxygen (DO), biological and chemical oxygen demand (BOD and COD), turbidity levels, the concentration of different ions (e.g., Mg2á and Ca2á), concentrations of heavy metals or other pollutants, pH, temperature, and more, can be observed or predicted. Authors mentioned that many studies have been conducted, but very few of the methods for predicting river WQ are applied extensively in engineering. DTs, RFs, SVMs, artificial neural networks (ANNs), and Deep Learning (DL) are some of the guided learning methods that make up the models. It is possible to use new hybrid algorithms to make estimates more accurate. However, the accuracy of the prediction depends on more than just the algorithm that was used. It also depends on the WQ metrics that were present, how they were chosen, and the input data that were used to teach the ML model.

Abbas et al. (2024) focused on utilizing the WQI to access the suitability of water for various applications. Nevertheless, the process of generating a precise WQI can be time-consuming due to the possibility of inaccuracies in the computations of sub-indices. To address this difficulty, an AI forecasting model was developed with the goal of making the process more efficient and reducing errors. To fully explore the WQI characteristics of the area, 422 data samples were collected from the city of Mirpurkash in the province of Sindh. Additionally, the study sought to look at the connections between various elements in the physiochemical analysis of water. When WQI was predicted using a variety of ML classifiers, the findings revealed that RF and gradient boosting had the greatest accuracy rates, at 95% and 96%, respectively, with SVM coming in second at 92%. The accuracy rate of KNN is 84%, while decision trees obtain 77%. Conventional methods for evaluating WQ are inefficient and prone to mistakes. However, a revolutionary strategy that utilizes AI and ML overcomes these drawbacks. The researchers not only predicted the WQI but also performed an uncertainty analysis of the models using the R-factor. This analysis offered valuable information on the reliability and consistency of the forecasts.

Lowe et al. (2022) demonstrated the water-based applications that have been combined with AI/ML, including river-level monitoring, water quality index monitoring, adsorption, membrane filtering, automation and monitoring for aquaponics and hydroponics, and river-level monitoring. As AI techniques, ML models, and intelligent technologies like sensors, systems, and the IoTs discussed in this review have shown success in control, optimization, and modeling, they also face common and widespread challenges and limitations. It is critical to solve challenges including poor data management, limited explainability, insufficient model repeatability and standardization, and a lack of academic transparency for these intelligent applications to be adopted successfully. Solutions to the difficulties of explainability, data management, repeatability, and models are provided to aid in the effective use of these powerful tools.

Shams et al. (2024) described that predicting the WQI is a challenge, which is a crucial indicator of the integrity of the water. This work employs parameter tuning and optimization to raise the accuracy of multiple ML models. ML-based techniques predict the WQC and WQI. Grid search is an essential method to enhance and fine-tune the parameters of four classification and regression models. The classification models used to predict WQC include RF, extreme gradient boosting (XGboost), gradient boosting, and adaptive boosting (Ada-Boost). WQI is predicted using regression techniques including multi-layer perceptron regressor models, DT, support vector regressor, and K-nearest neighbor (KNN).

Ataei et al. (2024) introduced an advanced decision support system (DSS) designed to address the disconnect between theory and practice in groundwater management. This study is of essential relevance due to the constant demand for improved systems that can analyze large volumes of data to enable sustainable groundwater decision-making. Data from six randomly chosen wells were utilized to create an extensive database of groundwater pumping parameters to address this issue. Flow rate, pressure, and current intensity were these variables. Water pressure and electrical current threshold values were established because of statistical study of these characteristics. Moreover, the RF approach was used to construct a soft sensor that allowed for the real-time prediction of important elements. This was achieved by continuously comparing pump design parameters and the results of routine field testing with real-time telemetry data. The authors have suggested an ML approach that guarantees reliable empirical monitoring of the health of wells and pumps. In addition, the extensive operational expertize of water management specialists, obtained using a classical Delphi technique, was smoothly incorporated.

Jayaraman, Nagarajan, Partheeban, & Krishnamurthy (2024) discussed the several sensors employed in WQM, focusing especially on the WQI, which considers different physical, chemical, and biological factors. The authors focus on the use of IoT-based sensors for remote monitoring of water quality indicators. These sensors use low-power WAN technology to transmit the data they have collected. The pH, turbidity, TDS, and temperature were monitored by the IoT system with a 95% accuracy rate, compared to merely 85% accuracy with the previous technique. This study also examined a number of AI-based methods for evaluating WQ, such as SVMs, deep neural networks, KNNs, and traditional ML methods. ML and DL have the potential to greatly increase the accuracy of groundwater quality assessments when compared to conventional techniques.

Popescu et al. (2024) examined current progress in utilizing AI, sensors, and IoT for monitoring environmental pollution, considering the challenges of forecasting and monitoring pollution fluctuations caused by the ever-changing environment. Although applying ML techniques might revolutionize environmental research, there are challenges involved as well. Finding a balance between the model's performance and interpretability, understanding the ML model's needs, selecting appropriate models, and addressing data-sharing constraints are important considerations. This paper aims to analyze these difficulties and emphasize the most recent developments in utilizing AI and the IoT for monitoring environmental contamination.

Biraghi, Loftian, Carrion, & Brovelli (2021) explored the feasibility of utilizing AI to facilitate water monitoring. The focus of this study is to examine the accuracy and dependability of the data collected through Citizen Science. The researchers focused on creating a prefiltering system for volunteer geographic information related to worldwide lake water monitoring. They utilized the resources and data from the SIMILE project. The objective is to minimize the duration needed for human assessment of the contributions by automatically detecting detrimental occurrences (such as foams and algae) in the photographs that citizen scientists provide. The diverse range of data, which encompasses geotagged photographs captured without specific guidance, poses a challenge in the task. To achieve this objective, many advanced DL approaches, such as convolutional neural network (CNN), quicker region-based CNN algorithm for object recognition, the Clarifai platform, and other state-of-the-art methods, have been evaluated. The original dataset, consisting of observations from the SIMILE—Lake Monitoring program, has been enhanced with data collected through web crawling of Flickr and findings from keyword and picture searches on search engines such as Google and Bing. The algorithms' capability to accurately recognize and categorize the occurrence is showcased, along with potential tactics for further enhancement.

Krishnan et al. (2022) discussed several key domains of application necessary for effective water management. These are current developments in the recycling of wastewater, distribution of water, collection of rainwater, and control of irrigation, all utilizing different AI models. The information gathered for these uses is distinctive and varies according to category. To provide answers for each of these applications, it is necessary to employ a model or algorithm that is universally applicable. AI and DL methods combined with IoT architecture may be used to construct an intelligent water management system that promotes the sustainable use of water from natural sources. This study investigated the integration of AI/DL with the IoT system and looked at several approaches to water management. To develop a framework for more successful water management, statistical analysis and case studies were also incorporated.

Muharemi, Logoftu, & Leon (2019) represented the various methods for detecting changes or irregularities in time series data related to WQ. This work also addressed and presented a resolution to certain obstacles encountered while working with time series data. Many models, such as logistic regression, linear discriminant analysis, SVM, ANN, deep neural networks, recurrent neural networks (RNN), and long short-term memory (LSTM), are used to analyze the WQ data. The F-score measure is used to evaluate the performance. The F-score measure is used in simulation research to access each algorithm's performance. To address imbalanced data, intentional adjustments must be made to the data to get convincing rather than accurate results. The findings show that all algorithms are vulnerable; however, the relative levels of susceptibility for SVM, ANN, and logistic regression are lower than those of DNN, RNN, and LSTM.

Miller et al. (2024) explored and examined the incorporation of ML in environmental monitoring, emphasizing its superiority over conventional statistical techniques in managing extensive and complex data streams. This study involved a thorough review of different ML algorithms designed to accurately forecast WQ. Both supervised and unsupervised learning models were used in these algorithms. Along with discussing these inherent challenges, the paper also projects prospects for this rapidly developing field of ML, including data quality and model interpretability. The potential of ML to transform environmental policy-making and resource management through enhanced prediction skills underscores the transformative impact of ML in environmental research.

Kamyab et al. (2023) described the current uses of Big Data analytics (BDA) and AI in waste resource management, highlighting how these technologies can overcome existing limitations. Examining AI technologies, including DL and ML, and their applications in forecasting water demand, allocating water resources, and WQM are all part of the project. The study also looked at the application of BDA for the management of water resources. It included comprehensive details on the many data sources that can be used, such as social media, IoT devices, and remote sensing.

Shaikh and Birajdar (2024) demonstrated the applications of AI, highlighting its various layers, discussing the obstacles it presents, and exploring its future possibilities. AI-powered advancements aim to transform groundwater management by offering accurate predictive modeling, live monitoring, and seamless data integration. Nevertheless, these advancements encounter obstacles such as difficulties in understanding, the need for specific technical knowledge, and constraints in the quality and quantity of data, which might hinder the effectiveness of AI models. The authors also noted that AI has great potential for improving groundwater management in the future. Advanced AI models have the capability to generate more accurate forecasts of groundwater behavior, detect locations that are susceptible to contamination and depletion, initiate preemptive interventions, and facilitate collaborative platforms including scientists, policymakers, and local communities. AI-driven collaborative platforms have the potential to facilitate synergistic involvement among scientists, policymakers, and local populations, jointly driving the management of groundwater resources. It is crucial to acknowledge and tackle the issues associated with AI while fully embracing its potential to ensure sustainable and resilient groundwater management techniques. The evolution of groundwater resource management will persist by embracing the potential of AI while simultaneously resolving its challenges.

Rane, Choudhary, & Rane (2024) introduced a cutting-edge water quality analysis system that utilizes sophisticated ML algorithms to conduct efficient and precise evaluations of water conditions. Conventional approaches that depend on manual testing and occasional sampling frequently fail to deliver immediate and up-to-date findings. The suggested system utilizes regression and classification models to analyze several water quality indicators, such as pH, dissolved oxygen, turbidity, and chemical concentrations. The system's main characteristics encompassed the creation of regression models to continuously monitor, classification models to comprehensively water conditions, real-time monitoring abilities, effortless integration with sensor networks for frequent data collection, and the production of insights that contribute to a DSS. Through rigorous training on large datasets, the ML system can recognize patterns that are linked to changes in WQ. This allows for preventative measures to control pollution and make educated decisions regarding water resource management. This method represented a notable progress in the evaluation of WQ, offering prompt and precise evaluations that contribute to the overall health of ecosystems and human populations (Table 2.1).

2.3 Discussions

Water makes up around 71% of the Earth's surface, which means that it is necessary for all known life forms to exist. It is imperative that water supply companies throughout the world guarantee the quality of their drinking water since many toxins pose a serious threat. Almost every aspect of life on Earth is impacted by WQ management, and having access to clean water is a necessity. Observing various factors, a survey study discovered that pH, temperature, and humidity are the most observed variables for WQ. These elements are essential for assessing the health of aquatic ecosystems, spotting pollution, and guaranteeing that rules are followed (Fig. 2.1)

It becomes possible to identify changes in WQ in real time by keeping an eye on these metrics. This improves the control of WQ and enables a timely response to likely contamination incidents. Conversely, certain parameters may be evaluated less often due to considerations such as the cost and availability of sensors or their minimal impact on WQ. Monitoring metals and metalloids can incur high costs, and the availability of sensors for these properties may be limited compared to other qualities. In certain agricultural regions, where runoff poses a potential hazard, it is crucial to closely monitor the levels of

Table 2.1: Summary of AI and ML models used in water treatment and water quality monitoring.

Technique	Water treatment and monitoring applications	Advantages	Drawbacks
Random Forest (RF) (Miller et al., 2024)	– Modeling the percentage removal of adsorption processes – Uncomplicated and blended dissolved oxygen modeling	– Intuitive model architecture – Inputs that are categorical or continuous can be processed by the system even in the presence of missing values or data. It is less impacted by noise and outliers and is generally stable. – The bagging algorithm reduces variance and minimizes overfitting.	– The accuracy and robustness of decision trees are determined by their density. – Higher density levels lead to substantial increases in model intricacy, duration of model training, and processing power needed.
Support vector machines/ regressions (SVM/SVR) (Muharemi et al., 2019; Nallakaruppan et al., 2024; Shams et al., 2024)	– Modeling of disinfection by-products (DBPs) – Modeling of membrane processes – Modeling of parameters – River dissolved oxygen modeling, including biological oxygen demand (BOD) and chemical oxygen demand (COD) modeling – Modeling the development rate of aquaponics – Classification of growth stages in aquaponics.	– Proficient in managing datasets with many inputs and a lesser number of outputs. – Proficient in managing tiny changes in datasets. – Capable of processing both linear and nonlinear data.	– The process of selecting a kernel can be initially challenging and time-consuming. – Modeling with SVM/SVR takes significant computer capacity, which makes it less suitable for larger datasets. – Prone to being affected by noise in datasets – Training times are relatively lengthy.
Artificial neural network (ANN)-general (Muharemi et al., 2019)	– Chlorine dosage or set-point – Modeling the generation of DBP – Modeling the parameters of the adsorption process – Modeling of dissolved	– Able to effectively process datasets with many dimensions – Efficiently obtained modeling and prediction findings within a realistic timeframe. – Forward propagation enables	– The backward propagation step is characterized by its high processing power. – Certain models and architectural designs can be challenging to comprehend.

(*Continued*)

Table 2.1: (Continued)

Technique	Water treatment and monitoring applications	Advantages	Drawbacks
	oxygen concentration; control, categorization of hydroponic systems; and modeling of membrane process parameters	efficient and rapid computation at a low cost. – Refer to the following for precise advantages of the ANN model.	
K-nearest neighbor (KNN) (Abbas et al., 2024; Shams et al., 2024)	– Aquaponics growth stage classification	– Simple to integrate without requiring extensive training time – Able to effectively manage the incorporation of new data	– Inadequate efficiency while dealing with extensive datasets and datasets with a high number of dimensions – Prone to being affected by noise, incomplete data, and extreme values.
Recurrent neural network (RNN)/long short-term memory (LSTM) (Rana et al., 2023)	– Modeling of parameters in membrane processes. – Modeling the concentration of dissolved oxygen.	– Suitable for sequential datasets and modeling and time series datasets. – The length of the dataset inputs is not limited.	– Demands significant computational resources – Training is challenging due to the need for extensive and varied datasets.
Convolutional neural network (CNN) (Biraghi et al., 2021)	– DBP formation modeling	– Data are generally considered to be highly precise. – The model operates simultaneously, allowing for fast acquisition of data. – It excels at addressing problems that involve image-based inputs.	– The model and architecture themselves are intricate and complex. – It necessitates significant computational capacity.

AI, Artificial intelligence; *ML*, machine learning.

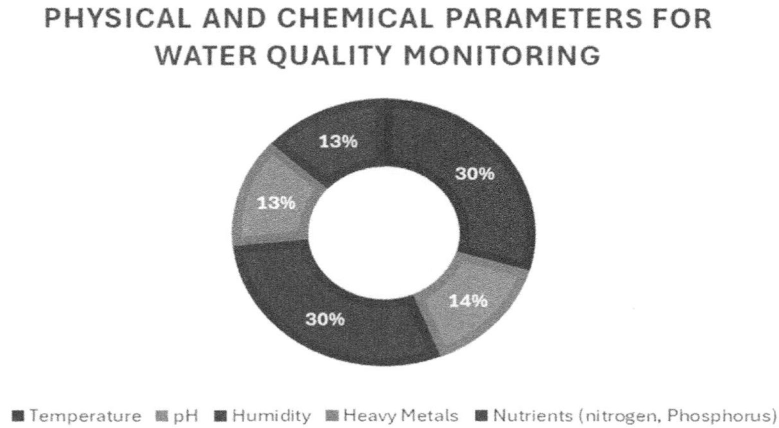

Figure 2.1
Parameters for water quality evaluation.

nitrate and phosphate. However, in other circumstances, this monitoring may not hold the same degree of significance. In the study of managing WQ, emerging pollutants such as microplastics, medications, and personal care products are becoming more and more concerning. These contaminants are usually present in trace concentrations, and little is known about how they affect the environment and human health. This makes it more difficult to access the possible risk and choose effective surveillance techniques. However, conventional monitoring methods have inherent constraints. These methods are only implemented in regions and monitor a restricted range of metrics, which may not offer a thorough comprehension of WQ and potential contaminants. In addition, the frequency of monitoring may fail to detect temporary changes that can have significant impacts on the environment and human well-being. The significant cost associated with monitoring can provide a hindrance, particularly in impoverished regions and emerging countries where the establishment of monitoring systems may be challenging. Poorly calibrated sensors in water quality monitors might result in inadequate or imprecise data, leading to erroneous evaluations of WQ. Poor-quality sensors can cause false alarms or missed detections in addition to being unreliable and requiring more regular calibration or replacement. This may cause a delay in the creation of efficient management plans meant to reduce water pollution and guarantee the security of water supplies. The applications of ML and AI in WQ management have demonstrated significant effectiveness, completely transforming the accuracy, efficiency, and effectiveness of monitoring, analyzing, and predicting WQ. In this section, we will analyze and synthesize the findings from the evaluated research on developments in WQM using ML and AI for environmental management, based on the literature survey.

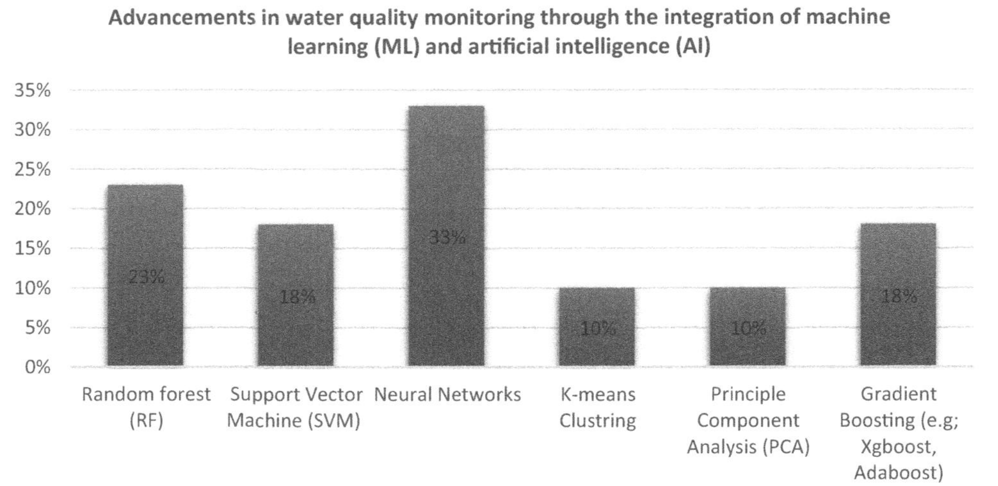

Figure 2.2
Integration of AI/ML algorithms for water quality monitoring. *AI*, Artificial intelligence; *ML*, machine learning.

2.4 Conclusion

Cutting-edge technologies ML and AI-based WQM systems have the potential to greatly transform the methods of monitoring, managing, and preserving water resources. These devices utilize AI and ML algorithms to offer immediate, precise, and practical information regarding WQ factors. Intelligent water systems are designed to swiftly identify and handle any changes in WQ, protecting and preserving the water used for agriculture and maintained for the environment. The prompt detection of pollution problems supports data-driven decision-making and aid in environmental protection initiatives (Fig 2.2).

The combination of ML and AI in WQM shows great potential for future research in environmental management. Although challenges such as data accuracy, security, scalability, and integration exist, ongoing technological advancements and increasing adoption of ML and AI in water management are paving the way for a future where intelligent monitoring systems are vital in ensuring clean and sustainable water sources. This review paper provides a comprehensive analysis of the recent advancements in ML and AI in the context of predicting WQ. To summarize, this study showcases the potential of combining ML and AI for WQM to enhance environmental practices.

References

Abbas, F., Cai, Z., Shoaib, M., Iqbal, J., Ismail, M., Alrefaei, A. F., & Albeshr, M. F. (2024). Machine learning models for water quality prediction: A comprehensive analysis and uncertainty assessment in Mirpurkhas, Sindh, Pakistan. *Water*, *16*(7), 941.

Aggarwal, N., Tripathi, B., Chottani, D., & Tiwana, P. S. (2021). Attacks opportunities in the cyber physical space and the role of cybersecurity. In *2021 second international conference on computational methods in science & technology (ICCMST)* (pp. 276–280). IEEE.

Ataei, P., Takhtravan, A., Gheibi, M., Chahkandi, B., Faramarz, M. G., Wacławek, S., & Behzadian, K. (2024). An intelligent decision support system for groundwater supply management and electromechanical infrastructure controls. *Heliyon*, *10*(3).

Biraghi, C. A., Loftian, M., Carrion, D., & Brovelli, M. A. (2021). AI in support to water quality monitoring. In *Proceedings of XXIV ISPRS congress international society for photogrammetry and remote sensing*, 5–9 July 2021, Digital Edition.

Cojbasic, S., Dmitrasinovic, S., Kostic, M., Turk Sekulic, M., Radonic, J., Dodig, A., & Stojkovic, M. (2023). Application of machine learning in river water quality management: A review. *Water Science & Technology*, *88*(9), 2297–2308.

Essamlali, I., Nhaila, H., & El Khaili, M. (2024). Advances in machine learning and IoT for water quality monitoring: A comprehensive review. *Heliyon*, *10*, e27920. Available from https://doi.org/10.1016/j.heliyon.2024.e27920.

Fu, Z. (2020). Water quality prediction based on machine learning techniques. University of Nevada, Las Vegas. Available from https://doi.org/10.34917/22110053.

Kadam, A. J., Sunny, A., Admuthe, M., Bhosale, A., & Bhujbal, A. (2023). Water quality monitoring model using machine learning. *Computer Integrated Manufacturing Systems*, *29*(6), 113–124.

Jayaraman, P., Nagarajan, K. K., Partheeban, P., & Krishnamurthy, V. (2024). Critical review on water quality analysis using IoT and machine learning models. *International Journal of Information Management Data Insights*, *4*(1), 100210.

Kamyab, H., Khademi, T., Chelliapan, S., SaberiKamarposhti, M., Rezania, S., Yusuf, M., & Ahn, Y. (2023). The latest innovative avenues for the utilization of artificial Intelligence and big data analytics in water resource management. *Results in Engineering*, 101566.

Krishnan, S. R., Nallakaruppan, M. K., Chengoden, R., Koppu, S., Iyapparaja, M., Sadhasivam, J., & Sethuraman, S. (2022). Smart water resource management using artificial intelligence—A review. *Sustainability*, *14*(20), 13384.

Lowe, M., Qin, R., & Mao, X. (2022). A review on machine learning, artificial intelligence, and smart technology in water treatment and monitoring. *Water*, *14*(9), 1384. Available from https://doi.org/10.3390/w14091384.

Maroju, R. G., Choudhari, S. G., Shaikh, M. K., Borkar, S. K., & Mendhe, H. (2023). Application of artificial intelligence in the management of drinking. *Cureus*, *15*(11), Wwater: aA nNarrative rReview.

Miller, T., Łobodzińska, A., Kozlovska, P., Lewita, K., Kaczanowska, O., & Durlik, I. (2024). Advancing water quality prediction: The role of machine learning in environmental science. *Grail of Science*, *36*, 519–525.

Muharemi, F., Logoftu, D., & Leon, F. (2019). Machine learning approaches for anomaly detection of water quality on a real-world data set. *Journal of Information and Telecommunication*, *3*(3), 294–307.

Nallakaruppan, M. K., Gangadevi, E., Shri, M. L., Balusamy, B., Bhattacharya, S., & Selvarajan, S. (2024). Reliable water quality prediction and parametric analysis using explainable AI models. *Scientific Reports*, *14*(1), 7520.

Popescu, S. M., Mansoor, S., Wani, O. A., Kumar, S. S., Sharma, V., Sharma, A., & Chung, Y. S. (2024). Artificial intelligence and IoT driven technologies for environmental pollution monitoring and management. *Frontiers in Environmental Science*, *12*, 1336088.

Rajitha, A., Aravinda, K., Nagpal, A., Kalra, R., Maan, P., Kumar, A., & Abdul-Zahra, D. S. (2024). Machine learning and AI-driven water quality monitoring and treatment. In E3S web of conferences (505, p. 03012). EDP Sciences.

Rana, R., Kalia, A., Boora, A., Alfaisal, F. M., Alharbi, R. S., Berwal, P., & Qamar, O. (2023). Artificial intelligence for surface water quality evaluation, monitoring and assessment. *Water*, *15*(22), 3919.

Rane, N., Choudhary, S., & Rane, J. (2024). Enhancing water and air pollution monitoring and control through ChatGPT and similar generative artificial intelligence implementation. Available at SSRN 4681733.

Shaikh, M., & Birajdar, F. (2024). Artificial intelligence in groundwater management: Innovations, challenges, and prospects. *International Journal of Science and Research Archive*, *11*(1), 502–512.

Shams, M. Y., Elshewey, A. M., El-Kenawy, E. S. M., Ibrahim, A., Talaat, F. M., & Tarek, Z. (2024). Water quality prediction using machine learning models based on grid search method. *Multimedia Tools and Applications*, *83*(12), 35307–35334.

Tiwana, P. S., & Singh, J. (2023). Load optimization accessions, ramification the QoS in software defined networking. *2023 International Conference on Artificial Intelligence and Smart Communication (AISC)*. Available from https://doi.org/10.1109/AISC56616.2023.10085328.

CHAPTER 3

Revolutionizing water quality management the impact of machine learning and artificial intelligence

Richa Sharma[1], Aparna Satapathy[2], Vaishnavi Srivastava[3] and Richa Saxena[4]

[1]Department of Microbiology, Mahatma Gandhi University of Medical Sciences & Technology, Jaipur, India, [2]Department of Zoology, MDS University, Ajmer, Rajasthan, India, [3]School of Life Sciences and Biotechnology, Chhatrapati Shahu Ji Maharaj University, Kanpur, Uttar Pradesh, India, [4]Department of Biotechnology, Invertis University, Bareilly, Uttar Pradesh, India

3.1 Introduction to machine learningand artificial intelligence in water quality monitoring

In recent years, machine learning (ML) and artificial intelligence (AI) have been two such transformative technologies in many varied fields, incuding environmental monitoring and management. More specifically, their application for water quality monitoring is a quantum mandate toward the efficient, accurate, and timely assessment of water resources.

Conventional water quality monitoring relies on periodic sampling and laboratory analysis, which are resource-intensive and time-consuming (Chen, Tan, Peng, Yang, & Li, 2024). In addition, most of the techniques provide spatial and temporal coverage and therefore are not able to understand the changes in water quality dynamics. That is where ML and AI technologies come into the picture.

3.2 Overview of machine learning and artificial intelligence technologies

Machine learning refers to algorithms and statistical models that enable computer systems to progressively improve their performance on a specific task without being explicitly programmed. On the other hand, AI encompasses broader concepts where machines mimic cognitive functions associated with human minds, like learning and problem-solving.

ML and AI techniques in water quality monitoring are applied to a large dataset usually drawn from many sources, including remote sensors, satellites, and on-site probes. The technologies can discover complex patterns within the data, offer anomaly detection,

Computational Automation for Water Security.
DOI: https://doi.org/10.1016/B978-0-443-33321-7.00016-0

and predict future trends (Diro et al., 2024). Common ML methods one could use in this domain include supervised learning for classification tasks, such as identifying pollutants, and unsupervised learning for clustering, such as grouping water bodies of similar characteristics. Reinforcement learning would operate in refinement of sampling over time.

3.3 Revolutionizing water management

The implication of ML and AI adoption in water quality monitoring is very deep because of the following impacts on water management:

3.3.1 Real-time monitoring

Real-time data analysis by the ML models provide instant warnings against water quality deviations. These may signify an early detection of either contaminants or alterations in water conditions that could have an impact on public health and ecosystem safety.

3.3.2 Higher accuracy and efficiency

Large datasets and complex algorithms—processing capabilities enabled by ML and AI—can substantially improve the accuracy of a water quality assessment (Jayaraman, Nagarajan, Partheeban, & Krishnamurthy, 2024). This reduces reliance on manual sampling and laboratory analyses, saving costs with faster decision-making processes.

3.3.3 Predictive capabilities

ML makes it possible to make use of historical data and the environmental drivers in predicting water quality trends. This prediction can help in proactive planning and management of water resources, for instance, algal bloom estimates or climate change assessments.

3.3.4 Data integration and scalability

By using AI/ML methods, it is considerably easier to integrate source data from sensor networks, satellite images, and hydrological models (Sharifi et al., 2024). As a result, the holistic approach would offer scalability, increasing coverage, and a comprehensive picture of water quality across a wide geographic area.

3.3.5 Policy support and decision making

Reliable data generated from the ML and AI models support evidence-based policy-making processes at both central and state levels, in compliance with regulatory provisions. These insights will also be of help to governments and environmental agencies in setting standards, efficiently allocating resources, and monitoring intervention effectiveness.

3.4 Automation and accuracy in monitoring processes

3.4.1 Advantages of machine learning and artificial intelligence in automating water quality analysis

3.4.1.1 Continuous monitoring

Traditional water quality monitoring depends on periodic sampling, which may completely miss transient events of pollution. Continuous monitoring in real-time is made possible with the advent of ML and AI through automated systems working round the clock. Water quality parameters can hence continuously be assessed by AI-enabled sensors, triggering alarm signals at the very start of deviation (Bian, Liu, Zhou, Lukowicz, & Magno, 2024).

3.4.1.2 Scalability

ML and AI, being essentially automation technologies, could cover large areas and at numerous sampling points in an uninterrupted manner. Scalability in this regard is, therefore, incumbent for full-scale watershed management and monitoring of lakes, rivers, and oceans.

3.4.1.3 Less human intervention

Automated processes reduce the occurrence of manual data collection and analysis, hence cutting down on labor costs and human error. This is particularly ideal for remote or hazardous environments where human access may prove to be difficult or dangerous.

3.4.1.4 Speed and efficiency

High-speed data processing and analysis within the automated systems are possible. AI algorithms identify patterns and anomalies with speed, offering timely information for decision-making (Chataut, Nankya, & Akl, 2024). This faster response ability is very critical for preventing or mitigating a pollution event.

3.4.1.5 Resource optimization

This frees up resources for more complex analysis tasks or strategic planning activities, optimizing both personal and financial resource use at the water management agencies and organizations.

3.4.1.6 IoT device integration

This simply means that ML and AI can be incorporated together with IoT devices, allowing multiple devices to connect with one another and forming a smart sensor network. From such a networked approach, it becomes possible to collect and analyze data in a much more comprehensive way, obtaining full overview of water quality based on different locations/depths (Zamani et al., 2024).

3.4.2 Improved accuracy and reliability of data

3.4.2.1 Advanced data analytics

Probably the principal strength of ML algorithms is in processing complex datasets for subtle pattern identification, often eluding traditional techniques of analysis. In this regard, these algorithms process large volumes of data from several sources, which leads to better accuracy in assessments related to water quality.

3.4.2.2 Anomaly detection

AI systems are adept at detecting anomalies in data. These systems can highlight unusual readings that may indicate a pollution event, equipment failure, or other problems. Early anomaly detection by such systems helps prevent the escalation of potential issues.

3.4.2.3 Predictive modeling

Such ML models can predict the future trends of water quality with historical data and other environmental factors. This predictive capability enables advanced management to take timely interventions, thus reducing the risk of deleterious events.

3.4.2.4 Improved Calibration

AI continuously calibrates sensors against environmental variables to ensure that, at all times, the data being collected are accurate. This reduces issues with drift and degradation of sensor accuracy, which in some cases is a problem for long-term monitoring (Koziel, Pietrenko-Dabrowska, Wojcikowski, & Pankiewicz, 2024).

3.4.2.5 Data fusion

AI techniques can fuse data from in-situ sensors, remote sensing satellites, and hydrological models. This type of data integration provides an enriched and more accurate view of water quality through the fusing of independent datasets and thus reduces the limitations suggested by individual sources.

3.4.2.6 Noise reduction

The noise from raw data can be cleaned by ML algorithms to get a clearer signal. This refines the reliability of the data by taking care that irrelevant or incorrect information and inauthentic data are not used for further analysis and decision-making (Bukhari et al., 2024).

3.4.2.7 Adaptive learning

AI systems can learn and self-adjust to changing conditions over time. If new data are continuously collected, models may be updated to portray the latest trends and insights for these systems, ensuring that they remain accurate and relevant in changing environments.

3.5 Future predictive capabilities of water quality

3.5.1 How machine learning and artificial intelligence make predictions of a change in water quality

3.5.1.1 Historical data

The interpretation of the historical data on water quality by ML models constrains patterns and trends. They learn how different factors have impacted their quality over time to make predictions of future variations. Historical data may comprise past events of pollution, seasonal changes, and long-term trends due to climate change or other industrial activities.

3.5.1.2 Environmental and meteorological data integration

AI systems make use of environment and meteorology data sources such as rainfall, temperature, and humidity (Narayana et al., 2024). All these factors bear an impact on water quality considerations: nutrient loading, sedimentation, and microbial growth. Using ML models, these variables can be correlated to arrive at water quality data and predict how future weather may affect such quality.

3.5.1.3 Hydrological and hydrodynamic models

These advanced ML algorithms can be coupled with process-based hydrological and hydrodynamic models simulating water movement and distribution (Liu et al., 2024). Such models allow for the prediction of pollutant spread in a water body, changes in water flow due to precipitation or drought, and dynamics occurring in overall water quality.

3.5.1.4 Remote sensing and satellite data

Satellite imagery and remote sensing data cover huge spatial and temporal extents over water bodies. AI processes these data to detect changes in water color, temperature, and vegetation, which might indicate the shifts in quality. This insight will be further useful in predicting trends such as algal blooms or sediment plumes.

3.5.1.5 Real-time sensor data

On the other side, IoT-enabled sensors constantly report real-time data related to various water quality parameters such as pH, dissolved oxygen, turbidity, and specific contaminants. The ML models run an endless analysis of these data to detect an early sign of change and determine what future conditions will be like given present trends.

3.5.1.6 Anomaly detection and event prediction

AI systems are good at identifying anomalies in water quality data that would otherwise point to the rise of pollution events or other problems (Qiu et al., 2024). Spotting these anomalies in time, predictive models would be able to project the likelihood and potential impact of such events; hence, interventions would take place in optimal timing.

3.5.1.7 Machine-learning techniques

It exploits techniques such as time series analysis, regression models, neural networks, and ensemble methods for predicting future water quality. For example, time series analysis helps in understanding periodic trends, while regression models predict specific outcomes based on multiple input variables.

3.5.2 Implications on proactive management decisions

3.5.2.1 Early warning systems

Predictive capabilities can be developed into early warning systems that will notify the appropriate authorities and stakeholders of the eventuating water quality problem. This allows for time to execute remedial measures against risks to human health and ecosystems with an advantage.

3.5.2.2 Resource allocation

In this respect, accurate predictions will enable the concerned water management agencies to plan their resources more effectively. They can then give more attention to those areas prone to higher risks of pollution or water quality degradation and ensure that monitoring and remediation efforts are focused where needed (Ukhurebor et al., 2024).

3.5.2.3 Preventive measures

Predictive insights allow for preventive measures to be taken before things get out of hand. For example, if a model predicted an algal bloom, the authorities would be able to take measures to reduce nutrient runoff from farmlands, which is typically the cause.

3.5.2.4 Compliance with regulations

Predictive models facilitate compliance with environmental regulations by predicting violations. The proactive approach helps industries and municipalities adjust their operations so that they do not cross the set limits of pollutant discharge.

3.5.2.5 Public health protection

Predicting changes in water quality could aid in proactive management decisions for the protection of drinking water sources, and thereby public health.

This could include timely advisories, switching to alternative sources, or putting in place enhanced treatment processes in a timely manner.

3.5.2.6 Sustainable water management

Long-term views of trends in water quality enable predictive analytics that support sustainable water management practices. This will build up strategies that work in line with human needs and ecological health. These examples include calibration of water use in agriculture or resorting to management of reservoirs to sustain ecological flow (Fu et al., 2021).

3.5.2.7 Climate change adaptation

Finally, predictive models can adapt management strategies in response to climate change impacts on water resources. For example, more frequent or more severe droughts predicted under climate change will better plan the timing of water conservation measures and infrastructure improvements.

3.5.2.8 Community engagement and education

Predictive information about water can be shared with communities; this increases public awareness and engagement even more. An informed community would definitely be supportive and responsive to safety initiatives regarding protection of water resources (Tables 3.1 and 3.2).

3.6 Enhancing efficiency and resource management

3.6.1 Optimizing operations and reducing waste

Inclusion of ML and AI in the water quality monitoring and management systems has dramatically improved operational efficiency and reduced wastage. Such technologies are capable of automatically completing routine monitoring tasks with minimal need for labor and associated high costs. It analyzes huge datasets to identify inefficiencies in water treatment processes and distribution networks so that these can be set right with optimization in operation using AI algorithms (Achite et al., 2024). For example, predictive maintenance models may anticipate equipment failure before it actually happens, curtailing downtime and preventing water loss

Table 3.1: Predictive capabilities for future water quality using machine learning and artificial intelligence.

Application area	Description
Historical data analysis	ML models analyze historical water quality data to identify patterns and trends. This helps predict future changes by understanding past influences such as pollution events, seasonal variations, and long-term trends related to climate change or industrial activities.
Environmental and meteorological data integration	AI integrates environmental and meteorological data like rainfall, temperature, and humidity, which significantly affect water quality. By correlating these variables with water quality data, ML models forecast how future weather conditions might impact water quality.
Hydrological and hydrodynamic models	Advanced ML algorithms coupled with hydrological and hydrodynamic models simulate water movement and distribution, predicting pollutant spread, changes in water flow due to precipitation or drought, and overall impacts on water quality.
Remote sensing and satellite data	Satellite imagery and remote sensing data provide extensive spatial and temporal coverage of water bodies. AI processes these data to detect changes in water color, temperature, and vegetation, indicating shifts in water quality and predicting trends such as algal blooms or sediment plumes.
Real-time sensor data	IoT-enabled sensors provide continuous real-time data on water quality parameters like pH, dissolved oxygen, turbidity, and contaminants (Sarminingsih, Juliani, Budihardjo, Puspita, & Mirhan, 2024). ML models analyze these data in real-time, detecting early signs of changes and predicting future conditions based on current trends.
Anomaly detection and event prediction	AI systems identify anomalies in water quality data that may signal pollution events or other issues. Predictive models forecast the likelihood and impact of such events, allowing for timely interventions.
Machine learning techniques	Techniques such as time series analysis, regression models, neural networks, and ensemble methods are used to predict future water quality. Time series analysis helps understand periodic trends, while regression models predict specific outcomes based on multiple input variables.

associated with an unplanned failure event. The AI systems will also optimize chemical dosing at treatment plants in real-time through dynamic control, avoiding under- or over-application, therefore reducing costs and environmental impact. Real-time data analytics thus helps to identify leakages and unauthorized water usage in the distribution network, which then significantly cuts down on water loss and increases the overall efficiency of operations. To achieve sustainability in the consideration of using water, ML and AI also save resources and extend infrastructure lifetime.

3.6.2 Impact on sustainable water resource management

The applications of ML and AI technologies help significantly in sustainable water resource management. These technologies may be used for full appraisal of water quality data related to identifying the pollution emission source and its respective impacts on aqua ecosystems (Ejaz et al., 2024). These ML and AI systems help in the design of some mitigation strategies through

Table 3.2: Implications for proactive management decisions.

Application area	Description
Early warning systems	Predictive capabilities enable early warning systems that alert authorities about impending water quality issues, allowing for timely corrective measures to protect public health and ecosystems.
Resource allocation	Accurate predictions enable efficient resource allocation, prioritizing areas at higher risk of pollution or degradation, ensuring focused monitoring and remediation efforts.
Preventive measures	Predictive insights allow for preventive measures before issues escalate. For example, if an algal bloom is predicted, authorities can reduce nutrient runoff from agricultural lands, preventing the bloom from occurring.
Regulatory compliance	Predictive models help ensure compliance with environmental regulations by anticipating potential violations, allowing industries and municipalities to adjust operations to avoid exceeding pollutant discharge limits.
Public health protection	Predicting water quality changes that affect drinking water sources enables proactive management to safeguard public health, including timely advisories and enhancing treatment processes.
Sustainable water management	Predictive analytics support sustainable practices by providing long-term views of water quality trends, aiding in developing strategies that balance human needs with ecological health, such as optimizing agricultural water use or managing reservoirs for ecological flow (Alshami, Ali, Elsayed, Eltoukhy, & Zayed, 2024).
Climate change adaptation	Predictive models help adapt management strategies to changing conditions caused by climate change, such as anticipating more frequent or severe droughts and planning water conservation measures accordingly.
Community engagement and education	Providing communities with predictive water quality information enhances public awareness and engagement, encouraging support and participation in water protection initiatives.

regulatory policies by providing the predictions supported by trend analyses. It provides a way to know beforehand the effects that, for instance, agricultural runoff, urbanization, and industrial discharges have on water bodies. As a result, stakeholders can take proper measures to reduce contamination of these water bodies and thus promote the health of an ecosystem. Even AI-powered decision support systems now help in striking a balance in the allocations between competing agricultural, industrial, and domestic uses while maintaining ecological integrity and allowing for equitable distribution. By enhancing our ability to monitor, predict, and manage water quality, ML and AI contribute in great measure to the achievement of several sustainable development goals that are associated with clean water and sanitation, fostering a future where management of the water resources is carried out in a way that sustains all. Not only are these systems scalable to systems of various sizes, but they also adjust well to different settings.

The scalability and adaptability of ML and AI technologies make them appropriate for such extensive realization in water resource management frameworks, from local communities to large-scale regional systems (Adelani, Okafor, Jacks, & Ajala, 2024). These are technologies applicable and adaptive to meet any need that arises in setups as small as a town or in rural places, where inexpensive and reliable water quality monitoring solutions are needed. Low-cost IoT sensors, worked out with AI, can deliver real-time insights into these communities' water systems, enabling timely interventions and resource-efficient management. On a bigger level, ML and AI can handle the complexities of volumes of data generated by urban water networks in largely populated cities and river basins. They facilitate the combination of several data sources, including remote sensing, hydrological models, and in-situ sensors, to produce an overall spatial view of water quality in regions. This flexibility of application will ensure the deployment of ML and AI solutions at any given scale, which supports local authorities, in addition to large utilities, to enhance their practice in the management of water (Sharifi et al., 2024).

3.6.3 Applicability from local to large-scale water management

In this respect, ML and AI technologies are uniquely fitted for handling water management requirements at different scales. These technologies are granular in monitoring and management, which are feasible at very small, local settings such as those of municipal water supply systems or small watershed areas. For example, community water systems can utilize AI to detect early threats from contaminants and provide safe drinking water for residents. For locally applied AI, one can use it in the optimization of the irrigation scheduling at farms, hence saving the water used and improving crop production. In this regard, the regional and global scale use of ML and AI can go a long way in administering shared water resources by regional and international entities. These could include the complexities of modeling hydrological system, predicting the climate change impact, assisting nations in reaching cooperative agreements on water sharing, and much more. In this respect, ML and AI enable customized solutions to be issued for the different scales in water management while still addressing the specific needs and challenges related to the proper and sustainable use of water resources (Lowe, Qin, & Mao, 2022).

3.6.4 Global gains and adaptability of the solutions from machine learning and artificial intelligence

The benefits and adaptability of ML and AI solutions in monitoring water quality are very wide and cut across different geographical and socio-economic contexts with identifiable advantages. These technologies usher in low-cost, scalable solutions for water quality monitoring and management in developing countries, where clean water is already a critical concern. Coupled with low-cost sensors and cloud-based analytics, communities can access real-time insights into their water systems to, in turn, act immediately toward contamination and improvement in water safety (Nag et al., 2024). ML and AI in the developed regions bring added sophistication to the practice of water management by integrating the state-of-the-art analytics into existing infrastructure and ensuring that increasingly stringent environmental regulations are being complied with while optimizing resource use. The adaptability of such technologies further indicates the ability to be localized to any particular setting, whether addressing the challenges of water scarcity in arid regions or managing flood risks in areas with a high frequency of heavy rainfall. It will globalize ML and AI by supporting joint water management through shared data sharing platforms, thereby transitioning to a more sustainable approach to climate change. Eventually, better water security, public health, and environmental sustainability could be achieved with wide ML and AI penetration in water management.

Addition of ML and AI to conventional methods of water quality monitoring is a very strong combination. These two additions produce a sure increase in efficiency in water resource management. Manual sampling and laboratory analysis of traditional methods give high-accuracy data that serve as the critical baseline for the calibration and validation of models under AI and ML (Prajapati et al., 2024). As such, these constructed models could further be deciphered about large sets of data in a better way regarding the pattern detection and prediction of the future described more efficiently than manual means. Established on these approaches, the absolute knowledge of water quality will then comprise traditional techniques for accuracy and reliability, while ML and AI apply for scale and real-time analytics. Furthermore, local knowledge and community engagement are crucial in putting data into perspective in the face of regional environmental conditions and the socio-economic context. That is how the integration of a high-tech solution and traditional knowledge leads to scientifically valid and culturally appropriate water management strategies, resulting in more effective and sustainable outcomes (Gao, Li, & Shi, 2024).

3.6.5 Development of resilient water quality management strategies

Combination of strengths from traditional methods of water quality management and advanced technologies of ML and AI is required to formulate robust strategies. Traditional techniques provide a foundational understanding of water ecosystems through direct observation and empirical analysis (Tang, Riley, Manzoni, & Maggi, 2024). When integrated with AI-driven predictive analytics and ML algorithms, these methods can significantly enhance the depth and breadth of water quality assessments. For example, ML models can digest historical data from more traditional methods to recognize long-term trends and possible future risks. This eventually allows for better, more informed decision-making. AI systems can automate routine monitoring tasks, thus freeing resources for a more in-depth type of analysis or planning strategies. This enables the integration of a multifaceted approach into water management, consisting of real-time monitoring, predictive forecasting, and adaptive response strategies. Water managers can thus develop holistic strategies that, by marrying state-of-the-art technology with traditional knowledge, meet existing challenges and foresee future requirements in a resilient and sustainable manner within water quality management (Ahmed et al., 2024).

3.7 Applications of machine learning and artificial intelligence in automating water quality monitoring, analysis, and management

- Data Collection: Real-time water quality-related data collection is done by IoT devices and sensors.
- Data Preprocessing: Cleaning and transformation of raw data for noise and inconsistency avoidance to carry out the correct analysis.
- Anomaly Detection: Identify abnormal fluctuations in indicators of water quality by using ML algorithms, such as SVM and Isolation Forest.
- Predictive Modeling: Forecast the trends in water quality based on historical data, environmental conditions, and predictive variables.
- Quality Assessment: Automated assessment of compliance in water quality related to regulatory standards with AI and flagged deviations.
- Optimization Algorithms: Optimization of water treatment processes for efficiency and cost-effectiveness through genetic algorithms and neural networks.
- Pattern Recognition: AI has been used to recognize certain patterns in water quality data attributable to particular pollutants or other environmental factors of interest.
- Decision Support Systems: Real-time advisory and decision-making in water management is exercised in view of the current situation and forecast by AI.
- Risk Assessment: With ML models of historical incidents correlated with environmental factors in assessing risks associated with contamination events.
- Resource Allocation: Establishing priority-based resource allocation on water monitoring and management with the help of AI-based systems.
- Remote Sensing Integration: Monitoring the status of the large water areas with the help of satellite imaging as well as integrating remote sensing data with AI algorithms for distant pollution source detection.
- Maintenance Scheduling: It can predict maintenance requirements of the water infrastructure usage pattern, environmental factors, and degradation models through AI algorithms.
- Natural Language Processing: Study of reports, social media, and citizen feedback in a manner of understanding the issues in water quality from the public's perspective.
- Visualization and Reporting: Develop visual dashboards and reports summarizing the trends, anomalies, and status of water quality compliance by stakeholders using AI-powered tools (Fig. 3.1).

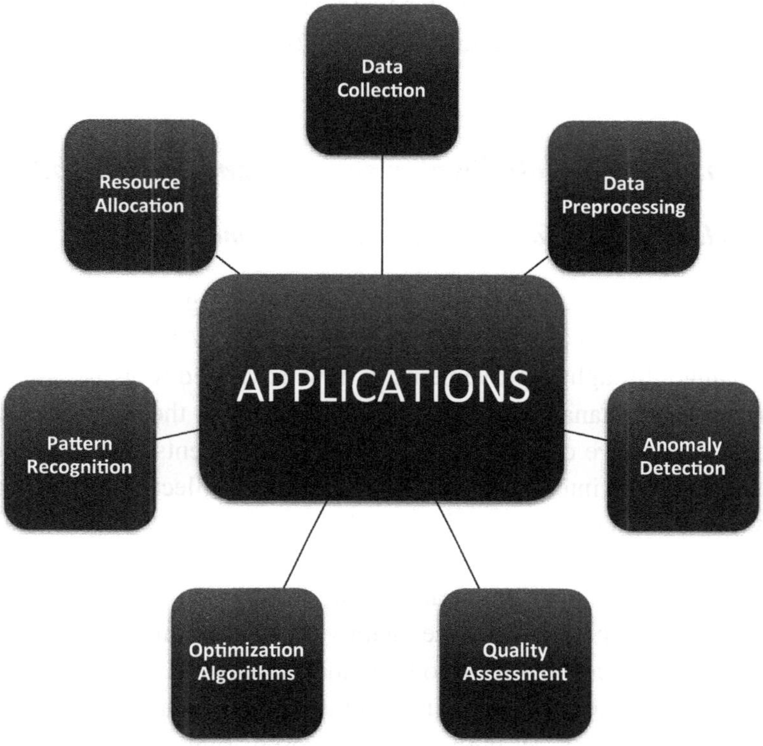

Figure 3.1
Applications of machine learning and artificial intelligence in water quality monitoring.

3.8 Discussion

ML and AI are transforming water quality management by providing advanced tools for monitoring, predicting, and improving water quality. Traditional methods of water quality assessment often involve manual sampling and laboratory analysis, which can be time-consuming and expensive. ML and AI, however, enable real-time data collection and analysis through smart sensors and IoT devices, offering a more efficient and accurate approach. AI algorithms can analyze vast amounts of data from various sources, identifying patterns and anomalies that might indicate pollution or other issues. ML models can predict future water quality trends based on historical data, allowing for proactive management and timely interventions. This predictive capability is crucial for preventing water contamination and ensuring safe drinking water. Moreover, AI-driven systems can optimize water treatment processes by adjusting chemical dosing and filtration settings in real-time, leading to more effective and cost-efficient operations. The integration of AI and ML into water quality management also facilitates better decision-making by providing actionable insights to stakeholders. Overall, the application of AI and ML in water quality management holds significant promise for enhancing the safety, reliability, and sustainability of water resources

3.9 Conclusion: toward global water security and sustainability

3.9.1 Role of collaborative efforts and technological innovation

Global water security and sustainability require both technological innovation and collaborative action across sectors and stakeholders. ML and AI have the potential for transformative change through improvements in new, advanced tools for increased monitoring, analysis, and management of water resources, but their potential can only be effectively explored in active collaboration between governments, industries, academia, and local communities; it is the intersection of technology and collective action that can effectively drive what it takes to bring about global water security and sustainability. Collaborative efforts are only a requirement to ensure that the development and implementation of water management strategies are technically sound and socially inclusive in nature. International cooperation is once again very vital whenever there are transboundary water resources that have to be managed and shared data, as well as common endeavors that defuse conflicts and give an impetus toward sustainable use. Thereafter, public–private cooperation can get the innovative works going by leveraging technical know-how and resources from the two sectors to develop and deploy new technologies. By creating a cooperative environment, the whole bundle of knowledge and capabilities required to address complex water security and sustainability challenges can be unlocked.

3.9.2 Potential for ensuring global water security

ML and AI in water quality management have very high potential for worldwide water security. The technologies provide real-time monitoring, which forms the basis of any predictive capabilities considered critical for proactive water management (Fu, Jin, Sun, Yuan, & Butler, 2022). In reinforcing this statement, ML and AI assure more resilient and adaptive water systems in view of predicting and mitigating a series of risks arising from pollution events, water shortages, and climate change impacts. The scales of these technologies range from local community to large transboundary basin. This adaptability makes these technologies relevant for both developed and developing countries, where conditions are generally expected to differ, hence equating the access to clean and safe water (Zhalmurziyeva, Tokbolat, Durdyev, Mustafa, & Karaca, 2024). This could also help in achieving international water objectives, such as the United Nations' Millennium Development Goals, particularly No. 6, which focus on the availability and sustainable management of water and sanitation for all. With much innovation, collaboration, and commitment to sustainability, ML and AI can play a very important role in ensuring water resources for generations to come, promoting global water security, and improving environmental sustainability.

References

Achite, M., Farzin, S., Elshaboury, N., Valikhan Anaraki, M., Amamra, M., & Toubal, A. K. (2024). Modeling the optimal dosage of coagulants in water treatment plants using various machine learning models. *Environment, Development and Sustainability*, *26*(2), 3395–3421.

Adelani, F. A., Okafor, E. S., Jacks, B. S., & Ajala, O. A. (2024). Theoretical frameworks for the role of ai and machine learning in water cybersecurity: Insights from African and US applications. *Computer Science & IT Research Journal*, *5*(3), 681–692.

Ahmed, N., Zhang, B., Deng, L., Bozdar, B., Li, J., Chachar, S., ... Tu, P. (2024). Advancing horizons in vegetable cultivation: A journey from ageold practices to high-tech greenhouse cultivation—A review. *Frontiers in Plant Science*, *15*, 1357153.

Alshami, A., Ali, E., Elsayed, M., Eltoukhy, A. E., & Zayed, T. (2024). IoT innovations in sustainable water and wastewater management and water quality monitoring: A comprehensive review of advancements, implications, and future directions. *IEEE Access*.

Bian, S., Liu, M., Zhou, B., Lukowicz, P., & Magno, M. (2024). Body-area capacitive or electric field sensing for human activity recognition and human-computer interaction: A comprehensive survey. arXiv preprint arXiv:2401.06000.

Bukhari, S. M. S., Zafar, M. H., Abou Houran, M., Moosavi, S. K. R., Mansoor, M., Muaaz, M., & Sanfilippo, F. (2024). Secure and privacy-preserving intrusion detection in wireless sensor networks: Federated learning with SCNN-Bi-LSTM for enhanced reliability. *Ad Hoc Networks*, *155*, 103407.

Chataut, R., Nankya, M., & Akl, R. (2024). 6G networks and the AI revolution—Exploring technologies, applications, and emerging challenges. *Sensors*, *24*(6), 1888.

Chen, H., Tan, C., Peng, H., Yang, W., & Li, L. (2024). A qualitative study of water quality using Landsat 8 and station water quality-monitoring data to support SDG 6.3. 2 evaluations: A case study of Deqing, China. *Water*, *16*(10), 1319.

Diro, A., Kaisar, S., Vasilakos, A. V., Anwar, A., Nasirian, A., & Olani, G. (2024). Anomaly detection for space information networks: A survey of challenges, techniques, and future directions. *Computers & Security*, *139*, 103705.

Ejaz, U., Khan, S. M., Jehangir, S., Ahmad, Z., Abdullah, A., Iqbal, M., ... Svenning, J. C. (2024). Monitoring the Industrial waste polluted stream-Integrated analytics and machine learning for water quality index assessment. *Journal of Cleaner Production*, *450*, 141877.

Fu, G., Jin, Y., Sun, S., Yuan, Z., & Butler, D. (2022). The role of deep learning in urban water management: A critical review. *Water Research*, *223*, 118973.

Fu, Y., Leng, J., Zhao, J., Na, Y., Zou, Y., Yu, B., ... Wu, W. (2021). Quantitative calculation and optimized applications of ecological flow based on nature-based solutions. *Journal of Hydrology*, *598*, 126216.

Gao, P., Li, Z., & Shi, R. (2024). Impact of natural resource dependence on green technology development: Role of digital governance in mitigating resource-curse using big data. *Resources Policy*, *92*, 105023.

Jayaraman, P., Nagarajan, K. K., Partheeban, P., & Krishnamurthy, V. (2024). Critical review on water quality analysis using IoT and machine learning models. *International Journal of Information Management Data Insights*, *4*(1), 100210.

Koziel, S., Pietrenko-Dabrowska, A., Wojcikowski, M., & Pankiewicz, B. (2024). Statistical data pre-processing and time series incorporation for high-efficacy calibration of low-cost NO_2 sensor using machine learning. *Scientific Reports*, *14*(1), 9152.

Liu, C., Xie, T., Li, W., Hu, C., Jiang, Y., Li, R., & Song, Q. (2024). Research on machine learning hybrid framework by coupling grid-based runoff generation model and runoff process vectorization for flood forecasting. *Journal of Environmental Management*, *364*, 121466.

Lowe, M., Qin, R., & Mao, X. (2022). A review on machine learning, artificial intelligence, and smart technology in water treatment and monitoring. *Water, 14*(9), 1384.

Nag, A., Hassan, M. M., Das, A., Sinha, A., Chand, N., Kar, A., ... Alkhayyat, A. (2024). Exploring the applications and security threats of Internet of Thing in the cloud computing paradigm: A comprehensive study on the cloud of things. *Transactions on Emerging Telecommunications Technologies*, *35*(4)e4897.

Narayana, T. L., Venkatesh, C., Kiran, A., Kumar, A., Khan, S. B., Almusharraf, A., & Quasim, M. T. (2024). Advances in real time smart monitoring of environmental parameters using IoT and sensors. *Heliyon*, *10*(7).

Prajapati, B., Dharaiya, V., Sahu, M., Venkatraman, C., Biswas, P., Yadav, K., ... Jehangir, A. (2024). Development of a physics-based method for calibration of low-cost particulate matter sensors and comparison with machine learning models. *Journal of Aerosol Science*, *175*, 106284.

Qiu, C., Shao, G., Zhang, Z., Zhou, C., Hou, Y., Zhao, E., ... Guan, X. (2024). Unsupervised real time and early anomalies detection method for sewer networks systems. *IEEE Access*.

Sarminingsih, A., Juliani, H., Budihardjo, M. A., Puspita, A. S., & Mirhan, S. A. A. (2024). Water quality monitoring system for temperature, pH, turbidity, DO, BOD, and COD parameters based on Internet of Things in the Garang watershed. *Ecological Engineering & Environmental Technology (EEET)*, *25*(2).

Sharifi, A., Beris, A. T., Javidi, A. S., Nouri, M. S., Lonbar, A. G., & Ahmadi, M. (2024). Application of artificial intelligence in digital twin models for stormwater infrastructure systems in smart cities. *Advanced Engineering Informatics*, *61*, 102485.

Tang, J., Riley, W. J., Manzoni, S., & Maggi, F. (2024). Feasibility of formulating ecosystem biogeochemical models from established physical rules. *Journal of Geophysical Research: Biogeosciences*, *129*(6), e2023JG007674.

Ukhurebor, K. E., Hossain, I., Pal, K., Jokthan, G., Osang, F., Ebrima, F., & Katal, D. (2024). Applications and contemporary issues with adsorption for water monitoring and remediation: A facile review. *Topics in Catalysis*, *67*(1), 140−155.

Zamani, M. G., Nikoo, M. R., Al-Rawas, G., Nazari, R., Rastad, D., & Gandomi, A. H. (2024). Hybrid WT−CNN−GRU-based model for the estimation of reservoir water quality variables considering spatio-temporal features. *Journal of Environmental Management*, *358*, 120756.

Zhalmurziyeva, K., Tokbolat, S., Durdyev, S., Mustafa, M. Y., & Karaca, F. (2024). Assessment of sustainability indicators for urban water infrastructure in a developing country. *International Journal of Building Pathology and Adaptation*, *42*(3), 337−351.

CHAPTER 4

Enhancing water quality management: the role of predictive modeling and IoT in monitoring, analysis, and intervention

Kartavya Mathur[1], Paridhi Sharma[1], Nisha Gaur[1] and Shubham Kumar[2]

[1]Gautam Buddha University, Greater Noida, Uttar Pradesh, India, [2]University of Minnesota, Minneapolis, MN, United States

4.1 Introduction

Water is one of the greatest imperative assets vivacious for the wherewithal of all the lifeforms (Zainab, Amina, Abdulmuhaimin, Sadiku, & Baballe, 2023). The milieu, commerce, floras, and faunas, as well as we people, all profoundly depend on the eminence of the water. During the past few years, effluence and adulteration have disturbed water quality. Water pollution sinks the eminence of the water, which distresses the wellbeing of marine life and, subsequently, that of creatures who exploit it. Because of this, it is critical to oversee the excellence of the water and assure the subsistence of maritime life. It is also vital to twig complications and apprehensions associated with water quality to alleviate and control water effluence. Many administrations throughout the ecosphere have commenced developing green water supervision plans to better apprehend the state of the maritime bionetwork. Many people die every year as a corollary of soiled water, derisory cleanliness, as well as tainted settings; besides, virtually one billion people lack admittance to unsoiled drinking water. Consequently, upholding the quality of freshwater is elemental (Nasir et al., 2022; Shams et al., 2024).

The water excellence varies spatiotemporally, i.e., it is dependent upon location as well as the time at which the water sample is analyzed. For illustration, solar radiation distresses rivulet temperatures throughout the day, riparian vegetation cover, as well as natural climate disparities result in differences in stream temperature, and the headwaters of streams at higher elevations are characteristically colder than wide-ranging streams at lower elevations. Fires, windthrows, and even debris floods are examples of disturbances that can distress the temperature and turbidity, besides other characteristics of water quality in streams. Climate, geomorphology, and geology all affect the quality of water. Effluent from

Computational Automation for Water Security.
DOI: https://doi.org/10.1016/B978-0-443-33321-7.00008-1

an industrial outlet, or an end-of-pipe discharge, that is absorbed straight into a stream is an example of point source pollution.

Broadly, water quality strictures are convened into corporeal, chemical, and biological constraints. The corporeal qualities of water are reliant on the physical state as well as the condition of water. These parameters include temperature, turbidity, vibrant hue, total dissolved solids (TDS), and electrical conductivity. The chemical parameters explore the chemical makeup of the streams and reservoirs, including anthropogenically and naturally occurring contaminants. These parameters include dissolved oxygen (DO), nutrients, heavy metals, and chloramines, besides the pH of the water. Biological parameters quantitate the presence of any biological constituents like harmful algal blooms, microorganisms like fecal microflora and viruses, and planktons, among various others. These parameters and their apparent effect on water quality are thoroughly reviewed in different studies. At times even ascetic parameters like any odor, color, taints, or floating matter may be considered a valuable water quality parameter (Srivastava, Gaur, Ohri, Srivastava, & Singh, 2024; Zainab et al., 2023).

Water quality analysis and comprehension are indispensable to safeguarding the longstanding use and administration of this priceless supply. An extensively used indicator that stipulates a thorough evaluation of water grade centered on numerous features is the water quality index (WQI). The WQI assesses the grade of water for consumption, industrial usage, and the thriving of aquatic creatures. Another widely used indicator for water quality estimation is water quality classification (WQC). WQC catalogs the water bodies into somewhat soiled or highly contaminated (Shams et al., 2024). Estimating both the indices (WQI and WQC) requires regular monitoring of the water resources. However, lately, direct measurement of all parameters has become impractical due to high associated costs and labor, besides time complexity. Studies have suggested water quality can be evaluated promptly by building precise and efficient prediction models that enable real-time monitoring of these parameters. This would enable taking provocative steps to prevent damages as well as safeguard water resources by identifying potential threats or degradation in water quality before time (Chidiac, El Najjar, Ouaini, El Rayess, & El Azzi, 2023; Shams et al., 2024).

A wide range of studies have utilized machine learning (ML) as well as deep learning techniques utilizing prominent algorithms like support vector networks (SVNs), regression analysis, artificial neural networks (ANNs), and gray systems, among various others. Besides this, the introduction of remote sensing has prompted the use of computer vision for comprehending water quality by analyzing the imaging data obtained from satellites through remote sensing (Shams et al., 2024). Previously, Liu, Wang, Sangaiah, Xie, and Yin (2019) secondhanded a long short-term memory (LSTM) system to envisage drinking water grade in the Yangtze River Basin. This algorithm was trained on DO, pH, and chemical oxygen demand (COD), besides the presence of ammonium ions. Similarly, a study by Khullar and Singh (2022) Khullar and Singh used the Bi-LSTM archetypal grounded on deep learning (DLBL-WQA) to foresee the water grading factors of the Yamuna River in India. These are a few among many illustrations for foreseeing water quality constraints.

ML as well as artificial intelligence (AI) possesses a wide range of applications in analyzing, monitoring, and managing water quality. Some of these areas include data collection and monitoring of water quality, predicting contaminants like algal blooms, organic matter, etc., detection of minute variations in water quality, optimizing treatment processes, besides development of high-end decision support systems for managing water quality. We shall be exploring strategies for the expansion as well as deployment of predictive learning for water grade valuation, monitoring, and management. Furthermore, we shall be reviewing different applications of AI/ML in improving forecasts of water quality.

4.2 Developing a machine learning model for assessing and monitoring water quality parameters

Predictive learning is being widely applied for enhanced estimation of water quality parameters and indices. Both classifications as well as regression-based models have been evenly applied for this task. Recently, through a systematic review, Ighalo, Adeniyi, and Marques (2021) reconnoitered the contribution of AI in surface water grade supervising and evaluation. They noted ML methods have been applied to both surface water and groundwater quality assessment (Ighalo et al., 2021). Nordin et al. (2021) appraised the four most utilized predictive modeling techniques used in groundwater assessment, viz., ANN, adaptive network-based fuzzy inference system (ANFIS), evolutionary algorithm, and support vector networks (SVC/SVR). Their review suggested ANN performs better in analyzing large datasets as well as deciphering complex relationships between nonlinear data (Che Nordin et al., 2021). ANN has also shown promising results in assessing the water grade of irrigation water from Ele River Nnewi in Anambra State by monitoring pH, total dispersed solids, and electric conduction besides the presence of sodium ions (Ubah et al., 2021). Recently, in their review, Fu, Jin, Sun, Yuan, and Butler (2022) suggested predictive learning practices are still in a preliminary phase, using mostly benchmarked systems, artificial data, workroom as well and experimental arrangements to assess the accomplishment of predictive modeling procedures with insignificant real-world implementation (Fu et al., 2022). Bagheri, Farshforoush, Bagheri, and Shemirani (2023) also recently reviewed the utility of TinyML mockups for in-situ instantaneous ecological examining of information (Bagheri et al., 2023).

The basic steps involved in the expansion of a high-efficiency model for predicting water grade parameters involve data acquisition/sampling, data preprocessing, feature engineering and selection, model development, model testing, and evaluation followed by performance evaluation (Fig. 4.1). (1) Data Acquisition/Sampling: Traditionally, water samples were

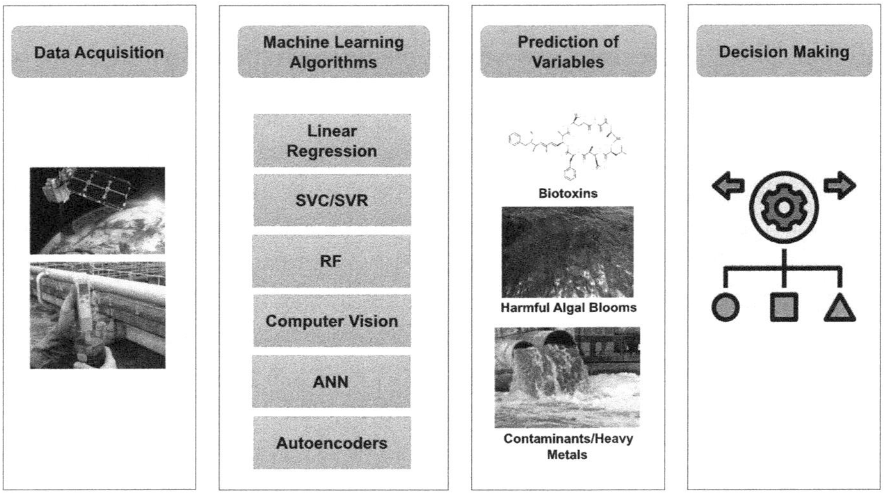

Figure 4.1
Steps elaborate in the expansion of AI/ML archetypal for water grade extrapolation. *AI*, Artificial intelligence; *ML*, machine learning.

acquired using manual sampling. However, the development of various sensors like pH sensors, thermal sensors, colorimetric sensors, hyperspectral as well as multispectral sensors, along with unmanned aerial vehicles (UAVs), have eased the sample acquisition process by providing real-time data. Publicly available data for water quality monitoring have been added in Kaggle and other repositories like GemStat (https://gemstat.org/data-gemstat/data-portal/), US EPA (www.epa.gov/waterdata/water-quality-data), etc. Recently, Naderian et al. (2024) established a water quality archive for global lakes. Similarly, Thorslund and van Vliet (2020) established a global dataset for the measurement of salinity from subterranean water as well as superficial water. (2) Model Development: The ML processes applied for envisaging water quality metrics include widely used classification, clustering, and regression-based approaches. Kanyama, Bhunu Shava, Gamundani, and Hartmann (2024) highlighted the use of SVNs, K-nearest neighbor (KNN), autoencoders, density-based spatial clustering of application with noise, random forests (RFs), decision trees (DT), and neural networks, among various others. (3) Performance Evaluation: Different performance metrics are available for measuring the performances of predictive learning models. Majorly, the use of performance scores for evaluating models depends upon the type of algorithm used. The classification algorithms generally estimate confusion matrix by comparing the class of predicted variable with its actual class. These embrace true positives, false positives, true negatives, and false negatives. This measure can additionally be used secondhand to cipher the accuracy score, correctness, sensitivity, and specificity of the algorithms. Regression algorithms, on the other hand, compute error rates like R2 and mean square error, besides root mean square error (RMSE) (Table 4.1).

Table 4.1: Methods for performance measurement of predictive learning models in water quality assessment.

Performance measurement of predictive learning		
Confusion matrix		
	Predicted class	
Actual class	True positives (TP)	False negative (FN)
	False positive (FP)	True negative (TN)
Accuracy	$\frac{(TP + TN)}{(TP + TN + FP + FN)}$	
Precision	$\frac{TP}{FP + TP}$	
Recall or sensitivity	$\frac{(TP)}{TP + FN}$	
Specificity	$\frac{TN}{TN + FP}$	
R^2	$\left[\left(n\Sigma xy-(\Sigma x)(\Sigma y)\right)/\left(\sqrt{n\Sigma x^2-(\Sigma x)^2}\times\sqrt{n\Sigma y^2-(\Sigma y)^2}\right)\right]^2$ x and y represent two variables	
Mean square error (MSE)	$\Sigma(P_i - O_i)2/n$	
Root mean square error (RMSE)	$\text{Sqrt}\left(\Sigma(P_i - O_i)2/n\right)$	

4.3 Strategies for data acquisition: remote sensing, Internet of Things, and unmanned aerial vehicles

As mentioned earlier, water quality varies both with physical location as well as the time-point at which the samples were collected from the water surface. Traditionally, in-situ sensors or manual labor were the methods for acquiring samples. These methods are inherently time-consuming. The advancement in technologies, development in remote sensing technology, the arena of computer vision, and ML/AI have enabled easy as well as fast analysis of samples. However, the deployment of these models in the field poses inherent difficulties. Recently, satellite-based remote imaging, devices based on Internet of Things (IoT), and UAVs have provided means to deploy these algorithms into remote areas for fast data acquisition, monitoring, and analysis (Srivastava et al., 2024; Wasehun, Hashemi Beni, & Di Vittorio, 2024). IoT, as the name suggests, formulates a network of sensors as well as devices rooted all over the water substructure. These instruments, together with smart meters, leak detectors, and water eminence radars, assemble information on dissimilar water quality strictures in real-time, thus providing very high spatial and temporal tenacity of the information. The information hence obtained is steadily conveyed to a principal cloud-based platform for stowing as well as investigation. The analyzed data can then be visualized by individuals using predesigned dashboards to gain insight into water leaks and water usage, besides water quality measures. This enables making informed decisions regarding infrastructure and resource allocation as well as identifying regions for conservation and optimization of treatment processes (Konde & Deosarkar, 2020; Miller, Kisiel,

Cembrowska-Lech, Durlik, & Miller, 2023; Mukta, Islam, Barman, Reza, & Hossain Khan, 2019; Pasika & Gandla, 2020). These devices have been suggested as capable of monitoring purposes, especially in trivial local water bodies, viz., lagoons, ponds, brooks, swamplands, etc. (Srivastava et al., 2024; Wasehun et al., 2024).

Remote sensing has been demonstrated to be an operative scheme for apprehending water quality. It measures the spectral signature returned after the scattering of light through multiple sensors. The quantity of scattered backscattered spectrum from water estimating alteration in the parameter being monitored. A few key factors monitored using remote sensing that affect water quality are presence of algae [viz., carotenoids, chlorophyll-a (Chl-a)], suspended sediments (viz., turbidity), presence of organic matters, presence of heavy metals and chemicals (viz., lead, nutrients, pesticides), pathogenic organisms (viz., bacteria like *Enterococcus* and *Escherichia coli*, fungi, viruses, and microalgae), aquatic plants, and oil leaks. These features alter energy spectra of reflected and emitted radiations from different riverines. Thus enabling spatiotemporal analysis of water bodies. The spectral reflectance measures inherent (viz., absorption coefficient, scattering coefficient, volume scattering, refraction index, beam attenuation, etc.) and apparent optical properties (viz., medium in which light travels). Major types of sensors used in remote sensing include airborne sensors, space-borne sensors, visible region spectral, near-infrared spectral, short-wave infrared spectral, mid-infrared spectral, hyperspectral, and multispectral sensors. Hyperspectral imaging sensors use continuous and contiguous ranges of wavelength, while multispectral imaging sensors divide spectra into different bands (Srivastava et al., 2024; Wasehun et al., 2024).

Previously, Kim et al. (2014) used a Geostationary Ocean Color Imager to compute the presence of Chl-a along with suspended particulate matter (SPM) in seaside water in South Korea. They were able to correlate the GOCI-derived vivacity information with Chl-a and SPM presence. They had identified the proportion of band 2 to band 4 as well as band 6 to band 5 to have provided more influential variables in the discrimination of Chl-a as well as SPM levels. Furthermore, GOCI-derived hourly images helped in comprehending spatiotemporal variations in data (Kim, Im, Ha, Choi, & Ha, 2014). Likewise, Zhu et al. (2022) also used a secondhanded, distant perception for analyzing water quality parameters. They used images retrieved from the Sentinel-2 satellite from Shenzhen Bay, China. The finest features for water quality were carefully chosen from eight spectral bands along with seven spectral catalogs. Using a shapely additive elucidation approach, the features were quantitated from the images, and the impact of Chl-a, turbidity, and DO were measured. Band 3 provided the maximum positive contribution to Chl-a, and Band 7 had the maximum influence on DO. These parameters helped in comprehending spatiotemporal variations in water grading strictures as well as the effect of industrial effluent (Zhu et al., 2022). Recently, Tian et al. used a multispectral instrument installed on Sentinel-2 with high spatial resolution for monitoring highly productive waters. They analyzed discrepancies in water grading constraints like the presence of Chl-a, DO, and NH_3-N in water systems over 3 years (2018–20) both annually and seasonally (Tian et al., 2023). Quang, Dinh, Dien, and Son (2023) also used surface-reflected radiance imaging patterns obtained from multispectral instruments attached to Sentinel-2 for water quality parameter estimation. This enabled analysis of spatiotemporal alterations in water eminence (Quang et al., 2023).

The incorporation of IoT devices, sensors, and entrenched systems has empowered stowage, examining, and altering water quality change. By including ML algorithms, these sensors and devices can gather information and engender alarms in actuality, which would improve fault detection as well as reduce the infrastructural load for management. Diverse ML prototypes like RF, XGBoost, and Naïve Bayes (NB) have been used to analyze trends and management architecture. RF attained an accuracy of 88%, XGBoost accomplished an accuracy of 85%, and NB reached 49% correctness (Bhardwaj et al., 2022). IoT seems to offer an appreciated means for the expansion of smart water metering networks (SWMN) along with smart water quality monitoring (SWQM). Lakshmikantha et al. (2021) thoroughly reviewed the recent works for smart monitoring of water pollution (Lakshmikantha et al., 2021). For instance, Pasika and Gandla (2020) used the application of IoT-based ThinkSpeak to measure constraints like turbidity, pH, water level, and succulence, besides temperature. Similarly, Mukta et al. (2019) used Arduino Uno in conjunction with four sensors to monitor pH, temperature, turbidity, and electrical conductivity. In another study, Konde and Deosarkar (2020) presented an SWQM system using sensors like turf programmable gate array and ZigBee radiocommunication components to predict WQI using constraints like turbidity, pH, moistness, water level, water temperature, and CO_2 in actual. The specific solicitations of AI/ML representations in water eminence evaluation, scrutinizing, as well as administration are provided in subsequent sections.

4.4 Applications of machine learning and artificial intelligence in water quality assessment

Predictive modeling techniques have a varied range of solicitations in water eminence estimation as well as examination. It has been well comprehended that water quality is influenced by both ecological and anthropogenic factors. Pollution in the marine environment due to sewage discharge, rainfall runoff, water circulation, animal excreta, etc. can cause a rapid change in water quality. Researchers have used ML algorithms for forecasting the dilution of nutrients like nitrogen and phosphorus in water. These nutrients often released into the water as agricultural run-offs may result in harmful algal blooms and eutrophication, causing increased Chl-a levels. It has become feasible for the prediction of single parameters like the presence or absence of bacterial species, prediction of Chl-a levels, prediction of salinity, DO, etc., as well as multiple parameters (Yan et al., 2024).

Predictive algorithms trained on land use and weather patterns besides previous nutrient levels have shown high accuracy in forecasting future values. In many countries, like Singapore and the United States, ML models have been installed for regularly examining and forecasting the eminence of urban water systems. Broadly, the applications of ML and AI can be grouped into data collection and monitoring of water quality, forecasting risks, algal blooms and contaminations, detection of anomalies, optimization of treatment processes, and development of decision support systems (Fig. 4.2). We shall be exploring these applications in detail in this section.

4.4.1 Data collection and monitoring

One of the major applications of ML and AI in water quality assessment is data collection and surveillance. Different types of sensors and imaging methods are utilized for capturing real-time data for analysis of water quality parameters. AI/ML can easily comprehend complex relationships between alternating parameters like salinity, DO, turbidity, and their predicted outcomes (Yan et al., 2024). Recently, Kumar, Afzal, and Ahmad (2022) used a combination of predictive learning models to estimate turbidity in the nautical environment of Hong Kong. Turbidity showcases a nonlinear time-series behavior, which is difficult to

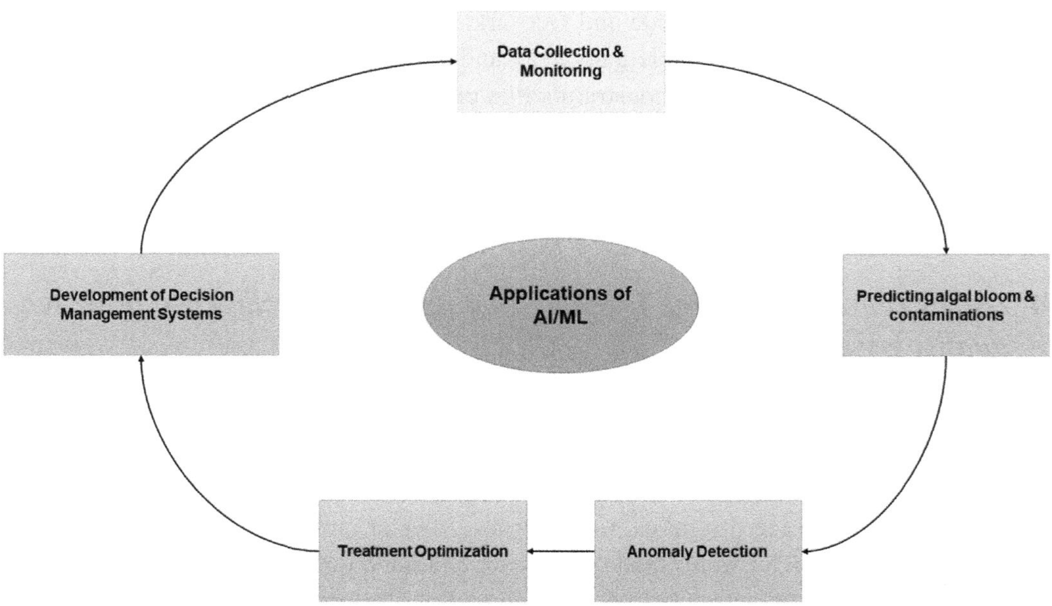

Figure 4.2
Applications of AI/ML in data collection and monitoring of water quality, prediction of contaminations and algal blooms, detection of anomalies, optimization of water treatment, besides development of decision support systems. *AI*, Artificial intelligence; *ML*, machine learning.

predict. They used ANN, LSTM-RNN, along with SVR to forecast turbidity. LSTM-RNN outperformed ANN and SVR with an accuracy of 88.45% (Kumar et al., 2022). Biological oxygen demand (BOD) at a span of 5 years is a critical indicator that is again complex to decipher. Ma, Ding, Cheng, Jiang, and Xu (2020) investigated the utility of predictive modeling techniques to estimate BOD. They applied deep matrix factorization as well as deep neural network (DNN) on data collected from New York anchorage water grounded on 32,323 samples. This method presented 11.54%–17.23% reduced RMSE in assessment to conformist matrix completion method, besides 19.2%–25.16% reduced RMSE in evaluation to standard ML processes (Ma et al., 2020). Salinity is an added decisive feature determining water eminence. It can influence the physical, chemical, as well as biological processes of water bodies. Salt affects the density of the water reservoir, thus influencing the progress as well as imitation of aquatic creatures. Diverse studies have tried to feign nonlinear and complex associations between brininess and input strictures, viz., tide-induced free-surface advancement, river emancipations, besides wind speed (Yan et al., 2024). For instance, Chawla et al. (2023) explored the effect of fertilizers and pesticides due to agricultural runoff on the water quality of the Salton Sea. They applied different ML algorithms like LR, RF, SVN, and LSTM. It predicted water qualities like salinity, pH, and DO (Chawla et al., 2023). In another study, Lal and Datta predicted salinity in groundwater. They applied ANNs, genetic programming, and SVR, besides Gaussian process regression, to data obtained from a small Pacific island coastal aquifer (Lal & Datta, 2020). Few studies have also aimed at predicting DO levels using predictive modeling. DO is an imperative pointer of water quality, as it provides a descriptive extent of molecular oxygen present in aquatic environments. The concentration of DO may vary with alterations in temperature, salinity, and atmospheric pressure. While a decreased atmospheric pressure enhances DO in water, high temperature and salinity may reduce DO (Yan et al., 2024). Malarkodi, Tarakeswari, and Tomy (2021) also presented the utility of deep learning systems for the calculation of water eminence. They combined IoT-based data acquisition with deep learning techniques. Using LSTM, they achieved water quality cataloging on total dispersed solids, turbidity, and water temperature. They also addressed the challenges of latency and data loss in IoT networks (Malarkodi et al., 2021). In an alternative study, Alfwzan et al. also proposed a combination of bi-directional LSTM grids and computational fluid dynamics (CFD) for enhancing water quality estimation. They used water quality constraints viz. COD and BOD along with contamination movement and diffusion on water samples collected from the Yamuna River. They observed the proposed combination of Bi-LSTM and CFD outperformed LSTM, GRU, and recurrent neural network (RNN). This process attained a recall of 98.25% and specificity of 96.48%, along with an exactness of 97.56% and precision of 97.76% (Alfwzan, Selim, Almalki, & Alharbi, 2024). Similarly, Rahu et al. (2024) suggested WQI and WQC prediction on data obtained from multiple sensors from Rohri Canal and Gajrawah Canal for efficient water resource management in agricultural irrigation (Rahu et al., 2024). Wu and Wang (2022) proposed a hybrid model for forecasting water quality parameters using ANN, wavelet transforms, as well as long short-term memory by analyzing Jinjiang River basin water samples (Wu & Wang, 2022). Recently, Quinn, Tansey, and Lu (2021) compared deterministic and algebraic prototypes for water quality acquiescence using data obtained from the San Joaquin River Basin, California. Table 4.2 enlists studies using AI/ML techniques to forecast multiple water quality constraints—WQI and WQC.

4.4.2 Forecasting risks, algal blooms, and contaminations

Predictive modeling enables the detection of microorganisms like *Enterococcus* and *Escherichia coli* that may act as contaminants in aquatic systems. It can also be used to forecast the presence of heavy metals like lead and antimony. Predictive modeling can also be used for forecasting the markers of eutrophication. Eutrophication refers to the aging of water bodies due to increased nutrients and overabundance of microalgae and plankton, which eventually utilize large amounts of CO_2. Detrimental algal blooms (DABs) are a naturally happening miracle in the majority of maritime settings, whether fresh, brackish, or marine. These are majorly considered by the fast propagation of noxious phytoplankton. The regularity as well as strength of DAB incidents are snowballing in retort to global warming and other climatic alterations. The shellfish that feed upon microalgae accumulate biotoxins, which may cause a threat to seafood consumers. Great efforts have been put to model and forecast the persistence and spread of these microalgae. Multivariate time-series data have been used to predict the contamination of shellfish with the toxins. ML approaches, viz., autoregressive models like autoregressive integrated moving average (ARIMA), SVC/SVR, RF, probabilistic graphical models, ANNs like FFNNs, convolutional neural networks (CNN), and RNN have been applied to predict algal bloom (Cruz, Reis Costa, Vinga, Krippahl, & Lopes, 2021; Guan, Bao, Lou, Zhou, & Yin, 2022; Litchman, 2023). Lately, Liu et al. (2022) anticipated a fusion tactic—long short-term memory outperforming DNN and ARIMA (Liu et al., 2022). Recently, Hannan and Anmala (2021) streamlined an algorithm to classify and predict the presence of fecal coliform in stream waters of Upper Green River Watershed, Kentucky, US, by using the DT algorithm (Hannan & Anmala, 2021). Table 4.3 explores the applications of AI/ML in predicting risks, algal blooms, and contamination in water.

4.4.3 Detection of anomalies

The large amount of data obtained from different water monitoring amenities based on a variety of water grading strictures has permitted the sleuthing of latent atypical vicissitudes along with early warnings. Anomalies constitute the detection of instances that do not align with anticipated patterns in a dataset. Generally, these instances can be outliers, noises, deviations, or exceptional values. In terms of the water quality environment, these

Table 4.2: Selected studies highlighting the relevance of ML and DL in data collection and examination/prediction of water quality.

S. No.	Water source	Quality parameter	Prediction index	ML algorithm	Best model	Performance	References
1	Indian rivers	pH, DO, BOD, conductivity	Weighted arithmetic water quality	SVM, DT, NB	DT	Accuracy = 98.50%	Radhakrishnan and Pillai (2020)
2	Rivers in India data submitted in Kaggle	pH, temperature, DO	Water quality index	RF, SVM, KNN	RF	Accuracy = 92.13%	Jain, Shah, Mehta, Lodaria, and Kurup (2021)
3	Rivers in India data submitted in Kaggle	DO, pH, conductivity, BOD, nitrate, fecal coliform, temperature	Water quality classification Water quality index	Neuro-fuzzy inference (ANFIS), KNN, Fed-Forward Neural Network (FFNN)	ANFIS for WQI FFNN for WQC	Accuracy (ANFIS) = 96.17% Accuracy (FFNN) = 100%	Hmoud Al-Adhaileh and Waselallah Alsaade (2021)
4	Kelantan River Basin	BOD, DO, COD, NH_4^+, pH	Water quality classification	DT, NB, GB, KNN, ANN, RF, SVM	GB	Accuracy = 94.9%	Malek, Wan Yaacob, Md Nasir, and Shaadan (2022)
5	Gulshan Lake	pH, DO, COD, TDS, turbidity, chloride, SS, alkalinity, ECI, WQI	Water quality index Water quality status	PCR, GB	GB	Accuracy (GBoost) = 100%	Islam Khan, Islam, Uddin, Islam, and Nasir (2022)
6	La Buong River	Temperature, pH, DO, BOD, COD, NH_4^+, PO_4^{3-}, TSS, TUR, coliform	Water quality index	AdaBoost, GB, HGBoost, LGBoost, XGBoost, DT, ET, RF, MLP, RBF, DFFNN, CNN	XGBoost	$R^2 = 0.989$ RMSE = 0.107	Khoi, Quan, Linh, Nhi, and Thuy (2022)
7	Indian water quality dataset at Kaggle	DO, BOD, pH, conductivity, nitrate, fecal coliform, total coliform	Water quality index Water quality classification	RF, XGB, GBM, AdaBoost, MLP, DT, KNN, SVR	GBM (WQC) MLP (WQI)	Accuracy (WQC) = 99.5% R^2 (WQI) = 99.8%	Shams et al. (2024)
8	Lake Paijanne, Finland	COD, total phosphorus	Water quality index	SVR, neural network, LSTM	LSTM	RMSE (TP) = 0.91 RMSE (COD) = 0.11	Kim et al. (2024)

(Continued)

Table 4.2: (Continued)

S. No.	Water source	Quality parameter	Prediction index	ML algorithm	Best model	Performance	References
9	–	TDS, turbidity, temperature	Water quality classification	LSTM, ANN. ARIMA	LSTM	Dissolved oxidants: $R^2 = 0.984$ MSE = 0.0068 RMSE = 0.0782	Malarkodi et al. (2021)
10	Johor River Basin	NH_3-N, suspended solids, pH	Water quality index	ANFIS, RBF-ANN, MLP-ANN, WDT-ANFIS	WDT-ANFIS	$R^2 > 0.9$	Najah Ahmed et al. (2019)
11	Eagle River Nnewi	pH, TDS, electrical conductivity, sodium	Water quality parameters	ANN-feed forward multilayer perceptron	ANN-FFMLP	pH: $R^2 = 0.945-0.968$ RMSE = 0.022−0.088 TDS: $R^2 = 0.946-0.968$ RMSE: 0.012−0.087 EC: $R^2 = 0.944-0.967$ RMSE = 0.015−0.085 Na: $R^2 = 0.949-0.965$ RMSE = 0.014−0.084	Ubah et al. (2021)

Note: *ANFIS*, Adaptive network-based fuzzy inference system; *ANN*, artificial neural network; *BOD*, biological oxygen demand; *CNN*, convolutional neural network; *DFFNN*, deep feed forward neural network; *DT*, decision trees; *ECI*, ecological co-inhabitance index; *ET*. Extra Trees; *GB*, gradient boosting; *FFMLP*, feed forward multiple layer perceptron; *GBM*, gradient boosting machine; *KNN*, K-nearest neighbor; *LSTM*, long short-term memory; MLP, multiple layer perceptron; *MSE*, mean square error; *NB*, Naïve Bayes; *PCR*, principal component regression; *RBF*, radial bias function; *RF*, random forests; *RMSE*, root mean square error; *SVM*, support vector machines; *SS*, suspended solids, *SVR*, support vector regression; *TP*, true positives; *TTS*, total suspended solids; *TSS*, total suspended solids; *TUR*, turbidity; *WDT-ANFIS*, wavelet decision tree-adaptive neuro fuzzy inference system; *WQC*, water quality classification; *WQI*, water quality index; *XGB*, Extra Gradient Boost.

Table 4.3: Selected studies highlighting the application of ML and DL in predicting risks, algal bloom, and contamination in water.

S. No.	Water source	Country	Predicted parameter	ML algorithm applied	Validation method	Best model	Performance	References
1	Coastal water	South Korea	Chlorophyll-a, suspended particulate matter	RF, Cubist, SVR	Leave one-out cross-validation	SVR	Chl-a: $R^2 = 0.91$ RMSE = 1.74 mg/m^3 SPM: $R^2 = 0.98$ RMSE = 11.42 mg/m^3	Kim et al. (2014)
2	Inland waters	China	Chl-a, DO, NH_3-N	XGBoost, SVR, RF, ANN	Fivefold cross-validation	XGBoost	Chl-a: $R^2 = 0.73$, RMSE = 0.2558, MAPE = 7.59, Bias = − 3.17 DO: $R^2 = 0.90$, RMSE = 0.1379, MAPE = 0.71, Bias = 0.07 NH_3-N: $R^2 = 0.82$, RMSE = 0.0850, MAPE = 28.6, Bias = − 21.8	Tian et al. (2023)
3	Mar Menor Lagoon	Spain	Chl-a, temperature, pH, suspended solids, turbidity, DO, salinity, nitrogen, and phosphorus	Multilayer neural networks, SVR	Recursive feature elimination (RFE) feature selection	SVR	Chl-a:$R^2 \sim 0.7$	Jimeno-Sáez, Senent-Aparicio, Cecilia, and Pérez-Sánchez (2020)

(*Continued*)

Table 4.3: (Continued)

S. No.	Water source	Country	Predicted parameter	ML algorithm applied	Validation method	Best model	Performance	References
4	Coastal water	China	Chl-a, turbidity, DO	XGBoost, SVR, MLP, mixture density networks	Shapely addictive explanations (SHAP)	All	Chl-a: Error: 1.7% Turbidity: Error: 1.5% DO: Error: 0.02%	Zhu et al. (2022)
5	Coastal zone of Binh Dinh	Vietnam	Total suspended solids, Chl-a, COD, DO	DT, RF, gradient augmented regression, Ada augmented regression		RF	Mean $R^2 = 0.56$ (original) Mean $R^2 = 0.70$ (calibrated) Chl-a: $R^2 = 0.74$	Quang et al. (2023)
6	Aitoutan watershed	China	Phosphorus, nitrogen, NH_3-N	RF, SVN, backpropagation neural network	K-fold cross-validation	RF	Phosphorus: 79 + 1.3% Total nitrogen: 84 + 0.9% NH_4^+-N: 75 + 1.3%	(Chen, Zhang, Lin, and Huang (2022))

Note: *MAPE*, mean absolute percentage error; *ML*, Machine learning; *RF*, random forests; *RMSE*, root mean square error; *SPM*, suspended particulate matter; *ANN*, artificial neural network; *Chl-a*, chlorophyll-a; *SVN*, support vector network.

anomalies can be caused either due to alterations in background data, process operations, or water pollutants. Water pollution has been considered the most frequent form of anomaly researchers decipher during water quality valuation and investigation (Ni, Cao, Zhao, Yuan, & Tan, 2024; Patabendige, Cardell-Oliver, Wang, & Liu, 2018). Water quality incongruity recognition majorly encompasses gauging a diverse indicator for discerning nonconformities from archetypal water quality circumstances. The research into anomaly detection has been on a surge. Previously, different mechanistic and hydrodynamic approaches, statistical modeling approaches, as well as predictive modeling, have been applied for sleuthing incongruities in water quality. These methods are somewhat based on setting certain threshold values from given data and predictive performance on test data (Ni et al., 2024). Currently, predictive modeling is the most advanced method for identifying anomalies. Studies have previously used both ML and AI to serve the purpose. Previously, Kanyama, Nyirenda, and Temaneh Nyah (2017) suggested the use of KNN, cluster-based local outlier factor besides histogram-based outlier score, for the development of SWMN for anomaly detection. They used performance metrics such as false-positivity rate (FPR) and detection rate for estimating the functioning of ML representations. The study disclosed that KNN achieved the best FPR and the worst DT, while the cluster-based method showed the opposite trend. The histogram-based method achieved an intermediate performance (Kanyama et al., 2017). Patabendige et al. (2018) suggested both supervised and unsupervised learning approaches are not trustworthy for water systems. While labeling in supervised learning is time-consuming and needs to be done for each, identifying relevant patterns through unsupervised learning algorithms has shown low performance in classification. They have suggested time-based/calendar context-aware to be beneficial for anomaly detection (Patabendige et al., 2018). In another report, Fang, Sun, and Huang (2019) detected an anomaly in water quality from the Ministry of Water Resources of China. They used one-class SVN on one-dimensional, three-dimensional, and seven-dimensional datasets. They identified cases having values 10 times more than the median and zero entries with approximately large values to have extreme aberration (Fang et al., 2019). Recently, Harsha Vardhan, Subramanyam, Lakshmi Reddy, Gowtham Reddy, and Ramesh (2021) also displayed the practice of SVM and DT algorithms in deciphering anomalies in the Tunisian water distribution system (Harsha Vardhan et al., 2021). Kanyama et al. (2024) also swotted the applications of ML in detecting anomalies in smart water metering systems (Kanyama et al., 2024).

4.4.4 Optimization of treatment processes

Conventionally, desalinization and reverse osmosis are the two techniques being widely used for water treatment and purification. Despite advances in reducing overall cost, high recovery, advancement in pretreatment processes, and reserve osmosis membrane modules, the procedure suffers from performance volatility triggered by disparities in forage rivulet grading as well as extreme vivacity feasting. Recently, predictive modeling has turned out to be the center of attraction for water conduct as well as seawater detoxification commerce. It finds vast application in self-adaptive fluidity augmentation as well as instantaneous optimization in the water sector. However, for the application of AI, relevant and most discriminating features are selected, which requires precise examination. AI in the arena of water grading tends to mend data quality, attain scalability, recuperate efficiency, and attain steadiness as well as suitability, besides augmenting recall and specificity of recognition (Ray, Verma, Singh, Ganesapillai, & Kwon, 2023). A variety of ML algorithms as well as AI architectures, viz., SVNs, LR, KNN, DT, AdaBoost, ANN, CNN, fuzzy inference systems, along with genetic algorithms, have previously been applied to forecast various water treatment processes, besides increasing cost-effectiveness.

4.4.5 Decision support systems

Presently, the environmental changes enforced by climate change and natural disasters due to human-induced problems have put water-reliant ecosystems in a state of crisis. This has led to the depletion of aquifers, land collapsing, periodic aeration of major rivers, waterlogging, and salinization of the aquifer and water bodies, besides the impact on human health due to the overuse of fertilizers, pesticides, and household chemicals. All of these problems are somewhere connected with water quality. Therefore there is a pressing need for fundamental reconsidering of resource administration strategies and the development of newfangled tools that support experts and controllers craft innovative resolutions (Quinn, Dinar, & Sridharan, 2022; Wardropper & Brookfield, 2022). This is possible by developing computer-aided tools enabling decision-makers to make informed verdicts by providing them with relevant evidence, facts, as well as examination tools. These tools help in environmental management by supporting policymaking on disputes related to water resources, expansion, as well as effluence hegemony besides fortification (Wang, 2024). The decision support systems are suggested to play an essential role in leading socio-economic decision-making by assisting multisectoral suppliers to make brand-informed pronouncements; advising technical strategy as well as funding investments and controlling research to recognize knowledge gaps; consenting supervisory agencies to trail improvement on the way to accomplishing water quality goals and abridging strategy administration; supporting administrators and experts to make evidence-based water supervision judgments; as well as helping as a mouthpiece to the community, affording resources for influencing inhabitant science ingenuities (Quinn et al., 2022). However, preparation, management, and augmenting water quality examination programs are not as simple as they seem (Behmel, Damour, Ludwig, & Rodriguez, 2021). They had earlier presented an intellectual decision sustenance scheme (IDSS) grounded on tacit acquaintance from around 44 connoisseurs and two case studies, along with an explicit understanding of challenges in water quality monitoring. IDSS could enable planning, management, and optimization of the monitoring programs to facilitate good and prompt decision-making (Behmel et al., 2021). Over the years, many significant developments have been made to develop decision support systems for improving water quality. Previously, Teodsiu et al. had provided an overview of decision support systems and their functioning. It also presented some early examples of DSS like MULINO-DSS, a multisectoral, cohesive, and effective decision sustenance scheme for supportable usage of water possessions at the catchment measure; NOSTRUM-DSS, a network on ascendency science and technology for maintainable water supplies management in Mediterranean area; RIBASIM, river basin imitation archetypal; WEAP, a water valuation as well as scheming structure; MIKE BASIN, MODSIM, WBalMo, ELBE-DSS, MulinoDSS Belgium, a Nethen catchment, among various others (Teodosiu, Ardeleanu, & Lupu, 2009). Recently, Bornstein, Dayan, Cahn, Wells, and Housh (2022) demonstrated the utility of the CE-QUAL-W2 archetypal for an environment-based decision sustenance structure. They proposed a web-based service based on Docker, Kubernetes, Helm, and cloud computing. This service enabled adaptive management, community engagement, as well as long-term resilience planning (Bornstein et al., 2022). Dinar and Quinn also proposed a decision sustenance scheme for provincial agrarian nonpoint salinity effluence administration. They used an updateable Watershed Analysis Risk Management Framework to forecast salt concentration assimilating capacity in the San Joaquin River. They showed how a policy permutation of well-crafted river brininess purposes by the supervisor and the relevance of an easy-to-use besides justifiable decision sustenance tool by patrons can assist in reducing saltiness levels over a two-decade exploration (Dinar & Quinn, 2022).

4.5 Real-life applications of artificial intelligence in water quality management

Many applications as well as models developed using the employment of predictive modeling techniques for water grade examining and analysis, besides management, are still in the developmental phase. A few noteworthy illustrations exist that have found their way to the market. These include CyanoLakes, Aquarius, OptiNimbus, WaterHub, Project Premonition by Microsoft, Prognosis, and Fluidion Alert System, among a few others. CyanoLakes, formulated by Mark Matthews as Copernicus Masters Idea Challenge, is one of the first services based entirely upon satellite-enabled remote sensing. It is based on the idea that cyanobacteria blooms enhance the operational risk by alleviating the water treatment capital and pose a hazard to hominoid wellbeing. It improves the knowledge of raw water resources by monitoring the algal bloom. It detects Chl-a levels in real-time. The tool is developed on publicly available data and is available as a mobile application for us. CyanoLakes allows users to compare new data, view historical records of the water resource, and view high-resolution images. (https://www.cyanolakes.com/). Aquarius by

Aquatic Informatics is a software platform utilized for monitoring water quality globally. It enables data management by acquiring, processing, modeling, and publishing water information in real-time. It integrates data from multiple resources to offer the greatest straightforward as well as reliable outcomes. It enables visualization of water eminence indexes through the use of powerful charts and representations. It enables streamlining of sample acquisition from labs and fields. It is viable for analyzing floods, water quality, groundwater levels, etc. (https://aquaticinformatics.com/). OptiNimbus, developed by Cobalt Water Global, is another innovative system combining the IoT with remote sensing. It has been established at Brooklyn Botanic Park, New York. It tackles the challenge of collective sewer runoff drop while upholding esthetics as well as saving water. This software gathers information from the forecasts of the National Weather Service and National Oceanic and Atmospheric Administration and then pairs it with an algorithm to adjust water levels to the forecasted rainfall. At the time of storm events, it preemptively releases water from the pond (https://www.iotone.com/). WaterHub is a decision-making system that applies predictive modeling methods to analyze data from water utilities, regulatory bodies, and other sources to provide insights into water quality trends and potential risks. The platform utilizes ML and AI-based analytics to envisage water quality issues, permitting upbeat actions for averting pollution and guaranteeing regulatory acquiescence. Clean Water AI is another exquisite tool developed by Peter Ma on an Intel Movidius-based AI platform. It analyzes water bodies for the presence of contaminants like bacteria, organic matter, or lead particles in real-time using a custom-developed CNN architecture and computer vision. It detects the shape of the molecules using a microscope camera and maps the identified structures to alert local people and administrators or users about water quality (https://cleanwaterai.com/). Fluidion Alert System is another high-end tool developed by Fluidion. It enables water quality and environmental surveillance through a cutting-edge, fully validated monitoring system, supporting decision-making. Like Clean Water AI, it also applies predictive modeling to real-time monitoring of contaminants, viz., pathogens and other organic materials (https://fluidion.com/) (Angelescu, Rerolle, Vassart, & Wong, 2023). The real-life applications of AI and predictive modeling are not limited to the illustrations provided above. Newer technologies are being developed, modified, and tested to ensure accurate and reliable water quality analysis and management.

4.6 Discussion

In this study, we explored in depth how predictive modeling approaches as well as IoT-based sensing can be pertained to improving water eminence monitoring, assessment, as well as management by illustrating some noteworthy applications. We noted maintaining water grading is of chief interest for ensuring the survival of all the dependent lifeforms. The water eminence of every water resource is not stringent and can vary concerning the time and place of sampling. The traditional methods like manual sampling and deployment of real-time monitoring systems used for investigating water quality parameters are costly as well as laborious, causing access to a sparse dataset. Moreover, these methods were able to capture very few parameters, such as water level, temperature, and conductivity, despite having seen unprecedented technological advancements (Kulwant & Yadav, 2024). We found that predictive modeling methods were able to counter these challenges by finding unusual patterns in the dataset, thus considerably reducing the time and labor. This enabled predicting the behavior of a water body and any discrepancies, thus making informed decisions. We also noted that the advent of IoT and the development of UAVs together with satellite-based remote sensing led to the easy deployment of predictive models into real-world applications. We also explored the possible requirements for developing water quality prediction models. We identified a variety of algorithms ranging from classical to neural network-based approaches, applied in WQI, namely, support vector, regressions, AdaBoost, KNN, RF, DT, CNN, gradient boosting, and NB, among a few others, which were captured in detail by Sangwan and Bhardwaj (2024) along with different performance metrics. We understand the performance of these representations plays a pivotal function in comprehending the reliability of the water eminence forecast systems. Additionally, in this review, we also discussed a few recently developed water quality management systems, viz., Aquarius, CyanoLakes, OptiNimbus, Clean Water AI, and WaterHub. These decision support platforms have been shown to scrutinize water grades in accordance with the presence of microbiota and Chl-a using the integration of computer vision and neural networks. These newer platforms are certainly different from preexisting and benchmarked water quality monitoring models (viz., MIKE, Delft3D, AQUATOX, EFDC, and WASP) used in developed countries like the USA, France, Germany, Russia, and China. These models were dependent upon manual data collection for easily accessible sites and wireless sensors for complex sites and utilized information entropy, multivariate statistics (principal component analysis, discriminant analysis, clustering), geostatistics, as well as topological methods like geographical information systems to identify warning and source identification of water contamination. Each of these approaches had its limitations working on small water reservoirs like pools, lakes, or ponds; could only analyze locations with low variability and a good number of water quality parameters; required a large amount of spatially scaled data; and required historical data of water reservoirs (Talukdar, Kumar, & Kulkarni, 2023).

We also noted that predictive learning algorithms were able to minimize these suggested limitations by selecting an optimal set of water quality parameters by reducing the variations in the data using different activation functions and hidden layers and providing high prediction accuracy. We observed majority of studies related to water quality analysis using predictive modeling have been carried out on surface water reservoirs like lakes, ponds, and rivers. These algorithms have shown promising results on groundwater reservoirs as well. While ANNs and deep learning architectures coupled with computer vision algorithms have been found to provide better performances for groundwater quality monitoring, classical and collaborative ML procedures like DTs, RF, regression, and SVM, and a few neural network approaches have been found to better suit surface water reservoirs. Based on this evidence, we can establish that predictive learning approaches as well as IoT-based sensing have a high potential for improving the analysis, monitoring, and management of water eminence.

Although the methods for water eminence monitoring, assessment, and management have immensely benefited from the predictive learning approaches as well as IoT, it has been presented with its own sets of challenges and limitations. These challenges include the unavailability of large datasets for training the models to achieve high accuracy and reliability. Although a substantial amount of data is available for designing the workflow, it often contains gaps and missing values, which renders the effectiveness of ML models or a diverse range of environmental conditions. One of the foremost complications most ML models face is overfitting, i.e., the model is unable to differentiate between data whose pattern is not similar to that of training data. Hence, ensuring the models to be generalized for a diverse data point is one of the major challenges. Another challenge of ML algorithms may be ethical and privacy considerations of the available data. Since data is outsourced from public or community sources, it may create a privacy violation situation. Most importantly, the ML model must not have any inherent biases in prediction, which may lead to harm to consumers due to misidentification of uncontaminated or contaminated locations (Miller et al., 2024). Similarly, IoT devices and sensors also have their challenges due to their complicated design. IoT devices are nonstandardized. One of the key complications IoT devices face is related to the integration of different sensors and devices, as the majority of these are incompatible with each other. Although IoT devices tend to reduce costs on a larger scale, these devices themselves are costly and difficult to install and maintain. Another important challenge associated with IoT devices is security and privacy. Since the data collected and stored by IoT devices is highly sensitive, there is a potentially high risk associated with its misuse and unauthorized access by individuals and organizations. A wide range of mitigation methods are being established and instigated to diminish the vulnerabilities associated with cyberattacks and potential misuse. Some suggested measures include using secure passwords, keeping devices regularly updated to diminish vulnerabilities, and using encryption systems (viz., secure sockets layer, internet protocol securities, secure shell, advanced encryption system, and two-factor authentication systems) (Miller et al., 2023).

Despite all the challenges and limitations, AI/ML has shown promising results in water quality valuation and supervision. If used diligently, it can revolutionize the sector of water quality monitoring. Integration of predictive modeling with IoT seems the most promising approach to enhancing water quality scrutinizing and estimation. With a wide network of sensors, it would enable the analysis of real-time data on highly diverse parameters, facilitating timely intervention to water quality alterations. Moreover, this would help policymakers to manage the environment efficiently, thus speeding up the decision-making process. Furthermore, we can aim to develop new repositories for sharing the data and algorithms. This requirement arises because no central channel exists globally for exchanging water quality data except for freshwater quality (GEMStat, United Nations Environment Programme). Although few repositories are available, like the Central Pollution Control Board (CPCB) (https://rtwqmsdb1.cpcb.gov.in/), USGS (Riskin & Lee, 2021), and European Environment Agency (EEA), they lack historical evidence on the water eminence of dissimilar water bodies. The availability of large volumes of historical data on water bodies would allow for capturing more diverse data, enabling the development of a more robust forecasting model. This would help researchers accelerate the innovation in IoT sensing, improving model accuracy, and ensuring transparency. Thus facilitating interdisciplinary collaborations.

4.7 Conclusions

In this chapter, we provided an impression of predictive learning procedures that have been applied to water eminence assessment, monitoring, and supervision, besides how modern techniques are enabling its implementation. Different methods like classification, pattern recognition, computer vision, and remote sensing have been applied to comprehend water quality. Here, we emphasized versatile applications of predictive modeling in data organization and surveillance, forecasting algal blooms, pollutants, and other contaminants in water, detection of anomalies, besides development of decision systems for effective management of water resources. These algorithms have helped in reducing time and labor for large-scale analysis considerably. However, despite the advancement of the techniques, water grading investigation, as well as organization, is difficult due to the high variability of water grading strictures, unavailability of large datasets, presence of gaps and missing values, incompatibility between IoT sensors and devices, and security risks. More efficient methods and algorithms are required to be developed for comprehending these enigmatic patterns from the water samples as well as for tackling highlighted challenges to enhance the water quality prediction models. Furthermore, a centralized data-sharing portal supplemented with historical and instantaneous water eminence data is required for fostering collaboration and innovation and ensuring transparency, enabling enhanced water quality monitoring and management.

References

Alfwzan, W. F., Selim, M. M., Almalki, A. S., & Alharbi, I. S. (2024). Water quality assessment using Bi-LSTM and computational fluid dynamics (CFD) techniques. *Alexandria Engineering Journal, 97*, 346–359. Available from https://doi.org/10.1016/j.aej.2024.04.030.

Angelescu, D. E., Rerolle, V., Vassart, L., & Wong, J. (2023). Light in the water: Single-fiber spectroscopy for next-generation microfluidic water quality sensors. In Optica Sensing Congress 2023 (AIS, FTS, HISE, Sensors, ES) (2023), Paper AM3A.1, AM3A.1. Available from https://doi.org/10.1364/AIS.2023.AM3A.1.

Bagheri, M., Farshforoush, N., Bagheri, K., & Shemirani, A. I. (2023). Applications of artificial intelligence technologies in water environments: From basic techniques to novel tiny machine learning systems. *Process Safety and Environmental Protection*, *180*, 10–22. Available from https://doi.org/10.1016/j.psep.2023.09.072.

Behmel, S., Damour, M., Ludwig, R., & Rodriguez, M. J. (2021). Intelligent decision-support system to plan, manage and optimize water quality monitoring programs: Design of a conceptual framework. *Journal of Environmental Planning and Management*, *64*(4), 703–733. Available from https://doi.org/10.1080/09640568.2020.1782858.

Bhardwaj, A., Dagar, V., Khan, M. O., Aggarwal, A., Alvarado, R., Kumar, M., ... Proshad, R. (2022). Smart IoT and machine learning-based framework for water quality assessment and device component monitoring. *Environmental Science and Pollution Research*, *29*(30), 46018–46036. Available from https://doi.org/10.1007/s11356-022-19014-3.

Bornstein, Y., Dayan, B., Cahn, A., Wells, S., & Housh, M. (2022). Environmental decision support systems as a service: demonstration on CE-QUAL-W2 model. *Water*, *14*(6). Available from https://doi.org/10.3390/w14060885, Article 6.

Chawla, P., Cao, X., Fu, Y., Hu, C., Wang, M., Wang, S., & Gao, J. Z. (2023). Water quality prediction of the Salton Sea using machine learning and big data techniques. *International Journal of Environmental Analytical Chemistry*, *103*(18), 6835–6858. Available from https://doi.org/10.1080/03067319.2021.1963713.

Che Nordin, N. F., Mohd, N. S., Koting, S., Ismail, Z., Sherif, M., & El-Shafie, A. (2021). Groundwater quality forecasting modelling using artificial intelligence: A review. *Groundwater for Sustainable Development*, *14*, 100643. Available from https://doi.org/10.1016/j.gsd.2021.100643.

Chen, S., Zhang, Z., Lin, J., & Huang, J. (2022). Machine learning-based estimation of riverine nutrient concentrations and associated uncertainties caused by sampling frequencies. *PLoS One*, *17*(7), e0271458. Available from https://doi.org/10.1371/journal.pone.0271458.

Chidiac, S., El Najjar, P., Ouaini, N., El Rayess, Y., & El Azzi, D. (2023). A comprehensive review of water quality indices (WQIs): History, models, attempts and perspectives. *Reviews in Environmental Science and Bio/Technology*, *22*(2), 349–395. Available from https://doi.org/10.1007/s11157-023-09650-7.

Cruz, R. C., Reis Costa, P., Vinga, S., Krippahl, L., & Lopes, M. B. (2021). A review of recent machine learning advances for forecasting harmful algal blooms and shellfish contamination. *Journal of Marine Science and Engineering*, *9*(3). Available from https://doi.org/10.3390/jmse9030283, Article 3.

Dinar, A., & Quinn, N. W. T. (2022). Developing a decision support system for regional agricultural nonpoint salinity pollution management: Application to the San Joaquin River, California. *Water*, *14*(15). Available from https://doi.org/10.3390/w14152384, Article 15.

Fang, S., Sun, W., & Huang, L. (2019). Anomaly detection for water supply data using machine learning technique. *Journal of Physics: Conference Series*, *1345*(2), 022054. Available from https://doi.org/10.1088/1742-6596/1345/2/022054.

Fu, G., Jin, Y., Sun, S., Yuan, Z., & Butler, D. (2022). The role of deep learning in urban water management: A critical review. *Water Research*, *223*, 118973. Available from https://doi.org/10.1016/j.watres.2022.118973.

Guan, W., Bao, M., Lou, X., Zhou, Z., & Yin, K. (2022). Monitoring, modeling and projection of harmful algal blooms in China. *Harmful Algae*, *111*, 102164. Available from https://doi.org/10.1016/j.hal.2021.102164.

Hannan, A., & Anmala, J. (2021). Classification and prediction of fecal coliform in stream waters using decision trees (DTs) for Upper Green River Watershed, Kentucky, USA. *Water*, *13*(19). Available from https://doi.org/10.3390/w13192790, Article 19.

Harsha Vardhan, A., Subramanyam, B., Lakshmi Reddy, M., Gowtham Reddy, G., & Ramesh, A. (2021). Anomaly detection in water distribution systems. In *Fourth Smart Cities Symposium (SCS 2021), 2021* (pp. 219–222). Available from https://doi.org/10.1049/icp.2022.0344.

Hmoud Al-Adhaileh, M., & Waselallah Alsaade, F. (2021). Modelling and prediction of water quality by using artificial intelligence. *Sustainability*, *13*(8), 4259. Available from https://doi.org/10.3390/su13084259.

Ighalo, J. O., Adeniyi, A. G., & Marques, G. (2021). Artificial intelligence for surface water quality monitoring and assessment: A systematic literature analysis. *Modeling Earth Systems and Environment*, *7*(2), 669–681. Available from https://doi.org/10.1007/s40808-020-01041-z.

Islam Khan, M. S., Islam, N., Uddin, J., Islam, S., & Nasir, M. K. (2022). Water quality prediction and classification based on principal component regression and gradient boosting classifier approach. *Journal of King Saud University - Computer and Information Sciences*, *34*(8), 4773–4781. Available from https://doi.org/10.1016/j.jksuci.2021.06.003.

Jain, D., Shah, S., Mehta, H., Lodaria, A., & Kurup, L. (2021). A machine learning approach to analyze marine life sustainability. In A. P. Pandian, R. Palanisamy, & K. Ntalianis (Eds.), *Proceedings of International Conference on Intelligent Computing, Information and Control Systems* (pp. 619–632). Springer. Available from https://doi.org/10.1007/978-981-15-8443-5_53.

Jimeno-Sáez, P., Senent-Aparicio, J., Cecilia, J. M., & Pérez-Sánchez, J. (2020). Using machine-learning algorithms for eutrophication modeling: Case study of Mar Menor Lagoon (Spain). *International Journal of Environmental Research and Public Health*, *17*(4). Available from https://doi.org/10.3390/ijerph17041189, Article 4.

Kanyama, M., Nyirenda, C., & Temaneh Nyah, C. (2017). Anomaly detection in smart water metering networks.

Kanyama, M. N., Bhunu Shava, F., Gamundani, A. M., & Hartmann, A. (2024). Machine learning applications for anomaly detection in smart water metering networks: A systematic review. *Physics and Chemistry of the Earth, Parts A/B/C*, *134*, 103558. Available from https://doi.org/10.1016/j.pce.2024.103558.

Khoi, D. N., Quan, N. T., Linh, D. Q., Nhi, P. T. T., & Thuy, N. T. D. (2022). Using machine learning models for predicting the water quality index in the La Buong River, Vietnam. *Water*, *14*(10), 1552. Available from https://doi.org/10.3390/w14101552.

Khullar, S., & Singh, N. (2022). Water quality assessment of a river using deep learning Bi-LSTM methodology: Forecasting and validation. *Environmental Science and Pollution Research*, *29*(9), 12875–12889. Available from https://doi.org/10.1007/s11356-021-13875-w.

Kim, H. I., Kim, D., Mahdian, M., Salamattalab, M. M., Bateni, S. M., & Noori, R. (2024). Incorporation of water quality index models with machine learning-based techniques for real-time assessment of aquatic ecosystems. *Environmental Pollution*, *355*, 124242. Available from https://doi.org/10.1016/j.envpol.2024.124242.

Kim, Y. H., Im, J., Ha, H. K., Choi, J.-K., & Ha, S. (2014). Machine learning approaches to coastal water quality monitoring using GOCI satellite data. *GIScience & Remote Sensing*, *51*(2), 158–174. Available from https://doi.org/10.1080/15481603.2014.900983.

Konde, S., & Deosarkar, D. S. (2020). IOT based water quality monitoring system. *SSRN Electronic Journal*. Available from https://doi.org/10.2139/ssrn.3645467.

Kulwant, M., & Yadav, A. K. (2024). Challenges and opportunities for water quality monitoring and management in India. In A. K. Yadav, K. Yadav, & V. P. Singh (Eds.), *Integrated management of water resources in India: A computational approach: Optimizing for sustainability and planning* (pp. 121–137). Switzerland: Springer Nature. Available from https://doi.org/10.1007/978-3-031-62079-9_7.

Kumar, L., Afzal, M. S., & Ahmad, A. (2022). Prediction of water turbidity in a marine environment using machine learning: A case study of Hong Kong. *Regional Studies in Marine Science*, *52*, 102260. Available from https://doi.org/10.1016/j.rsma.2022.102260.

Lakshmikantha, V., Hiriyannagowda, A., Manjunath, A., Patted, A., Basavaiah, J., & Anthony, A. A. (2021). IoT based smart water quality monitoring system. *Global Transitions Proceedings*, *2*(2), 181–186. Available from https://doi.org/10.1016/j.gltp.2021.08.062, https://www.sciencedirect.com/science/article/pii/S2666285X2100090X.

Lal, A., & Datta, B. (2020). Performance evaluation of homogeneous and heterogeneous ensemble models for groundwater salinity predictions: A regional-scale comparison study. *Water, Air, & Soil Pollution*, *231*(6), 320. Available from https://doi.org/10.1007/s11270-020-04693-w.

Litchman, E. (2023). Understanding and predicting harmful algal blooms in a changing climate: A trait-based framework. *Limnology and Oceanography Letters*, *8*(2), 229–246. Available from https://doi.org/10.1002/lol2.10294.

Liu, M., He, J., Huang, Y., Tang, T., Hu, J., & Xiao, X. (2022). Algal bloom forecasting with time-frequency analysis: A hybrid deep learning approach. *Water Research*, *219*, 118591. Available from https://doi.org/10.1016/j.watres.2022.118591.

Liu, P., Wang, J., Sangaiah, A., Xie, Y., & Yin, X. (2019). Analysis and prediction of water quality using LSTM deep neural networks in IoT environment. *Sustainability*, *11*(7), 2058. Available from https://doi.org/10.3390/su11072058.

Ma, J., Ding, Y., Cheng, J. C. P., Jiang, F., & Xu, Z. (2020). Soft detection of 5-day BOD with sparse matrix in city harbor water using deep learning techniques. *Water Research*, *170*, 115350. Available from https://doi.org/10.1016/j.watres.2019.115350.

Malarkodi, B., Tarakeswari, P., & Tomy, J. (2021). A deep learning strategy for water quality monitoring. In 2021 second International Conference on Secure Cyber Computing and Communications (ICSCCC), 555–559. Available from https://doi.org/10.1109/ICSCCC51823.2021.9478174.

Malek, N. H. A., Wan Yaacob, W. F., Md Nasir, S. A., & Shaadan, N. (2022). Prediction of water quality classification of the Kelantan River Basin, Malaysia, using machine learning techniques. *Water*, *14*(7), 1067. Available from https://doi.org/10.3390/w14071067.

Miller, M., Kisiel, A., Cembrowska-Lech, D., Durlik, I., & Miller, T. (2023). IoT in water quality monitoring—are we really here? *Sensors*, *23*(2). Available from https://doi.org/10.3390/s23020960, Article 2.

Miller, T., Łobodzińska, A., Kozlovska, P., Lewita, K., Kaczanowska, O., & Durlik, I. (2024). Advancing water quality prediction: The role of machine learning in environmental science. *Grail of Science*, *36*. Available from https://doi.org/10.36074/grail-of-science.16.02.2024.092, Article 36.

Mukta, M., Islam, S., Barman, S. D., Reza, A. W., & Hossain Khan, M. S. (2019). IoT based smart water quality monitoring system. In *2019 IEEE fourth International Conference on Computer and Communication Systems (ICCCS)* (pp. 669–673). Available from https://doi.org/10.1109/CCOMS.2019.8821742.

Naderian, D., Noori, R., Heggy, E., Bateni, S. M., Bhattarai, R., Nohegar, A., & Sharma, S. (2024). A water quality database for global lakes. *Resources, Conservation and Recycling*, *202*, 107401. Available from https://doi.org/10.1016/j.resconrec.2023.107401.

Najah Ahmed, A., Binti Othman, F., Abdulmohsin Afan, H., Khaleel Ibrahim, R., Ming Fai, C., Shabbir Hossain, M., . . . Elshafie, A. (2019). Machine learning methods for better water quality prediction. *Journal of Hydrology*, *578*, 124084. Available from https://doi.org/10.1016/j.jhydrol.2019.124084.

Nasir, N., Kansal, A., Alshaltone, O., Barneih, F., Sameer, M., Shanableh, A., & Al-Shamma'a, A. (2022). Water quality classification using machine learning algorithms. *Journal of Water Process Engineering*, *48*, 102920. Available from https://doi.org/10.1016/j.jwpe.2022.102920.

Ni, Q., Cao, X., Zhao, Z., Yuan, J., & Tan, C. (2024). An unsupervised water quality anomaly detection method based on a combination of time-frequency analysis and clustering. *Environmental Science and Pollution Research*, *31*(10), 15920–15931. Available from https://doi.org/10.1007/s11356-024-32170-y.

Pasika, S., & Gandla, S. T. (2020). Smart water quality monitoring system with cost-effective using IoT. *Heliyon*, *6*(7), e04096. Available from https://doi.org/10.1016/j.heliyon.2020.e04096.

Patabendige, S., Cardell-Oliver, R., Wang, R., & Liu, W. (2018). Detection and interpretation of anomalous water use for non-residential customers. *Environmental Modelling & Software*, *100*, 291–301. Available from https://doi.org/10.1016/j.envsoft.2017.11.028.

Quang, N. H., Dinh, N. T., Dien, N. T., & Son, L. T. (2023). Calibration of Sentinel-2 surface reflectance for water quality modelling in Binh Dinh's coastal zone of Vietnam. *Sustainability*, *15*(2). Available from https://doi.org/10.3390/su15021410, Article 2.

Quinn, N. W. T., Dinar, A., & Sridharan, V. (2022). Decision support tools for water quality management. *Water*, *14*(22). Available from https://doi.org/10.3390/w14223644, Article 22.

Quinn, N. W. T., Tansey, M. K., & Lu, J. (2021). Comparison of deterministic and statistical models for water quality compliance forecasting in the San Joaquin River Basin, California. *Water*, *13*(19). Available from https://doi.org/10.3390/w13192661, Article 19.

Radhakrishnan, N., & Pillai, A.S. (2020). Comparison of water quality classification models using machine learning. In *2020 Fifth International Conference on Communication and Electronics Systems (ICCES)* (pp. 1183−1188). Available from https://doi.org/10.1109/ICCES48766.2020.9137903.

Rahu, M. A., Shaikh, M. M., Karim, S., Soomro, S. A., Hussain, D., & Ali, S. M. (2024). Water quality monitoring and assessment for efficient water resource management through Internet of Things and machine learning approaches for agricultural irrigation. *Water Resources Management*. Available from https://doi.org/10.1007/s11269-024-03899-5.

Ray, S. S., Verma, R. K., Singh, A., Ganesapillai, M., & Kwon, Y.-N. (2023). A holistic review on how artificial intelligence has redefined water treatment and seawater desalination processes. *Desalination*, *546*, 116221. Available from https://doi.org/10.1016/j.desal.2022.116221.

Riskin, M. L., & Lee, C. J. (2021). USGS National Water Quality Monitoring Network. In Fact Sheet (2021−3019). U.S. Geological Survey. https://doi.org/10.3133/fs20213019.

Sangwan, V., & Bhardwaj, R. (2024). Application of machine learning model for assessing water quality index. In A. K. Yadav, K. Yadav, & V. P. Singh (Eds.), *Integrated management of water resources in India: A computational approach: optimizing for sustainability and planning* (pp. 291−308). Switzerland: Springer Nature. Available from https://doi.org/10.1007/978-3-031-62079-9_16.

Shams, M. Y., Elshewey, A. M., El-kenawy, E.-S. M., Ibrahim, A., Talaat, F. M., & Tarek, Z. (2024). Water quality prediction using machine learning models based on grid search method. *Multimedia Tools and Applications*, *83*(12), 35307−35334. Available from https://doi.org/10.1007/s11042-023-16737-4.

Srivastava, M. K., Gaur, S., Ohri, A., Srivastava, P. K., & Singh, N. (2024). Chapter 12—Applications of remote sensing in water quality assessment. In S. Lamine, P. K. Srivastava, A. Kayad, F. Muñoz-Arriola, & P. C. Pandey (Eds.), *Remotesensing in precision agriculture.* (pp. 217−236). Academic Press. Available from https://doi.org/10.1016/B978-0-323-91068-2.00019-9.

Talukdar, P., Kumar, B., & Kulkarni, V. V. (2023). A review of water quality models and monitoring methods for capabilities of pollutant source identification, classification, and transport simulation. *Reviews in Environmental Science and Bio/Technology*, *22*(3), 653−677. Available from https://doi.org/10.1007/s11157-023-09658-z.

Teodosiu, C., Ardeleanu, C., & Lupu, L. (2009). An overview of decision support systems for integrated water resources management. *Environmental Engineering and Management Journal*, *8*(1), 153−162. Available from https://doi.org/10.30638/eemj.2009.023.

Thorslund, J., & van Vliet, M. T. H. (2020). A global dataset of surface water and groundwater salinity measurements from 1980−2019. *Scientific Data*, *7*(1), 231. Available from https://doi.org/10.1038/s41597-020-0562-z.

Tian, S., Guo, H., Xu, W., Zhu, X., Wang, B., Zeng, Q., . . . Huang, J. J. (2023). Remote sensing retrieval of inland water quality parameters using Sentinel-2 and multiple machine learning algorithms. *Environmental Science and Pollution Research*, *30*(7), 18617−18630. Available from https://doi.org/10.1007/s11356-022-23431-9.

Ubah, J. I., Orakwe, L. C., Ogbu, K. N., Awu, J. I., Ahaneku, I. E., & Chukwuma, E. C. (2021). Forecasting water quality parameters using artificial neural network for irrigation purposes. *Scientific Reports*, *11*(1), 24438. Available from https://doi.org/10.1038/s41598-021-04062-5.

Wang, F. (2024). Design and practice of decision support system for integrated water resources management. In J. C. Hung, N. Yen, & J.-W. Chang (Eds.), *Frontier computing on industrial applications* (Volume 2, pp. 329−335). Springer Nature. Available from https://doi.org/10.1007/978-981-99-9538-7_51.

Wardropper, C., & Brookfield, A. (2022). Decision-support systems for water management. *Journal of Hydrology*, *610*, 127928. Available from https://doi.org/10.1016/j.jhydrol.2022.127928.

Wasehun, E. T., Hashemi Beni, L., & Di Vittorio, C. A. (2024). UAV and satellite remote sensing for inland water quality assessments: A literature review. *Environmental Monitoring and Assessment*, *196*(3), 277. Available from https://doi.org/10.1007/s10661-024-12342-6.

Wu, J., & Wang, Z. (2022). A hybrid model for water quality prediction based on an artificial neural network, wavelet transform, and long short-term memory. *Water*, *14*(4). Available from https://doi.org/10.3390/w14040610, Article 4.

Yan, X., Zhang, T., Du, W., Meng, Q., Xu, X., & Zhao, X. (2024). A comprehensive review of machine learning for water quality prediction over the past five years. *Journal of Marine Science and Engineering*, *12*(1). Available from https://doi.org/10.3390/jmse12010159, Article 1.

Zainab, A., Amina, I., Abdulmuhaimin, M., Sadiku, A.S., & Baballe, M.A. (2023). The water monitoring system's disadvantages. Available from https://doi.org/10.5281/ZENODO.8161049.

Zhu, X., Guo, H., Huang, J. J., Tian, S., Xu, W., & Mai, Y. (2022). An ensemble machine learning model for water quality estimation in coastal area based on remote sensing imagery. *Journal of Environmental Management*, *323*, 116187. Available from https://doi.org/10.1016/j.jenvman.2022.116187.

CHAPTER 5

Enhancing water quality management through artificial intelligence and machine learning technologies

Aakriti Chauhan[1], Purnima Mehta[2] and Arun Lal Srivastav[3]

[1]Department of Agriculture, Vivekananda Global University Jagatpura, Jaipur, Rajasthan, India, [2]Dr. Yashwant Singh Parmar University of Horticulture and Forestry, Solan, Himachal Pradesh, India, [3]Department of Applied Sciences, Chitkara University School of Engineering and Technology, Chitkara University, Solan, Himachal Pradesh, India

5.1 Introduction

Water covers over 70% of Earth's surface, with only 3% being freshwater found in ice caps, glaciers, groundwater, rivers, lakes, and more. Of this, only about 0.5% is usable for various life forms (Oğuz & Ertuğrul, 2023). As water resources face mounting challenges, the importance of monitoring water quality (WQ) is growing significantly. This monitoring is vital for maintaining the health and sustainability of Earth's water resources. It is essential to monitor and evaluate water samples from resources such as lakes, ponds, and rivers to preserve and enhance the quality of the environment. Traditional methods for assessing indicators of WQ include in situ measurements and/or the collection of water samples for subsequent laboratory analysis (Ritchiel & Schiebe, 2000). This approach has both advantages and disadvantages. It is a costly and time-consuming method. Moreover, it might not be the most economical course of action. Consequently, cutting-edge approaches like artificial intelligence (AI) and machine learning (ML) are employed to address this issue. The chemical approach of evaluating WQ criteria is eliminated by AI and ML, which is also potentially, frequently, and economically advantageous (Ragi, Holla & Manju, 2019). In recent years, AI technology has developed rapidly (Zhou et al., 2020). AI describes computer programs that can carry out activities that normally need intellect from humans. ML, a component of AI, involves algorithms that improve through experience, allowing systems to adjust their behavior based on new information (Riedl, 2019). ML constructs computer models from data, enabling predictions without further programming.

AI is primarily used to monitor various WQ parameters, including physical, chemical, and biological ones (Malviya & Jaspal, 2021). AI algorithms have advanced capabilities for

Computational Automation for Water Security.
DOI: https://doi.org/10.1016/B978-0-443-33321-7.00022-6

assessing WQ, reducing the need for manual, time-consuming analysis and enabling efficient, rapid assessments (Kaddoura, 2022). ML has revolutionized WQ monitoring through algorithms and statistical models that learn from data, identify patterns, and make predictions without specific instructions, optimizing numerous parameters (Baratchi et al., 2024). Several ML algorithms applied in WQ evaluation of surface river waters include artificial neural networks (ANN), autoregressive integrated moving average (ARIMA), polynomial neural network (PNN), cascade correlation neural network, deep neural network (DNN), convolutional neural network (CNN), and adaptive neuro-fuzzy inference systems (ANFIS) (Sharbiani et al., 2022). Common types also include decision trees, support vector machines (SVM), random forests, neural networks, logistic regression, gradient boosting (GB) machines, XGBoost, and k-nearest neighbors, all of which efficiently predict the water quality index (WQI) with high accuracy (Hussein et al., 2023; Khoi, Quan, Linh, Nhi, & Thuy, 2022a; Nasir et al., 2022). Dissolved oxygen (DO) concentration can be accurately analyzed using ANN, specifically the radial basis function algorithm (Ay & Kisi, 2012). The attention-based recurrent neural network model is particularly efficient in predicting DO levels (Liu, Zhang, Song, & Chen, 2019). In hypoxic river systems, where DO concentration is low, many models fail to predict accurately, but the SVM is found to be the best for analyzing DO in such conditions (Ji, Shang, Dahlgren, & Zhang, 2017).

Several steps are involved in obtaining data on WQ parameters using AI and ML. This process starts with the collection and identification of data on water pollution from various sources (Koelmans et al., 2019; Moroni et al., 2016). It includes the acquisition and preprocessing of remote sensing data as well as the measurement and analysis of WQ parameters at the ground level. These parameters can be categorized into physical, chemical, and biological aspects (Chauhan & Verma, 2015; Siyang & Kerdcharoen, 2016). The collected data are then fitted into a selected model, which is tested for accuracy by identifying and correcting missing or duplicate values to reduce noise and enhance performance (Emmanuel et al., 2021).

WQ issues persist globally, with many regions experiencing deteriorating conditions. Clean, freshwater is essential for both terrestrial and aquatic life. Various human activities, such as agricultural runoff, sewage discharge, and industrial waste, contribute to the nutrient enrichment of freshwater (Kling et al., 2014). ML can be applied to surface WQ monitoring, wastewater treatment, and water quality management systems. This methodology involves real-time water quality monitoring, tracking pollutant sources, analyzing their concentrations, allocating the water resources, and optimizing wastewater treatment methods (Zhu et al., 2022). Numerous WQ analysis models have evolved recently, involving several steps (Sharma, Sharma, & Jindal, 2021; Tiyasha et al., 2020). Therefore the book chapter will mainly emphasize on different AI approaches for WQ monitoring and assessment.

5.1.1 Need of water quality analysis

River WQ in India has been declining over time due to increase in the pollution from industrial processes, agricultural land, domestic sewage discharge, erosion due to deforestation, etc. Various surveys have shown incredible pollution of surface water bodies (Chawla, Kumar, Singh, Hasteer, & Ghanshyam, 2015). The swift increase in water pollution is a significant cause for alarm as it deteriorates the quality of the water and makes it unfit for any purpose (Khullar & Singh, 2022). Water is a very important resource, not only for the life on the Earth but also for the environment and economic development. When natural composition of the water is deteriorated due to entry of various contaminants into water bodies (streams, lakes, ponds, rivers, groundwater bodies, etc.), such condition is referred to as pollution. The primary contaminants found in water are acids, alkalies, detergents, fertilizers, pesticides, oil, heat from industrial processes, heavy metals, nutrients, pathogens, etc. (Göksu & Ziya, 2015). The water's fitness for drinking and potential health effects are examined based on the WQI calculation and comparison with guidelines, which is only possible with the help of WQ analysis (Kılıç, 2020).

5.1.2 Need of water quality analysis by artificial intelligence and machine learning technologies

Testing water for quality checks involves tedious and step-wise chemical examinations; this makes the process continuous and exhausting (Nanjundan, George, & Vij, 2022). Conventional laboratory examination is costly, and it consumes a lot of time. Algorithms employed in ML techniques are developed to forecast WQ in a short span of time (Pal, 2022). ML-powered real-time monitoring systems quickly identify abnormalities in water samples so that timely action can be taken; this system works across wide aquatic systems (Perera, Karunasiri, Rajapaksha, & Sirisena, 2024).

5.1.3 A case study of water quality in the Chambal River, Rajasthan

Rajasthan is the largest and most water-scarce state in India. The state is home to a large number of freshwater lakes and dams (Meena & Sharma, 2023). Except for Chambal and Mahi, all of the current rivers are seasonal, and the region's 39.65% of land has almost minimal drainage. Due to water scarcity, most people get their water from groundwater for their water requirements. This case study focuses on the Chambal River WQ, the longest river in Rajasthan, which flows through Kota City, a famous industrial and educational center in Rajasthan state, India. The Chambal River water is used for agriculture, drinking, residential, and industrial purposes by the local communities. In the absence of a functioning sewage infrastructure, the entire city of Kota dumps garbage and polluted water into the Chambal River (Gupta, Nafees, Jain, & Kalpana, 2011).

The case study below examines the Chambal River's WQ metrics at two sites in Rajasthan, India: Keshoraipatan in Bundi District and Rangpur in Kota District. The data are presented based on the most current year. The data for this study were taken from the Rajasthan Pollution Control Board website. The data cover Rangpur, Kota District, Rajasthan, from June 2024 to January 2020 (Fig. 5.1) and December 2019 to January 2014 (Fig. 5.2). Fig. 5.3 depicts statistics for Keshoraipatan in Bundi District from May 2024 to January 2020, whereas Fig. 5.4 provides data from December 2019 to January 2014. Ten years of data were collected for all parameters, and their mean values were used in this case study. The WQ metrics of the Chambal River at Rangpur fluctuate over time (2014–24), showing changes in the river's ecological health. While some metrics, such as pH, temperature, and DO, have remained reasonably consistent, others, such as conductivity, turbidity, BOD, COD, and fecal coliform, have fluctuated significantly. These changes underline the importance of constant monitoring and management in addressing future WQ issues and ensuring the river ecosystem's sustainability.

The pH levels were relatively steady, with only small changes. The lowest pH was measured in 2019 (8.11) and the highest in 2024 (8.76). Conductivity has fluctuated significantly over the years, indicating variations in the concentration of dissolved ions. Conductivity fluctuated, peaking at 767.42 μmho/cm in 2022 and dropping to 543.63 μmho/cm in 2020. Turbidity has been growing, with noteworthy maxima in 2019 (7.61 JTU/NTU) and 2023 (16.26 JTU/NTU),

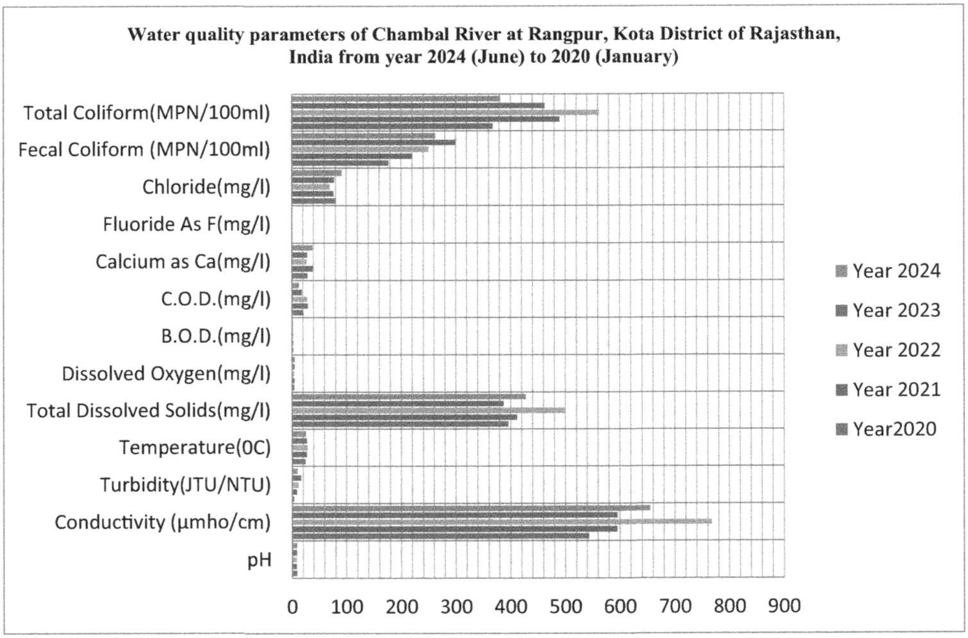

Figure 5.1
Water quality parameters of the Chambal River at Rangpur, Kota District of Rajasthan, India, from 2024 (June) to 2020 (January).

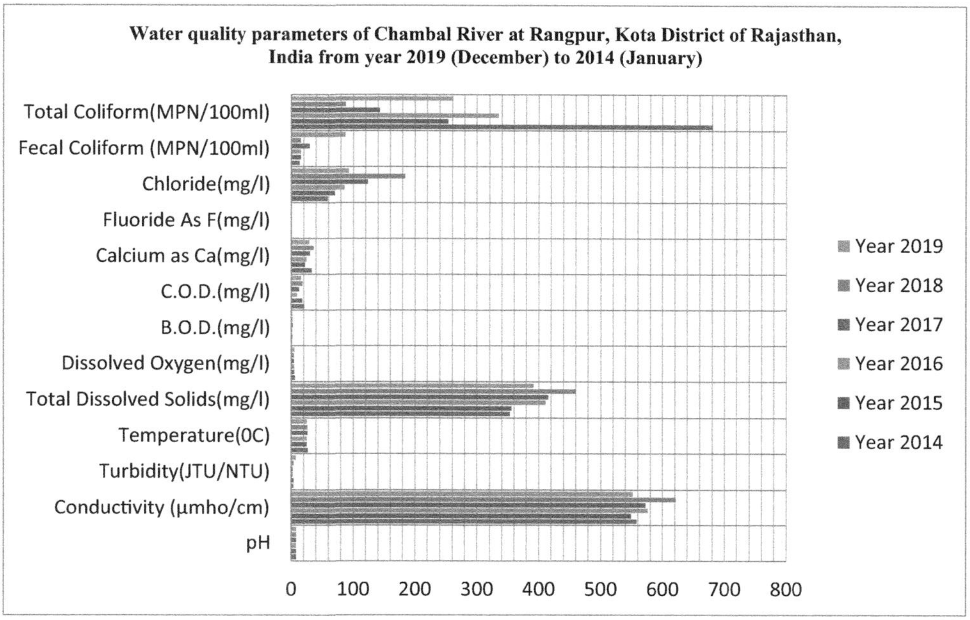

Figure 5.2
Water quality parameters of the Chambal River at Rangpur, Kota District of Rajasthan, India, from 2019 (December) to 2014 (January).

indicating probable river disturbances. Temperature has displayed usual seasonal changes, with no noticeable long-term trend, peaking in 2022 (28.91°C).

TDS levels have fluctuated significantly, peaking in 2018 (460 mg/L) and 2022 (498.5 mg/L), reflecting fluctuations in the concentration of dissolved minerals and salts. DO levels have fluctuated somewhat, with no notable long-term trend, indicating moderate to good WQ. The lowest DO level was measured in 2022 (4.25 mg/L). BOD levels fluctuated, peaking in 2017 (2.71 mg/L). BOD levels show the amount of organic matter in the water, and higher readings imply more pollution. COD levels have fluctuated significantly, reaching peaks in 2021 and 2018 (29.57 and 18.42 mg/L, respectively). Calcium levels have varied, with the highest in 2024 (37.76 mg/L). Calcium levels have shown fluctuations, with peaks in 2018 and 2024, indicating changes in mineral content. Fluoride levels have been steadily declining, reaching their lowest point in 2023 (0.35 mg/L). Chloride levels have fluctuated, peaking in 2018 (184.33 mg/L). These differences could be attributed to changes in the river's flow patterns or inputs from nearby locations. Fecal coliform levels peaked in 2023 (300 MPN/100 mL), indicating possible bacterial infection, whereas total coliform levels fluctuated. Total coliform levels varied, with the highest in 2014 (680.25 MPN/100 mL).

The WQ of the Chambal River near Keshoraipatan has fluctuated in several parameters over the past decade (2014–24) (Figs. 5.3 and 5.4). While pH and DO levels remained reasonably

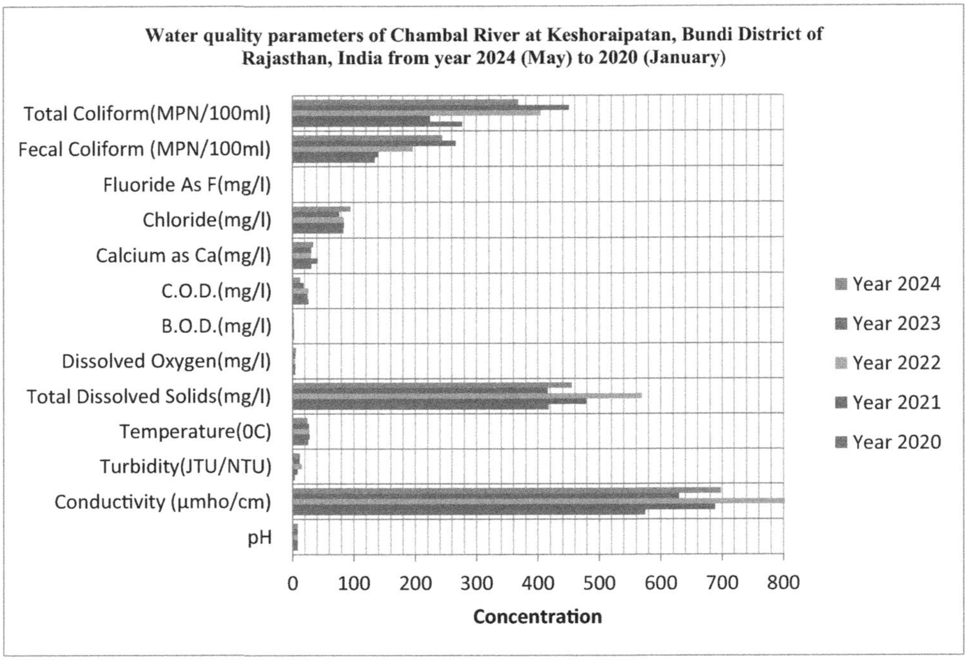

Figure 5.3
Water quality parameters of the Chambal River at Keshoraipatan, Bundi District of Rajasthan, India, from 2024 (May) to 2020 (January).

steady, conductivity, total dissolved solids (TDS), and coliform counts all changed significantly. These findings underline the importance of ongoing monitoring and management to ensure the river's WQ while also protecting its biodiversity and agricultural uses. The pH levels ranged between 8.23 and 9.12 over the years, indicating an alkaline nature. There were no notable departures from this range, indicating a pH-stable environment. Conductivity increased from 553.01 μmho/cm in 2014 to 698.12 μmho/cm by 2024.

This rising trend indicates an increase in dissolved ions in the water, which could be attributed to a variety of factors such as pollution and increased mineral content. Turbidity varied from 3.10 to 15.12 JTU/NTU. Higher turbidity levels were recorded in 2014, 2019, and 2021, indicating periods of decreasing water clarity. This could be attributed to an increase in silt load or suspended particles in the water. Temperature changes were minimal, with the greatest recorded in 2021 (28.00°C) and the lowest in 2024 (24.10°C).

TDS increased from 351 mg/L in 2014 to 443.21 mg/L in 2018, before significantly decreasing to 407.12 mg/L in 2019. TDS increased significantly, reaching 570.34 mg/L in 2022 before falling to 455.11 mg/L in 2024. DO levels vary between 4 and 6 mg/L. While there were significant deviations, the general trend was rather consistent, indicating sufficient oxygen levels for aquatic life. BOD levels remained low and showed no

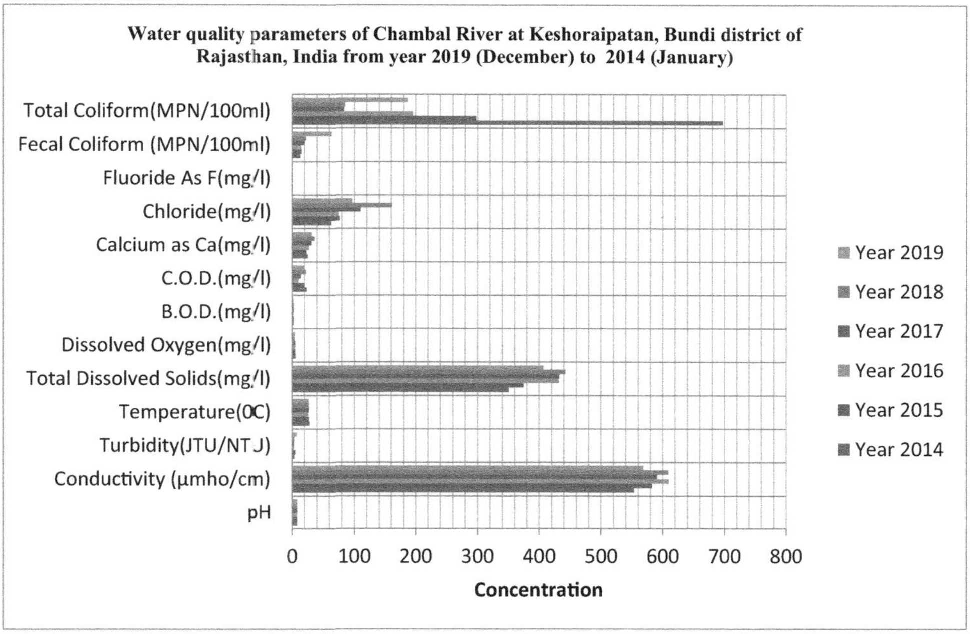

Figure 5.4
Water quality parameters of the Chambal River at Keshoraipatan, Bundi District of Rajasthan, India, from 2019 (December) to 2014 (January).

notable changes. The BOD levels were 2.89 mg/L in 2024, 2.57 mg/L in 2023, 2.67 mg/L in 2022, 2.45 mg/L in 2021, and 3.12 mg/L in 2020. COD values were greater between 2018 and 2020 (27.32 mg/L), indicating periods of increased chemical pollution. Calcium levels vary, with 35.23 mg/L in 2024, 42.22 mg/L in 2021, 31.56 mg/L in 2019, 36.34 mg/L in 2018, and 25.21 mg/L in 2014, whereas chloride levels ranged from 63.54 to 161.57 mg/l for the selected decade. Fluoride levels remained steady at approximately 1 mg/L. Fecal and total coliform levels fluctuated significantly between 2014 and 2024, with fecal coliform ranging from 13 MPN/100 mL to 244 MPN/100 mL and total coliform ranging from 84 MPN/100 mL to 451 MPN/100 mL.

5.2 Major artificial intelligence techniques used in water quality analysis, monitoring, and management

In recent years, researchers have increasingly turned to AI techniques to predict WQ. The basis ML flow chart is shown in Fig. 5.5. These methods include ANN model, multivariate linear regression (MLR) model, ANFIS, long short-term memory (LSTM), and fuzzy time series (FTS). These models excel at capturing both the straightforward and complex relationships within WQ data.

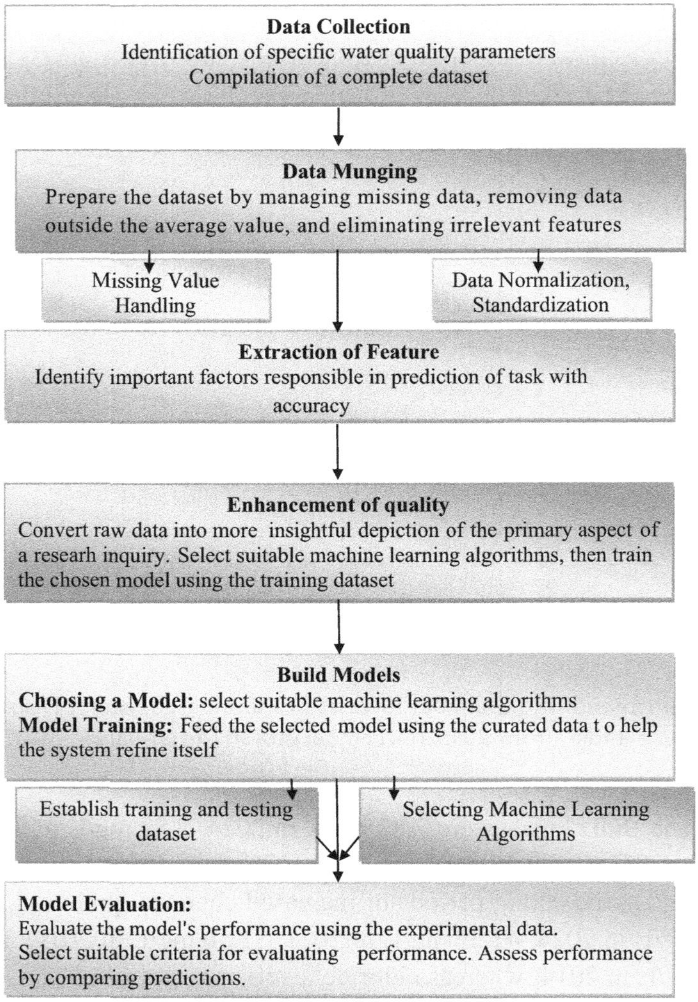

Figure 5.5
Steps in building models through machine learning.

5.2.1 Artificial neural network

The model is a sort of ML algorithm that can identify complicated patterns in data, which makes it helpful for identifying variations and trends within data on WQ (Rana et al., 2023). An ANN model consists of an input, hidden, and an output layer for data output. It is composed of interconnected neurons that process input data mathematically and transmit the results to neurons in subsequent layers, ultimately producing the output. For example, the DO concentration in surface water can be analyzed using a two-layer ANN model, as demonstrated in a study of the Yamuna River in India (Sarkar & Pandey, 2015). ANNs can also predict other

WQ parameters such as water temperature, conductivity, TDS, and turbidity in streams (Piotrowski, Napiorkowski, Napiorkowski, & Osuch, 2015). The study of Rana et al. (2023) concluded that various AI approaches were used to analyze different WQ factors at Ashwani Khad, Solan, Himachal Pradesh. They found that the MSE and root mean square error (RMSE) for ANN and LSTM approaches ranged from 0.52 to 0.6 and 0.04 to 0.21, respectively. The LSTM approach demonstrated a higher accuracy of 95% for surface WQ monitoring.

5.2.2 Multivariate linear regression

MLR is a statistical tool that involves the use of comprehensive multivariate statistical techniques such as correlation analysis and multiple linear regression models (Chowdhury & Zahrani, 2014). Its speed and simplicity contribute to its widespread adoption in analyzing large and complex WQ datasets (Kadam, Wagh, Muley, Umrikar, & Sankhua, 2019; Liu et al., 2018; Zare, 2014). The PCA, factor analysis (FA), hierarchical cluster analysis (HCA), and discriminant analysis are the most widely utilized multivariate analytic techniques. The goal of HCA is to cluster water samples according to their squared Euclidean distance, the most widely used unit of measurement for distance. Higher levels of heterogeneity are seen in samples from distinct clusters compared to samples from the same cluster. In hydrogeochemical research, PCA and FA are mainly utilized to describe linear combinations between various variables using a correlation-centered methodology. PCA utilizes a few variables, or principal components, to keep the same level of variance. FA, on the other hand, estimates factors that represent underlying structures not directly quantifiable. These multivariate techniques allow for efficient and insightful analysis of complex WQ data, enabling researchers to identify patterns, relationships, and underlying factors that impact WQ.

5.2.3 Adaptive neuro-fuzzy inference system

An innovative AI method termed the ANFIS models complicated, nonlinear relationships in data by combining the fuzzy logic with neural networks. The model can handle the variability and imprecision associated with WQ measurements by using fuzzy logic. Fuzzy logic creates procedures that can cope with incomplete data by utilizing functions of membership and linguistic variables. The data input, fuzzification, rule, normalization, and defuzzification layers are five different layers that comprise the ANFIS model. In Fig. 5.6, the significance of these strata is illustrated. A fuzzy inference system (FIS) and ANN are successfully combined in the ANFIS model. Fuzzy logic is used in FIS to create a mapping from input to output (Chanal, Steiner, Petrone, Chamagne, & Péra, 2022). ANN interconnectivity with anthropomorphic thought is incorporated in ANFIS (Walia, Sinh, & Sharma, 2015). The best parameters produced using a genetic algorithm can be used to make ANFIS more efficient and optimum (Jangir and Satavalekar, 2022).

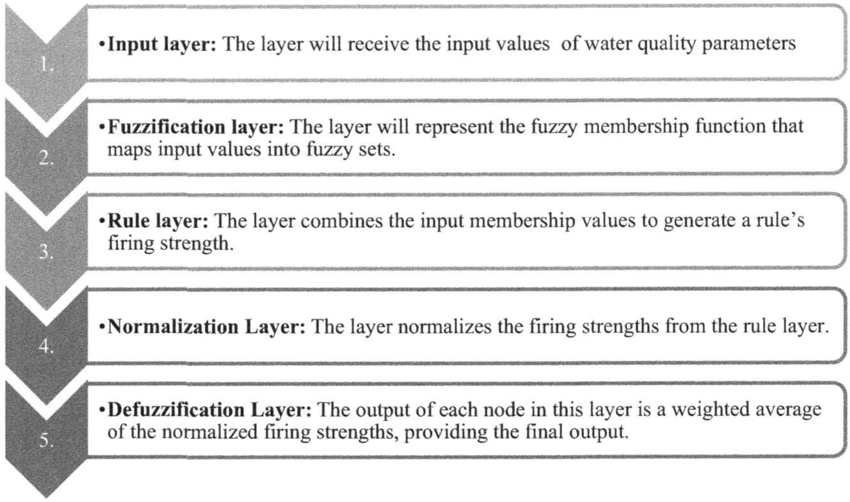

Figure 5.6
Different layers of adaptive neuro-fuzzy inference system.

5.2.4 Fuzzy time series

Over time, numerous techniques have been explored to enhance forecasting accuracy, including traditional statistical models as well as soft computing and artificial intelligence techniques (Huarng, 2001;Sadaei, Enayatifar, Guimarães, Mahmud, & Alzamil, 2015). In FTS technique, fuzzy relational equations are employed (Song and Chissom, 1993). In numerous studies, they are reported for their fine accuracy, simplicity, accountability, and low calculation complexity (Lucas, Orang, Silva, Mendes, & Guimaraes, 2021).

In recent years, FTS methods have gained more attention due to their good accuracy. Soft computing techniques such as FTS generate nonparametric, easily comprehensible, computationally accessible, and data-driven models for time series analysis and forecasting. In addition, they do not require a large dataset as in statistical models and deep learning models (Silva, 2019).

5.3 Prediction of various water quality parameters

5.3.1 Prediction of dissolved oxygen

The different ML models such as extreme gradient boosting (Xgboost), support vector regression (SVR), and mixture density networks, and according to their study, ensemble machine learning model performed the best for analyzing DO concentration in water (Zhu et al., 2022). According to Tian et al. (2022), they used various algorithms such as extreme gradient boost, SVR, ANN, and RF for DO prediction in the water and found that extreme gradient boost was the best method among all other methods. Wang et al. (2013) used ANN, ARIMA, BANN (bootstrapped artificial neural network), and BWNN (bootstrapped wavelet neural network) to analyze DO concentration in the water. According to their comparative study, the hierarchy of performance of different models was in the following decreasing order: BWNN > BANN > WNN > ANN > ARIMA. The study conducted by Siljic Tomic, Antanasijevic, Ristic, Peric-Grujic, and Pocajt (2018) used a PNN for DO analysis with the input parameters such as BOD, COD, EC, calcium, and magnesium and found that the model performed very well with near about 74% of accuracy. According to Zhi et al. (2021), tested the LSTM model for DO prediction and reported that the algorithm's performance was up to the mark.

5.3.2 Prediction of multiple water quality parameters

An ANFIS model can predict the total phosphorus and total nitrogen content in the water samples, which makes it more reliable (Khadr and Elshemy, 2016). It can also be used to calculate approximately the BOD of surface water; one of the examples is the Surma River in Bangladesh (Ahmed and Shah, 2017). ANIS model can predict the EC and TDS in the surface water (Najah, El-Shafie, Karim, & Jaafar, 2010). Wavelet-ANFIS model can also be used to assess the groundwater level (Yan, Ma, Song, & Zhou, 2015). DNN, SVR, and RF are reliable for BOD estimation with input parameters such as TDS, pH, temperature, DO salinity, and total coliform (Ma, Ding, Cheng, Jiang, & Xu, 2020). ANFIS performed best in both quantitative and classification problems related to algal bloom in the surface water and includes physical, chemical, and biological input parameters (Sagan et al., 2020). The different approaches of AI, such as ANN, SVM, CNN, and RF, can be used for water pollution monitoring, according to Pu, Ding, Chao, Yu, and Xu (2019) and Wu, Zhang, Xiao, and Feng (2020). Principal component analysis model is suited for heavy metal contamination assessment and can also predict different WQ parameters (Kumar et al., 2018; Tripathi and Singal, 2019). SVM performs better than ANN algorithm for parameter analysis such as EC, CO_3^{2-}, HCO_3^-, SO_4^{--}, Cl_2, TDS, Na, Mg, Ca, total nitrogen, and total phosphorous (Parsaie, Nasrolahi, & Haghiabi, 2018). The approaches such as RF and DT can also be used to predict WQ levels with different input parameters such as chlorophyll-a, temperature of H_2O, pH scale, DO, and flow rate, according to Castrillo and Garcia (2020) and Chen and Al (2020).

5.3.3 Prediction of water quality index using machine learning algorithms

In the last few decades, the WQ has been affected by contamination and different pollutants. One of the most used approaches for the assessment of quality is the WQI. It is derived from a single value ranging from 0 to 100 that combines physical, chemical, and biological factors. Earlier, the WQI was calculated on the basis of nine factors. The quality of drinking water can be predicted by knowing its WQI (Arjaria, Singh, Rathore, & Gupta, 2024). Due to the enhancement of AI approaches to calculate the WQI, there are about eight commonly used algorithms, including different models such as DT and RF techniques, SVM, Extra Tree, and Xgboost. Thus ExT and GXB models outperformed to calculate WQI with lower RMSE and MSE values (Uddin, Nash, Mahammad Diganta, Rahman, & Olbert, 2022). The flow chart of basic ML model is depicted in Fig. 5.7. According to Elvin and Wibowo (2024), XGBoost algorithm model produces an accurate prediction of WQ.

The formula is as follows:

$$A = \frac{\Sigma_{i=1} N q_i \times w_i}{\Sigma_{i=1} N w_i}$$

where q_i is the quality rating scale of a single parameter and W_i is weight (Chauhan and Verma, 2016).

$$q_i = 100 \times \left(\frac{V_i - V_{ideal}}{S_i - V_{ideal}} \right) \text{ and } W_i = \frac{K}{S_i}$$

where V_i is the observed value of variable (i) in the analyzed water samples, S_i is the suggested standard value of variable (i), and V_{Ideal} is the ideal value of variable (i) in pure water for each parameter with the exception of DO = 14:6 mg/L, pH = 7:0, and K is constant.

According to Aldhyani, Al-Yaari, Alkahtani, and Maashi (2020), various ML models, such as nonlinear auto-regressive neural network (NARNET), LSTM, and SVM approaches, are the best to calculate WQI. The SVM approach achieved about (97.01%) of the highest accuracy than the other said approaches, with a slight difference in coefficient of determination (NARNET = 96.17% and LSTM = 94.21%). ML is also helpful for predicting WQ classification. Few studies have been conducted to develop predictive models of WQI for rivers by using a more complex neutral network (Khuan, Hamzah, & Jailani, 2002). According to Shams et al. (2024), different approaches such as RF approach, Xgboost, GB, and adaptive boosting (AdaBoost) approach are used for WQC. They proved

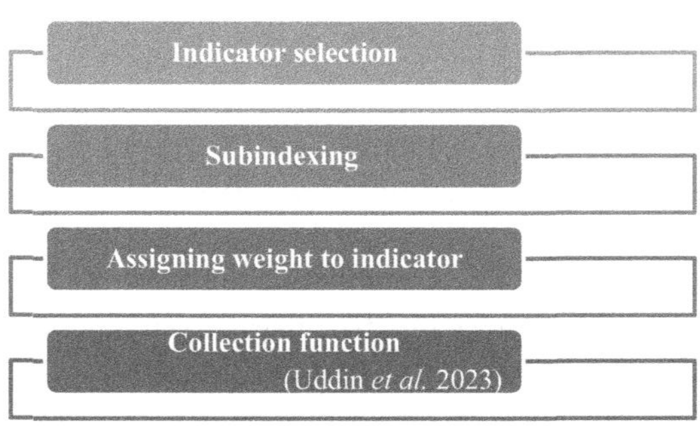

Figure 5.7
ML approach consists of different processes. *ML,* Machine learning.

that GB method, known as GB, predicts WQC more accurately with the highest accuracy of 99.50%, than other methods of ML algorithms. The MLP regressor algorithm outperformed with the highest accuracy of 99.8% in calculating WQI than KNN, DT, and SVM models. Khoi, Quan, Linh, Nhi, and Thuy (2022b) used 12 ML models to measure the quality of water. Ahmed et al. (2019) estimation of the WQI was done using supervised ML techniques. Kouadri, Elbeltagi, Islam, and Kateb (2021) predicted the WQI utilizing eight AI systems, including the correlation coefficient (*R*), mean absolute error (MAE), RMSE, relative absolute error, and root relative square error. Nair and Vijaya (2021) used a number of ML and Big Data prediction algorithms based on sensor networks (Fig. 5.8).

5.4 Discussion

Analyzing the WQ indicators for Rangpur (Kota District) and Keshoraipatan (Bundi District) indicates some key implications for the Chambal River's overall health and management. Both sites exhibit varied WQ measures, showing changes in the river's ecological health. Consistent monitoring and management are critical for addressing WQ issues and preserving the river's environment. Critical WQ metrics such as electrical conductivity (EC), TDS, and turbidity vary between sites. Such fluctuations frequently reflect changes in pollutant levels or river conditions. Rangpur has more dramatic fluctuations, whereas Keshoraipatan has an increasing tendency in conductivity and TDS. These trends point to increased pollution and mineral composition changes in the river, both

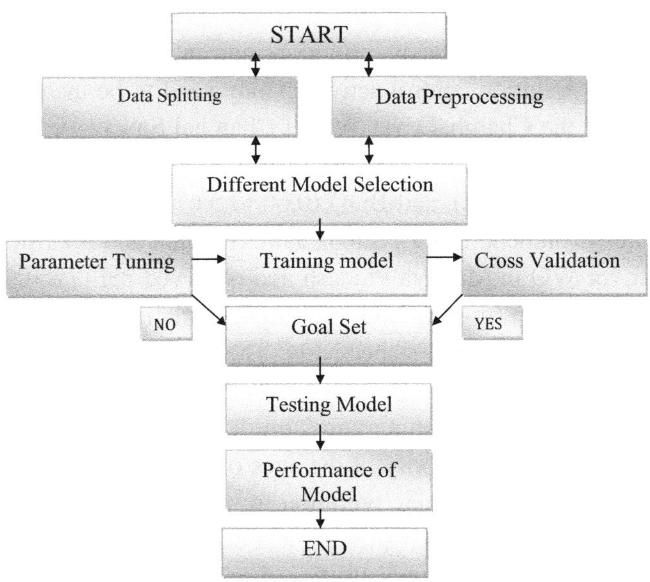

Figure 5.8
Machine learning models flow chart.

of which could have an impact on aquatic ecosystems. Both sites have changing BOD and COD levels, with peaks in particular years indicating increased organic and chemical pollution. Increased COD levels, in particular, indicate major instances of chemical pollution, which can damage WQ and kill aquatic life. Both locations retain moderate to good DO levels, with Rangpur showing more variability, including a low point in 2022. Adequate DO is critical for aquatic life, and substantial decreases can stress or even kill aquatic creatures. Consistent monitoring is required to ensure that DO levels remain within acceptable limits for ecosystem health. Turbidity at Rangpur is increasing, while turbidity at Keshoraipatan is changeable, indicating sediment load disturbances or shifts. High turbidity can restrict light penetration, harming aquatic plants and disturbing ecosystems. It may also suggest soil erosion or runoff problems that affect WQ. Both sites show variable fecal and total coliform levels, with noticeable peaks indicating bacterial contamination. Inadequate wastewater treatment or agricultural runoff is a common cause of this contamination. High levels of coliform bacteria endanger human health and highlight the need for better sanitation and pollution control. Variations in calcium, chloride, and fluoride levels at both sites indicate changes in mineral imports or pollution sources. Rising calcium and chloride levels, combined with variable fluoride levels, can affect water chemistry and aquatic life. Effective management techniques are required to meet these developments.

5.4.1 Comparative analysis and historical context

Keshoraipatan has more constant pH and DO levels, indicating a somewhat stable chemical environment, whereas Rangpur's measurements are more variable. Historical data show that, while some WQ measurements stay steady, pollution indicators such as COD and TDS are growing. This means that pollution control activities may need to be increased. Saksena, Garg, and Rao (2008) evaluated the Chambal river WQ at Chambal Sanctuary, Madhya Pradesh, in the year 2006, and they reported EC (145.60–884 μS/cm), TDS (260–500 mg/L), pH (7.60–9.33), DO (4.86–14.59 mg/L), and BOD (0.60 to 5.67 mg/L), respectively, and found a pollution-free aquatic environment. Shroti, Singh, Jain, Sharma, and Sharma (2021) also examined Chambal River WQ in Madhya Pradesh and observed pH (7.89), EC (430.91 μS/cm), TDS 356.25 (mg/L), B.O.D. 8.5 (mg/L), D.O. 6.2 (mg/L), calcium 32.32 (mg/L), chloride 37.52 (mg/L), and fluoride 0.33 (mg/L) respectively. They also assessed the WQI, which reported good WQ of the river Chambal flowing in Madhya Pradesh. Tiwari and Singh (2008) reported significant increases in COD (minimum 24mg/L to maximum 60 mg/L from October 2004 to July 2005) readings at the majority of the sites of the Chambal River in Dholpur District of Rajasthan and indicated organic pollution from industries and residential areas during the time period covered by the data. Both sites show a need for continuing monitoring and management to combat pollution and safeguard the river's environment. Historical data and new studies highlight the significance of taking proactive efforts to prevent additional degradation and maintain the river's health.

5.4.2 Limitations of the water quality analysis

The data only cover certain time periods and may not reflect seasonal variations or long-term trends. This constraint can result in an imperfect picture of the river's health over time. The analysis focuses solely on two sites. The Chambal River passes through several regions, and circumstances may differ dramatically in areas not addressed by this study. A more thorough examination, including additional sites, would provide a more accurate picture of the river's health. While key WQ indicators such as EC, TDS, turbidity, BOD, COD, DO, and coliform levels are addressed, other critical factors such as heavy metals, pesticides, and emerging pollutants are not. These can have a substantial impact on both WQ and ecosystem health. The procedures employed for sampling and analysis determine the data's accuracy and consistency. Any differences in data-gathering methods or laboratory procedures may have an impact on the results' reliability. The analysis fails to properly account for the impact of human activities (e.g., industrial discharge, agricultural practices) and environmental conditions (e.g., climate change, natural disasters) on WQ. Comparisons with historical data [e.g., studies by Saksena et al. (2008) and Shroti et al. (2021)] can be beneficial, although differences in research design, geographic focus, and measuring procedures may limit their utility. The investigation focuses on changes in WQ but does not thoroughly investigate the ecological consequences for aquatic life and biodiversity. Detailed ecological studies would be required to fully comprehend the ramifications for the river's environment.

5.4.3 Recommendations for management

5.4.3.1 Enhanced monitoring

Regular and extensive monitoring of WQ metrics at both sites is critical for detecting changes early and responding quickly.

5.4.3.2 Pollution control

Addressing pollution sources, such as wastewater discharge and agricultural runoff, is critical. Implementing stronger restrictions and improved waste management can help to reduce pollution.

5.4.3.3 Restoration efforts

Initiatives to restore and maintain riparian zones, reduce soil erosion, and regulate agricultural practices can all help to improve WQ and reduce turbidity.

5.4.3.4 Public awareness

Educating local communities about the effects of pollution and the need of water conservation can help improve management practices and reduce human-caused pollution.

5.5 Conclusion

Water resources are at risk of contamination; therefore assessing and maintaining WQ is vital for human population. As the population grows, pollution is depleting our water supplies. As a result, the analysis and prediction of WQ metrics have been completely transformed by advanced technologies like AI and ML. These technologies provide efficient and quick evaluations by facilitating the monitoring of several parameters. The WQI can potentially be predicted with a high degree of accuracy using different kinds of AI approaches. This helps early detection systems by identifying pollutants and processing spectrum data from several sources. These models have surpassed conventional methodologies, enabling estimates of WQ that are more accurate and timely. The AI approaches perform in a distinct area of WQ prediction. Furthermore, to fully realize the potential of these technologies, multidisciplinary cooperation between scientists, engineers, and local people is essential. The relevance of AI in tackling global WQ challenges is highlighted by its application in wastewater treatment, surface WQ monitoring, and management. Stakeholders may contribute to the sustainability and well-being of diverse water resources by investing in and implementing these cutting-edge technologies, which guarantee safer and cleaner water for all living forms.

References

Ahmed, A. A. M., & Shah, S. M. A. (2017). Application of adaptive neuro-fuzzy inference system (ANFIS) to estimate the biochemical oxygen demand (BOD) of Surma River. *Journal of King Saud University - Engineering Sciences*, *29*(3), 237–243.

Ahmed, U., Mumtaz, R., Anwar, H., Shah, A. A., Irfan, R., & García-Nieto, J. (2019). Efficient water quality prediction using supervised machine learning. *Water*, *11*. Available from https://doi.org/10.3390/w11112210.

Aldhyani, T. H. H., Al-Yaari, M., Alkahtani, H., & Maashi, M. (2020). Water quality prediction using artificial intelligence algorithms. *Applied Bionics and Biomechanics*, 2020. Available from https://doi.org/10.1155/2020/6659314.

Arjaria, S. K., Singh, S., Rathore, A. S., & Gupta, R. K.(2024). Assessing water quality with machine learning: A comprehensive analysis of prediction methods and performance evaluation (p. 17). Available from https://doi.org/10.4018/979-8-3693-2.351-9.ch001.

Ay, M., & Kisi, O. (2012). Modeling of dissolved oxygen concentration using diferent neural network techniques in Foundation Creek, El Paso County, Colorado. *Journal of Environment Engineering*, *138*, 654–662.

Baratchi, M., Wang, C., Limmer, S., Rijn, J. N. V., Hoos, H., Back, T., & Olhofer, M. (2024). Automated machine learning: Past, present and future. *Artifcial Intelligence Review*, *57*, 122. Available from https://doi.org/10.1007/s10462-024-10726-1.

Castrillo, M., & Garcia, A. L. (2020). Estimation of high frequency nutrient concentrations from water quality surrogates using machine learning methods. *Water Research*, *172*, 115490. Available from https://doi.org/10.1016/j.watres.2020.115490.

Chanal, D., Steiner, N. Y., Petrone, R., Chamagne, D., & Péra, M. C. (2022). Online diagnosis of PEM fuel cell by fuzzy C-means clustering. *Encyclopedia of Energy Storage*, *2*, 35. Available from https://doi.org/10.1016/B978-0-12-819723-3.00099.

Chauhan, A., & Verma, S. C. (2015). Impact of agriculture, urban and forest land use on physico-chemical properties of water: A review. *Internattional Journal of Current Microbiology and Applied Science*, *4*(10), 18–22.

Chauhan, A., & Verma, S. C. (2016). Impact of land uses and seasons on physico-chemical characteristics of surface water in Solan district of Himachal Pradesh. *Nature Environment and Pollution Technology*, *15*(2), 667–672.

Chawla, P., Kumar, P., Singh, M., Hasteer, N., & Ghanshyam, C. (2015). Prediction of pollution potential of Indian rivers using empirical equation consisting of water quality parameters. In *2015 IEEE technological innovation in ICT for agriculture and rural development (TIAR), Chennai, India* (pp. 214–219). Available from https://doi.org/10.1109/TIAR.2015.7358560.

Chen, K., & Al, E. (2020). Comparative analysis of surface water quality prediction performance and identification of key water parameters using different machine learning models based on big data. *Water Research*, *171*, 115454. Available from https://doi.org/10.1016/j.watres.2019.115454.

Chowdhury, S., & Zahrani, M. A. (2014). Water quality change in dam reservoir and shallow aquifer: Analysis on trend, seasonal variability and data reduction. *Environmental Monitoring and Assessment*, *186*, 6127–6143.

Elvin., & Wibowo, A. (2024). Forecasting water quality through machine learning and hyperparameter optimization. *Indonesian Journal of Electrical Engineering and Computer Science*, *33*(1), 496–506.

Emmanuel, T., Maupong, T., Mpoeleng, D., Semong, T., Mphago, B., & Tabona, O. (2021). A survey on missing data in machine learning. *Journal of Big Data*, *8*, 140. Available from https://doi.org/10.1186/s40537-021-00516-9.

Göksu, M., & Ziya, L. (2015). *Water pollution*. Ankara: Academician Bookstore.

Gupta, N., Nafees, S. M., Jain, M. K., & Kalpana, S. (2011). Physico-chemical assessment of water quality of river Chambal in Kota City area of Rajasthan state (India). *Rasayan Journal of Chemisry*, *4*(2), 295–298.

Huarng, K. H. (2001). Heuristic models of fuzzy time series for forecasting. *Fuzzy Sets and Systems*, *123*(3), 369–386. Available from https://doi.org/10.1016/S0165-0114(00)00093-2.

Hussein, E. E., Jat Baloch, M. Y., Nigar, A., Abualkhair, H. F., Aldawood, F. K., & Tageldin, E. (2023). Machine learning algorithms for predicting the water quality index. *Water*, *15*, 3540.

Jangir, H. K., & Satavalekar, R. (2022). Evaluating adaptive neuro-fuzzy inference system (ANFIS) to assess liquefaction potential and settlements using CPT test data. *Journal of Soft Computing in Civil Engineering*, *6*(3), 119–139.

Ji, X., Shang, X., Dahlgren, R. A., & Zhang, M. (2017). Prediction of dissolved oxygen concentration in hypoxic river systems using support vector machine: A case study of Wen-Rui Tang River, China. *Environmental Science and Polluion Reseach.*, *24*, 16062–16076.

Kadam, A. K., Wagh, M., Muley, A. A., Umrikar, B. N., & Sankhua, R. N. (2019). Prediction of water quality index using artificial neural network and multiple linear regression modelling approach in Shivganga River basin, India. *Modeling Earth Systems and Environment*, *5*(3), 951–996.

Kaddoura, S. (2022). Evaluation of machine learning algorithm on drinking water quality for better sustainability. *Sustainability*, *14*(18), 11478. Available from https://doi.org/10.3390/su141811478.

Khadr, M., & Elshemy, M. (2016). Data-driven modeling for water quality prediction case study: The drains system associated with Manzala Lake, Egypt. *Ain Shams Engineering Journal*, 1–9.

Khoi, D. N., Quan, N. T., Linh, D. Q., Nhi, P. T. T., & Thuy, N. T. D. (2022a). Using machine learning models for predicting the water quality index in the La Buong River, Vietnam. *Water*, *14*, 1552.

Khoi, D. N., Quan, N. T., Linh, D. Q., Nhi, P. T. T., & Thuy, N. T. D. (2022b). Using machine learning models for predicting the water quality index in the La buong river, Vietnam. *Water*, *14*, 10. Available from https://doi.org/10.3390/w14101552.

Khuan, L.Y., Hamzah, N., & Jailani, R. (2002). Prediction of water quality index (WQI) based on artificial neural network (ANN). In *Proceedings of the student conference on research and development, Shah Alam, Malaysia.*

Khullar, S., & Singh, N. (2022). River water quality classification using a hybrid machine learning technique. In *2022 ninth international conference on computing for sustainable global development (INDIACom), New Delhi, India* (pp. 808–813). doi: 10.23919/INDIACom54597.2022.9763301.

Kılıç, Z. (2020). The importance of water and conscious use of water. *International Journal of Hydrology*, *4*(5), 239–241. Available from https://doi.org/10.15406/ijh.2020.04.00250.

Kling, C. L., Panagopoulos, Y., Rabotyagov, S. S., Valcu, A. M., Gassman, P. W., Campbell, T., . . . Rabalais, N. N. (2014). LUMINATE: linking agricultural land use, local water quality and Gulf of Mexico hypoxia. *European Review of Agricultural Economics*, *41*(3), 431–459.

Koelmans, A. A., Mohamed Nor, N. H., Hermsen, E., Kooi, M., Mintenig, S. M., & De France, J. (2019). Microplastics in freshwaters and drinking water: Critical review and assessment of data quality. Water. *Reseach*, *155*, 410–422.

Kouadri, S., Elbeltagi, A., Islam, A. R. M. T., & Kateb, S. (2021). Performance of machine learning methods in predicting water quality index based on irregular data set: Application on Illizi region (Algerian southeast). *Applied Water Science*, *11*, 12. Available from https://doi.org/10.1007/s13201-021-01528-9.

Kumar, V., Sharma, A., Kumar, R., Bhardwaj, R., Kumar Thukral, A., & RodrigoComino, J. (2018). Assessment of heavy-metal pollution in three different Indian water bodies by combination of multivariate analysis and water pollution indices. *Human Ecology and Risk Assessment*, *26*, 1–16. Available from https://doi.org/10.1080/10807039.2018.1497946.

Liu, Y., Zhang, Q., Song, L., & Chen, Y. (2019). Attention-based recurrent neural networks for accurate short-term and long-term dissolved oxygen prediction. *Computer and Electronics in Agriculture*, *165*, 104964.

Liu, Z., Joo, J. C., Choi, S. H., Heo, N., Jang, J., & Hur, J. W. (2018). Assessment of surface water quality in Geum River Basin, Korea using multivariate statistical techniques. *International Journal of Applied Engineering Research*, *13*(9), 6723–6732.

Lucas, P. O., Orang, O., Silva, P. C. L., Mendes, E. M. A. M., & Guimaraes, F. G. (2021). A tutorial on fuzzy time series forecasting models: Recent advances and challenges, learning and nonlinear models. *Journal of the Brazilian Society on Computational Intelligence*, *19*(2), 29–50.

Ma, J., Ding, Y., Cheng, J. C. P., Jiang, F., & Xu, Z. (2020). Soft detection of 5-day BOD with sparse matrix in city harbor water using deep learning techniques. *Water Research*, *170*, 115350. Available from https://doi.org/10.1016/j.watres.2019.115350.

Malviya, A., & Jaspal, D. (2021). Artificial intelligence as an upcoming technology in wastewater treatment: A comprehensive review. *Environment Technology Reviews*, *10*(1), 177–187.

Meena, R., & Sharma, V. K. (2023). Assessment of surface water quality and physicochemical analysis of Mamchari Dam in District Karauli (Rajasthan), India. *Biological Forum – An International Journal*, *15*(5), 859–869.

Moroni, D., Pieri, G., Salvetti, O., Tampucci, M., Domenici, C., & Tonacci, A. (2016). Sensorized buoy for oil spill early detection. *Methods Oceanography*, *17*, 221–231.

Nair, J. P., & Vijaya, M. S. (2021). Predictive models for river water quality using machine learning and big data techniques - A survey. In *Proceedings of the international conference on artificial intelligence and smart systems (ICAIS)*, March 2021, Coimbatore, India. IEEE.

Najah, A. A., El-Shafie, A., Karim, O. A., & Jaafar, O. (2010). Water quality prediction model utilizing integrated wavelet-ANFIS model with cross-validation. *Neural Computing and Applications*, *21*(5), 833–841.

Nanjundan, P., George, J. P., & Vij, A. (2022). A reliable method of predicting water quality using supervised machine learning model. In *2022 IEEE international conference on data science and information system (ICDSIS)*, Hassan, India (pp. 1–5). Available from https://doi.org/10.1109/ICDSIS55133.2022.9915819.

Nasir, N., Kansal, A., Alshaltone, O., Barneih, F., Sameer, M., Shanableh, A., & Al-Shamma'a, A. (2022). Water quality classification using machine learning algorithms. *Jounal of Water Process Engineeing*, *48*, 102920.

Oğuz, A., & Ertuğrul, O. F. (2023). A survey on applications of machine learning algorithms in water quality assessment and water supply and management. *Water Supply*, *23*(2), 895–922.

Pal, K. O. (2022). The quality of drinkable water using machine learning techniques. *International Journal of Advanced Engineering Research and Science*, *9*(6), 16–23.

Parsaie, A., Nasrolahi, A. H., & Haghiabi, A. H. (2018). Water quality prediction using machine learning methods. *Water Quality Research Journal*, *53*, 3–13. Available from https://doi.org/10.2166/wqrj.2018.025.

erera, N., Karunasiri, W., Rajapaksha, R. M. L. S., & Sirisena, K. (2024). AI-based machine learning algorithms for water quality analysis: A review. In *14th International conference on sustainable built environment*, Sri Lanka, Vol. 2.

Piotrowski, A. P., Napiorkowski, M. J., Napiorkowski, J. J., & Osuch, M. (2015). Comparing various artificial neural network types for water temperature prediction in rivers. *Journal of Hydrology*, *529*(1), 302–315.

Pu, F., Ding, C., Chao, Z., Yu, Y., & Xu, X. (2019). Water-quality classification of Inland lakes using landsat8 images by convolutional neural networks. *Remote Sensing*, *11*, 141674. Available from https://doi.org/10.3390/rs11141674.

Ragi, N. M., Holla, R., & Manju, G. (2019). Predicting water quality parameters using machine learning. *Four*th ınternational conference on recent trends on electronics, ınformation, communication & technology (RTEICT) (pp. 1109–1)112. IEEE.

Rana, R., Kalia, A., Boora, A., Alfaisal, F. M., Alharbi, R. S., Berwal, P., ... Qamar, O. (2023). Artificial intelligence for surface water quality evaluation, monitoring and assessment. *Water*, *15*(22), 3919. Available from https://doi.org/10.3390/w15223919.

Ritchiel, J. C., & Schiebe, F. R. (2000). Water quality. *Remote sensing in hydrology and water management* (pp. 287–303). Springer.

Riedl, M. O. (2019). Human-centered artificial intelligence and machine learning. *Human Behaviour and Emerging Technology*, *1*, 33–36.

Sadaei, H. J., Enayatifar, R., Guimarães, F. G., Mahmud, M., & Alzamil, Z. (2015). Combining ARFIMA models and fuzzy time series for the forecast of long memory time series. *Neurocomputing*, *175*. Available from https://doi.org/10.1016/j.neucom.2015.10.079.

Sagan, V., Peterson, K. T., Maimaitijiang, M., Sidike, P., Sloan, J., Greeling, B. A., ... Adams, C. (2020). Monitoring inland water quality using remote sensing: Potential and limitations of spectral indices, bio-optical simulations, machine learning, and cloud computing. *Earth Science Review*, *205*, 103187. Available from https://doi.org/10.1016/j.earscirev.2020.103187.

Saksena, D. N., Garg, R. K., & Rao, R. J. (2008). Water quality and pollution status of Chambal river in National Chambal sanctuary, Madhya Pradesh. *Journal of Environmental Biology*, *29*(5), 701–710.

Sarkar, A., & Pandey, P. (2015). River water quality modelling using artificial neural network technique. *Aquatic Procedia*, *4*, 1070–1077.

Shams, M. Y., Elshewey, A. M., El-kenawy, E. S. M., Ibrahim, A., Talaat, F. M., & Tarek, Z. (2024). Water quality prediction using machine learning models based on grid search method. *Multimedia Tools and Applications*, *83*, 35307–35334.

Sharabiani, V. R., Kaveh, M., Taghinezhad, E., Abbaszadeh, R., Khalife, E., Szymanek, M., & Dziwulska-Hunek, A. (2022). Application of artificial neural networks, support vector, adaptive neuro-fuzzy inference systems for the moisture ratio of parboiled hulls. *Applied Science*, *12*(4), 1771. Available from https://doi.org/10.3390/app12041771.

Sharma, N., Sharma, R., & Jindal, N. (2021). Machine learning and deep learning applications- A vision. *Global Transitions Proceedings*, *2*(1), 24–28.

Shroti, K., Singh, M. K., Jain, V. K., Sharma, R., & Sharma, S. (2021). Water quality assessment and pollution analysis of one of the major tributary rivers of Yamuna in Central India: The Chambal River. *Bulletin of Environment, Pharmacology and Life Sciences*, *11*(1), 52–60.

Siljic Tomic, A., Antanasijevic, D., Ristic, M., Peric-Grujic, A., & Pocajt, V. (2018). A linear and non-linear polynomial neural network modeling of dissolved oxygen content in surface water: Inter- and extrapolation performance with inputs' significance analysis. *Science of Total Environment*, *610–611*, 1038–1046. Available from https://doi.org/10.1016/j.scitotenv.2017.08.192.

Silva P. C. D. L. (2019). Scalable models for probabilistic forecasting with fuzzy time series. Thesis.

Siyang, S., & Kerdcharoen, T. (2016). Development of unmanned surface vehicle for smart water quality inspector. In *Proceedings of the 2016 13th international conference on electrical engineering/electronics, computer, telecommunications and information technology (ECTI-CON)*, Chiang Mai, Thailand, 28 June–1 July 2016 (pp. 1–5).

Song, Q., & Chissom, B. S. (1993). Fuzzy time series and its model. *Fuzzy Sets and Systems, 54*(3), 269–277. Available from https://doi.org/10.1016/0165-0114(93)90372-O.

Tian, S., Guo, H., Xu, W., Zhu, X., Wang, B., Zeng, Q., ... Huang, J. J. (2022). Remote sensing retrieval of inland water quality parameters using Sentinel-2 and multiple machine learning algorithms. *Environment Science and Pollution Research, 30*, 18617–18630.

Tiwari, P. S., & Singh, M. K. (2008). Physico-chemical study of Chambal river water in Dholpur (Rajasthan). *Current World Environment, 3*(2), 289–292.

Tiyasha., Tung, T. M., & Yaseen, Z. M. (2020). A survey on river water quality modelling using artificial intelligence models: 2000–2020. *Journal of Hydrology, 585*. Available from https://doi.org/10.1016/j.jhydrol.2020.124670.

Tripathi, M., & Singal, S. K. (2019). Use of principal component analysis for parameter selection for development of a novel water quality index: a case study of river Ganga India. *Ecologgical Indicator, 96*, 430–436. Available from https://doi.org/10.1016/j.ecolind.2018.09.025.

Uddin, M. G., Nash, S., Mahammad Diganta, M. T., Rahman, A., & Olbert, A. I. (2022). Robust machine learning algorithms for predicting coastal water quality index. *Journal of Environmental Management*, 321–115923. Available from https://doi.org/10.1016/j.jenvman.2022.115923.

Walia, N., Sinh, H., & Sharma, A. (2015). ANFIS: Adaptive neuro-fuzzy inference system- A survey. *International Journal of Computer Applications, 123*, 32–38.

Wang, Y., Zheng, T., Zhao, Y., Jiang, J., Wang, Y., Guo, L., & Wang, P. (2013). Monthly water quality forecasting and uncertainty assessment via bootstrapped wavelet neural networks under missing data for Harbin, China. *Environment Science and Pollution Research, 20*, 8909–8923. Available from https://doi.org/10.1007/s11356-013-1874-8.

Wu, Y., Zhang, X., Xiao, Y., & Feng, J. (2020). Attention neural network for water image classification under IoT environment. *Applied Science, 10*, 030909. Available from https://doi.org/10.3390/app10030909.

Yan, Q., Ma, C., Song, Y., & Zhou, W. (2015). Wavelet ANFIS combination model for. *Revista Tecnica De La Facultad De Ingenieria Universidad Del Zulia, 39*(2), 317–328.

Zare, A. H. (2014). Evaluation of multivariate linear regression and artificial neural networks in prediction of water quality parameters. *Journal of Environmental Health Science & Engineering, 12*(1), 1–8.

Zhi, W., Feng, D., Tsai, W. P., Sterle, G., Harpold, A., Shen, C., & Li, L. (2021). From hydrometeorology to river water quality: Can a deep learning model predict dissolved oxygen at the continental scale? *Environment Science and Technology, 55*, 2357–2368. Available from https://doi.org/10.1021/acs.est.0c06783.

Zhou, X., Li, C., Rahaman, M. M., Yao, Y., Ai, S., Sun, C., Wang, Q., Zhang, Y., Li, M., Li, X., et al. (2020). A comprehensive review for breast histopathology image analysis using classical and deep neural networks. *IEEE Access, 8*, 90931–90956.

Zhu, M., Wang, J., Yang, X., Zhang, Y., Zhang, L., Ren, H., ... Ye, L. (2022). A review of the application of machine learning in water quality evaluation. *Ecology Environment and Health*, 107–116.

Zhu, X., Guo, H., Huang, J. J., Tian, S., Xu, W., & Mai, Y. (2022). An ensemble machine learning model for water quality estimation in coastal area based on remote sensing imagery. *Journal of Environment Management, 323*, 116187.

CHAPTER 6

Machine learning and artificial intelligence application in automotive water quality monitoring, analysis, and management

Arvind Kumar, Abdul Gaffar Sheik, Faizal Bux and Sheena Kumari

Institute for Water and Wastewater Technology, Durban University of Technology, Durban, South Africa

Nomenclature

AdBoost	Adaptive boosting
ANFIS	Adaptive neuro-fuzzy inference system
ANN	Artificial neural network
AI	Artificial intelligence
ARIMA	Autoregressive integrated moving average
BOD	Biological oxygen demand
CatBoost	Categorical gradient boosting
COD	Chemical oxygen demand
CEEMDAN	Complete ensemble empirical mode decomposition with adaptive noise
CVPS	Cross-validation parameter selection
DT	Decision tree
ERT	Extremely randomized tree
DO	Dissolved oxygen
GA	Genetic algorithm
GBT	Gradient boosted trees
GEP	Gene expression programming
GMDH	Group method of data handling
GPR	Gaussian process regression
GWO	Gray wolf optimization
LR	Logistic regression
LRM	Linear regression model
LSTM	Long short-term memory
MCS	Monte carlo simulation
MLPNN	Multi-layer perceptron neural network
MLR	Multiple linear regression
MLPNN	Multi-layer perceptron neural network
NB	Naive Bayes
NAS	Neural architecture search

Computational Automation for Water Security.
DOI: https://doi.org/10.1016/B978-0-443-33321-7.00004-4

NN	Neural network
TDS	Total dissolved solids
TH	Total hardness
TON	Total organic nitrogen
RBF-ANN	Radial basis function neural networks
REPT	Reduced error pruning tree
RT	Random tree
REPT	Reduced error pruning tree
RFC	Randomizable filtered classification
SAR	Sodium absorption ratio
SMOTE	Synthetic minority oversampling technique
SSA	Sparrow search algorithm
WDT-ANFIS	Neuro-fuzzy inference system based augmented wavelet de-noising technique
XGBoost	Extreme gradient boosting
URMS	Unweighted root means squared aggregation function
VFA	Volatile fatty acid

6.1 Introduction

Water is one of the most valuable resources necessary for all life. Water pollution lowers the quality of the water, which affects the health of marine life and, consequently, that of humans who utilize it. Because of this, it is imperative to monitor water quality (WQ) and guarantee the survival of aquatic species (Jain, Shah, & Mehta, 2021). The long-term feasibility of a diversion plan depends on the quality of the water. Since resources must be reallocated to fix water distribution infrastructure whenever a problem arises, low-quality water may also be expensive. To ensure that drinking water is safe and affordable, there is an increasing need for improved water management and quality control. Organizational monitoring concerns, disposal methods, and freshwater evaluations must all be thoroughly examined to address these problems (Hu, Zhang, & Zhao, 2019). The process of predicting WQ involves predicting the features of fluctuations in the state of a water system at a given point in time. Planning and regulating WQ depend heavily on the results of WO assessments. By projecting future updates in water purity at different levels of pollution and developing practical water pollution prevention and control procedures, methods for avoiding and regulating water pollution may be improved. Plans for the diversion of water ought to assess the overall consistency of the water. To address common drinking challenges, a significant amount of water is transported. Therefore research should be done on ways to predict WQ today (Zhou, Wang, & Xiao, 2018). Nowadays, security concerns require the application of artificial intelligence (AI) and machine learning (ML) innovations, which map the relationship between system inputs and their outcomes rather than emphasizing intricate operational techniques (Waqas et al., 2022).

Conventional methods of evaluating WQ, such as laboratory analysis and computation using quantifiable criteria, can be expensive, time-consuming, and limited in their ability to provide real-time data. (Ahmed, Mumtaz, Anwar, Mumtaz, & Qamar, 2020). An alternative approach is provided by predictive modeling, which uses ML techniques to estimate WQ based on available data. Based on predicted WQ categories, decision-makers can target automotive WQ monitoring, and analysis and optimize treatment options (Sundui et al., 2021). Early detection systems for problems with WQ can be developed using predictive models as their base of operation (Lowe, Qin, & Mao, 2022). The fields of AI and ML-driven WQ monitoring and automation are currently converging at a promising and dynamic point. AI is being used for many creative purposes, such as creating predictive models for WQ, building neural networks to identify specific pollutants such as nitrogen and phosphorous, and automating water treatment processes with the use of AI-powered decision-making tools. The fields of AI-driven WQ monitoring and treatment are currently converging at a promising and dynamic point. AI is being used for many creative purposes, such as creating predictive models for water processes, building neural networks to identify specific pollutants like heavy metals and infections, and automating water treatment processes with the use of AI-powered decision-making aids (Godavarthi, Nalajala, & Ganapuram, 2017). Alongside the aforementioned expansion, the Internet of Things (IoT) is becoming more widely used. IoT devices produce continuous data streams that are used by better AI models to achieve increased accuracy and speedy analysis. In the field of water process engineering, researchers are currently focusing on removing obstacles related to data security, privacy, and ethical application of AI (Kumari et al., 2021). The use of AI and ML in WQ treatment and automation is a stunning example of how far technology has come. Furthermore, it denotes a commitment to advancing environmental sustainability and protecting public health, thus portending a future in which technology and ecology will be together harmoniously.

The remaining sections of the chapter are arranged as follows: Real-time water quality monitoring and prediction are included in Section 6.2. Section 6.3 provides contaminant detection and source identification. Automated water quality management systems are reported in Section 6.4. Data-driven decision making and integration of IoT are elucidated in Section 6.5. Lastly, recommendations and conclusions are drawn from the chapter.

6.1.1 Bibliographic analysis

After conducting a search on Google Scholar (GS), PUBMED (PM), Scopus, and Web of Science (WofS) databases, the number of publications for the keywords "water quality prediction using AI and ML, automated water quality management, data-driven decision making for WWTP, Internet of Things, contaminant detection, and source identification by employing ML in WWTP" is shown in Fig. 6.1A, respectively: GS (18500), PM (16), Scopus (120), and WofS (6); GS (88000), PM (241), Scopus (596), and WofS (485); GS (15600), Scopus (10), PM (5), and WofS (13); GS (18100), PM (20), Scopus (174), and WofS (52); GS (17000), PM (6), Scopus (43), and WofS (12). In conclusion, there were

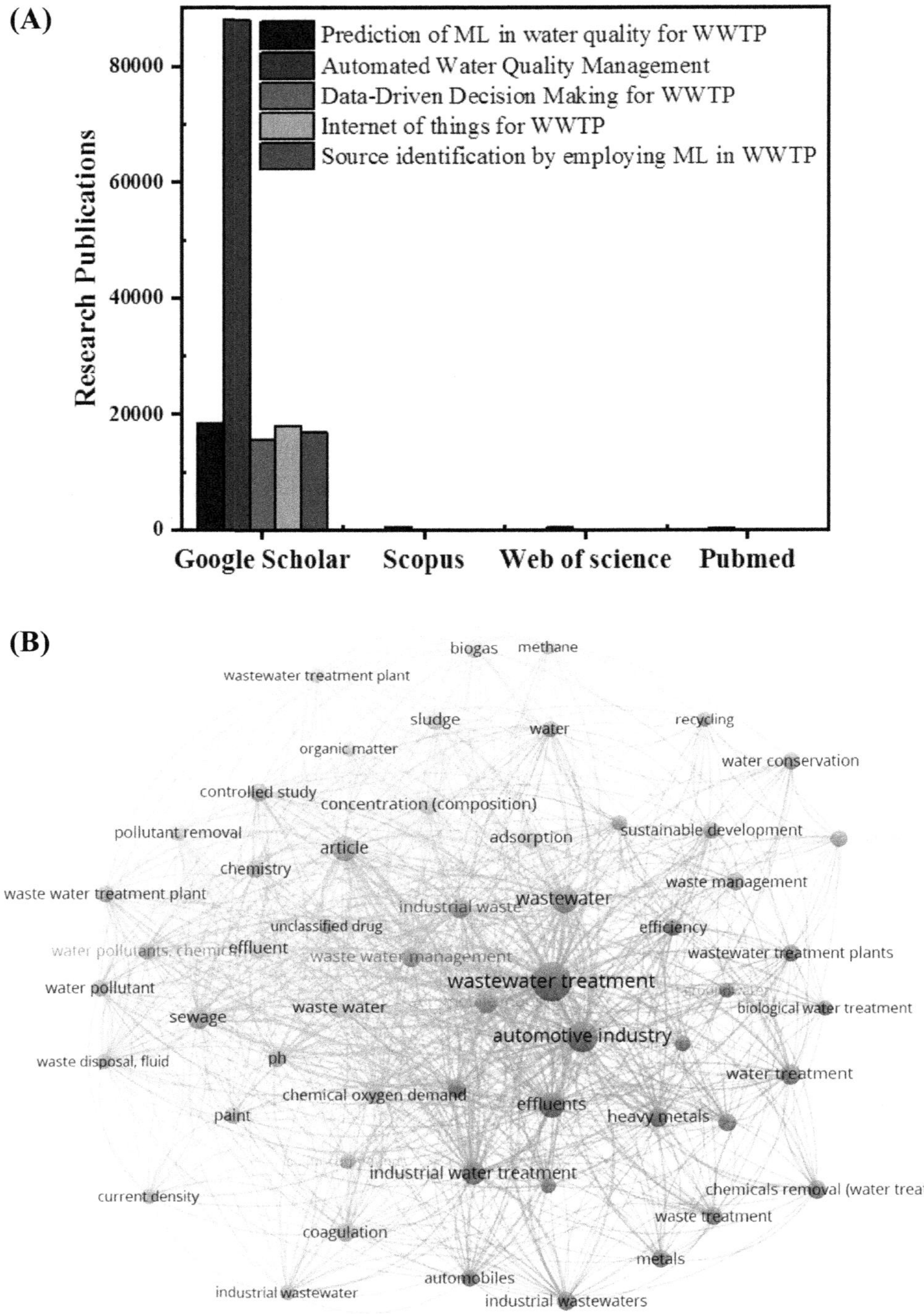

Figure 6.1

(A) Bar graph representation of various applications of AI and ML in WWTP; (B) representation of co-occurrence map with keyword of wastewater treatment and automation. *AI*, Artificial intelligence; *ML*, machine learning.

differences in the quantity of articles for the given keywords between the four databases. Nevertheless, a number of the articles from the listed sources are pertinent to the keywords and offer information about automation, IoT, WQ, AI and ML, and WWTP. Data were collected from 2013 to 2024; every term was examined separately. Fig. 6.1A represents the bar graph of various applications of AI and ML in WWTP.

The study starts with a collection of terms (keywords) associated with phosphorus, wastewater, and other relevant subjects. The terms "automation AND wastewater AND treatment" were employed in this investigation. In addition, this study's main data source is the Scopus database, which is among the most reliable scientific databases. With each search round, no more than 1000 "new" OR "highly cited" articles may be found and converted to CSV files. Information is assessed using the free and open-source VOS viewer bibliometric tool. This study used more than 500 papers on automation, wastewater, and related subjects to gather data and conduct research. Less than 100 publications are covering a certain issue, such as wastewater treatment and digitization, between 2013 and 2024. Data filtration was started once the data were loaded into VOS viewer to exclude repetitive and unrelated keywords (like paper, study, etc.). Different combinations of terms (like wastewater and automation) were combined and recorded as a single unique phrase. Fig. 6.1B displays the initial data mapping result. Three groupings (colors) represent the research on automation and wastewater. Keywords like wastewater, pollutants removed, and nutrients are highlighted and repeated with large circles. Furthermore, it is implied that there is overlap between two keywords the closer two items are to each other and the more prominently they interact. According to Fig. 6.1B, distinct clusters (depicted by color) were produced by the author keywords (from the Scopus database) assigned to modeling and wastewater treatment papers. These clusters represented the worldwide hotspots in the field of wastewater treatment.

6.2 Discussion section

This section discusses all the ML and AI applications in automotive WQ monitoring, analysis, and management as follows:

6.2.1 Real-time water quality monitoring and prediction using AI and ML

AI and ML are popular methods for data analysis that are used to find patterns or predict outcomes from large amounts of data gathered from various WQ scenarios. Data collection, the proper algorithm choice, training models, and model testing must be done before AI and ML are put into practice (Ahmed et al., 2019). Selecting an appropriate algorithm is essential for these procedures. Two primary categories of ML algorithms are supervised and unsupervised learning (Berry, Mohamed, & Yap, 2019). Supervised learning has been employed for data classification and regression, and various techniques are available, such as LR, artificial neural network (ANN), DT, SVM, NB, k-nearest neighbor (KNN), RF, and so on (Nasir et al., 2022). On the other hand, unsupervised learning uses clustering and dimensionality reduction to categorize the training data into distinct groups based on their unique features. Typically used unsupervised ML methods include K-means, KNN, and others (Zamri et al., 2022). Furthermore, another category of ML method is called reinforcement learning, which describes a machine's capacity to generalize and provide accurate answers to issues that it has not been taught. It is rarely used in the subject of aquatic environments, nevertheless, in contrast to the other two ML classes (Hasan et al., 2019). Due to their great precision and accuracy in estimating linear, nonlinear, and complicated systems, a variety of ML methods were used for modeling and optimizing various water and WWT systems (Mahadeva, Manik, Goel, & Dhakal, 2019). Wastewater treatment system modeling and optimization have been studied using tools such as SVM, ANN (Ismail, 2024), fuzzy networks (Srivastava, Handa, & Madhulika, 2022), IoT (Asli, 2025), deep belief networks with event-triggered learning (Wang, Jia, Zhou, Bi, & Qiao, 2021; Wang, Thunéll, et al., 2021), wireless sensors and support vector machines (SVMs) (Alprol, Mansour, Ibrahim, & Ashour, 2024), decision tree, multivariate adaptive regression splines (Jumber, Damtie, & Tegegne, 2024), and long short-term memory (LSTM) (Inbar and Avisar, 2024). ML models have the significant benefit of reflecting actual reaction/ process scenarios as opposed to preformulated processes based on fundamental principles. Because many of the mechanisms underlying wastewater treatment are still unknown, they must be strong and thorough (Nadiri, Shokri, Tsai, & Moghaddam, 2018). Consequently, ML modeling is frequently utilized as an alternative to mechanistic modeling for WWTPs (Liu, Zhang, & Zhang, 2020). This is an opportunity for researchers, as understanding how operational parameters impact effluent quality is crucial for developing engineering scenarios that enhance WWTP processes. Tree-based models have an advantage over ANN-based models in terms of interpretability, but they also have a similar variation captured (called "accuracy/precision" in prediction scenarios) (Chowdhury, Rahman, Rahman, & Mahdy, 2020; Kumari & Toshniwal, 2021). This is significant because it improves the accuracy of model interpretation and analysis by capturing more variation. The WQ parameters of wastewater treatment with applications of ML are reported in Table 6.1.

6.2.2 Contaminant detection and source identification by employing ML

In general, factors like pH, temperature, concentrations of substrates, and pollutants can all have an impact on a WWTP's WQ. This is so because microorganisms employ their metabolism to remediate wastewater. Nonetheless, biological treatment displays extremely nonlinear and time-varying properties that are influenced by known and unknown parameters (Mjalli, Al-Asheh, & Alfadala, 2007). Owing to these intricate aspects, a mathematical model for process simulation and control was used in numerous prior researches to assess and

Table 6.1: Comparative study on water quality parameters of wastewater treatment with applications of ML.

S. No.	Models used	Parameters	Remarks	References
1	ANN, GMDH, and SVM	Temperature (*T*), pH, conductivity, bicarbonate, sulfates, chlorides, TDS, sodium, magnesium, calcium	SVM had a smaller data dispersion than the other two when the performance of GMDH, SVM, and ANN was compared using water quality parameters. Their parameters are nearly equal, even though model GMDH's accuracy is lower than model SVM's. Additionally, an analysis of the applied models' performance revealed that GMDH and SVM model results were more trustworthy than ANN's.	Haghiabi, Nasrolahi, and Parsaie (2018)
2	MLP-ANN, ANFIS, RBF-ANN, WDT-ANFIS	*T*, conductivity, salinity, nitrate, turbidity, phosphate, chloride, potassium, sodium, magnesium, iron, and *Escherichia coli* (*E. coli*)	The prediction accuracy of the WDT-ANFIS model was significantly improved, outperforming all other suggested models for every parameter related to water quality.	Ahmed et al. (2019)
3	XGBoost and RF, CEEMDAN	*T*, DO, pH value, specific conductance, and turbidity.	The models with the lowest average MAPEs—3.90% and 3.71%, respectively—are CEEMDAN-RF and CEEMMDAN-XGBoost, suggesting that their overall prediction accuracy is the best.	Lu and Ma (2020)
5	GEP, ANN, and LRM	TDS and specific conductivity	The ANN approach came in second to the GEP's performance, which proved to be the most accurate. For a given set of inputs, both GEP and ANN can model the water quality parameters. It would be simple to anticipate monthly TDS and EC using the GEP formulas for TDS and EC levels.	Shah et al. (2021)
6	MLR, M5P model tree, SVR, and RF	TDS, SAR, and TH	When it came to TDS prediction (R^2 = 0.98, RMSE = 70.50), SVR prediction (R^2 = 0.99, RMSE = 0.04), and TH prediction (R^2 = 0.99, RMSE = 1.54), the RFR model performed the best. The comparison of the data showed that all stations' TDS, SAR, and TH could be correctly estimated by the ML models.	Nouraki, Alavi, Golabi, and Albaji (2021)

(*Continued*)

Table 6.1: (Continued)

S. No.	Models used	Parameters	Remarks	References
7	RF, M5P, and REPT, bagging, CVPS, and RFC	pH, DO, TDS, conductivity, salinity, chloride, total alkalinity, TH, sulfate, and nitrate	RT-ANN hybrid could not effectively predict severe WQI values. Nearly all algorithms overestimated WQI values except BA-RF, RF, BA-REPT, REPT, RFC-M5P, RFC-REPT, and ANN-ANFIS.	Aslam et al. (2022)
8	SVR, RF, LR, DT, CATBoost, XGBoost, and MLP.	DO, pH, dissolved oxygen, conductivity, BOD, nitrate, and fecal and total coliform	As the top-performing classifier and stacking model to improve accuracy, CATBoost was evaluated. A variety of datasets can be utilized with the CATBoost classifier due to its high-performance metrics.	Nasir et al. (2022)
10	NB, LR, kNN, DT and ensemble techniques, and SMOTE	20 water quality parameters are used	With an accuracy and recall of 98.1%, precision of 100%, and an AUC of 99.9%, the stacking classification model following SMOTE with 10-fold cross-validation performs better than the other models, according to the data.	Dritsas and Trigka (2023)
11	MLPNN, SVM, RF, ERT, XGBoost, and CatBoost	COD, TP, T, pH, total Kjeldahl nitrogen (TON), Tur, Chl-a, DO, oil and grease (OG), TDS, nitrite-nitrogen, orthophosphates, conductivity, chloride, fluoride, copper, zinc, and arsenic.	With R2 values of 0.917, 0.906, 0.901, and 0.937 and RMSE values of 1.978, 0.843, 2.020, and 1.572, respectively, MLPNN-SVM, ET-CatBoost, MLPNN-CatBoost, and CaBboost demonstrated the best performance in Clusters 1 through 4. According to the sensitivity analysis results, fewer WQ parameters can be used to get acceptable local prediction results.	Wang et al. (2023)
12	MCS, GPR, and URMS	Ammonia, transparency, TON, *T*, DO, ammoniacal nitrogen, pH, salinity, BOD, and chlorophyll	Higher prediction accuracy was seen in the WQM and RMS models, suggesting that these WQI models could be useful and trustworthy instruments for low-uncertainty evaluation of the state of the water quality.	Uddin, Nash, Rahman, and Olbert (2023)
13	DT, RF, GBT, ANN, and SVM	pH, electrical conductivity, DO, BOD, nitrate-nitrogen, and total coliform	The modeling tests demonstrate that GBT can predict WQI classes with >80% of performance metrics using just three input factors. GBT, RF, and ANN all have only two input parameters and can all produce	Singh, Das, and Sharma (2024)

(Continued)

Table 6.1: (Continued)

S. No.	Models used	Parameters	Remarks	References
			trustworthy estimates. Our research shows that ML models can be used by water resource policymakers as tools to assist in their decision making.	
14	AdaBoost, KNN, GBT, RF, Sup SVR, and BR	COD, TOC, oil and grease, ammonia, nitrogen, arsenic, nickel, and zinc	The GB model, which uses a combination of just seven parameters, outperformed other models in the comparative evaluation of predictive models for the water quality by accomplishing higher accuracy in both training (R^2 = 0.88, RMSE = 7.24) and testing (R^2 = 0.85, RMSE = 8.67).	Ejaz et al. (2024)
15	SSA, GWO, GA, and SVR	DO and COD	The GWO-SVR model has a stronger prediction effect than the PSO-SVR model; however, the prediction outcomes of both models are not that different. Compared to the GWO-SVR model, the SSA-SVR model has a superior prediction effect.	Hu, Lyu, Wang, Zhou, and Fang (2024)

diagnose WWTP performance (Sheik, Tejaswini, Seepana, & Ambati, 2023; Solís, Guisasola, Flores-Alsina, Jeppsson, & Baeza, 2022). Recently, new approaches for tracking pollution from point sources and detection have been made possible by data mining and ML techniques (Wang et al., 2024). Distinct pollutants from different municipal industries interact differentially with river interactions, hydrological events, and transformation processes (Budi et al., 2024). Numerous studies have been carried out to use data mining techniques like semisupervised learning and classification to extract hidden predictive information from data (Huang, Ma, Ma, Huangfu, & He, 2021). Recurrent neural networks and LSTMs are two examples of ML time-series models that are particularly good at identifying temporal sequences and correlations in water quality source identification (Wang et al., 2019). The flow chart of AI and ML applications for WQ monitoring is represented in Fig. 6.2.

6.2.3 Automated WQ management systems

The automation of applying ML techniques to real-world problems is known as automated ML or Auto ML. It was created to address problems with AI and the increased difficulty of

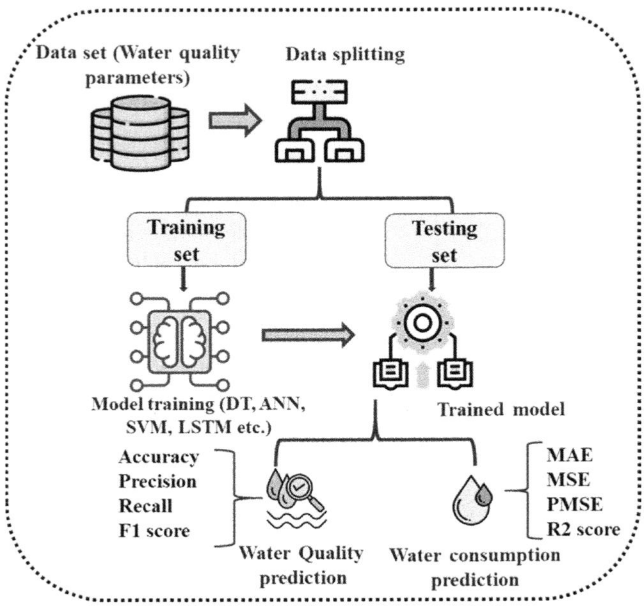

Figure 6.2
Flow chart of AI and ML applications for WQ monitoring. *AI*, Artificial intelligence; *ML*, machine learning; *WQ*, water quality.

implementing ML (Thornton, Hutter, Hoos, & Leyton-Brown, 2013). The primary innovation employed by Auto ML is the search for and optimization of hyperparameters, which are used for preliminary component processing and model selection. It makes use of optimization algorithms such as genetic algorithm platforms, which use their essential processes like mutation, crossover, genome coding, and choosing to search for appropriate hyperparameters and NN concepts automatically (Caruana, Niculescu-Mizil, Crew, & Ksikes, 2004). Bayesian optimization is a continuous design plan that considers global optimization of the hyperparameters (Mockus, 2005). NAS is an optimization used for the automated design of neural networks (Elsken, Metzen, & Hutter, 2019). Escalante (2021) analysis of Auto ML explores the technology's evolutionary history. To obtain insights, the key findings are evaluated and summarized. The pipeline of Auto ML techniques, including feature engineering, data preparation, hyperparameter optimization, and NAS, is reviewed by He, Zhao, and Chu (2021). This essay offers insights into the procedure and the gaps found during it. Zöller and Huber (2019) use 137 datasets from well-known Auto ML benchmark suits to apply a few Auto ML frameworks. Second, the subject of study on automated ML is still fairly new. This creative concept was initially applied in this study project. Despite producing effective outcomes, Prasad et al. (2021) AI and WQ are still in their infancy as the platform is only a few years old. Although Auto ML facilitates the user's automation of the entire pipeline, its applicability and effectiveness for practical issues such as WQ assessments are currently being investigated. Identifying areas for growth and performance lapses is essential for Advance Auto ML, and this will be our driving force.

6.2.4 Data-driven decision making

Process engineering and data science expertize are needed to comprehend, evaluate, and get useful information from the created data. Even though there are a large number of AI models in WWTPs, critical AI examinations of applications in the simulation and optimization of WWTPs are particularly scarce (Bhagat, Tung, & Yaseen, 2020; Singh et al., 2023). The literature has studied a variety of neural network topologies; the most employed one is feed-forward neural networks that are trained using backpropagation methods (Golzar, Nilsson, & Martin, 2020). Kazemi, Bengoa, Steyer, and Giralt (2021) developed a soft sensor for controlling VFA in anaerobic digestion using SVM. The findings show that SVM is better than ANFIS at detecting VFA faults and producing residual errors that are more accurate. LS-SVM was employed by Liu, Wan, Ma, and Wang (2019) to estimate effluent COD online in an anaerobic WWTP. Out of all the SVM studies that were reviewed, 72% employed SVM for predicting and 28% for optimization. SVM was a helpful optimization technique because of its ability to detect and distinguish. In contrast to ANN, SVM necessitates subjective interpretations and has not been computed for the local optima (Bahramian, Dereli, Zhao, Giberti, & Casey, 2023). RF was used by Zhou et al. (2019) to anticipate influent flow rates in WWTPs. Rahimian and Behnam (2020) categorized the quantity of operating pumps in a full-scale treatment plant using trees. In a similar vein, Byliński, Sobecki, and Gębicki (2019) used DTs to forecast the characteristics and strength of the odor in a biological WWTP. According to Revollar, Vilanova, Francisco, & Vega (2018), ammonium concentration indicates effluent quality, nitrate concentration at the end of the anoxic zone indicates emissions, and DO concentration indicates energy consumption by employing fuzzy logic (FL). Sheik et al. (2024) assessed the WQ in the plant's effluent using the EQ and the operational cost to measure energy performance by applying different control strategies. Overall, there is a lot of evidence to support FL's advantage when modeling procedures have partial data. Additionally, FL controllers enable the design of the processes to be controlled to incorporate expert knowledge. ANFIS and ARIMA models were integrated by Lotfi et al. (2020) to forecast the wastewater influent/effluent BOD, COD, and TSS. Mahjouri et al. (2017) identified the critical indicators for choosing the best wastewater treatment method using a novel fuzzy Delphi and fuzzy analytic hierarchy process. To train the MLP created to mimic the removal of alkylphenols in a moving bed biofilm reactor, Mohammadi, Bina, Karimi, Rahimi, and Yavari (2020) assessed the effectiveness of Levenberg Marquardt and particle swarm methods. To reduce energy consumption in a full-scale treatment facility, Zhang, Han, and Qiao (2020) tracked the achieved optimal DO set points using a mix of fuzzy neural networks and adaptive multiobjective particle swarm optimization.

A further novel concept to guarantee data quality in WWTPs is the in-situ monitoring system. Using this technique, sensors spread throughout the WWTP are connected to a data processor that is integrated with a cloud storage unit and a library of real-time past information. The signal is transferred to the cloud storage unit by the data processor after it has compared the incoming signal to comparable previous values in the library and determined that there is no substantial statistical abnormality. Soft sensors are additional useful uses of data-driven approaches in WWTPs. These sensors use several ML methods to approximate the WQ indicators reasonably. Chemical oxygen demand soft sensors are one such instance. Another illustration of an effective approach in usage is FL, which is utilized for monitoring greenhouse gas emissions in addition to detecting and treating control (Santín, Barbu, Pedret, & Vilanova, 2018). The advantages of data-driven models in decision-making are shown in Fig. 6.3.

The research community should be able to suggest novel approaches, and practitioners should be able to provide the highest level of data quality assurance to implement such dependable and sophisticated methodologies. Deeper communication between scholars with various backgrounds is also required. Successful use of innovative methodologies in industry would require strong collaboration between researchers with various backgrounds to translate knowledge into practice.

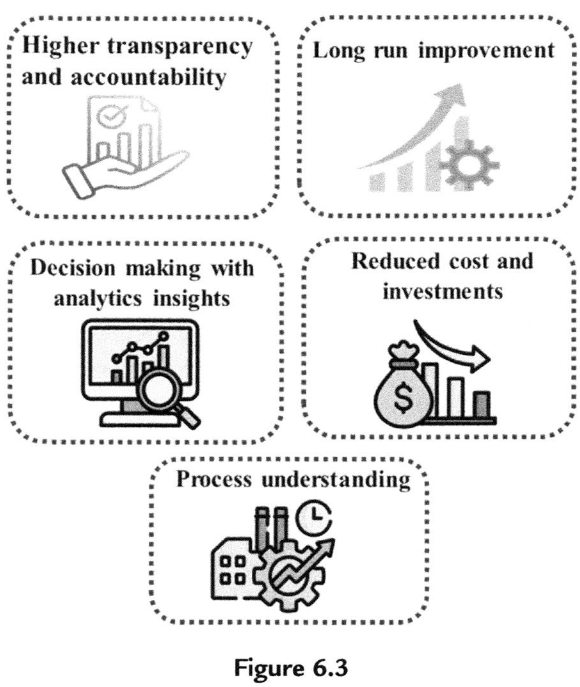

Figure 6.3
Advantages of data-driven models in decision-making.

6.2.5 Internet of Things

Connectivity between devices and sensors to gather and share data is known as the IoT. IoT devices in wastewater treatment create a full data ecosystem by offering real-time information on multiple factors. The schematic of industrial applications of IoT in WWTP is presented in Fig. 6.4. AI and ML algorithms have the potential to increase system performance by optimizing resource utilization, enhancing decision-making processes, and more (Rane, Choudhary, & Rane, 2023). However, AI gives smart water systems the ability to make better decisions and do sophisticated analytics. The vast datasets generated by IoT devices are analyzed by ML methods within AI systems (Himeur et al., 2023). They convert unprocessed data into useful insights by identifying trends and formulating predictions. Smart water management systems can anticipate possible problems, optimize operations, and adjust to changing environmental circumstances thanks to this synergy. IoT technologies are essential to improving water infrastructure monitoring and control. Smart sensors are used in water treatment plants to continuously measure important factors such as chemical concentrations, pH levels, and turbidity (Park, Kim, & Lee, 2020). Real-time data on water flow, pressure, and identification of leaks is provided via IoT-enabled devices across the distribution network. Smart meters that are IoT-enabled allow for accurate tracking of water usage by the user, which makes invoicing more effective and promotes water conservation. Remote monitoring and control are made possible by the IoT's smooth interaction with water infrastructure, which also increases data collection's timeliness and accuracy. Preventive maintenance, quick reaction to emergencies, and early anomaly identification make this capability especially useful for bolstering the water management system's overall resilience (Syrmos et al., 2023). According to Yang, Cao, Meng, and Si (2021), adaptive learning algorithms allow the system to continually enhance its

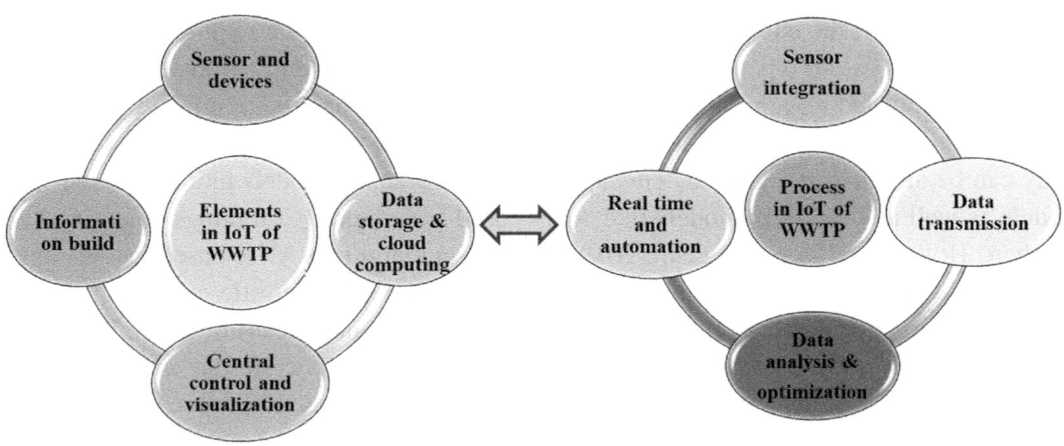

Figure 6.4
Industrial applications of IoT in WWTP. *IoT*, Internet of Things.

performance using real-time feedback, enabling a more sustainable and adaptable approach to water treatment. Predictive modeling powered by AI is also capable of anticipating equipment breakdowns, which lowers maintenance expenses and downtime. AI systems are able to detect anomalies in system behavior or water quality, which can notify operators of any problems before they get worse. The dependability and efficiency of water treatment procedures are improved by this proactive approach (Nong & Tarek, 2023; Sahu, Kaur, Singh, & Arya, 2023). When IoT and AI technologies are smoothly combined to create a comprehensive and adaptive system, smart water management's entire potential becomes apparent. AI algorithms use the constant data streams from IoT devices as inputs to produce insights and useful recommendations. Water management systems may make decisions in real-time and adapt flexibly to changing conditions thanks to this closed-loop connection.

6.2.6 Comparative analysis

The comparative analysis of different AI and ML models used in WQ monitoring reveals notable distinctions in their performance and applicability. SVM and tree-based models, such as decision trees and random forests, demonstrate superior accuracy and interpretability compared to more complex models like ANNs. SVM and tree-based models are adept at handling nonlinear relationships and can provide clear decision paths, making them easier to understand and trust for stakeholders in water management. In contrast, ANNs, while powerful and capable of capturing intricate patterns in data, often act as "black boxes," offering limited insight into their decision-making processes. This lack of transparency can be a significant barrier to their widespread adoption in critical applications where understanding the rationale behind predictions is crucial. Additionally, the computational efficiency of SVM and tree-based models makes them more suitable for real-time WQ monitoring and treatment scenarios, where quick and reliable decisions are essential. However, the choice of model also depends on the specific requirements of the application, such as the type of WQ parameters being monitored and the availability of data. Most published research has concentrated on prediction or developing soft sensors using nonlinear ML models without interpreting the models to learn how the targets under study can be affected or managed. This applies to interpretable models like tree-based models as well as black-box models like ANN-based techniques (Corominas et al., 2018; Newhart, Holloway, Hering, & Cath, 2019). There is a claim that ANN-based models are hard to understand since the weights or ANN structures reveal very little regarding the function's usage (Guidotti et al., 2018). According to Chen and Guestrin (2016), Ke et al. (2017), and other authors, tree-based models have an intrinsic structure that makes them highly interpretable. Considering the algorithms that were evaluated between conventional and Auto ML systems, it is reasonable to say that Auto ML is just as capable as a conventional model. The outcomes demonstrate a notable improvement in algorithm performance. It was also decided that each system had a few options for expansions and customization within reason. For instance, from the beginning, data shown to significantly influence both models and analysis is still being studied. AI algorithms can initiate automated responses, such as modifying treatment procedures or notifying operators of manual intervention, if IoT sensors discover an anomaly in WQ (Adeyemi, Grove, Peets, & Norton, 2017; Jansen et al., 2009). Water management has become a proactive, data-driven, and intelligent process that can precisely and swiftly handle problems thanks to the synergy between IoT and AI. The requirement for specialized knowledge and skills is one issue that is frequently disregarded when adopting smart water technologies. Water utilities are responsible for hiring and keeping staff members who are knowledgeable about cutting-edge technologies like IoT, AI, and data analytics, as well as water management (Geetha & Bhanu, 2018). Overall, the comparative analysis underscores the importance of selecting the right AI/ML model to balance accuracy, interpretability, and computational efficiency for effective WQ management.

6.2.7 Summary of key findings

This chapter explored the integration of ML and AI in automotive WQ management systems. The primary findings include:

- It is important to evaluate all of these arguments in the context of the fact that Auto ML requires remarkably little time investment and academic experience. In general, Auto ML can be referred to as a superior tool. Auto ML has the drawback of being more effective in static environments than dynamic ones.
- It has been reported that very few research articles specifically emphasize using ML algorithms in contaminant detection and source identification cases, even though there is a great deal of potential for control and prediction with such tools.
- Establishing and sustaining smart water systems requires bridging the knowledge gap between conventional water management techniques and state-of-the-art technologies. Additionally, developing interdisciplinary cooperation is essential to developing a workforce capable of navigating the intricacies of technological innovation and water management.

6.2.8 Limitations, challenges, and implication

A major obstacle is the availability and caliber of data. Even though a lot of data may have been gathered from sensors and other sources, it might be difficult to guarantee that it is reliable and relevant for precise analysis. Water systems in automotive can be complicated, involving a number of variables like flow rates, pressure, temperature, and pollutants. The complexities of these systems may be difficult for ML models to adequately represent. Deep learning models in particular are sometimes regarded as "black boxes," making it challenging to understand the conclusions they make. This lack of interpretability can be a major drawback, particularly in crucial applications such as monitoring water quality. Scalability issues may arise when ML/AI solutions are implemented in automotive water systems, especially when dealing with extensive automotive manufacturing facilities. It may be difficult for ML models to adjust in real-time to shifting environmental factors or newly identified pollutants, necessitating regular updates and recalibration. The implications of these findings are profound for both environmental sustainability and public health. Improved WQ monitoring and treatment processes can lead to safer drinking water and reduced environmental pollution, enhancing overall public health outcomes. The use of AI and ML in water management underscores the potential of technological advancements to address critical environmental challenges. These findings also highlight the necessity for continued investment in research and development to refine AI and ML technologies, expand their applications, and ensure their effective integration into water management practices. Overall, these advancements represent a significant step toward more efficient, reliable, and sustainable water management systems.

6.2.9 Recommendations and future directions

The real-time monitoring capabilities of IoT sensors can be improved by further integrating ML/AI. This integration can help with proactive management of water quality issues by offering a multitude of data for study. Accuracy and interpretability can be increased by creating hybrid models that incorporate ML approaches with the advantages of physics-based models. These models can take advantage of ML's data-driven capabilities while also utilizing domain expertise. One can adopt explainable artificial intelligence (XAI) research to improve the interpretability and transparency of ML models. Better decision making and more trust can result from the application of XAI techniques in automotive WQ monitoring. The goal of future research should be to create ML models that are resilient to shifting circumstances. This covers methods for self-calibration and ongoing learning to sustain performance over time. Advancement in this sector can be fueled by cooperation between automotive manufacturers, specialists in water management, ML/AI researchers, and regulatory agencies. Multidisciplinary methods can result in comprehensive fixes for problems that deal with technical and legal issues. ML/AI applications should prioritize guaranteeing compliance with pertinent norms and laws, given the growing legislation pertaining to environmental preservation and WQ. To do this, modeling systems may need to directly incorporate regulatory requirements. A further noteworthy distinction is that time gaps between process steps in WWTP processes are typically either completely ignored or treated with inadequate rigor in models. When trying to govern processes based on cause-and-effect correlations found through such analyses and interpretations, this could result in inaccurate or misleading interpretations and analyses of the model outputs (Wang, Jia, et al., 2021; Wang, Thunéll, et al., 2021).

6.3 Conclusion

This chapter elaborates on the various challenges in WQ smart management. Real-time WQ monitoring and prediction, as well as contaminants and source identification, are all challenges that AI and ML address. Additionally, this chapter describes the usage of automated WQ management systems, data-driven decision-making, and IoT. The fusion of AI and ML with the above platforms and with their feasibility toward WQ management. This chapter also reviews the drawbacks and future scope for the development of WQ by using advanced computer tools. Overall, the application of ML and AI in automotive WQ monitoring, analysis, and management offers significant benefits in terms of efficiency, cost savings, and reliability. These technologies enable real-time monitoring, predictive maintenance, and automated management, ensuring high WQ standards and effective resource management. As AI and ML technologies continue to advance, their integration into automotive WQ systems will likely become even more sophisticated and impactful.

References

Adeyemi, O., Grove, I., Peets, S., & Norton, T. (2017). Advanced monitoring and management systems for improving sustainability in precision irrigation. *Sustainability*, *9*(3), 353.

Ahmed, A. N., Othman, F. B., Afan, H. A., Ibrahim, R. K., Fai, C. M., Hossain, M. S., ... Elshafie, A. (2019). Machine learning methods for better water quality prediction. *Journal of Hydrology*, *578*, 124084.

Ahmed, U., Mumtaz, R., Anwar, H., Mumtaz, S., & Qamar, A. M. (2020). Water quality monitoring: From conventional to emerging technologies. *Water Supply*, *20*(1), 28–45.

Alprol, A. E., Mansour, A. T., Ibrahim, M. E. E. D., & Ashour, M. (2024). Artificial intelligence technologies revolutionizing wastewater treatment: Current trends and future prospective. *Water*, *16*(2), 314.

Aslam, B., Maqsoom, A., Cheema, A. H., Ullah, F., Alharbi, A., & Imran, M. (2022). Water quality management using hybrid machine learning and data mining algorithms: An indexing approach. *IEEE Access*, *10*, 119692–119705.

Asli, H. H. (2025). Applications of networked sensors and Internet of Things (IoT) for wastewater treatment. In *Sustainable water treatment and ecosystem protection strategies* (pp. 3–42). Apple Academic Press.

Bahramian, M., Dereli, R. K., Zhao, W., Giberti, M., & Casey, E. (2023). Data to intelligence: The role of data-driven models in wastewater treatment. *Expert Systems with Applications*, *217*, 119453.

Berry, M. W., Mohamed, A., & Yap, B. W. (Eds.), (2019). *Supervised and unsupervised learning for data science*. Springer Nature.

Bhagat, S. K., Tung, T. M., & Yaseen, Z. M. (2020). Development of artificial intelligence for modeling wastewater heavy metal removal: State of the art, application assessment and possible future research. *Journal of Cleaner Production*, *250*, 119473.

Budi, H. S., Catalan Opulencia, M. J., Afra, A., Abdelbasset, W. K., Abdullaev, D., Majdi, A., ... Mohammadi, M. J. (2024). Source, toxicity and carcinogenic health risk assessment of heavy metals. *Reviews on Environmental Health*, *39*(1), 77–90.

Byliński, H., Sobecki, A., & Gębicki, J. (2019). The use of artificial neural networks and decision trees to predict the degree of odor nuisance of post-digestion sludge in the sewage treatment plant process. *Sustainability*, *11*(16), 4407.

Caruana, R., Niculescu-Mizil, A., Crew, G., & Ksikes, A. (2004). Ensemble selection from libraries of models. In *Proceedings of the twenty-first international conference on machine learning* (p. 18).

Chen, T., Guestrin, C. (2016). Xgboost: A scalable tree boosting system. In *Proceedings of the 22nd acm sigkdd international conference on knowledge discovery and data mining* (pp. 785–794).

Chowdhury, R., Rahman, M. A., Rahman, M. S., & Mahdy, M. R. C. (2020). An approach to predict and forecast the price of constituents and index of cryptocurrency using machine learning. *Physica A: Statistical Mechanics and its Applications*, *551*, 124569.

Corominas, L., Garrido-Baserba, M., Villez, K., Olsson, G., Cortés, U., & Poch, M. (2018). Transforming data into knowledge for improved wastewater treatment operation: A critical review of techniques. *Environmental Modelling & Software*, *106*, 89–103.

Dritsas, E., & Trigka, M. (2023). Efficient data-driven machine learning models for water quality prediction. *Computation*, *11*(2), 16.

Ejaz, U., Khan, S. M., Jehangir, S., Ahmad, Z., Abdullah, A., Iqbal, M., … Svenning, J. C. (2024). Monitoring the Industrial waste polluted stream-Integrated analytics and machine learning for water quality index assessment. *Journal of Cleaner Production*, *450*, 141877.

Elsken, T., Metzen, J. H., & Hutter, F. (2019). Neural architecture search: A survey. *Journal of Machine Learning Research*, *20*(55), 1–21.

Escalante, H. J. (2021). Automated machine learning—A brief review at the end of the early years. In Automated design of machine learning and search algorithms (pp. 11–28).

Geetha, R., & Bhanu, S. R. D. (2018). Recruitment through artificial intelligence: A conceptual study. *International Journal of Mechanical Engineering and Technology*, *9*(7), 63–70.

Godavarthi, B., Nalajala, P., & Ganapuram, V. (2017). Design and implementation of the vehicle navigation system in urban environments using internet of things (IoT). *IOP Conference Series: Materials Science and Engineering*, *225*(1), 012262.

Golzar, F., Nilsson, D., & Martin, V. (2020). Forecasting wastewater temperature based on artificial neural network (ANN) technique and monte carlo sensitivity analysis. *Sustainability*, *12*(16), 6386.

Guidotti, R., Monreale, A., Ruggieri, S., Turini, F., Giannotti, F., & Pedreschi, D. (2018). A survey of methods for explaining black box models. *ACM Computing Surveys (CSUR)*, *51*(5), 1–42.

Haghiabi, A. H., Nasrolahi, A. H., & Parsaie, A. (2018). Water quality prediction using machine learning methods. *Water Quality Research Journal*, *53*(1), 3–13.

Hasan, M. M., Lwin, K., Imani, M., Shabut, A., Bittencourt, L. F., & Hossain, M. A. (2019). Dynamic multi-objective optimisation using deep reinforcement learning: Benchmark, algorithm and an application to identify vulnerable zones based on water quality. *Engineering Applications of Artificial Intelligence*, *86*, 107–135.

He, X., Zhao, K., & Chu, X. (2021). AutoML: A survey of the state-of-the-art. *Knowledge-based systems*, *212*, 106622.

Himeur, Y., Elnour, M., Fadli, F., Meskin, N., Petri, I., Rezgui, Y., …… Amira, A. (2023). AI-big data analytics for building automation and management systems: A survey, actual challenges and future perspectives. *Artificial Intelligence Review*, *56*(6), 4929–5021.

Hu, Y., Lyu, L., Wang, N., Zhou, X., & Fang, M. (2024). Application of machine learning model optimized by improved sparrow search algorithm in water quality index time series prediction. *Multimedia Tools and Applications*, *83*(6), 16097–16120.

Hu, Z., Zhang, Y., Zhao, Y., et al. (2019). A water quality prediction method based on the deep LSTM network considering correlation in smart mariculture. *Sensors*, *19*, 1420.

Huang, R., Ma, C., Ma, J., Huangfu, X., & He, Q. (2021). Machine learning in natural and engineered water systems. *Water Research*, *205*, 117666.

Inbar, O., & Avisar, D. (2024). Enhancing wastewater treatment through artificial intelligence: A comprehensive study on nutrient removal and effluent quality prediction. *Journal of Water Process Engineering*, *61*, 105212.

Ismail, R. (2024). Improving wastewater treatment plant performance: An ann-based predictive model for managing average daily overflow and resource allocation optimization using tabu search. *Asian Journal of Civil Engineering*, *25*(2), 1427–1441.

Jain, D., Shah, S., Mehta, H., et al. (2021). A machine learning approach to analyze marine life sustainability. In *Proceedings of international conference on intelligent computing, information and control systems* (pp. 619–632). Springer.

Jansen, J.-D., Douma, S., Brouwer, D. R., Van den Hof, P., Bosgra, O., & Heemink, A. (2009). Closed-loop reservoir management. In *Paper presented at the SPE reservoir simulation conference?.*

Jumber, M. B., Damtie, M. T., & Tegegne, D. (2024). Integration of multivariate adaptive regression splines and weighted arithmetic water quality index methods for drinking water quality analysis. *Water Conservation Science and Engineering, 9*(1), 6.

Kazemi, P., Bengoa, C., Steyer, J. P., & Giralt, J. (2021). Data-driven techniques for fault detection in anaerobic digestion process. *Process Safety and Environmental Protection, 146*, 905–915.

Ke, G., Meng, Q., Finley, T., Wang, T., Chen, W., Ma, W., … Liu, T. Y. (2017). Lightgbm: A highly efficient gradient boosting decision tree. *Advances in Neural Information Processing Systems, 30.*

Kumari, P., & Toshniwal, D. (2021). Extreme gradient boosting and deep neural network based ensemble learning approach to forecast hourly solar irradiance. *Journal of Cleaner Production, 279*, 123285.

Liu, H., Zhang, Y., & Zhang, H. (2020). Prediction of effluent quality in papermaking wastewater treatment processes using dynamic kernel-based extreme learning machine. *Process Biochemistry, 97*, 72–79.

Liu, Z. J., Wan, J. Q., Ma, Y. W., & Wang, Y. (2019). Online prediction of effluent COD in the anaerobic wastewater treatment system based on PCA-LSSVM algorithm. *Environmental Science and Pollution Research, 26*(13), 12828–12841.

Lotfi, K., Bonakdari, H., Ebtehaj, I., Delatolla, R., Zinatizadeh, A. A., & Gharabaghi, B. (2020). A novel stochastic wastewater quality modeling based on fuzzy techniques. *Journal of Environmental Health Science and Engineering, 18*, 1099–1120.

Lowe, M., Qin, R., & Mao, X. (2022). A review on machine learning, artificial intelligence, and smart technology in water treatment and monitoring. *Water, 14*(9), 1384.

Lu, H., & Ma, X. (2020). Hybrid decision tree-based machine learning models for short-term water quality prediction. *Chemosphere, 249*, 126169.

Mahadeva, R., Manik, G., Goel, A., & Dhakal, N. (2019). A review of the artificial neural network based modelling and simulation approaches applied to optimize reverse osmosis desalination techniques. *Desalination and Water Treatment, 156*(April 2018), 245–256.

Mahjouri, M., Ishak, M. B., Torabian, A., Abd Manaf, L., Halimoon, N., & Ghoddusi, J. (2017). Optimal selection of iron and steel wastewater treatment technology using integrated multi-criteria decision-making techniques and fuzzy logic. *Process Safety and Environmental Protection, 107*, 54–68.

Mockus, J. (2005). The Bayesian approach to global optimization. In *System modeling and optimization: proceedings of the 10th IFIP conference,* New York City, USA, August 31–September 4, 1981 (pp. 473–481). Berlin, Heidelberg: Springer.

Mjalli, F. S., Al-Asheh, S., & Alfadala, H. E. (2007). Use of artificial neural network black-box modeling for the prediction of wastewater treatment plants performance. *Journal of Environmental Management, 83*(3), 329–338.

Mohammadi, F., Bina, B., Karimi, H., Rahimi, S., & Yavari, Z. (2020). Modeling and sensitivity analysis of the alkylphenols removal via moving bed biofilm reactor using artificial neural networks: Comparison of levenberg marquardt and particle swarm optimization training algorithms. *Biochemical Engineering Journal, 161*, 107685.

Nadiri, A. A., Shokri, S., Tsai, F. T. C., & Moghaddam, A. A. (2018). Prediction of effluent quality parameters of a wastewater treatment plant using a supervised committee fuzzy logic model. *Journal of Cleaner Production, 180*, 539–549.

Nasir, N., Kansal, A., Alshaltone, O., Barneih, F., Sameer, M., Shanableh, A., & Al-Shamma'a, A. (2022). Water quality classification using machine learning algorithms. *Journal of Water Process Engineering, 48*, 102920.

Newhart, K. B., Holloway, R. W., Hering, A. S., & Cath, T. Y. (2019). Data-driven performance analyses of wastewater treatment plants: A review. *Water Research, 157*, 498–513.

Nong, N. B., & Tarek, M. (2023). Surveillance approaches to intrusion detection, crop health, and disease prevention in agriculture. *Quarterly Journal of Emerging Technologies and Innovations*, *8*(3), 1–17.

Nouraki, A., Alavi, M., Golabi, M., & Albaji, M. (2021). Prediction of water quality parameters using machine learning models: A case study of the Karun River, Iran. *Environmental Science and Pollution Research*, *28*(40), 57060–57072.

Park, J., Kim, K. T., & Lee, W. H. (2020). Recent advances in information and communications technology (ICT) and sensor technology for monitoring water quality. *Water*, *12*(2), 510.

Prasad, D. V. V., Kumar, P. S., Venkataramana, L. Y., Prasannamedha, G., Harshana, S., Srividya, S. J., ... Indraganti, S. (2021). Automating water quality analysis using ML and auto ML techniques. *Environmental Research*, *202*, 111720.

Rahimian, P., & Behnam, S. (2020). A novel data-driven and feature-based forecasting framework for wastewater optimization of network pressure management system. *International Journal of Industrial Engineering & Production Research*, *31*(3), 423–433.

Rane, N., Choudhary, S., & Rane, J. (2023). Leading-edge Artificial Intelligence (AI), Machine Learning (ML), Blockchain, and Internet of Things (IoT) technologies for enhanced wastewater treatment systems (October 31, 2023).

Revollar, S., Vilanova, R., Francisco, M., & Vega, P. (2018). PI Dissolved Oxygen control in wastewater treatment plants for plantwide nitrogen removal efficiency. *IFAC-PapersOnLine*, *51*(4), 450–455.

Sahu, S., Kaur, A., Singh, G., & Arya, S. K. (2023). Harnessing the potential of microalgae-bacteria interaction for eco-friendly wastewater treatment: A review on new strategies involving machine learning and artificial intelligence. *Journal of Environmental Management*, *346*, 119004.

Santín, I., Barbu, M., Pedret, C., & Vilanova, R. (2018). Fuzzy logic for plant-wide control of biological wastewater treatment process including greenhouse gas emissions. *ISA Transactions*, *77*, 146–166.

Shah, M. I., Alaloul, W. S., Alqahtani, A., Aldrees, A., Musarat, M. A., & Javed, M. F. (2021). Predictive modeling approach for surface water quality: Development and comparison of machine learning models. *Sustainability*, *13*(14), 7515.

Sheik, A. G., Krishna, S. B. N., Patnaik, R., Ambati, S. R., Bux, F., & Kumari, S. (2024). Digitalization of phosphorous removal process in biological wastewater treatment systems: Challenges, and way forward. *Environmental Research*, 119133.

Sheik, A. G., Tejaswini, E. S. S., Seepana, M. M., & Ambati, S. R. (2023). Control of anaerobic-anoxic-aerobic (A2/O) processes in wastewater treatment: A detailed review. *Environmental Technology Reviews*, *12*(1), 420–440.

Singh, N. K., Yadav, M., Singh, V., Padhiyar, H., Kumar, V., Bhatia, S. K., & Show, P. L. (2023). Artificial intelligence and machine learning-based monitoring and design of biological wastewater treatment systems. *Bioresource Technology*, *369*, 128486.

Singh, S., Das, A., & Sharma, P. (2024). Predictive modeling of water quality index (WQI) classes in Indian rivers: Insights from the application of multiple machine learning (ML) models on a decennial dataset. *Stochastic Environmental Research and Risk Assessment*, 1–18.

Solís, B., Guisasola, A., Flores-Alsina, X., Jeppsson, U., & Baeza, J. A. (2022). A plant-wide model describing GHG emissions and nutrient recovery options for water resource recovery facilities. *Water Research*, *215*, 118223.

Srivastava, K., Handa, C., & Madhulika. (2022). A proposed fuzzy logic model for waste water treatment analysis. In *International conference on computational intelligence in pattern recognition* (pp. 403–415). Singapore: Springer Nature.

Sundui, B., Ramirez Calderon, O. A., Abdeldayem, O. M., Lázaro-Gil, J., Rene, E. R., & Sambuu, U. (2021). Applications of machine learning algorithms for biological wastewater treatment: Updates and perspectives. *Clean Technologies and Environmental Policy*, *23*, 127–143.

Syrmos, E., Sidiropoulos, V., Bechtsis, D., Stergiopoulos, F., Aivazidou, E., Vrakas, D., & Vlahavas, I. (2023). An intelligent modular water monitoring iot system for real-time quantitative and qualitative measurements. *Sustainability*, *15*(3), 2127.

Thornton, C., Hutter, F., Hoos, H. H., & Leyton-Brown, K. (2013). Auto-WEKA: Combined selection and hyperparameter optimization of classification algorithms. In *Proceedings of the 19th ACM SIGKDD international conference on Knowledge discovery and data mining* (pp. 847–855).

Uddin, M. G., Nash, S., Rahman, A., & Olbert, A. I. (2023). A novel approach for estimating and predicting uncertainty in water quality index model using machine learning approaches. *Water Research*, *229*, 119422.

Wang, D., Thunéll, S., Lindberg, U., Jiang, L., Trygg, J., Tysklind, M., & Souihi, N. (2021). A machine learning framework to improve effluent quality control in wastewater treatment plants. *Science of the Total Environment*, *784*, 147138.

Wang, G., Jia, Q. S., Zhou, M., Bi, J., & Qiao, J. (2021). Soft-sensing of wastewater treatment process via deep belief network with event-triggered learning. *Neurocomputing*, *436*, 103–113.

Wang, J., Xue, B., Wang, Y., Yinglan, A., Wang, G., & Han, D. (2024). Identification of pollution source and prediction of water quality based on deep learning techniques. *Journal of Contaminant Hydrology*, *261*, 104287.

Wang, P., Yao, J., Wang, G., Hao, F., Shrestha, S., Xue, B., ... Peng, Y. (2019). Exploring the application of artificial intelligence technology for identification of water pollution characteristics and tracing the source of water quality pollutants. *Science of the Total Environment*, *693*, 133440.

Wang, Q., Li, Z., Cai, J., Zhang, M., Liu, Z., Xu, Y., & Li, R. (2023). Spatially adaptive machine learning models for predicting water quality in Hong Kong. *Journal of Hydrology*, *622*, 129649.

Waqas, S., Harun, N. Y., Sambudi, N. S., Arshad, U., Nordin, N. A. H. M., Bilad, M. R., ... Malik, A. A. (2022). SVM and ANN modelling approach for the optimization of membrane permeability of a membrane rotating biological contactor for wastewater treatment. *Membranes*, *12*(9), 821.

Yang, Q., Cao, W., Meng, W., & Si, J. (2021). Reinforcement-learning-based tracking control of waste water treatment process under realistic system conditions and control performance requirements. *IEEE Transactions on Systems, Man, and Cybernetics: Systems*, *52*(8), 5284–5294.

Zamri, N., Pairan, M. A., Azman, W. N. A. W., Abas, S. S., Abdullah, L., Naim, S., ... Gao, M. (2022). A comparison of unsupervised and supervised machine learning algorithms to predict water pollutions. *Procedia Computer Science*, *204*, 172–179.

Zhang, L., Han, H., & Qiao, J. (2020). Dynamic multiobjective optimal control with knowledge-decision for wastewater treatment process. In *2020 seventh international conference on information, cybernetics, and computational social systems (ICCSS)* (pp. 72–78). IEEE.

Zhou, J., Wang, Y., Xiao, F., et al. (2018). Water quality prediction method based on IGRA and LSTM. *Water*, *10*, 1148.

Zhou, P., Li, Z., Snowling, S., Baetz, B. W., Na, D., & Boyd, G. (2019). A random forest model for inflow prediction at wastewater treatment plants. *Stochastic Environmental Research and Risk Assessment*, *33*, 1781–1792.

Zöller, M.-A., & Huber, M. F. (2019). Survey on automated machine learning. arXiv preprint arXiv, 1904.12054 (9), 844.

CHAPTER 7

Integrating AI, machine learning, and nanotechnology: shaping the future of water quality management

Manish Pant and Jabrinder Singh
Department of Chemistry- SoPS, DIT University, Dehradun Uttarakhand, India

7.1 Introduction

Water quality management is a complex process that includes resource allocation, harvesting, conservation, and planning. This calls for the creation and implementation of consistent rules and processes, typically within a fragmented regulatory framework. Adopting sustainable techniques is required to guarantee the long-term availability of resources because traditional approaches have proven ineffective. Pollution lowers the availability of drinkable water because, due to salinity, approximately 97% of the world's water is unfit for human consumption. Major causes of pollution include mining, intensive agriculture, wastewater, industrial activity, and untreated urban runoff (Water, 2011; Berthet, Vincent, & Fleury, 2021). Utilizing different water sources efficiently is essential, yet traditional methods are expensive and do not readily integrate with modern information and communication technologies (ICTs) (Koech & Langat, 2018).

Ensuring universal access to safe drinking water is a vital goal outlined in the United Nations Sustainable Development Goals (SDGs), and it is one of the most important concerns of the modern era (Pant, Singhal, & Singh, 2023; UN, 2015). Regarding performance assessment and efficiency enhancement, a great deal of attention has been paid to the application of different models and optimization technologies. Artificial intelligence (AI) presents a chance to get beyond the challenges associated with current methods for treating water (Wang et al., 2023). Water industries are expected to invest $6.3 billion in AI by 2030. AI is expected to minimize operational expenses by 20%–30% by optimizing chemical usage and reducing expenditures (WFM Staff, 2020).

Due to its simplicity in design, ease of installation, and versatility, AI is an attractive alternative for water treatment applications. This quickly developing field has practical uses

Computational Automation for Water Security.
DOI: https://doi.org/10.1016/B978-0-443-33321-7.00006-8

in banking, e-commerce, smart cities, healthcare, and academia (Allam & Dhunny, 2019). AI techniques are also capable of analyzing increasingly more complicated environmental data, such as assessments of water quality predictions and evaluations (Georgescu et al., 2023) or spatiotemporal environmental data prediction (Amato, Guignard, Robert, & Kanevski, 2020). Industry 4.0, blockchain, cloud computing, IoT, and machine learning (ML) are just a few of the applications that require the expertize of AI subfields like data analytics, ML, and deep learning (Lu, 2019). The field of water quality management is entering a transformative era due to the advancements in ML and AI. This has resulted in a significant change in the approaches used for data analysis and interpretation. According to Naushad, Sharma, and Alothman (2019), ML algorithms have demonstrated exceptional usefulness in anticipating contamination events, examining the sources of pollution, and improving treatment procedures. The ability of different algorithms to effectively handle big datasets and identify intricate patterns significantly advances research in a variety of fields.

AI-powered systems are increasingly being used in water quality management for tasks like forecasting and response actions, in addition to basic monitoring. According to Paschen, Pitt, and Kietzmann (2020), proactive oversight made possible by these systems reduces risks to ecosystems and public health by enabling early detection of possible threats. Several studies have shown how different AI tools can be used to effectively model and optimize water treatment processes, including pollutant removal (Bagheri, Akbari, & Mirbagheri, 2019; Zhao et al., 2020). However, major obstacles still stand in the way of the widespread use of AI in water filtration. Additionally, the role of ML in managing water quality is examined, emphasizing how it can enhance data processing, identify patterns, and provide predictive maintenance for water systems.

Modern needs are being met by innovative water purification methods made possible by the latest advancements in nanoscale science and engineering. Designing, characterizing, producing, and utilizing systems, devices, and structures at the nanoscale with precise control is known as nanotechnology. According to Shah and Ahmad (2010), properties of materials and phenomena studied in nanoscience are very different from those observed at larger sizes because of the atomic, molecular, and macromolecular levels (Shah & Ahmad, 2010). The potential use of nanostructured materials as catalysts or adsorbents to remove dangerous chemicals from wastewater has drawn a lot of attention lately. Specifically, higher surface-to-volume ratios, enhanced magnetic properties, and unique catalytic characteristics distinguish single and multimetal or doped metal oxides (Gupta et al., 2011).

The integration of nanotechnology, ML, and AI is then explored, with a focus on how these technologies could cooperate to improve the efficacy and efficiency of water quality control systems. Future directions for overcoming these barriers are also offered, along with an examination of the difficulties in implementing this state-of-the-art technology for water management. This study examines the individual and combined contributions of AI, ML, and nanotechnology in an effort to present a complete strategy to impact the future of water quality management.

7.2 Fundamentals of water quality management

The monitoring of water quality is a crucial component of environmental management, with the main objective being to ascertain its appropriateness for different uses, including industrial, drinking, and ecosystem preservation (Tripathi, Agarwal, Awasthi, & Arun, 2022). This test comprises an extensive range of criteria divided into topics related to chemistry, physics, and biology. Each of these characteristics offers vital information regarding the condition of the water. Examining physical attributes like wetness, temperature, and coloring is necessary to analyze the visual and sensory aspects of water. Chemical parameters, which include pH, dissolved oxygen (DO), phosphorus, nitrogen, various dangerous metals, and organic pollutants, are highly variable and comprise a wide range of components and compounds (Singh et al., 2023; Sudhakar, 2023). Important details regarding the chemical balance of water and the possible presence of dangerous substances are provided by these chemical indicators.

Determining the quality of water requires analyzing microorganisms, including bacteria, viruses, and phytoplankton (Reddy, Roopa, Kovvada Rajeev, & Nandan, 2020). The assessment of the general health and safety of water is contingent upon these elements, particularly with respect to its suitability for human consumption and its impact on the environment. As Table 7.1 illustrates, traditional approaches to water quality monitoring have primarily relied on manual processes that necessitate significant human intervention. According to Agrawal, Singh, Saxena,

Table 7.1: Comparative charts on traditional, current, and nanotechnology monitoring techniques.

Aspect	Traditional monitoring techniques	Current monitoring techniques	Nanotechnology techniques	References
Data collection	Manual sampling and laboratory analysis	Real-time sensor networks and IoT integration	Nanosensors for highly accurate and selective identification at the molecular level	Lu (2019)
Sampling frequency	Periodic and infrequent sampling	Continuous and high-frequency monitoring	Constant real-time monitoring with nanodevices	Zhuiykov (2012)
Parameter monitoring	Limited parameters (e.g., pH, turbidity)	Monitoring of multiparameters (e.g., pH, DO, EC, conductivity)	Monitoring of a wide range of parameters encompassing chemical and biological agents	Silva et al. (2022)
Data processing and analysis	Manual data entry and analysis	Automated data aggregation and AI-based analysis	Nanoscale data analytics and their possible combination with quantum computing for increased processing power	Madhura et al. (2019)

(*Continued*)

Table 7.1: (Continued)

Aspect	Traditional monitoring techniques	Current monitoring techniques	Nanotechnology techniques	References
Detection and alerting	Reactive response to contamination events	Predictive modeling for early detection	Nanomaterial-based sensors enable highly sensitive and proactive detection	Wang, Cao, and Barati Farimani (2021)
Cost and resource requirements	Labor-intensive and costly	Cost-effective and resource-efficient	Initially expensive, but potentially cost-effective with scale nanomaterial production.	Qu, Brame, Li, and Alvarez (2013)
Scalability and coverage	Limited spatial coverage and resolution	Broad spatial coverage and high resolution	High spatial resolution nanosensor networks provide extensive coverage.	Vikesland (2018)
Integration with treatment systems	Disconnected from treatment processes	Integration with AI-enabled treatment systems	Advanced integration using nanocatalysts and targeted nanodelivery systems for improved treatment efficiency	Yunus, Harwin, Kurniawan, Adityawarman, and Indarto (2012)
Adaptability to environmental changes	Limited adaptability to dynamic conditions	Adaptive monitoring and response to changes	High flexibility using smart and responsive nanomaterials.	Qu et al. (2013)
Durability and lifespan	Frequent maintenance and calibration needed	Improved durability but still needs regular maintenance.	Because of their robustness, nanosensors have the potential to last for a long time with minimal maintenance.	Fulekar, Pathak, and Kale (2014)
Sensitivity and accuracy	Moderate sensitivity and accuracy	Advanced sensors ensure high sensitivity and accuracy	Nanoscale detection capabilities enable extremely high sensitivity and precision.	Kumar (2023)
Environmental impact	Chemical reagents and waste have the capability to make a significant environmental impact.	Reduced environmental impact through greener technologies	Low environmental effect by developing eco-friendly nanoparticles and producing minimum trash.	Kumar (2023)
Data security	Minimal issues due to manual data management.	Data security measures are becoming increasingly important as a result of digital data transfer.	With potentially large volumes of sensitive data being collected and shared, data security is extremely important.	Steffens, Steffens, Graboski, Manzoli, and Leite (2017)

and Buddhi (2023), these processes usually entail collecting water samples from various locations and sending them to specialized labs for examination. These labs employ a range of tools and methods, like microscopy to identify microbes and spectroscopy for chemical analysis. Traditional methods have serious shortcomings, notwithstanding their historical significance in the assessment of water quality. A notable challenge is the delay that occurs between the collection of samples and the processing of results, which might hinder the timely identification of contamination incidents (Arun, Shukla, Singh, & Upadhyay, 2015). When prompt action is necessary to save the public's health or prevent environmental harm, such delays can have disastrous consequences. Moreover, traditional methods are sometimes expensive and require specialized knowledge, which limits their use and monitoring frequency, particularly in environments with limited resources.

Furthermore, because standard processes are irregular in nature, they frequently fall short of providing a comprehensive evaluation of water quality. The methods now in use are inadequate for detecting the seasonal fluctuations in water quality that may arise from changes in the seasons, meteorological events, or infrequent pollution incidents (SudhirSastry, Krishna, & Budarapu, 2015). Inconsistent data might cause knowledge gaps and make it more difficult to manage water resources effectively. Furthermore, the spatial range of these techniques is limited as they often focus on specific areas of interest that may not accurately reflect the general features of the entire aquatic environment. This limitation makes it challenging to accurately evaluate the general health of extensive water supply systems or whole aquatic ecosystems. Consequently, the application of conventional techniques for water quality monitoring has produced insightful results. However, the limitations of these methods highlight the need for more advanced, real-time monitoring technologies to be developed. Such methods would offer a more comprehensive and dynamic understanding of the water quality (Geetha et al., 2023).

7.3 Role of AI in water quality management

7.3.1 Introduction to artificial intelligence and its applications in various fields

Water quality management is critical to assuring the supply of safe and clean water for human use, agriculture, and health of ecosystems. Given the growing issues caused by pollution, climate change, and population increase, new solutions are more important than ever. According to an IBM business policy paper (IBM, 2015), technological breakthroughs are propelling us towards a vision of a "smart planet," with the goal of developing autonomous, intelligent systems.

ChatGPT has emerged as one of the most advanced AI systems as of 2023. It functions as a powerful language model integrated into a massive neural network that is extensively trained with significant amounts of internet text data. ChatGPT excels at comprehending spoken language and producing responses that closely mimic human speech. However, it is only one example of the sophisticated AI systems in use today. Another noteworthy AI is AlphaGo, which was developed by DeepMind, an Alphabet Inc. subsidiary. The use of AI is currently ubiquitous. According to Accenture and Frontier Economics research, AI has the potential to increase productivity growth in industrialized countries by up to 40% by 2035 (Buchholz, 2020). AI is also used creatively to attain SDGs related to water quality. Safe Water AI, for example, leverages convolutional neural networks and Internet of Things (IoT) technology created in the United States to enable real-time monitoring and identification of pollutants such as bacteria even in the absence of an internet connection. This system is built using low-cost commercial off-the-shelf components, and the Safe Water AI package is currently priced at approximately USD 500. Further price cuts are projected as AI technology progresses and use grows (Hmoud Al-Adhaileh & Waselallah Alsaade, 2021; Olatunde, Adelani, & Sikhakhane, 2024).

AI has proven to be a useful tool in addressing water quality-related concerns because of its ability to do real-time monitoring, predictive modeling, and advanced analytics (Mahardhika & Putriani, 2023). Fig. 7.1 illustrates the most widely used AI-based water treatment techniques. By means of prediction, diagnosis, assessment, and simulation, these strategies support water reuse, conservation efforts, and cost reduction in the management of wastewater treatment operations.

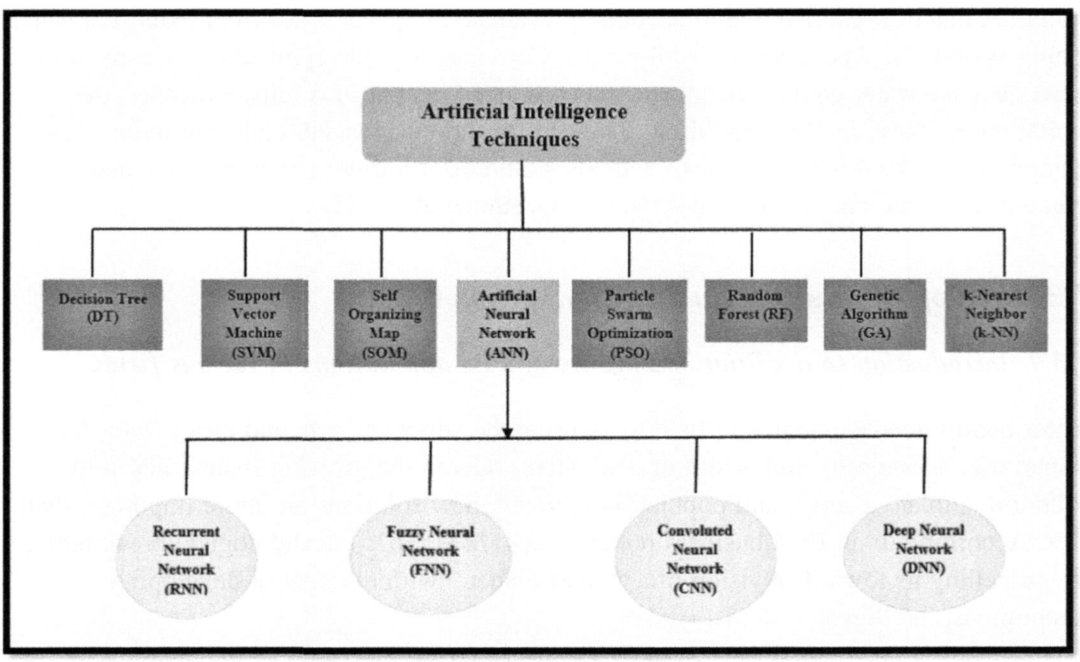

Figure 7.1
Classification of AI techniques. *AI*, Artificial intelligence.

7.3.1.1 AI applications in water quality management

AI is proving to be a promising new tool for water management. With its advanced algorithms and data-driven insights, AI has unparalleled capacity to enhance water security, optimize resource allocation, and mitigate the consequences of climate change. The effects of AI on water quality and management are discussed below (Bahramian, Dereli, Zhao, Giberti, & Casey, 2023; Konya & Nematzadeh, 2024).

7.3.1.1.1 Data analysis and prediction

AI systems are capable of going through massive datasets that include satellite imaging, weather forecasts, and sensor networks to generate precise projections regarding water availability and any shortages. Because of AI's analytical capabilities, authorities can create proactive regulations in reaction to changing climate circumstances.

7.3.1.1.2 Efficient irrigation systems

AI-driven smart irrigation systems will revolutionize agriculture. These gadgets will promote industry-wide water efficiency by continuously monitoring crop demands, soil moisture, and weather conditions. AI-powered agriculture increases agricultural yields while preserving water, which supports sustainable farming practices.

7.3.1.1.3 Leak detection and infrastructure monitoring

AI processes data from water distribution networks in real-time, enabling early leak identification. It can do this by monitoring variations in pressure and flow patterns, which minimizes water loss, aids in infrastructure maintenance, and extends the life of water supply networks (Nova, 2023).

7.3.1.1.4 Optimization of water treatment processes

AI technologies are revolutionizing water treatment systems through the ability to predict changes in demand and adjust treatment protocols accordingly. Singh, Pant, Lakhera, and Singhal (2023) claim that the intelligent control systems raise efficiency, reduce energy consumption, and enhance overall water treatment standards.

7.3.1.1.5 Consumer water management

The development of AI-powered smart meters allows consumers to monitor and manage their water consumption. Programs driven by AI provide personalized water-saving guidance, encouraging individuals to lead water-conscious lives.

7.3.1.1.6 Reservoir management and flood control

AI models also assist in the analysis of current and historical data, as well as the effective management of reservoir levels, which have a significant impact on the water quality. To reduce the negative effects of extreme weather events on communities, predictive analytics is essential for putting proactive flood control measures into place (Mahardhika & Putriani, 2023).

7.3.1.1.7 Remote sensing and GIS

Combining AI with remote sensing and geographical information systems (GIS) can improve water quality monitoring over large areas. AI can process satellite imagery and sensory data to identify pollution sources and track changes in water bodies, supporting large-scale environmental management efforts (Najafzadeh & Basirian, 2023).

7.3.1.1.8 Decision support systems for policy making

AI-driven decision support systems help the government and policymakers manage water resources by offering data-driven insights. Authorities can assess different policies using scenario modeling and simulations, enabling well-informed decision-making for sustainable water management (Konya & Nematzadeh, 2024).

7.3.2 Case studies and examples of AI applications in water management

Current developments that demonstrate the deployment of intelligent water management systems in real-time scenarios.

7.3.2.1 Using artificial intelligence for smart water management systems

ICTs have spread throughout many industries, including water management. Intelligent data analysis can increase water distribution efficiency, resulting in cost savings. AI approaches are critical in making intelligent water-use decisions for a variety of applications. Integrating ICT and AI provides a route to achieving SDGs for water management and sanitation. Furthermore, AI applications in water management help to address water shortage by accounting for factors such as population density and developing ways to reduce water leakage (Jenny, Alonso, Wang, & Minguez, 2020).

7.3.2.2 Smart water management—case study report

The International Water Resources Association (IWRA) and Korea Water Resources Corporation worked together to create a comprehensive smart water management (SWM) system. With this state-of-the-art system, IWRA can respond strategically to global water challenges by receiving real-time water data using ICT. For flood control, leak detection, irrigation optimization, and water quality monitoring, it offers AI-powered solutions. The system offers complete insights by merging real-time satellite data, GIS monitoring engines, and IoT devices. Additionally, AI training increases the system's capacity for automated services across a range of scenarios and boosts the effectiveness of decision-making (Rocher, 2018).

7.3.2.3 Grid intelligence water case study

Using modern technologies to manage water resources is a difficult task. During dry spells, the fair allocation of water resources to different industries requires large financial outlays. Installing precise and effective water metering systems in urban areas is essential. A cloud-based intelligent water metering system designed specifically for small cities in the Southeast has been unveiled by Verizon Grid Wide Intelligent Water Solutions. An IoT gateway provides safe multipoint connectivity, and water meter sensors are integrated into the system to monitor and control water usage. Water leaks and anomalies in usage of water patterns are found using ML.

7.3.2.4 Smart water management: the way to (artificially) intelligent water management

Many countries have adopted SWM systems, which use a variety of policies and technologies to fulfill the needs of modern water management. A summary paper by Nickum, Kuisma, Bjornlund, and Stephan (2020) highlights multiple SWMs from various countries. Mexico, Korea, and France, for example, have created smart flood monitoring systems that use IoT networks and AI for predictive analysis. In Mexico, the "PUMAGUA" initiative aims to improve water quality and reduce overall water usage by monitoring data within an intelligent water resource network. Similarly, South Korean researchers have developed the Hydro Intelligent Toolkit, an advanced SWM that uses a variety of characteristics such as hydrological data, rainfall forecasts, flood assessments, and groundwater levels obtained from intelligent IoT network data analysis.

7.3.2.5 Smart water management toward future water sustainable networks

In recent years, the water sector has faced significant problems in increasing efficiency and maintaining long-term performance across social, technical, and environmental dimensions. The integration of smart technologies sets the path for the creation of water-smart cities while also contributing to the energy nexus. Smart cities benefit from technological improvements that save costs, increase service quality, and optimize system performance.

In a detailed case study, the use of technology for monitoring and water loss control led to a significant boost in efficiency, especially when it came to lowering water losses and related expenses. The application of creative thinking has helped these cities move up the list of the world's most efficient cities from 20 to 5. In addition, the analysis highlights the considerable potential for technically and financially feasible micro-hydropower projects in the water sector, which could lead to lower CO_2 emissions and more energy efficiency.

Over a 12-year period, the case study shows significant savings of 57 GWh and 100 Mm^3, resulting in a decrease of 47,385 t CO_2-eq in CO_2 emissions. Furthermore, there are optimistic projections for energy output in Portugal (Ramos, McNabola, López-Jiménez, & Pérez-Sánchez, 2019).

7.4 Application of machine learning technology in water quality management

The management of water quality needs to ensure both ecological sustainability and the availability of safe and clean water for human consumption. A sophisticated set of instruments and technologies is desperately needed to successfully monitor and control water quality given the complexity of water systems, the increasing challenges posed by pollution, and the effects of climate change. The field of ML has gained significant traction in the management of water quality as it has developed into a potent tool for pattern recognition, large-scale dataset analysis, and predictive modeling.

Although ML models can successfully model DBP concentrations and crucial parameters for membrane filtration and adsorption processes, AI techniques have proven successful in managing chlorination. To assess the outcomes, a variety of statistical measures are frequently used, such as the correlation coefficient (R), coefficient of determination (R^2), mean square error, root mean square error, and relative error. It has been shown that using ML models to anticipate and model water quality-management elements is advantageous.

Monitoring the impact of human activities on river and other waterway water quality is essential to preserving these ecosystems. The dissolved oxygen concentration is a frequently monitored measure that is used to assess the condition of a river or other waterway. As a result of rising nutrient loading and global warming, hypoxia, or the absence of DO in streams, is becoming more common (Collins et al., 2019). Because of the interactions between DO, pollution, and human activity, it is becoming increasingly important to monitor DO concentrations to foresee and perhaps prevent hypoxic zones, which cause substantial harm to these aquatic ecosystems. Accurate and real-time data are often the most beneficial since slight changes in DO can have potentially fatal repercussions in certain species (Breitburg et al., 2018). In certain situations, environmental factors can make it difficult to detect DO, which highlights the present need for models and methods that can solve the shortcomings of traditional sensing methodologies (Wei et al., 2019). Table 7.2 presents a summary of research on the use of ML to model DO and other water-quality-management parameters. Future river water levels and/or the more general water quality index are the focus of other investigations.

Table 7.2: Assessment of machine learning techniques used in water quality evaluation.

Study area	Water quality parameters	ML technique	Output	References
Sutlej River	Ammonia, nitrates, chlorides, and fecal coliforms	Fuzzy C-means (FCM-ANFIS) and subtractive clustering- based adaptive neuro-fuzzy interference system (SC-ANFIS)	WQI	Tiwari, Babbar, and Kaur (2018)
Rawal watershed, Pakistan	Turbidity, temperature, pH	TDS-supervised prediction algorithms	DO, pH	Ahmed, Mumtaz, Baig, and Zaidi (2022)
Ebinur Lake Watershed	pH, BOD, NH^{3+}, iron, zinc, copper, DO, TDS	WQI method, RI, DI, and SVM model	pH, WQ	Wang et al. (2017)
Yamuna River	pH, COD, BOD, water temperature, flow rate, and NH_3 concentration	ANN, ANFIS, SVM	DO concentration	Elkiran, Nourani, and Abba (2019)
Johor State, Malaysia (Johor River)	Conductivity, temperature, nitrate, salinity, turbidity, chloride, potassium, phosphate, sodium, magnesium, iron, and *E. Coli* concentrations	ANFIS	pH, suspended solids, and ammoniacal nitrogen	Sillberg, Kullavanijaya, and Chavalparit (2021)
An Kim Hai irrigation system	DO, turbidity, TSS, and coliform	Multilinear regression (MLR) and support vector machine (SVM)	WQI	Lap et al. (2023)
Mathura River, India	Water temperature, pH, BOD, and dissolved oxygen (DO) (independent variables for each input are collected at upstream, midstream, and downstream stations)	Artificial neural network (ANN), ANFIS	Concentration of DO in upstream, midstream, and downstream	Elkiran, Nourani, Abba, and Abdullahi (2018)
Chao Phraya River, Thailand	DO, BOD, ammonia concentration, salinity, fecal coliform bacteria, and total coliform bacteria	SVM with varying kernel functions	Fecal coliform bacteria, BOD, DO, salinity, ammonia concentration, and total coliform bacteria	Najah Ahmed et al. (2019)

AI techniques are also capable of analyzing even more complicated environmental data, including water quality evaluation and forecasting (Georgescu et al., 2023) or spatiotemporal environmental data prediction (Amato et al., 2020). While ANNs are the main methodology utilized in studies of modeling aids for managing water quality using

ML techniques, other approaches have also been investigated, including ANFIS, RNN, EML, random forest (RF), SVM, HW, and hybrid ML models that combine elements of the ML techniques with RF models. Most experiments' models produced correct predictions, albeit this ultimately depends on the setting. With R^2 values more than 0.999, the ANN and ANFIS models both generated precise estimates of water levels (Lowe, Qin, & Mao, 2022).

7.4.1 Real-time data analysis using the integration of sensors, IoT, and machine learning algorithms

This chapter examines how sensor technologies and the IoT are increasingly combining, highlighting how this might revolutionize data gathering, analysis, and application in a variety of sectors. It evaluates existing installations, looks into the technological aspects of this integration, and projects future trends and difficulties. The chapter highlights the importance of IoT and sensor technology in the current digital world, including innovations that have made it possible to reduce costs, miniaturize sensors, and get IoT devices widely accepted (Kim, Im, Ha, Choi, & Ha, 2014).

7.4.1.1 Technological evolution and components

This section explores the shift from individual sensors to cohesive IoT networks, emphasizing important infrastructural and technology developments. It offers information on several sensor kinds, including motion, pressure, and temperature sensors, and examines how they integrate into IoT devices. Additionally, this section explores data transmission techniques, communication protocols, and the function of edge computing and cloud computing in this ecosystem (Kamyab et al., 2023).

7.4.1.2 Practical applications

A thorough analysis of real-world sensor and IoT integration applications is provided, with an emphasis on the manufacturing, healthcare, smart city, and agricultural sectors. According to Mabuwa and Msomi (2020), case studies illustrate the useful advantages of these integrations, such as improved operational efficiency, accuracy, and well-informed decision-making.

7.4.1.3 Challenges and solutions

The main challenges of incorporating sensor technologies into IoT systems are covered in this section, with a focus on interoperability, scalability, and data security and privacy. It suggests several strategies and methods to deal with these issues in an efficient manner, such as improved encryption methods, standardized protocols, and scalable design frameworks (Guo, Huang, Chen, Guo, & Singh, 2021).

7.4.1.4 Current and future developments

This section looks at recent and upcoming developments in sensor and IoT integration. It focuses on research topics that are still being investigated, like improvements in sensor technology and IoT connectivity, like 5G. It also looks at how AI and ML techniques are being integrated to enhance data processing capabilities.

7.4.1.5 Case study: water quality monitoring

Sensor integration is a crucial technique in many applications that involves combining data from several sensors to create more accurate and reliable monitoring findings. For example, in the subject of water quality monitoring, several sensors are employed to evaluate temperature, turbidity, pH, and chemical composition. With the use of ML algorithms, a multiparameter approach provides a comprehensive assessment of the quality of the water. ML approaches outperform conventional single-parameter approaches in the following domains: integrated data processing, anomaly detection, and pattern identification. Water quality management requires real-time data transmission and analysis to enable prompt decision-making and action. IoT sensors constantly provide data to centralized systems for AI-powered analytics. This swift processing ensures that any contamination or deterioration in quality is found fast, enabling prompt action. AI models may be able to predict future changes in water quality, which could help with preventive actions.

The monitoring of water quality has changed significantly with the advent of smart sensor networks. These networks combine intelligent and networked sensors to give extensive coverage over large areas. When the environment changes or abnormalities are detected, they can self-adjust sampling rates or parameters. AI analysis of the data from these networks yields significant insights about trends in contamination, the causes of pollution, and the overall health of the ecosystem. IoT, machine intelligence, and sensor technology together have the power to transform a number of industries. An extensive summary of the technological elements, real-world applications, issues, and potential future developments in this field is given in this chapter. In real-time analysis applications, the combination of sensors, IoT, and AI will spur innovation, improve productivity, and allow for better-informed decision-making as technology develops (Alprol, Mansour, Ibrahim, & Ashour, 2024).

7.4.2 Challenges of employing machine learning technology in water management

There are still issues with data scarcity, model interpretability, and scalability with ML in water quality management, despite its potential advantages. Next-generation ML-based water management solutions should be developed with an emphasis on resolving these problems. By offering advanced tools for data analysis, modeling, and decision-making, ML has the potential to significantly change the management of water quality. It is possible for water managers to maintain and conserve freshwater resources by using ML to enhance their capacity to observe, predict, and address problems related to water quality.

7.5 Nanotechnology for water quality management

7.5.1 Introduction to nanotechnology and its significance in water treatment

Water filtration systems are changing as a result of advances in nanotechnology, which also increase efficiency and efficacy. To clean water using different methods, including filtration, adsorption, and photocatalysis, new nanomaterials like nanoparticles are being developed. These materials possess high surface area, chemical reactivity, and selectivity, which enable them to effectively remove a wide range of pollutants from water. It appears that photocatalysts and catalytic moist air oxidation are the most promising applications of nanomaterials for water treatment. They are able to effectively remove a variety of pollutants from water thanks to certain features. Moreover, real-time water quality detection and monitoring are made possible by nanotechnology-driven sensing and monitoring devices, offering innovative approaches to guarantee water safety. The combination of AI and nanotechnology promises to result in even more creative and efficient water treatment solutions. This combination could result in groundbreaking discoveries in the fight against problems with water quality, guaranteeing that people everywhere have access to clean, safe water (Mehta, Sharma, Haldar, & Kumar, 2023).

7.5.2 Nanomaterials for contaminant removal and water purification

Sustainability can be greatly enhanced by nanotechnology and advanced materials, particularly in fields like material recycling, CO_2 reduction, and water purification (Nagar & Pradeep, 2020). Nanomaterials are quickly emerging as strong substitutes for traditional water treatment methods, offering effectiveness at the lowest possible cost and with the least amount of downtime. For developing countries like Bangladesh and India, where the success of new removal procedures depends heavily on their cost-effectiveness, this action is especially advantageous. Fig. 7.2 shows how nanotechnology offers innovative ways to enhance the quality of water.

Traditional materials that need a lot of energy, resources, or are harmful to the environment may be replaced by nanomaterials. By producing nanoparticles using green chemistry principles, we can drastically cut down on waste generation and encourage safer methods of chemical synthesis. To find out if substituting nanoparticles lowers energy and material consumption without having unfavorable side effects, more quantitative data are necessary. Nanoparticle safety and environmental impact remain controversial topics. Although they can break down pollutants through catalysis, nanoparticles can be harmful to cells if consumed. Wider adoption is hampered by the dual nature of nanotechnology. However, over time, the potential financial savings offered by nanotechnology might make it a more viable option for cleaning up water contamination. Nanoparticles are used in the water treatment industry for a variety of purposes, including strong sorbents for separation, catalysts for the photochemical destruction of contaminants, nanosized zerovalent iron for

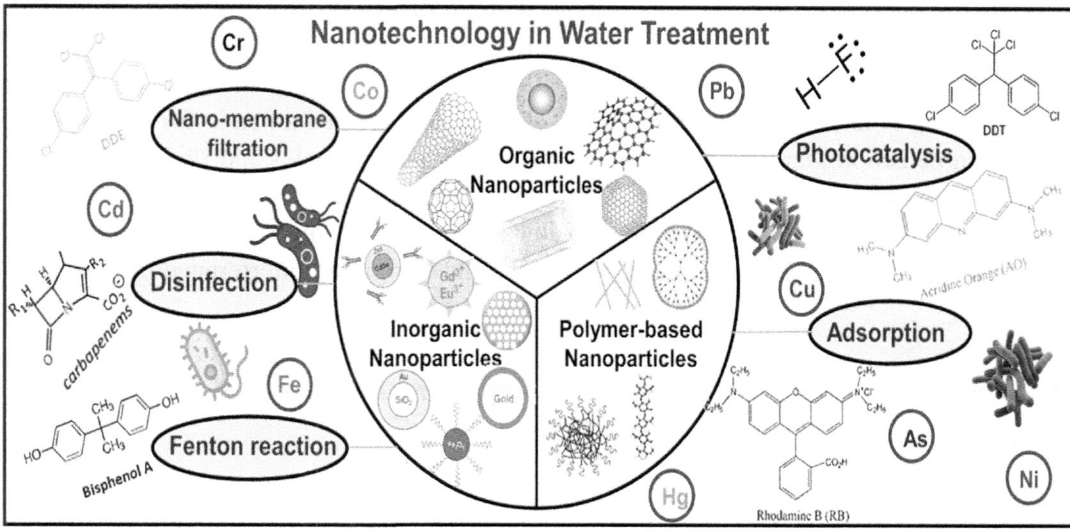

Figure 7.2

Nanotechnology in water treatment (Ajith et al., 2021). Source:*Reprinted from original article by Ajith, M.P., Aswathi, M., Priyadarshini, E., & Rajamani, P. (2021). Recent innovations of nanotechnology in water treatment: A comprehensive review.* Bioresource Technology, 342, *126000. https://doi.org/10.1016/j.biortech.2021.126000 with permission from Elsevier.*

the elimination of metals and organic molecules, and components of nanofiltration membranes (Rajan, 2011; Singh, Singh, & Rai, 2022). These uses highlight the versatility and promise of nanotechnology in addressing problems related to water quality, but they also highlight the necessity of carefully weighing the risks and benefits to the environment and public health.

7.5.3 Nanobased sensors for detecting pollutants in water

Three areas of nanotechnology applications in water treatment exist are pollution prevention, pollution detection, and contaminated water remediation (Yunus et al., 2012). Due to this, there is a greater need than ever for nanosensors that can detect micro, nano, and molecular pollutants with high sensitivity, selectivity, and precision (Ramezani et al., 2020). According to Steffens et al. (2017), these sensors use nanomaterials with nanostructures, such as quantum dots, nanorice, nanorods, and nanoparticles, which have a high binding affinity and a large surface area for target chemicals.

Nanosensors in monitoring of water quality are classified according to their detecting modes: optical, electrochemical, magnetic, or mechanical. Optical-based Fe_3O_4@Ag nanosensors, for example, have excellent sensitivity in detecting mercury ions (Hg^{2+}) at ultralow concentrations of 10 pM utilizing surface-enhanced Raman resonance scattering

(SERRS). The redox interaction between nanosilver and Hg^{2+} produces an Ag/Hg amalgam on the sensor's surface, which increases the SERRS signal intensity. However, different metallic divalent cations may undergo conjugation via the same redox reaction, offering a difficulty (Song et al., 2017). Similarly, Yuan et al. (2021) created a nanosensor for mercury ion detection that uses thymine-decorated gold nanoparticles and fiber-optic surface plasmon resonance stimulation. These new nanosensor devices demonstrate nanotechnology's promise to revolutionize water quality monitoring and pollution detection; however, hurdles remain in maintaining selectivity and mitigating interference from other ions.

7.6 Integration of AI, machine technology, and nanotechnology in water quality management

To address one of the most important problems facing the world today, which is environmental pollution, a holistic approach to controlling water quality is provided by the integration of AI, ML, and nanotechnologies. The accuracy and efficacy of selecting and optimizing adsorbents for the removal of pollutants have been greatly enhanced by the combination of these techniques (Zhao et al., 2020). Whereas AI enhances the accuracy of predictive modeling, ML approaches optimize sensor placement and data processing. According to Arellano Vidal and Govan (2024), high-performance detection and treatment capabilities are provided by nano-based sensors and remediation technologies, which aid in these endeavors. In particular, when applied to environmental cleanup projects, ML and nanotechnology yield creative solutions. A molecular level of contamination can be precisely targeted and neutralized using ML-created nanoparticles. The efficiency of eliminating toxins from soil, water, and the air could be significantly improved by this discovery. To improve environmental remediation efforts, for instance, ML algorithms can determine which nanoparticle designs are most effective at absorbing particular contaminants. Toxicology and Applied Pharmacology, in their paper "Towards a systematic exploration of nano-bio interactions," Bai et al. (2017) emphasize the significance of methodically altering the physicochemical properties of nanoparticles in conjunction with computational analysis to comprehend nano–bio interactions crucial for remediation.

7.6.1 Case studies and applications

Research on two-dimensional nanomaterials with remarkable physical characteristics, such as graphene, has been extensive. Graphene nanopores for effective water desalination are made with AI (Wang et al., 2021). To eliminate diazinon, a study used metal-organic frameworks and aminated graphitic carbonaceous structure. To predict the removal of pesticides, researchers employed artificial neural network (ANN) models and Box–Behnken Design (Yousefi et al., 2024). A study by Sadek, Fahmy, Nasr, and Mostafa (2023) claims that nano zerovalent aluminum was used to evaluate the removal of Cu(II) ions from aqueous solutions. Using the adsorption parameters, the Cu(II) removal efficiency (R%) was estimated and simulated. The most successful modeling technique for forecasting R% turned out to be ANN. To simulate the adsorbent's adsorption profile and model the adsorption process, ANNs have been used more and more in recent years (Mahato, Rawat, Gupta, & Yadav, 2024).

An important advancement in the field of adsorption science is the integration of AI and ML. These potent instruments provide answers to persistent problems, such as enhancing regeneration effectiveness and forecasting adsorption behavior in response to varying environmental circumstances. By utilizing ML and AI, scientists may now alter materials and procedures to create more intelligent adsorbents that adapt to their surroundings. This increases the environmental friendliness and performance of adsorption approaches and creates new opportunities for addressing challenging wastewater treatment problems (Satyam & Patra, 2024). Combining nanotechnology and AI technologies may also result in more inventive and effective water treatment solutions (Kumar, 2023).

7.7 Discussion

7.7.1 Key findings and interpretations

Technological advancement may be possible as a result of the prospective transformation of the water quality management field through the integration of nanotechnology, ML, and AI. Although the computational bases of AI and ML are similar, their integration with nanotechnology presents unique opportunities as well as challenges, highlighting the need for specific methodologies to reach their full potential. Machine intelligence-driven nanotechnology has the potential to solve complicated global issues and expand technological boundaries in the future. This synergy radically alters our understanding of what is possible in science and engineering, going beyond merely enhancing current procedures. It encourages the creation of improved nanoparticles for cleaning up the environment, more potent medications for focused treatments, and cutting-edge materials for electronics and energy storage. Because of this integration, humans are able to anticipate and govern occurrences in previously unimaginable ways, as well as observe previously invisible phenomena.

7.7.2 Impact and implications

7.7.2.1 Implications for ethics and society

When nanotechnology and AI are combined, it creates social and ethical questions that need to be resolved. Considerable attention needs to be paid to privacy issues, data security, and potential biases in AI systems. In addition, there should be transparent and responsible dialog regarding the risks, advantages, and possible social effects of AI systems powered by nanotechnology.

7.7.2.2 Safety and regulatory frameworks

Appropriate safety standards and regulatory frameworks must be created as nanotechnology develops to monitor its use in AI applications. When developing and putting into practice nanotechnology-enabled AI systems, it is crucial to take into account the safety of nanoscale components, their influence on the ecosystems and health of humans, and how to resolve any potential ethical issues.

7.7.3 Challenges and limitations

Notwithstanding the noteworthy advancements and prospective advantages associated with the amalgamation of nanotechnology and AI, certain obstacles must be surmounted to completely actualize the possibilities of this amalgamation. To successfully create and implement AI systems made possible by nanotechnology, it is essential to recognize these obstacles and focus on possible future developments.

7.7.3.1 Characterization and standardization of nanomaterials

The comprehensive characterization and standardization of nanomaterials used in AI applications is one of the most important problems. At the nanoscale, nanomaterials have unique features, yet characterizing them can be challenging and require specific tools. When assessing the effectiveness of nanomaterials in AI systems, it would be simpler to compare and assess them with confidence if there were standardized procedures for characterization and evaluation.

7.7.3.2 Manufacturing and scalability

For AI enabled by nanotechnology to be widely used, scalability is crucial. Although there has been promise in nanoscale fabrication techniques, large-scale manufacturing processes that can produce a large quantity of nanoscale devices and components must be developed. For nanotechnology to truly be used in AI systems, manufacturing challenges including yield optimization, repeatability, and cost-effectiveness must be resolved.

7.7.3.3 Arrangement and integration

Compatibility problems arise when integrating nanotechnology into current AI hardware and systems. It is necessary for nanoscale devices and components to blend seamlessly into the larger AI architecture while still being compatible with different hardware and software elements. Achieving optimal efficiency and collaboration requires bridging the divide between nanotechnology and AI frameworks, platforms, and algorithms.

7.7.3.4 Dependability and durability

For AI systems to be used practically, their nanoscale components must be dependable and long-lasting. Strong and long-lasting enough to withstand a variety of environmental factors and frequent use, nanomaterials and nanodevices are required. The key to the success and general acceptance of AI systems provided by nanotechnology is their long-term stability and durability.

7.7.4 Suggestions and recommendations

Traditional methods of water treatment and management include manual physicochemical analysis and data interpretations, which are unable to meet the current water resources deterioration. Nanotechnology applications help in monitoring a wide range of parameters encompassing chemical and biological agents with automated data aggregation and further AI-based analysis. The water industry is looking forward toward AI, which is expected to curtail operational expenses by 30% by reducing treatment expenses. In addition to basic monitoring, AI is also beneficial in forecasting precipitation rates for smooth harvesting management, pollution abatement, and providing predictive maintenance for water systems. The technique is cost-effective and resource-efficient. It reduces environmental impact through greener technologies. ANN, fuzzy, ANFIS, Hybrid-ML and SVM are the common tools employed for water resources treatment and monitoring purposes. However, the challenges and limitations discussed in the chapter must also be addressed while employing these tools.

7.8 Conclusion

The integration of nanotechnology, ML, and AI opens up new possibilities for managing water quality. The groundbreaking advantages of these technologies have been emphasized in this chapter, which also shows how they might be used to create more sustainable futures and enhance wastewater treatment systems. By utilizing AI and ML, wastewater treatment processes can achieve previously unheard-of levels of precision and efficiency. With the help of these technologies, treatment facilities may proactively anticipate and resolve problems with water quality. They also provide real-time monitoring, adaptive control systems, and predictive analytics. Furthermore, by providing innovative techniques for water treatment and purification, the use of nanotechnology enhances AI and ML capabilities. Because of their special properties, nanomaterials can be tailored to target certain contaminants, improving treatment efficacy and minimizing environmental impact. Utilizing AI, ML, and nanotechnology effectively is essential for long-term water management strategies, as we confront serious concerns about pollution and water scarcity. These cutting-edge technologies can help us optimize water treatment processes, protect priceless resources, and guarantee that clean water will be available for future generations.

References

Agrawal, R., Singh, S., Saxena, K. K., & Buddhi, D. (2023). A role of biomaterials in tissue engineering and drug encapsulation. *Proceedings of the Institution of Mechanical Engineers, Part E: Journal of Process Mechanical Engineering*. Available from https://doi.org/10.1177/09544089221150740.

Ahmed, M., Mumtaz, R., Baig, S., & Zaidi, S. M. H. (2022). Assessment of correlation amongst physico-chemical, topographical, geological, lithological and soil type parameters for measuring water quality of Rawal watershed using remote sensing. *Water Supply*, *22*(4), 3645–3660. Available from https://doi.org/10.2166/ws.2022.006.

Ajith, M. P., Aswathi, M., Priyadarshini, E., & Rajamani, P. (2021). Recent innovations of nanotechnology in water treatment: A comprehensive review. *Bioresource Technology*, *342*, 126000. Available from https://doi.org/10.1016/j.biortech.2021.126000.

Allam, Z., & Dhunny, Z. A. (2019). On big data, artificial intelligence and smart cities. *Cities (London, England)*, *89*, 80–91. Available from https://doi.org/10.1016/j.cities.2019.01.032.

Alprol, A. E., Mansour, A. T., Ibrahim, M. E. E. D., & Ashour, M. (2024). Artificial intelligence technologies revolutionizing wastewater treatment: current trends and future prospective. *Water*, *16*(2), 314. Available from https://doi.org/10.3390/w16020314.

Amato, F., Guignard, F., Robert, S., & Kanevski, M. (2020). A novel framework for spatiotemporal prediction of environmental data using deep learning. *Scientific Reports*, *10*, 22243. Available from https://doi.org/10.1038/s41598-020-79148-7.

Arellano Vidal, C. L., & Govan, J. E. (2024). Machine learning techniques for improving nanosensors in agroenvironmental applications. *Agronomy*, *14*(2), 341. Available from https://doi.org/10.3390/agronomy14020341.

Arun, V., Shukla, N. K., Singh, A. K., & Upadhyay, K. K. (2015). Design of all optical line selector based on SOA for data communication. In *Proceedings of the sixth international conference on computer and communication technology 2015* (pp. 281–285). Available from https://doi.org/10.1145/2818567.2818658.

Bagheri, M., Akbari, A., & Mirbagheri, S. A. (2019). Advanced control of membrane fouling in filtration systems using artificial intelligence and machine learning techniques: A critical review. *Process Safety and Environmental Protection*, *123*, 229–252. Available from https://doi.org/10.1016/j.psep.2019.01.013.

Bahramian, M., Dereli, R. K., Zhao, W., Giberti, M., & Casey, E. (2023). Data to intelligence: The role of data-driven models in wastewater treatment. *Expert Systems with Applications*, *217*, 119453. Available from https://doi.org/10.1016/j.eswa.2022.119453.

Bai, X., Liu, F., Liu, Y., Li, C., Wang, S., Zhou, H., ... Yan, B. (2017). Toward a systematic exploration of nano-bio interactions. *Toxicology and Applied Pharmacology*, *323*, 66–73. Available from https://doi.org/10.1016/j.taap.2017.03.011.

Berthet, A., Vincent, A., & Fleury, P. (2021). Water quality issues and agriculture: An international review of innovative policy schemes. *Land use Policy*, *109*, 105654. Available from https://doi.org/10.1016/j.landusepol.2021.105654.

Breitburg, D., Levin, L. A., Oschlies, A., Grégoire, M., Chavez, F. P., Conley, D. J., ... Zhang, J. (2018). Declining oxygen in the global ocean and coastal waters. *Science (New York, NY)*, *359*(6371). Available from https://doi.org/10.1126/science.aam7240.

Buchholz, K. (2020). Where AI is Aiding Productivity. Available from https://www.statista.com/chart/23779/ai-productivity-increase.

Collins, M., Tills, O., Turner, L. M., Clark, M. S., Spicer, J. I., & Truebano, M. (2019). Moderate reductions in dissolved oxygen may compromise performance in an ecologically-important estuarine invertebrate. *The Science of the Total Environment*, *693*, 133444. Available from https://doi.org/10.1016/j.scitotenv.2019.07.250.

Elkiran, G., Nourani, V., Abba, S., & Abdullahi, J. (2018). Artificial intelligence-based approaches for multi-station modelling of dissolve oxygen in river. *Global Journal of Environmental Science and Management*, *4*(4), 439–450. Available from https://doi.org/10.22034/gjesm.2018.04.005.

Elkiran, G., Nourani, V., & Abba, S. I. (2019). Multi-step ahead modelling of river water quality parameters using ensemble artificial intelligence-based approach. *Journal of Hydrology*, *577*, 123962. Available from https://doi.org/10.1016/j.jhydrol.2019.123962.

Fulekar, M. H., Pathak, B., & Kale, R. K. (2014). Nanotechnology: Perspective for environmental sustainability. *Environment and Sustainable Development*, 87–114. Available from https://doi.org/10.1007/978-81-322-1166-2_7.

Geetha, M., Bonthula, S., Al-Maadeed, S., Al-Lohedan, H., Rajabathar, J. R., Arokiyaraj, S., & Sadasivuni, K. K. (2023). Research trends in smart cost-effective water quality monitoring and modeling: Special focus on artificial intelligence. *Water*, *15*(18), 3293. Available from https://doi.org/10.3390/w15183293.

Georgescu, P. L., Moldovanu, S., Iticescu, C., Calmuc, M., Calmuc, V., Topa, C., & Moraru, L. (2023). Assessing and forecasting water quality in the Danube River by using neural network approaches. *Science of the Total Environment*, *879*, 162998. Available from https://doi.org/10.1016/j.scitotenv.2023.162998.

Guo, H., Huang, J. J., Chen, B., Guo, X., & Singh, V. P. (2021). A machine learning-based strategy for estimating non-optically active water quality parameters using Sentinel-2 imagery. *International Journal of Remote Sensing*, *42*(5), 1841–1866. Available from https://doi.org/10.1080/01431161.2020.1846222.

Gupta, K., Bhattacharya, S., Chattopadhyay, D., Mukhopadhyay, A., Biswas, H., Dutta, J., ... Ghosh, U. C. (2011). Ceria associated manganese oxide nanoparticles: Synthesis, characterization and arsenic(V) sorption behavior. *Chemical Engineering Journal*, *172*(1), 219–229. Available from https://doi.org/10.1016/j.cej.2011.05.092.

Hmoud Al-Adhaileh, M., & Waselallah Alsaade, F. (2021). Modelling and prediction of water quality by using artificial intelligence. *Sustainability*, *13*(8), 4259. Available from https://doi.org/10.3390/su13084259.

IBM. (2015). *IBM builds a smarter planet*. https://www.ibm.com/smarterplanet/us/en/. (Accessed 12.09.2023).

Jenny, H., Alonso, E. G., Wang, Y., & Minguez, R. (2020). *Using artificial intelligence for smart water management systems*. Available from https://doi.org/10.22617/BRF200191-2.

Kamyab, H., Khademi, T., Chelliapan, S., SaberiKamarposhti, M., Rezania, S., Yusuf, M., ... Ahn, Y. (2023). The latest innovative avenues for the utilization of artificial Intelligence and big data analytics in water resource management. *Results in Engineering*, 101566. Available from https://doi.org/10.1016/j.rineng.2023.101566.

Kim, Y. H., Im, J., Ha, H. K., Choi, J. K., & Ha, S. (2014). Machine learning approaches to coastal water quality monitoring using GOCI satellite data. *GIScience & Remote Sensing*, *51*(2), 158–174. Available from https://doi.org/10.1080/15481603.2014.900983.

Koech, R., & Langat, P. (2018). Improving irrigation water use efficiency: A review of advances, challenges and opportunities in the Australian context. *Water*, *10*(12), 1771. Available from https://doi.org/10.3390/w10121771.

Konya, A., & Nematzadeh, P. (2024). Recent applications of AI to environmental disciplines: A review. *Science of the Total Environment*, *906*, 167705. Available from https://doi.org/10.1016/j.scitotenv.2023.167705.

Kumar, S. (2023). Smart and innovative nanotechnology applications for water purification. *Hybrid Advances*, 100044. Available from https://doi.org/10.1016/j.hybadv.2023.100044.

Lap, B. Q., Du Nguyen, H., Hang, P. T., Phi, N. Q., Hoang, V. T., Linh, P. G., & Hang, B. T. T. (2023). Predicting Water Quality Index (WQI) by feature selection and machine learning: A case study of An Kim Hai irrigation system. *Ecological Informatics*, *74*, 101991. Available from https://doi.org/10.1016/j.ecoinf.2023.101991.

Lowe, M., Qin, R., & Mao, X. (2022). A review on machine learning, artificial intelligence, and smart technology in water treatment and monitoring. *Water*, *14*(9), 1384. Available from https://doi.org/10.3390/w14091384.

Lu, Y. (2019). Artificial intelligence: a survey on evolution, models, applications and future trends. *Journal of Management Analytics*, *6*(1), 1–29. Available from https://doi.org/10.1080/23270012.2019.1570365.

Mabuwa, S., & Msomi, V. (2020). The impact of submerged friction stir processing on the friction stir welded dissimilar joints. *Materials Research Express*, *7*(9), 096513. Available from https://doi.org/10.1088/2053-1591/abb6b6.

Madhura, L., Singh, S., Kanchi, S., Sabela, M., Bisetty, K., & Inamuddin. (2019). Nanotechnology-based water quality management for wastewater treatment. *Environmental Chemistry Letters*, *17*, 65–121. Available from https://doi.org/10.1007/s10311-018-0778-8.

Mahardhika, S. P., & Putriani, O. (2023). Deployment and use of artificial intelligence (AI) in water resources and water management. *IOP Conference Series: Earth and Environmental Science*, *1195*(1), 012056. Available from https://doi.org/10.1088/1755-1315/1195/1/012056, IOP Publishing.

Mahato, J. K., Rawat, S., Gupta, S. K., & Yadav, B. (2024). Adsorptive remediation of coal bed methane produced water (CBMW) using a novel bio-adsorbent and modern enable artificial intelligence modeling. *Chemical Engineering Research and Design*. Available from https://doi.org/10.1016/j.cherd.2024.05.029.

Mehta, K. P., Sharma, R., Haldar, S., & Kumar, A. (2023). Advancement in treatment of wastewater with nano technology. *Materials Today: Proceedings*, *80*, 3358–3362. Available from https://doi.org/10.1016/j.matpr.2021.07.253.

Nagar, A., & Pradeep, T. (2020). Clean water through nanotechnology: needs, gaps, and fulfillment. *ACS Nano*, *14*(6), 6420–6435. Available from https://doi.org/10.1021/acsnano.9b01730.

Najafzadeh, M., & Basirian, S. (2023). Evaluation of river water quality index using remote sensing and artificial intelligence models. *Remote Sensing*, *15*(9), 2359. Available from https://doi.org/10.3390/rs15092359.

Najah Ahmed, A., Binti Othman, F., Abdulmohsin Afan, H., Khaleel Ibrahim, R., Ming Fai, C., Shabbir Hossain, M., ... Elshafie, A. (2019). Machine learning methods for better water quality prediction. *Journal of Hydrology*, *578*, 124084. Available from https://doi.org/10.1016/j.jhydrol.2019.124084, vol.

Naushad, M., Sharma, G., & Alothman, Z. A. (2019). Photodegradation of toxic dye using Gum Arabic-crosslinked-poly (acrylamide)/$Ni(OH)_2$/FeOOH nanocomposites hydrogel. *Journal of Cleaner Production*, *241*, 118263. Available from https://doi.org/10.1016/j.jclepro.2019.118263.

Nickum, J. E., Kuisma, S., Bjornlund, H., & Stephan, R. M. (2020). Smart water management: The way to (artificially) intelligent water management, or just another pretty name? *Water International*, *45*(6), 515–519. Available from https://doi.org/10.1080/02508060.2020.1830581.

Nova, K. (2023). AI-enabled water management systems: an analysis of system components and interdependencies for water conservation. *Eigenpub Review of Science and Technology*, *7*(1), 105–124. Available from https://studies.eigenpub.com/index.php/erst/article/view/12.

Olatunde, T. M., Adelani, F. A., & Sikhakhane, Z. Q. (2024). A review of smart water management systems from Africa and the United States. *Engineering Science & Technology Journal*, *5*(4), 1231–1242. Available from https://doi.org/10.51594/estj.v5i4.1014.

Pant, M., Singhal, N., & Singh, J. (2023). Spatio-temporal variations in water quality of Rispana river in Dehradun, India. *Sustainable Water Resources Management*, *9*(4), 123. Available from https://doi.org/10.1007/s40899-023-00906-2.

Paschen, U., Pitt, C., & Kietzmann, J. (2020). Artificial intelligence: Building blocks and an innovation typology. *Business Horizons*, *63*(2), 147–155. Available from https://doi.org/10.1016/j.bushor.2019.10.004.

Qu, X., Brame, J., Li, Q., & Alvarez, P. J. (2013). Nanotechnology for a safe and sustainable water supply: Enabling integrated water treatment and reuse. *Accounts of Chemical Research*, *46*(3), 834–843. Available from https://doi.org/10.1021/ar300029v.

Rajan, C. S. (2011). Nanotechnology in groundwater remediation. *International Journal of Environmental Science and Development*, *2*(3), 182.

Ramezani, M., Taghdisi, S. M., Yazdian-Robati, R., Oroojalian, F., Abnous, K., & Alibolandi, M. (2020). *Nanosensors for water safety. Nanosensors for smart cities* (pp. 285–301). Elsevier. Available from http://doi.org/10.1016/B978-0-12-819870-4.00016-5.

Ramos, H. M., McNabola, A., López-Jiménez, P. A., & Pérez-Sánchez, M. (2019). Smart water management towards future water sustainable networks. *Water*, *12*(1), 58. Available from https://doi.org/10.3390/w12010058.

Reddy, K. S. P., Roopa, Y. M., Kovvada Rajeev, L. N., & Nandan, N. S. (2020). IoT based smart agriculture using machine learning. *2020 second international conference on inventive research in computing applications (ICIRCA)* (pp. 130–134). IEEE. Available from https://doi.org/10.1109/ICIRCA48905.2020.9183373.

Rocher, V. (2018). *Smart Water Management—Case Study Report;*. Daejeon,Korea: K-Water.

Sadek, A. H., Fahmy, O. M., Nasr, M., & Mostafa, M. K. (2023). Predicting Cu (II) adsorption from aqueous solutions onto nano zero-valent aluminum (nZVAl) by machine learning and artificial intelligence techniques. *Sustainability*, *15*(3), 2081. Available from https://doi.org/10.3390/su15032081.

Satyam, S., & Patra, S. (2024). Innovations and challenges in adsorption-based wastewater remediation: a comprehensive review. *Heliyon*. Available from https://doi.org/10.1016/j.heliyon.2024.e29573.

Shah, M. A., & Ahmad, T. (2010). *Principles of nanoscience and nanotechnology*. Alpha Science International.

Sillberg, C. V., Kullavanijaya, P., & Chavalparit, O. (2021). Water quality classification by integration of attribute-realization and support vector machine for the Chao Phraya River. *Journal of Ecological Engineering*, *22*(9), 70–86. Available from https://doi.org/10.12911/22998993/141364.

Silva, G. M. E., Campos, D. F., Brasil, J. A. T., Tremblay, M., Mendiondo, E. M., & Ghiglieno, F. (2022). Advances in technological research for online and in situ water quality monitoring—A review. *Sustainability*, *14*(9), 5059. Available from https://doi.org/10.3390/su14095059.

Singh, J., Pant, M., Lakhera, A., & Singhal, N. (2023). Water quality assessment of Rispana River at Dehradun (Uttarakhand) India with reference to WQI and microbial contamination. https://doi.org/10.21203/rs.3.rs-3321030/v1.

Singh, K. K., Singh, A., & Rai, S. (2022). A study on nanomaterials for water purification. *Materials Today: Proceedings*, *51*, 1157–1163. Available from https://doi.org/10.1016/j.matpr.2021.07.116.

Singh, N. K., Yadav, M., Singh, V., Padhiyar, H., Kumar, V., Bhatia, S. K., & Show, P. L. (2023). Artificial intelligence and machine learning-based monitoring and design of biological wastewater treatment systems. *Bioresource Technology*, *369*, 128486. Available from https://doi.org/10.1016/j.biortech.2022.128486.

Song, D., Yang, R., Wang, H., Li, W., Wang, H., Long, H., & Long, F. (2017). A label-free SERRS-based nanosensor for ultrasensitive detection of mercury ions in drinking water and wastewater effluent. *Analytical Methods*, *9*(1), 154–162. Available from https://doi.org/10.1039/c6ay02361d.

Steffens, C., Steffens, J., Graboski, A. M., Manzoli, A., & Leite, F. L. (2017). *Nanosensors for detection of pesticides in water. New pesticides and soil sensors* (pp. 595–635). Academic Press. Available from http://doi.org/10.1016/b978-0-12-804299-1.00017-5.

Sudhakar, M. (2023). Artificial intelligence applications in water treatment and water resource assessment: Challenges, innovations, and future directions. *Intelligent engineering applications and applied sciences for sustainability* (pp. 248–269). IGI Global.

SudhirSastry, Y. B., Krishna, Y., & Budarapu, P. R. (2015). Parametric studies on buckling of thin walled channel beams. *Computational Materials Science*, *96*, 416–424. Available from https://doi.org/10.1016/j.commatsci.2014.07.058.

Tiwari, S., Babbar, R., & Kaur, G. (2018). Performance evaluation of two ANFIS models for predicting water quality index of River Satluj (India). *Advances in Civil Engineering*, *2018*(1), 8971079. Available from https://doi.org/10.1155/2018/8971079.

Tripathi, G. P., Agarwal, S., Awasthi, A., & Arun, V. (2022). Artificial hip prostheses design and its evaluation by using ansys under static loading condition. *Biennial international conference on future learning aspects of mechanical engineering* (pp. 815–828). Singapore: Springer Nature Singapore. Available from https://doi.org/10.1007/978-981-99-3033-3_68.

UN, Department of Economic and Social Affairs Sustainable Development. (2015). https://sdgs.un.org/topics/water-and-sanitation. (Accessed 26.11.2020).

Vikesland, P. J. (2018). Nanosensors for water quality monitoring. *Nature Nanotechnology*, *13*(8), 651–660. Available from https://doi.org/10.1038/s41565-018-0209-9.

WFM Staff. (2020). Report: Data demand will drive $92 billion in investment by 2030, (2020). https://waterfm.com/report-data-demand-will-drive-92-billion-ininvestment-by-2030/ (Accessed November 24.11.2020).

Wang, J., Zhang, F., Kung, H. T., Ren, Y., Zhang, Y., & Yu, H. (2017). Linkage analysis of land use/cover patterns and hydro-chemical characteristics in different seasons in Ebinur Lake Watershed, China. *Water*, *9*(11), 888. Available from https://doi.org/10.3390/w9110888.

Wang, Y., Cao, Z., & Barati Farimani, A. (2021). Efficient water desalination with graphene nanopores obtained using artificial intelligence. *npj 2D Materials and Applications*, *5*(1), 66. Available from https://doi.org/10.1038/s41699-021-00246-9.

Wang, Y., Cheng, Y., Liu, H., Guo, Q., Dai, C., Zhao, M., & Liu, D. (2023). A review on applications of artificial intelligence in wastewater treatment. *Sustainability*, *15*(18), 13557. Available from https://doi.org/10.3390/su151813557.

Water, U. N. (2011). Policy brief: water quality. United Nations. http://www.unwater.org/downloads/waterquality_policybrief.pdf.

Wei, Y., Jiao, Y., An, D., Li, D., Li, W., & Wei, Q. (2019). Review of dissolved oxygen detection technology: From laboratory analysis to online intelligent detection. *Sensors*, *19*(18), 3995. Available from https://doi.org/10.3390/s19183995.

Yousefi, M., Akbari, H., Adibzadeh, A., Mohammadi, A. A., Baziar, M., Farajzadeh, M. A., & Akbari, H. (2024). Adsorption of diazinon from aqueous solution using metal organic framework and functionalized graphene: Comparison of BBD, ANN models. *Chemosphere*, *351*, 141222. Available from https://doi.org/10.1016/j.chemosphere.2024.141222.

Yuan, H., Sun, G., Peng, W., Ji, W., Chu, S., Liu, Q., & Liang, Y. (2021). Thymine-functionalized gold nanoparticles (Au NPs) for a highly sensitive fiber-optic surface plasmon resonance mercury ion nanosensor. *Nanomaterials*, *11*(2), 397. Available from https://doi.org/10.3390/nano11020397.

Yunus, I. S., Harwin., Kurniawan, A., Adityawarman, D., & Indarto, A. (2012). Nanotechnologies in water and air pollution treatment. *Environmental Technology Reviews*, *1*(1), 136–148. Available from https://doi.org/10.1080/21622515.2012.733966.

Zhao, L., Dai, T., Qiao, Z., Sun, P., Hao, J., & Yang, Y. (2020). Application of artificial intelligence to wastewater treatment: A bibliometric analysis and systematic review of technology, economy, management, and wastewater reuse. *Process Safety and Environmental Protection*, *133*, 169–182. Available from https://doi.org/10.1016/j.psep.2019.11.014.

Zhuiykov, S. (2012). Solid-state sensors monitoring parameters of water quality for the next generation of wireless sensor networks. *Sensors and Actuators B: Chemical*, *161*(1), 1–20. Available from https://doi.org/10.1016/j.snb.2011.10.078.

Further reading

Grid intelligence water case study. (2018). https://www.verizon.com/business/resources/articles/gridintelligence-water-case-study/. (Accessed 5.09.2022).

Lu, J. C., Jeng, S. L., & Wang, K. (2009). A review of statistical methods for quality improvement and control in nanotechnology. *Journal of Quality Technology*, *41*(2), 148–164. Available from https://doi.org/10.1080/00224065.2009.11917770.

Ramos, R., & Castro, R. (2020). Modeling renewable energy integration in the 2030 Portuguese power system: The role of energy storage. *Journal of Electrochemical Energy Conversion and Storage*, *17*(1), 011013. Available from https://doi.org/10.1115/1.4044958.

Where AI is aiding productivity. (2020). https://www.statista.com/chart/23779/ai-productivity-increase/. (Accessed 23.03.2023).

CHAPTER 8

Empowering water security: the synergy of automation and IoT integration

Neha Mumtaz[1], Tabish Izhar[1], Jazbi Izhar[2] and Syed Aqeel Ahmad[1]

[1]*Department of Civil Engineering, Integral University, Lucknow, Uttar Pradesh, India,* [2]*Applied A.I. Solutions Development, George Brown College's School of Computer Technology, Toronto, ON, Canada*

8.1 Introduction

Water security is a paramount global challenge, intricately linked to sustainable development, public health, and environmental preservation (Dell'Angelo, D'Odorico, & Rulli, 2017). As populations burgeon, climate change accelerates, and urbanization intensifies, the imperative to ensure reliable, safe, and equitable access to water resources has never been more pressing. In response to these complex challenges, the integration of automation and the Internet of Things (IoT) has emerged as a transformative approach to water management, promising to revolutionize how we monitor, manage, and safeguard water systems (Chen, Wu, & Ahmadi, 2022).

The background and context of this critical review delve into the evolving landscape of water security and the pivotal role of technology in addressing its multifaceted dimensions. Traditional water management approaches often face limitations in terms of real-time monitoring, predictive capabilities, resource optimization, and infrastructure resilience (Guna Priya, 2023). Integrating automation and IoT technologies offers a paradigm shift by enabling continuous data collection, advanced analytics, predictive modeling, and automated control mechanisms. This context sets the stage for understanding the significance and potential impacts of integrated automation and IoT solutions in empowering water security (Sharida, Hamdan, & AL-Hashimi, 2020).

The primary research objectives of this critical review are twofold. First, to comprehensively evaluate the benefits, challenges, and future prospects of integrating automation and IoT technologies in water security. This entails examining the potential impacts on real-time monitoring, data-driven decision making, resource optimization, infrastructure resilience, and safety measures within water systems. Second, to provide actionable insights, recommendations, and best practices for stakeholders, policymakers,

Computational Automation for Water Security.
DOI: https://doi.org/10.1016/B978-0-443-33321-7.00010-X

and researchers involved in implementing and leveraging integrated automation and IoT solutions in water management practices.

The scope of this chapter encompasses a wide range of topics related to the synergy of automation and IoT in water security. It covers conceptual frameworks, key technologies, benefits, challenges, case studies, future prospects, and recommendations. The structure of the review is organized into distinct sections, each focusing on critical aspects such as real-time monitoring, data-driven decision making, resource optimization, infrastructure resilience, challenges, future prospects, case studies, and conclusions. This structured approach provides a comprehensive analysis and synthesis of existing literature, industry practices, and technological advancements, offering a holistic perspective on the integration of automation and IoT in empowering water security. Fig. 8.1 depicts the framework of the relationship between IoT and water security.

8.1.1 Understanding water security

Understanding water security encompasses a multifaceted approach that goes beyond the mere availability of water to encompass various dimensions critical for human well-being and ecosystem sustainability. It involves ensuring reliable access to safe and sufficient

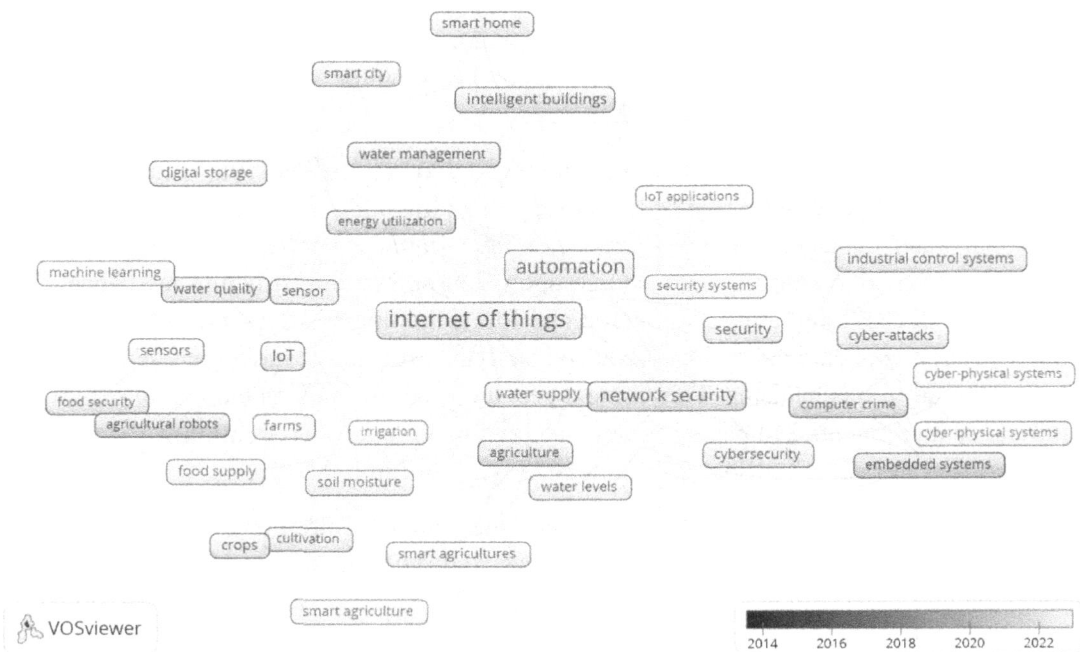

Figure 8.1
IoT and water security nexus. *IoT*, Internet of Things.

water for drinking, sanitation, hygiene, agriculture, industry, and environmental conservation (Hasan et al., 2023). Water security also entails addressing issues of affordability, equity, governance, and resilience to water-related risks such as floods, droughts, pollution, and climate change impacts (Rani, Jayalakshmi, & Baligar, 2020). Challenges related to water insecurity, including inadequate access to clean water, waterborne diseases, water scarcity, pollution, and conflicts over water allocation, have far-reaching implications for public health, economic development, food security, and environmental sustainability. Technology, including automation, IoT, data analytics, and innovative water treatment solutions, plays a crucial role in enhancing water security by enabling real-time monitoring, data-driven decision making, resource optimization, and stakeholder collaboration. Embracing a holistic approach to water security involves integrating technological innovations with effective governance, policy frameworks, community engagement, and sustainable water management practices to ensure the sustainable and equitable management of water resources for current and future generations (Harshitha, Manikanta Uma Srinivas, Eswar Sai, Kommuri, & Gopi Krishna, 2024). Water security encompasses a broad and multifaceted concept that goes beyond mere availability of water to include aspects of accessibility, quality, sustainability, and governance. At its core, water security is about ensuring that all individuals and communities have reliable access to safe and sufficient water resources for their basic needs and livelihoods (Vijayan, Rose, Arvindan, Revathy, & Amuthadevi, 2020). This includes access to drinking water, sanitation, hygiene, irrigation for agriculture, and water for industrial and environmental purposes. The dimensions of water security encompass not only the physical availability of water but also issues of affordability, equity, governance, and ecological sustainability.

Water insecurity poses significant challenges and implications at local, regional, and global scales. Inadequate access to clean water and sanitation leads to health risks, waterborne diseases, and economic burdens, particularly in marginalized communities and developing regions (Al-Mulla & Al-Badi, 2021). Water scarcity, exacerbated by population growth, climate change, and competing demands, intensifies pressure on water resources and ecosystems, leading to conflicts over water allocation and usage. Water pollution from industrial, agricultural, and urban sources further degrades water quality, threatens ecosystems, and compromises human health. The implications of water insecurity extend to food security, energy production, economic development, and environmental sustainability, making it a complex and urgent issue requiring holistic solutions (Sidhu, Kumar, Rana, & Jat, 2021).

Technology plays a pivotal role in enhancing water security by providing innovative solutions to address the challenges and dimensions of water insecurity. Automation and IoT technologies enable real-time monitoring of water systems, allowing for early detection of leaks, contamination events, and anomalies. Advanced sensors, data analytics, and machine learning (ML) algorithms enhance water quality monitoring, predictive modeling, and decision-making capabilities. Smart meters and digital platforms empower consumers to track and manage their water usage, promoting conservation and efficiency (Akkasaligar, Biradar, & Pujari, 2020). Water treatment technologies such as desalination, membrane filtration, and advanced oxidation processes improve water quality and expand water supply options. Moreover, technology facilitates integrated water management approaches, stakeholder collaboration, and data-driven policy interventions to enhance governance, resilience, and sustainability in water systems. By harnessing the potential of technology, water security can be strengthened, ensuring equitable access to clean and sustainable water resources for present and future generations.

8.2 The synergy of automation and IoT

The synergy of automation and the IoT represents a pivotal convergence in modern water management practices, promising transformative advancements in efficiency, sustainability, and resilience. As water systems grapple with escalating challenges such as population progression, climate change effects, old infrastructure, and supply constraints, the integration of automation and IoT emerges as a strategic imperative. This synergy entails the seamless integration of automated control mechanisms and intelligent IoT devices throughout water infrastructure, creating interconnected systems capable of real-time monitoring, data-driven decision making, and adaptive responses. In this critical review, we delve into the multifaceted dimensions of this synergy, exploring its key components, technologies, benefits, challenges, and prospects. By unlocking the potential of integrated automation and IoT solutions, water security can be strengthened, ensuring the sustainable management of water resources for present and future generations (Ibrahim & Al-Wadi, 2022).

The conceptual framework of integrating automation and the IoT in water management revolves around leveraging advanced technologies to create interconnected, intelligent, and responsive water systems. At its core, this framework involves deploying a network of IoT sensors and devices throughout water infrastructure to collect real-time data on water quality, quantity, flow rates, pressure levels, and environmental conditions. These data are then transmitted to centralized or cloud-based platforms for analysis, processing, and decision making. Automation mechanisms, including actuators, controllers, and smart algorithms, enable automated control and optimization of water processes based on data insights and predefined parameters. The integration of automation and IoT transforms traditional water management practices into dynamic, data-driven, and adaptive systems capable of proactively monitoring, managing, and optimizing water resources (Wang & Liu, 2024).

Several key components and technologies are integral to the integration of automation and IoT in water management:

1. IoT Sensors and Devices: These include various sensors such as water quality sensors, flow meters, pressure sensors, level sensors, temperature sensors, and environmental

sensors. IoT devices also encompass smart meters, valves, pumps, and actuators that interact with water infrastructure and collect data (Basharat & Bin Mohamad, 2022).

2. Communication Networks: IoT devices communicate over wired or wireless networks, including Wi-Fi, cellular, LoRaWAN, Zigbee, and Bluetooth. These networks facilitate data transmission between sensors, devices, control systems, and centralized platforms (Xie & Leung, 2021).
3. Data Analytics and Machine Learning: Advanced analytics and ML algorithms process the vast amount of data collected from IoT sensors. They analyze data patterns, detect anomalies, predict trends, and generate actionable insights for decision making and optimization.
4. Automation and Control Systems: Automated control mechanisms, such as supervisory control and data acquisition systems, programmable logic controllers (Stellios, Kotzanikolaou, & Psarakis, 2019), and distributed control systems, enable remote monitoring, control, and optimization of water processes. They adjust water flow, treatment parameters, and energy usage based on real-time data and predefined rules (Mijović, Tomašević, Janev, Stanojević, & Vraneš, 2011).
5. Cloud Computing and Edge Computing: Cloud-based platforms and edge computing technologies provide scalable, secure, and efficient data storage, processing, and analysis capabilities. Edge computing brings data processing closer to IoT devices, reducing latency, bandwidth usage, and dependency on centralized data centers (Kumar, Nayar, Mehta, & Sharma, 2021).

The integration of automation and IoT in water management offers a multitude of benefits and potential impacts on water security:

1. Real-time Monitoring and Early Detection: IoT sensors enable real-time monitoring of water systems, allowing for early detection of leaks, contamination events, equipment failures, and anomalies. This proactive approach minimizes downtime, reduces water losses, and enhances system reliability (Alotaibi, Karsou, Khan, Tohmeh, & Bashar, 2023).
2. Data-driven Decision Making: Advanced analytics and ML enable data-driven decision making by providing actionable insights into water usage patterns, infrastructure performance, and potential risks. This improves operational efficiency, supports predictive maintenance, and enhances resource allocation (Mendoza, Timbol, Samonte, & Blancaflor, 2020).
3. Resource Optimization: Automation mechanisms optimize water processes, such as water distribution, treatment, and irrigation, based on real-time data and predefined parameters. This leads to efficient resource usage, energy savings, cost reductions, and environmental benefits (Wang & Liu, 2024).
4. Enhanced Infrastructure Resilience: Automated control systems enhance the resilience of water infrastructure by facilitating rapid response to emergencies, adaptive control

strategies, and predictive maintenance. This improves system reliability, minimizes risks, and ensures the continuity of water services (Yuan et al., 2019).

5. Stakeholder Engagement and Transparency: IoT platforms and data analytics promote stakeholder engagement, transparency, and accountability by providing access to real-time data, performance metrics, and water usage insights. This fosters collaboration among water utilities, policymakers, researchers, and the public, leading to informed decision making and governance (Kaushik & Bagga, 2021).

Overall, the integration of automation and IoT in water management can revolutionize water security by creating smarter, more efficient, and resilient water systems capable of addressing current challenges and future uncertainties.

8.2.1 Real-time monitoring and control

Real-time monitoring and control stand as pillars of innovation in modern water management, enabled by the seamless integration of automation and the IoT (Allam & Jones, 2021). Through a network of IoT sensors and devices strategically deployed across water infrastructure, real-time data on water quality, flow rates, pressure levels, and environmental conditions are continuously collected and transmitted to centralized platforms. This influx of real-time data empowers water managers with a comprehensive and up-to-date view of system dynamics, allowing for swift detection of anomalies, leaks, contaminations, or equipment failures (Ahmed, Anifowose, & Khan, 2015). Automated control mechanisms, driven by this real-time data and preset parameters, enable proactive interventions, adjustments, and optimizations in water processes. The result is enhanced operational efficiency, reduced downtime, minimized water losses, and improved system reliability, laying the foundation for smarter, more responsive, and resilient water systems (Aslam, Tufail, Kim, Apong, & Raza, 2023).

Real-time monitoring holds immense importance in water management due to its ability to provide timely, accurate, and actionable insights into the status and performance of water systems. By continuously collecting data on parameters such as water quality, flow rates, pressure levels, and environmental conditions, real-time monitoring enables water managers to detect anomalies, leaks, contaminations, or equipment failures promptly. This proactive approach allows for swift intervention, mitigation measures, and optimizations in water processes, leading to enhanced operational efficiency, reduced downtime, minimized water losses, and improved system reliability. Implementation of real-time monitoring involves deploying a network of IoT sensors and devices throughout water infrastructure, establishing communication networks for data transmission, integrating data with centralized platforms for analysis, and utilizing automated control mechanisms for decision making and interventions.

Several case studies and examples demonstrate the efficacy and benefits of real-time control systems in water management. For instance, in water distribution networks, real-time monitoring of flow rates and pressure levels allows for immediate detection and isolation of leaks or bursts, preventing water losses and minimizing service disruptions. Automated control valves can adjust water flow based on real-time demand patterns, optimizing distribution and reducing energy consumption. In wastewater treatment plants, real-time monitoring of water quality parameters enables operators to adjust treatment processes, chemical dosing, and filtration rates in response to fluctuations, ensuring compliance with regulatory standards and maintaining effluent quality. Smart irrigation systems use real-time weather data and soil moisture sensors to optimize watering schedules, minimizing water usage and promoting water conservation in agriculture. These case studies highlight the versatility, efficiency, and effectiveness of real-time control systems in various water management applications (Dong, Werling, Cao, & Li, 2024).

Despite its benefits, real-time monitoring and control systems face several challenges and limitations. One of the primary challenges is ensuring data accuracy, reliability, and consistency across diverse IoT sensors and devices. Calibration, maintenance, and synchronization of sensors are crucial to mitigate data errors and discrepancies. Interoperability issues, compatibility constraints, and communication failures among IoT devices and platforms can hinder seamless data transmission and integration (Taylor & Amidy, 2020). Cybersecurity threats, data privacy concerns, and vulnerabilities in network infrastructure pose risks to data integrity and system security in real-time monitoring and control systems. Additionally, the complexity and cost of implementing and maintaining real-time systems, including hardware, software, training, and infrastructure upgrades, can be significant challenges for water utilities and organizations (Ainane, Ouzzif, & Bouragba, 2019). Overcoming these challenges requires a holistic approach that addresses technological, organizational, regulatory, and financial considerations to ensure the effective and sustainable deployment of real-time monitoring and control solutions in water management.

8.2.2 Data-driven decision making

Data-driven decision making stands as a cornerstone of innovation and efficiency in contemporary water management, driven by the integration of automation and the IoT. In an era where data have become the lifeblood of operations, the ability to harness, analyze, and derive actionable insights from vast datasets holds immense significance (Phasinam & Kassanuk, 2022). Data-driven decision making in water management involves leveraging advanced analytics, ML algorithms, and real-time data streams from IoT sensors to inform strategic, operational, and tactical decisions. By translating data into actionable intelligence, water managers can optimize resource allocation, improve operational efficiency, enhance system performance, and mitigate risks. In this critical review, we delve into the transformative impact of data-driven decision making, exploring its benefits, challenges, applications, and implications for water security and sustainability. Through a data-driven lens, water management can evolve into a dynamic, adaptive, and resilient domain, ensuring the effective stewardship of water resources for current and future generations (Delgado, Short, Roberts, & Vandenberg, 2019).

Data analytics and ML play a pivotal role in revolutionizing water management by providing advanced tools and techniques for extracting meaningful insights from complex datasets. In the context of water management, data analytics involves the use of statistical methods, algorithms, and visualization tools to analyze historical data, identify patterns, trends, and correlations, and derive actionable insights (Toh & Milojicic, 2021). ML, on the other hand, leverages algorithms and models to learn from data, make predictions, and automate decision-making processes based on patterns and trends identified in the data.

One key application of data analytics and ML in water management is predictive modeling. By analyzing historical data on water usage, weather patterns, infrastructure performance, and environmental conditions, predictive models can forecast water demand, identify potential risks such as leaks or equipment failures, optimize resource allocation, and improve operational efficiency. ML algorithms can also be used for anomaly detection, pattern recognition, and optimization of water treatment processes, leading to cost savings, energy efficiency, and improved water quality (Davenport, Guha, Grewal, & Bressgott, 2020).

Several case studies demonstrate the transformative impact of data-driven decision making in water management. For instance, in water distribution networks, data analytics can analyze flow data from sensors to detect anomalies indicative of leaks or bursts, allowing for proactive maintenance and intervention. Predictive modeling can optimize pump scheduling, valve operations, and pressure management to minimize energy consumption, reduce water losses, and improve system reliability. In wastewater treatment plants, data-driven decision making can optimize chemical dosing, filtration rates, and treatment processes based on real-time data, ensuring compliance with regulatory standards and minimizing environmental impact. These case studies highlight the efficacy of data analytics and ML in optimizing water management practices, enhancing operational efficiency, and achieving sustainability goals (Mumtaz & Izhar, 2023).

While data analytics and ML offer immense benefits in water management, they also raise ethical considerations and privacy concerns related to data usage. The collection, storage, and analysis of sensitive data, such as water usage patterns, customer information, and infrastructure details, require adherence to ethical standards, data protection regulations, and privacy policies. Water utilities and organizations must ensure data security, confidentiality, and anonymity; implement robust cybersecurity measures; obtain consent for data collection and usage; and establish transparency in data practices. Ethical considerations also encompass fairness, bias mitigation, and accountability in algorithmic decision-making processes to avoid unintended consequences or discriminatory outcomes (Ocaka, Briain, Davy, & Barrett, 2022). Balancing the benefits of data-driven decision making with ethical considerations and privacy concerns is essential for fostering trust, transparency, and responsible use of data in water management.

8.2.3 Resource optimization

Resource optimization stands as a critical imperative in modern water management, driven by the imperative to efficiently allocate and utilize finite water resources while minimizing waste, energy consumption, and environmental impact. The concept of resource optimization encompasses a multifaceted approach that integrates advanced technologies, data analytics, and strategic planning to maximize the efficiency, resilience, and sustainability of water systems. By optimizing water distribution, treatment processes, energy usage, and infrastructure maintenance, water managers can achieve significant cost savings, enhance operational performance, improve water quality, and mitigate risks. In this critical review, we delve into the intricacies of resource optimization in water management, exploring its key components, strategies, benefits, challenges, and potential impacts on water security and sustainability. Through a lens of resource optimization, water management can evolve into a dynamic and adaptive domain, ensuring the effective stewardship of water resources for present and future generations (Wang & Liu, 2024).

Optimizing water resources through automation and the IoT involves implementing a range of strategies aimed at maximizing efficiency, reducing waste, and enhancing sustainability in water management. One key strategy is real-time monitoring and control, where IoT sensors continuously collect data on water quality, flow rates, pressure levels, and environmental conditions. These data are then analyzed using data analytics and ML algorithms to identify patterns, trends, and anomalies. Automated control mechanisms, such as actuators and controllers, adjust water flow, treatment parameters, and energy usage based on real-time insights, optimizing resource utilization and minimizing inefficiencies. Another strategy is predictive maintenance, where ML algorithms analyze historical data to predict equipment failures, identify maintenance needs, and schedule proactive repairs, reducing downtime and extending asset lifespan. Additionally, demand forecasting and optimization use data analytics to predict water demand patterns, optimize water distribution, and prioritize resource allocation, ensuring efficient use of water resources (Davenport et al., 2020).

Implementing strategies for optimizing water resources through automation and IoT leads to significant efficiency gains, cost savings, and positive environmental impacts. By optimizing water distribution, treatment processes, and energy usage, water utilities can achieve operational efficiencies, reduce water losses, lower energy consumption, and decrease operational costs. Predictive maintenance reduces downtime, maintenance costs, and asset failures, improving system reliability and performance. Demand forecasting and optimization minimize overuse and wastage of water, promoting water conservation and sustainability. These efficiency gains and cost savings translate into tangible benefits for water utilities, stakeholders, and communities, while also reducing environmental footprint and resource depletion (Harshitha et al., 2024).

The scalability and adaptability of resource optimization solutions in water management are crucial for addressing evolving challenges, changing conditions, and future uncertainties. Scalability involves the ability of solutions to handle increasing data volumes, device deployments, and user requirements as water systems expand or face increased demand. Cloud-based platforms, edge computing, and modular architectures support scalability by providing flexible, scalable, and cost-effective solutions for data storage, processing, and analysis. Adaptability refers to the capability of solutions to adjust, evolve, and respond to changing environmental, regulatory, and operational conditions. ML algorithms, AI-driven decision support systems, and agile methodologies enhance adaptability by enabling dynamic adjustments, optimization strategies, and predictive modeling based on real-time data and feedback. By ensuring scalability and adaptability, resource optimization solutions can effectively address complex challenges, promote resilience, and support sustainable water management practices.

8.3 Enhanced infrastructure resilience

Enhanced infrastructure resilience stands as a cornerstone in the realm of water management, embodying the capacity of systems to withstand and recover from challenges, disruptions, and changing environmental conditions. As water systems face escalating pressures from population growth, climate change impacts, ageing infrastructure, and evolving regulatory frameworks, the imperative to bolster resilience has become increasingly pronounced. Enhanced infrastructure resilience involves integrating advanced technologies, risk management strategies, adaptive measures, and stakeholder collaboration to enhance system robustness, reliability, and responsiveness. In this critical review, we delve into the intricate facets of enhanced infrastructure resilience in water management, exploring its key components, strategies, benefits, challenges, and potential impacts on water security and sustainability. Through a lens of enhanced resilience, water systems can evolve into adaptive, sustainable, and resilient assets, ensuring the reliable and continuous delivery of water services for communities and ecosystems alike (Yuan et al., 2019).

Building resilient water infrastructure involves leveraging automation and the IoT to enhance system robustness, reliability, and adaptability. One key aspect is the deployment of IoT sensors throughout water infrastructure to enable real-time monitoring of critical parameters such as water quality, flow rates, pressure levels, and equipment status. This continuous data collection allows for the early detection of anomalies, leaks, equipment failures, and potential risks. Automated control mechanisms, driven by this real-time data and predictive analytics, can optimize water flow, treatment processes, energy usage, and asset performance to mitigate risks, reduce downtime, and enhance system resilience. Additionally, digital twins and simulation models can be used to create virtual replicas of water infrastructure, allowing for scenario analysis, predictive modeling, and risk assessment to inform decision making and planning (Dhavamani & Prem Priya, 2022).

Predictive maintenance, asset management, and risk mitigation strategies are crucial components of building resilient water infrastructure. Predictive maintenance uses data analytics and ML algorithms to predict equipment failures, identify maintenance needs, and schedule proactive repairs or replacements. This approach minimizes downtime, reduces maintenance costs, extends asset lifespan, and improves system reliability. Asset management strategies involve inventory management, life cycle analysis, condition assessment, and prioritization of critical assets based on risk and performance metrics. Risk mitigation encompasses identifying, assessing, and managing risks such as infrastructure vulnerabilities, climate change impacts, regulatory compliance, cybersecurity threats, and natural disasters. By integrating predictive maintenance, asset management, and risk mitigation strategies with automation and IoT technologies, water utilities can enhance infrastructure resilience, optimize resource allocation, and ensure the continuity of water services (Sobczak & Ziora, 2021).

Integrating resilience strategies with emergency response plans is essential for effectively managing and mitigating crises, disruptions, and unforeseen events in water systems. This involves developing comprehensive emergency response plans that incorporate resilience strategies, risk assessments, contingency plans, communication protocols, and stakeholder coordination. Automation and IoT technologies play a crucial role in enhancing emergency response capabilities by providing real-time data, situational awareness, and decision-support tools. Automated alerts, alarms, and notifications can be triggered based on predefined thresholds or abnormal conditions, enabling rapid response, mitigation measures, and resource mobilization during emergencies. Collaborative platforms and communication networks facilitate information sharing, coordination, and collaboration among stakeholders, emergency responders, government agencies, and communities. By integrating resilience strategies with emergency response plans, water utilities can improve preparedness, response times, recovery efforts, and overall resilience in the face of disruptions and crises (Rueda et al., 2023).

8.3.1 Challenges and considerations

Navigating the intricacies of water management, especially in the context of integrating automation and the IoT, brings forth a myriad of challenges and considerations. These challenges span technological, operational, regulatory, ethical, and societal dimensions, requiring a comprehensive approach to address and mitigate potential risks and obstacles. As water systems embrace advanced technologies and data-driven solutions, they must contend with issues such as data security and privacy, interoperability, scalability, and human factors. Moreover, ensuring the seamless integration of automation and IoT while maintaining reliability, resilience, and sustainability presents a complex landscape that demands careful planning, collaboration, and adaptation. In this critical review, we delve into the challenges and considerations inherent in the integration of automation and IoT in water management, exploring their impacts, implications, and strategies for overcoming these hurdles. By understanding and addressing these challenges, water utilities, stakeholders, and policymakers can navigate the path toward harnessing the full potential of automation and IoT technologies while ensuring the integrity, security, and effectiveness of water systems.

Addressing data security and privacy concerns is paramount in the integration of automation and IoT in water management. Water utilities must implement robust cybersecurity measures, encryption protocols, access controls, and data governance frameworks to protect sensitive data from unauthorized access, breaches, and cyber threats. Compliance with data protection regulations, such as GDPR (Zhang et al., 2021), HIPAA (Hussain, Dawood, & Al-Turjman, 2021), and CCPA (Bentotahewa, Hewage, & Williams, 2021), is essential to ensure the privacy rights of individuals and the confidentiality of data. Conducting regular security audits, vulnerability assessments, and risk assessments helps identify and mitigate potential security gaps and vulnerabilities in IoT devices, networks, and platforms. Building a culture of data security awareness, training employees on cybersecurity best practices, and fostering partnerships with cybersecurity experts and vendors contribute to a proactive approach to data security and privacy.

Overcoming interoperability challenges is crucial for ensuring seamless integration and communication among diverse IoT devices, systems, and platforms in water management. Standardization of protocols, data formats, and communication interfaces facilitates interoperability and compatibility among IoT devices from different manufacturers and vendors. Adopting open-source platforms, application programming interfaces, and middleware solutions enables data exchange, integration, and orchestration across heterogeneous systems. Collaborative efforts, industry partnerships, and consortiums promote interoperability standards and best practices, fostering an ecosystem of interconnected and interoperable IoT solutions in water management (Jan, Min-Allah, & Düştegör, 2021).

Scalability, maintenance, and upkeep of integrated systems pose challenges in sustaining the performance, reliability, and efficiency of automation and IoT solutions in water management. Scalability involves designing systems that can handle increasing data volumes, device deployments, and user requirements as water systems expand or face increased demand. Cloud-based platforms, edge computing, and modular architectures support scalability by providing flexible, scalable, and cost-effective solutions for data storage, processing, and analysis (Mani Sekhar, Tewari, Rahman, & Siddesh, 2020). Proactive maintenance strategies, including predictive maintenance, condition-based maintenance, and asset management, help optimize system performance, minimize downtime, and extend equipment lifespan. Regular software updates, firmware upgrades, and system patches ensure the security, stability, and functionality of integrated systems over time.

Human factors such as skills, training, and user acceptance play a critical role in the successful adoption and utilization of automation and IoT solutions in water management. Water utilities must invest in training programs, skill development initiatives, and knowledge transfer activities to equip employees with the necessary technical expertise, competencies, and capabilities to operate and maintain integrated systems effectively. User-centered design principles, usability testing, and feedback mechanisms help design intuitive interfaces, workflows, and user experiences that enhance user acceptance and adoption. Engaging stakeholders, fostering a culture of innovation, and addressing user concerns and feedback promote user acceptance and collaboration in leveraging automation and IoT technologies for water management (Wang & Liu, 2024).

The research gaps in the integration of automation and IoT for water security are mentioned in Table 8.1.

8.4 Future prospects and innovations

The future of water management is teeming with prospects and innovations driven by cutting-edge technologies and transformative approaches. As we navigate an era marked by rapid urbanization, climate change impacts, population growth, and resource constraints, the imperative to embrace innovative solutions has never been more pronounced. The convergence of automation, artificial intelligence (AI), machine learning, the IoT, and other emerging technologies holds immense promise in revolutionizing how we monitor, manage, and safeguard water resources. Future prospects in water management encompass a spectrum of possibilities, including AI-driven predictive modeling, blockchain-enabled traceability and transparency, edge computing for real-time analytics, and collaborative platforms for stakeholder engagement (Mumtaz, Izhar, Pandey, & Labhasetwar, 2022). These innovations herald a new era of resilience, efficiency, sustainability, and inclusivity in water management, paving the way for smarter, more adaptive, and resilient water systems that can meet the evolving needs of society and the environment. In this critical review, we explore the exciting prospects and innovations poised to shape the landscape of water management in the years to come.

Table 8.1: Research gaps in the integration of automation and IoT for water security.

Research area	Research gap	Solution
Data security and privacy (Chaudhari & Bansode, 2023)	• Insufficient comprehensive cybersecurity frameworks • Privacy protection mechanisms for sensitive water data	• Need for robust, scalable cybersecurity solutions tailored for IoT in water management. • Development of advanced data anonymization and encryption techniques.
Interoperability (Khan & Altayar, 2021)	Lack of standardized communication protocols Integration challenges with legacy systems	Need for universal standards to ensure compatibility among diverse IoT devices and systems. Strategies for seamless integration of new technologies with existing infrastructure.
Scalability (Madushanki, Halgamuge, Wirasagoda, & Syed, 2019)	Limited scalability of current IoT solutions Economic feasibility of scaling up IoT implementations	Research on scalable architectures to handle increasing data volumes and device numbers. Cost-benefit analysis and financial models for large-scale IoT deployments.
Predictive maintenance (Niedermaier, Hanka, Fischer, & Merli, 2021)	Inadequate predictive models for diverse water systems Limited real-world case studies demonstrating benefits	Development of machine learning models tailored to various types of water infrastructure. Comprehensive case studies and longitudinal studies to quantify impacts and benefits.
Data-driven decision making (Halgamuge et al., 2021)	Insufficient data analytics frameworks Integration of AI/ML with traditional decision-making processes	Advanced algorithms and frameworks for real-time data analysis and decision support. Research on effective integration methods and their impacts on decision accuracy and speed.
Resource optimization (Pavithra, Duraisamy, & Shankar, 2024)	Limited research on holistic resource optimization approaches Lack of adaptive resource management strategies	Comprehensive models that integrate water, energy, and environmental factors. Development of adaptive algorithms that can respond dynamically to changing conditions.
Human factors (Jaouhari, Palacios-Garcia, Anvari-Moghaddam, & Bouabdallah, 2019)	Need for training programs and skill development initiatives User acceptance and engagement strategies	Research on effective training methodologies and their impacts on user acceptance and system performance. Studies on psychological and sociocultural factors affecting technology adoption in water management.
Emerging technologies (Dhavamani & Prem Priya, 2022)	Impact of blockchain, edge computing, and other emerging technologies on water security Integration of digital twins and cyber-physical systems	Detailed studies on the applications, benefits, and limitations of these technologies in water management. Research on the practical implementation and advantages of digital twins in predictive modeling and system management.

(*Continued*)

Table 8.1: (Continued)

Research area	Research gap	Solution
Policy and regulation (Lu et al., 2023)	Gaps in regulatory frameworks addressing IoT and automation in water management Incentive structures for technology adoption	Analysis of existing policies and development of comprehensive regulatory guidelines. Research on effective incentive mechanisms to encourage the adoption of advanced technologies in water utilities.
Sustainability and environmental impact (Sharida et al., 2020)	Comprehensive assessment of environmental impacts of IoT implementations Strategies for achieving sustainable water management through technology	Life cycle assessments and studies on the environmental footprint of IoT devices and systems. Development of frameworks that align IoT and automation with sustainability goals and principles.

Note: *IoT*, Internet of Things; *AI*, artificial intelligence; *ML*, machine learning.

The advancements in AI, ML, and predictive analytics are poised to revolutionize water management by providing powerful tools for data analysis, decision making, and optimization. AI and ML algorithms can analyze vast amounts of data from IoT sensors, historical records, weather forecasts, and other sources to identify patterns, trends, correlations, and anomalies. Predictive analytics models can forecast water demand, detect leaks, predict equipment failures, optimize water distribution, and improve energy efficiency. These advancements enable water utilities to make data-driven decisions, automate routine tasks, optimize resource allocation, and enhance operational efficiency. Moreover, AI and ML technologies continue to evolve, offering opportunities for continuous learning, adaptive control, and predictive maintenance, thereby transforming water management practices and promoting resilience in water systems.

The potential of blockchain, edge computing, and collaborative platforms is significant in enhancing transparency, security, efficiency, and collaboration in water management. Blockchain technology enables secure, transparent, and immutable data recording, storage, and sharing, facilitating traceability, auditability, and accountability in water transactions, supply chains, and governance. Edge computing brings data processing closer to IoT devices, reducing latency, bandwidth usage, and dependency on centralized data centers, while enabling real-time analytics, faster decision making, and localized control. Collaborative platforms, including digital twins, smart grids, and data-sharing ecosystems, promote stakeholder engagement, information exchange, and collaborative decision making among water utilities, policymakers, researchers, and communities. These technologies unlock new possibilities for data-driven innovation, decentralized governance, and integrated solutions in water security and sustainability.

Several emerging trends and technologies are shaping the future of automation and IoT in water security. These include:

1. Digital Twins: Digital twins create virtual replicas of physical water systems, enabling simulation, modeling, and predictive analysis for optimizing operations, planning, and maintenance (Martínez et al., 2021).
2. Smart Sensors and IoT Devices: Advancements in smart sensors, IoT devices, and wireless communication technologies enhance real-time monitoring, data collection, and control capabilities in water infrastructure (Machado, Melina Nassif Mantovani Ribeiro, & Backx Noronha Viana, 2021).
3. AI-powered Optimization: AI-driven optimization algorithms optimize water treatment processes, energy usage, asset management, and resource allocation, leading to efficiency gains and cost savings (Tan, Chen, Yang, & Tan, 2023).
4. Robotic Process Automation (RPA): RPA automates repetitive tasks, workflows, and data entry processes, improving operational efficiency and reducing human errors in water management operations (Sobczak & Ziora, 2021).
5. Cyber-Physical Systems (CPS): CPS integrates physical processes with digital technologies, enabling cyber-physical monitoring, control, and automation of water systems for enhanced performance and resilience (Foschini, Martuscelli, Montanari, & Solimando, 2021).

These emerging trends and technologies herald a future of innovation, collaboration, and sustainability in automation and IoT, shaping the next generation of water security solutions.

8.4.1 Case studies and best practices

Success stories and lessons learned from implementations of automation and IoT in water management provide valuable insights and best practices for future deployments. For example, the implementation of real-time monitoring and control systems in water distribution networks has led to significant reductions in water losses, improved system reliability, and enhanced customer service (Naik, Gautam, Pachlor, & Patel, 2024). Case studies from cities like Barcelona, Singapore, and Copenhagen showcase successful implementations of smart water grids, IoT sensors, and predictive analytics, resulting in optimized resource allocation, energy savings, and improved water quality. Key lessons learned include the importance of stakeholder engagement, data quality assurance, cybersecurity measures, scalability planning, and continuous monitoring and evaluation. Success stories highlight the transformative impact of automation and IoT technologies in addressing water management challenges and achieving operational excellence (Sharifi, Allam, Bibri, & Khavarian-Garmsir, 2024).

Benchmarking and comparing different approaches in automation and IoT deployment allow water utilities and organizations to assess performance, identify best practices, and learn from peers and industry leaders. Benchmarking involves evaluating key performance indicators, such as water efficiency (Javaid, Pazil, Siong, & Ramli, 2022), energy consumption (Viswanathan, Sai Shibu, Rao, & Ramesh, 2017), cost savings (Roshandel & Eslamian, 2023), customer satisfaction (Helo, Shaker, & Abdul-Rahaim, 2021), and environmental impact (Parvathi Sangeetha et al., 2022), across different implementations and initiatives. Comparative analysis helps identify strengths, weaknesses, opportunities, and threats of different approaches, technologies, vendors, and strategies. Water utilities can benchmark their performance against industry benchmarks, standards, and benchmarks set by leading organizations and regulatory bodies (Bunting & Edwards, 2018). Comparative studies also enable knowledge sharing, collaboration, and continuous improvement in automation and IoT deployments, fostering innovation and excellence in water management practices.

Based on success stories, lessons learned, and benchmarking analyzes, recommendations for future deployments and initiatives in automation and IoT in water management include:

1. Holistic Approach: Adopt a holistic approach that integrates automation, IoT, data analytics, AI, and collaborative platforms to address multiple challenges and opportunities in water management (Sallam, Refaey, & Shami, 2019).
2. Stakeholder Engagement: Foster stakeholder engagement, collaboration, and partnerships among water utilities, technology providers, researchers, policymakers, and communities to co-create solutions, share knowledge, and drive innovation (Kumar et al., 2023).
3. Data Governance: Implement robust data governance frameworks, data quality assurance measures, and cybersecurity protocols to ensure data integrity, privacy, security, and compliance with regulations (Alkhani, 2020).
4. Scalability and Flexibility: Design solutions that are scalable, flexible, and adaptable to evolving needs, technologies, and regulatory requirements, ensuring long-term viability and sustainability (Wang & Liu, 2024).
5. Continuous Learning and Improvement: Embrace a culture of continuous learning, experimentation, and improvement by conducting pilot projects, monitoring performance metrics, gathering feedback, and iterating on solutions to achieve desired outcomes and impact (Achyuth & Manasa, 2021).

By following these recommendations, water utilities and organizations can optimize their approach to automation and IoT deployments, maximize benefits, and contribute to the advancement of water security, sustainability, and resilience.

8.5 Conclusion and recommendations

The conclusion and recommendations mark a pivotal juncture in our exploration of the integration of automation and the IoT for water security. As we reflect on the multifaceted dimensions, challenges, and innovations discussed throughout this critical review, we arrive at actionable insights and strategic recommendations that pave the way for future advancements and initiatives in water management. The conclusion encapsulates a synthesis of key findings, lessons learned, success stories, and emerging trends, highlighting the transformative potential of integrated automation and IoT solutions in enhancing water security, resilience, and sustainability. Furthermore, the recommendations offer a roadmap for stakeholders, policymakers, water utilities, and researchers to navigate the complexities of automation and IoT deployment, emphasizing the importance of holistic approaches, stakeholder engagement, data governance, scalability, continuous learning, and innovation. By embracing these recommendations, we can chart a course toward smarter, more efficient, and resilient water systems that safeguard water resources for present and future generations. The critical review of integrating automation and the IoT for water security has unearthed several key findings and insights. First, advancements in AI, ML, predictive analytics, and IoT technologies offer transformative potential in optimizing water management processes, enhancing operational efficiency, and mitigating risks. Real-time monitoring, predictive maintenance, asset management, and demand forecasting are key strategies driving efficiency gains and cost savings in water utilities. Additionally, collaborative platforms, data-sharing ecosystems, and digital twins facilitate stakeholder engagement, information exchange, and decision making in water management. Challenges such as data security, interoperability, scalability, and human factors require strategic planning, investments, and collaborations to address effectively.

Based on these key findings and insights, recommendations for policy, practice, and research are crucial for advancing the integration of automation and IoT in water security. Policy recommendations include the development of regulatory frameworks, standards, and incentives to promote data sharing, cybersecurity, interoperability, and sustainability in water management. Practice recommendations emphasize the adoption of holistic approaches, stakeholder engagement, capacity building, and continuous learning in implementing automation and IoT solutions. Research recommendations focus on filling knowledge gaps, conducting pilot projects, evaluating technologies, and fostering innovation in water management.

In conclusion, the way forward for empowering water security lies in embracing integrated automation and IoT solutions, guided by policy, practice, and research recommendations. By harnessing the potential of AI, ML, predictive analytics, IoT, and collaborative platforms, water utilities can optimize resource allocation, enhance operational resilience, and ensure sustainable water management practices. Collaboration among stakeholders, knowledge sharing, capacity building, and continuous monitoring and evaluation are essential elements in driving innovation, resilience, and efficiency in water security. The convergence of technology, policy, and stakeholder engagement holds the key to building smarter, more adaptive, and resilient water systems that meet the challenges of the future and safeguard water resources for generations to come.

References

Al-Mulla, Y. A., & Al-Badi, T. B. (2021). IoT and RS techniques for enhancing water use efficiency and achieving water security. In *ACMinternational conference proceeding series* (pp. 83−88). Available from https://doi.org/10.1145/3490725.3490738.

Achyuth, B., & Manasa, S. (2021). Implementation of recommendation system and technology for villages using machine learning and IoT. *Lecture notes on data engineering and communications technologies* (pp. 527−542). Springer Science and Business Media Deutschland GmbH.

Ahmed, Q., Anifowose, F. A., & Khan, F. (2015). System availability enhancement using computational intelligence-based decision tree predictive model. *Proceedings of the Institution of Mechanical Engineers, Part O: Journal of Risk and Reliability*, *229*(6), 612−626. Available from https://doi.org/10.1177/1748006X15595875.

Ainane, N., Ouzzif, M., & Bouragba, K. (2019). Data security of smart cities. *Lecture notes in intelligent transportation and infrastructure* (pp. 702−712). Part F1405.

Akkasaligar, P. T., Biradar, S., & Pujari, R. (2020). Internet of Things based smart secure home system. Lecture notes on data engineering and communications technologies (pp. 348−355). Springer Science and Business Media Deutschland GmbH.

Alkhani, R. (2020). Understanding private-sector engagement in sustainable urban development and delivering the climate agenda in northwestern europe—A case study of london and copenhagen. *Sustainability*, *12* (20), 1−35. Available from https://doi.org/10.3390/su12208431.

Allam, Z., & Jones, D. S. (2021). Future (post-COVID) digital, smart and sustainable cities in the wake of 6G: Digital twins, immersive realities and new urban economies. *Land Use Policy*, *101*. Available from https://doi.org/10.1016/j.landusepol.2020.105201.

Alotaibi, H., Karsou, W., Khan, S., Tohmeh, S., & Bashar, A. (2023). Bustani: A microcontroller-based automated hydroponic vertical farming solution. In *2023 IEEE international conference on agrosystem engineering, technology and applications, AGRETA 2023* (pp. 56−61). Available from https://doi.org/10.1109/AGRETA57740.2023.10262605.

Aslam, M. M., Tufail, A., Kim, K.-H., Apong, R. A. A. H. M., & Raza, M. T. (2023). A comprehensive study on cyber attacks in communication networks in water purification and distribution plants: challenges, vulnerabilities, and future prospects. *Sensors*, *23*(18). Available from https://doi.org/10.3390/s23187999.

Basharat, A., & Bin Mohamad, M. M. (2022) Security challenges and solutions for Internet of Things based smart agriculture: A review. In *Fourth international conference on smart sensors and application: Digitalization for societal well-being, ICSSA 2022* (pp. 102−107). Available from https://doi.org/10.1109/ICSSA54161.2022.9870979.

Bentotahewa, V., Hewage, C., & Williams, J. (2021). Security and privacy issues associated with coronavirus diagnosis and prognosis. In *Sixth EAI international conference on science and technologies for smart cities, SmartCity 2020*. Springer Science and Business Media Deutschland GmbH (pp. 108−116). Available from https://doi.org/10.1007/978-3-030-76063-2_8.

Bunting, S. W., & Edwards, P. (2018). *Global prospects for safe wastewater reuse through aquaculture*. Springer Singapore.

Chaudhari, A., & Bansode, R. (2023). A privacy preservation strategy using hybrid fully homomorphic encryption scheme in IoT. *International Journal of Cooperative Information Systems*, *32*(3). Available from https://doi.org/10.1142/S0218843023500077.

Chen, J.-L., Wu, Q.-A., & Ahmadi, C. (2022). AIoT@Water - An intelligent water resources management system. In*ICOIACT 2022 - fifth international conference on information and communications technology: A new way to make AI useful for everyone in the new normal era, proceeding* (pp. 199−204). Available from https://doi.org/10.1109/ICOIACT55506.2022.9971964.

Davenport, T., Guha, A., Grewal, D., & Bressgott, T. (2020). How artificial intelligence will change the future of marketing. *Journal of the Academy of Marketing Science*, *48*(1), 24−42. Available from https://doi.org/10.1007/s11747-019-00696-0.

Delgado, J. A., Short, N. M., Roberts, D. P., & Vandenberg, B. (2019). Big Data analysis for sustainable agriculture on a geospatial cloud framework. *Frontiers in Sustainable Food Systems*, *3*. Available from https://doi.org/10.3389/fsufs.2019.00054.

Dell'Angelo, J., D'Odorico, P., & Rulli, M. C. (2017). Threats to sustainable development posed by land and water grabbing. *Current Opinion in Environmental Sustainability*, *26–27*, 120–128. Available from https://doi.org/10.1016/j.cosust.2017.07.007.

Dhavamani, L., & Prem Priya, P. (2022). Energy-efficient and privacy-preserving approach for Internet of Things nodes using a novel hybrid fuzzy water cycle and evaporation strategy and matrix-based Rivest–Shamir–Adleman encryption algorithm. *Concurrency and Computation: Practice and Experience*, *34*(27). Available from https://doi.org/10.1002/cpe.7336.

Dong, Y., Werling, B., Cao, Z., & Li, G. (2024). Implementation of an in-field IoT system for precision irrigation management. *Frontiers in Water*, *6*. Available from https://doi.org/10.3389/frwa.2024.1353597.

Foschini, L., Martuscelli, G., Montanari, R., & Solimando, M. (2021). Edge-enabled mobile crowdsensing to support effective rewarding for data collection in pandemic events. *Journal of Grid Computing*, *19*(3). Available from https://doi.org/10.1007/s10723-021-09569-9.

Guna Priya, B. (2023). Whispers of the grid: Unraveling the mysteries of automated energy billing, fault detection, and raspberry Pi's power unplug system. In *Proceedings of the fourth international conference on smart electronics and communication, ICOSEC 2023* (pp. 95–100). Available from https://doi.org/10.1109/ICOSEC58147.2023.10275971.

Halgamuge, M. N., Bojovschi, A., Fisher, P. M. J., Le, T. C., Adeloju, S., & Murphy, S. (2021). Internet of Things and autonomous control for vertical cultivation walls towards smart food growing: A review. *Urban Forestry & Urban Greening*, *61*. Available from https://doi.org/10.1016/j.ufug.2021.127094.

Harshitha, A., Manikanta Uma Srinivas, C., Eswar Sai, M., Kommuri, K., & Gopi Krishna, P. (2024). Development of safety monitoring for an IoT-enabled smart environment. *Internet of Things* (pp. 281–292). Springer Science and Business Media Deutschland GmbH.

Hasan, M., et al. (2023). *IoT and sensor-based water and nutrient management for protected cultivation technology* In *Advances in water management under climate change* (pp. 168–188). CRC Press In.

Helo, M. O. A., Shaker, A., & Abdul-Rahaim, L. A. (2021). Design and implementation a cloud computing system for smart home automation. *Webology*, *18*, 879–893. Available from https://doi.org/10.14704/WEB/V18SI05/WEB18269.

Hussain, A. A., Dawood, B. A., & Al-Turjman, F. (2021). IoT and AI for COVID-19 in scalable smart cities. In *Sixth EAI international conference on science and technologies for smart cities, SmartCity 2020*, 372 (pp. 3–19). Springer Science and Business Media Deutschland GmbH. Available from https://doi.org/10.1007/978-3-030-76063-2_1.

Ibrahim, M., & Al-Wadi, A. (2022). Wastewater treatment plant security analysis. Available from https://doi.org/10.1109/ECAI54874.2022.9847468.

Jan, F., Min-Allah, N., & Düştegör, D. (2021). IoT based smart water quality monitoring: Recent techniques, trends and challenges for domestic applications. *Water (Switzerland)*, *13*(13). Available from https://doi.org/10.3390/w13131729.

Jaouhari, S. E. L., Palacios-Garcia, E. J., Anvari-Moghaddam, A., & Bouabdallah, A. (2019). Integrated management of energy, wellbeing and health in the next generation of smart homes. *Sensors (Switzerland)*, *19*(3). Available from https://doi.org/10.3390/s19030481.

Javaid, A., Pazil, A. F. M., Siong, C. H., & Ramli, N. (2022). GBI-based wireless home automation system using IoT. *Lecture notes in electrical engineering*, *865*, 109–117. Available from https://doi.org/10.1007/978-981-16-9781-4_12.

Kaushik, N., & Bagga, T. (2021). Smart cities using IoT. Available from https://doi.org/10.1109/ICRITO51393.2021.9596386.

Khan, S., & Altayar, M. (2021). Industrial internet of things: Investigation of the applications, issues, and challenges. *International Journal of Advances in Applied Sciences*, *8*(1), 104–113. Available from https://doi.org/10.21833/ijaas.2021.01.013.

Kumar, M., Nayar, N., Mehta, G., & Sharma, A. (2021). Application of IoT in current pandemic of COVID-19. In First international conference on computational research and data analytics, ICCRDA 2020 (Vol. 1022). Available from https://doi.org/10.1088/1757-899X/1022/1/012063.

Kumar, R., Kumar, P., Aljuhani, A., Islam, A. K. M. N., Jolfaei, A., & Garg, S. (2023). Deep learning and smart contract-assisted secure data sharing for iot-based intelligent agriculture. *IEEE Intelligent Systems*, *38* (4), 42–51. Available from https://doi.org/10.1109/MIS.2022.3201553.

Lu, H., et al. (2023). Application of Internet of Things in irrigation: A review. *J. Irrig. Drain*, *42*(1), 87–99. Available from https://doi.org/10.13522/j.cnki.ggps.2022402.

Machado, C., Jr., Melina Nassif Mantovani Ribeiro, D., & Backx Noronha Viana, A. (2021). Public health in times of crisis: An overlooked variable in city management theories? *Sustainable Cities and Society*, *66*. Available from https://doi.org/10.1016/j.scs.2020.102671.

Madushanki, A. A. R., Halgamuge, M. N., Wirasagoda, W. A. H. S., & Syed, A. (2019). Adoption of the Internet of Things (IoT) in agriculture and smart farming towards urban greening: A review. *International Journal of Advanced Computer Science and Applications*, *10*(4), 11–28. Available from https://doi.org/10.14569/ijacsa.2019.0100402.

Mani Sekhar, S. R., Tewari, S., Rahman, H., & Siddesh, G. M. (2020). *Data collection in fog data analytics. Studies in Big Data* (pp. 79–104). Springer Science and Business Media Deutschland GmbH.

Martínez, I., Zalba, B., Trillo-Lado, R., Blanco, T., Cambra, D., & Casas, R. (2021). Internet of things (IoT) as sustainable development goals (SDG) enabling technology towards smart readiness indicators (SRI) for university buildings. *Sustainability*, *13*(14). Available from https://doi.org/10.3390/su13147647.

Mendoza, I. C. P., Timbol, S. M., Samonte, M. J. C., Blancaflor, E. B. (2020). ImHome: An IoT for Smart Home Appliances. In *2020 IEEE seventh international conference on industrial engineering and applications, ICIEA 2020* (pp. 761–765). Available from https://doi.org/10.1109/ICIEA49774.2020.9101906.

Mijović, V., Tomašević, N., Janev, V., Stanojević, M., Vraneš, S. (2011). Ontology enabled decision support system for emergency management at airports. In *ACM international conference proceeding series* (pp. 163–166). Available from https://doi.org/10.1145/2063518.2063540.

Mumtaz, N., & Izhar, T. (2023). *COVID-19 impact on indian smart cities: A step toward build back better. Advances in construction safety* (pp. 319–327). Singapore: Springer Nature.

Mumtaz, N., Izhar, T., Pandey, G., & Labhasetwar, P. K. (2022). Utilizing artificial intelligence for environmental sustainability. In *Artificial intelligence for renewable energy systems*. Elsevier.

Naik, R. R., Gautam, S., Pachlor, R., & Patel, S. (2024). *The ecosystem of smart city surveillance using key elements. Green computing for sustainable smart cities: A data analytics applications perspective* (pp. 112–123). CRC Press.

Niedermaier, M., Hanka, T., Fischer, F., & Merli, D. (2021). A secure network scanner architecture for asset management in strongly segmented ics networks. In *International conference on information systems security and privacy* (pp. 347–355). Available from https://doi.org/10.5220/0010191603470355.

Ocaka, A., Briain, D. O., Davy, S., Barrett, K. (2022). Cybersecurity threats, vulnerabilities, mitigation measures in industrial control and automation systems: A technical review. Available from https://doi.org/10.1109/Cyber-RCI55324.2022.10032665.

Parvathi Sangeetha, B., Kumar, N., Ambalgi, A. P., Abdul Haleem, S. L., Thilagam, K., & Vijayakumar, P. (2022). IOT based smart irrigation management system for environmental sustainability in India. *Sustainable Energy Technologies and Assessments*, *52*. Available from https://doi.org/10.1016/j.seta.2022.101973.

Pavithra, M., Duraisamy, S., & Shankar, R. (2024). Illuminating agriculture: Crafting a strategy IoT-based architectural design for future growth. *Lecture Notes In Networks And Systems*, *878*, 77–93. Available from https://doi.org/10.1007/978-981-99-9489-2_8.

Phasinam, K., & Kassanuk, T. (2022). *Evaluation of vulnerabilities in IoT-based intelligent agriculture systems. Autonomous Vehicles: Smart Vehicles For Communication, Vol. 2* (pp. 237–258). Wiley.

Rani, D. S., Jayalakshmi, G. N., & Baligar, V. P. (2020). Low cost IoT based flood monitoring system using machine learning and neural networks: Flood alerting and rainfall prediction. In *Second international conference on innovative mechanisms for industry applications, ICIMIA 2020, Conference Proceedings* (pp. 261–267). Available from https://doi.org/10.1109/ICIMIA48430.2020.9074928.

Roshandel, S., & Eslamian, S. (2023). *Automation and Smart Irrigation. Handbook of irrigation hydrology and management: Irrigation methods* (pp. 275–293). CRC Press.

Rueda, P., et al. (2023). Smart production operations in a remote field in Latin America. Available from https://doi.org/10.2118/216827-MS.

Sallam, A., Refaey, A., & Shami, A. (2019). Securing smart home networks with software-defined perimeter. In *2019 15th international wireless communications and mobile computing conference, IWCMC 2019* (pp. 1989–1993). Available from https://doi.org/10.1109/IWCMC.2019.8766686.

Sharida, A., Hamdan, A., & AL-Hashimi, M. (2020). Smart cities: The next urban evolution in delivering a better quality of life. Studies in Computational Intelligence 846 (pp. 287–298). Springer Verlag.

Sharifi, A., Allam, Z., Bibri, S. E., & Khavarian-Garmsir, A. R. (2024). Smart cities and sustainable development goals (SDGs): A systematic literature review of co-benefits and trade-offs. *Cities (London, England), 146*, 104659. Available from https://doi.org/10.1016/j.cities.2023.104659.

Sidhu, R. K., Kumar, R., Rana, P. S., & Jat, M. L. (2021). Automation in drip irrigation for enhancing water use efficiency in cereal systems of South Asia: Status and prospects. In *Advances in agronomy* (Vol. 167) (pp. 247–300). Academic Press Inc.

Sobczak, A., & Ziora, L. (2021). The use of robotic process automation (RPA) as an element of smart city implementation: A case study of electricity billing document management at bydgoszcz city hall. *Energies, 14*(16). Available from https://doi.org/10.3390/en14165191.

Stellios, I., Kotzanikolaou, P., & Psarakis, M. (2019). *Advanced persistent threats and zero-day exploits in industrial Internet of Things. Advanced SCIENCES AND TECHNOLOGIES FOR SECURITY APPLICATIONS* (pp. 47–68). Springer.

Tan, X., Chen, W., Yang, J., & Tan, X. (2023). Application of a data-driven intelligent information system in infrastructure: Underwater tunnel case study. *Journal of Performance of Constructed Facilities, 37*(1). Available from https://doi.org/10.1061/JPCFEV.CFENG-4046.

Taylor, K., & Amidy, M. (2020). Data-driven agriculture for rural smallholdings. *Journal of Spatial Information Science, 20*, 125–135. Available from https://doi.org/10.5311/JOSIS.2020.20.669.

Toh, C. K., & Milojicic, D. (2021). Recent advances in smart cities. *IEEE Consumer Electronics Magazine, 10* (6), 67–68. Available from https://doi.org/10.1109/MCE.2021.3095705.

Vijayan, D. S., Rose, A. L., Arvindan, S., Revathy, J., & Amuthadevi, C. (2020). Automation systems in smart buildings: a review. *Journal of Ambient Intelligence and Humanized Computing*. Available from https://doi.org/10.1007/s12652-020-02666-9.

Viswanathan, A., Sai Shibu, N. B., Rao, S. N., Ramesh, M. V. (2018). Security challenges in the integration of IoT with WSN for smart grid applications. In *2017 IEEE International Conference on Computational Intelligence and Computing Research, ICCIC 2017*, Available from https://doi.org/10.1109/ICCIC.2017.8524233.

Wang, Y., & Liu, L. (2024). Research on sustainable green building space design model integrating IoT technology. *PLoS One, 19*(4). Available from https://doi.org/10.1371/journal.pone.0298982.

Xie, N., & Leung, H. (2021). Internet of Things (IoT) in Canadian smart cities: An overview. *IEEE Instrumentation & Measurement Magazine, 24*(3), 68–77. Available from https://doi.org/10.1109/MIM.2021.9436101.

Yuan, Z., et al. (2019). Sweating the assets – The role of instrumentation, control and automation in urban water systems. *Water Research, 155*, 381–402. Available from https://doi.org/10.1016/j.watres.2019.02.034.

Zhang, L., Shen, B., Barnawi, A., Xi, S., Kumar, N., & Wu, Y. (2021). FedDPGAN: Federated differentially private generative adversarial networks framework for the detection of COVID-19 pneumonia. *Information Systems Frontiers*. Available from https://doi.org/10.1007/s10796-021-10144-6.

CHAPTER 9

Cost-benefit analysis of automated water quality management: investment, savings, and strategic advantages

Nishtha Srivastava[1], Anamika Sen[2], Aastha Srivastava[3], Shaheen Efrah Ali[4] and Shubhra Khare[1]

[1]Department of Applied Sciences and Humanities, Invertis University, Bareilly, Uttar Pradesh, India, [2]Research Scholar, Department of Commerce, Bareilly College, Bareilly, Uttar Pradesh, India, [3]American Institute of Chemical Engineering, Den Haag, Zuid Holland, Netherlands, [4]Department of Management, Invertis University, Bareilly, Uttar Pradesh, India

9.1 Introduction

The existence of humans depends on water, a precious natural resource. 326 million trillion gallons is the estimated amount of water on Earth. Most freshwater on Earth is frozen in icebergs and glacier crowns, making up less than 3% of the total water supply.

Even though it is a natural source that is not scarce and can be found in plenty, only 0.04% of it can be utilized (Shafi, Mumtaz, Anwar, Qamar, & Khurshid, 2018). The quality of the water in water bodies shows the physical, chemical, and biological condition of the water bodies in question. Source water quality standards also vary depending on its use, agriculture, industry, recreation, fish and wildlife, and public supply water. For instance, the quality standards of water for human consumption are likely to be higher than those used for industrial and commercial or agricultural processes. Regarding water quality, 21% of the lakes are eutrophicated in the United States due to high concentrations of phosphorus and nitrogen loading while 46% of all the river and stream miles are classified as poor. Globalization of our globe is rising, and it is creating issues with the quality and availability of water. As suggested by these activities, some interference with quality can occur and therefore quality of water must be well addressed. It is important to point out that water quality parameters can be categorized into three classes. The first group of samples focuses on the physical characteristics such as warmth, color, smell, chroma, haziness, and electrical conductivity. pH, dissolved oxygen, chemical oxygen demand (COD), biochemical oxygen demand (BOD), complete inorganic carbon, heavy metal ions, and

Computational Automation for Water Security.
DOI: https://doi.org/10.1016/B978-0-443-33321-7.00021-4

nonmetallic poisons are part of the second group of chemical features. The third one is the microbiological category, which contains all the bacteria and coliform bacteria (AlMetwally, Hassan, & Mourad, 2020).

The definition of water quality criteria, the legal limit of contaminants or toxic pollutants, is another important activity for several national and state environmental protection agencies. Its purpose is to protect and rehabilitate the water quality as well as provide security of water supplies for the expanding population. The impact of degraded water quality on marketing communication can be observed at the regional and national level due to the number of deaths that may occur from waterborne illnesses such as diarrhea (Fernandez-Luqueno et al., 2013; Liu et al., 2012). Water infrastructures are primarily public organizations; most of them are financially unprofitable, environmentally, as well as socio-economically essential, and rely on disbursements. Apart from the communication aspect, the Internet of Things (IoT) today contributes significantly to data logging, archiving, and supervising as well as displaying in industrial processes. IoT systems open a new way of finding cheaper sources (Netinant & Rukhiran, 2020). The application of IoT technology in environmentally sensitive areas, including radiation poisoning, water pollution, air quality, and others, has risen recently (Daigavane & Gaikwad, 2017; Devalal & Karthikeyan, 2018; Ullo & Sinha, 2020).

These projects' financial, socioeconomic, and environmental advantages must therefore be justified to the funding agencies and decision-makers (Sartori, Latrônico, & Campos, 2014). Almost any field can be helped by CBA, which is often required as a step in the analytical process by the decision-makers who oversee granting awards. However, Sartori, Latrônico, & Campos (2014) pointed out the European Commission, World Bank, and Asian Development Bank. In this regard, CBA is a decision support tool often used to assess the impact of an infrastructure project in the society and frequently demanded by the funding agencies such as the national government and regional and international finance organizations. Most water infrastructures are being defined as public entities, which are typically financially nonprofit, but economically a critical element for the environment and society, heavily relying on expenditures. Numerous expenditure programs also require a qualitative comparison or an evaluation of the expenditure and the income together with an undertaking of a risk analysis all in one or a CBA and an expertize analysis all in another. Therefore enshrining awareness among the funding agencies and decision-makers on the financial socioeconomic and environmental gains of the projects is crucial (Sartori, Latrônico, & Campos, 2014). CBA is relevant in most fields, and it requires analysis by those who oversee payment decisions. Several disbursement programs require qualitative analysis or a composite analysis of CBA and either a risk analysis or expertize analysis, or an analysis from the perspective of utility to the environment, as highlighted by the European Commission, Asian Development Bank, and World Bank (Sartori, Latrônico, & Campos, 2014). This requirement is exemplified in the Norwegian regulations on impact assessments, which mandate a combination of monetary and nonmonetary analysis, indicating that CBA is just one of the evaluation tools utilized.

9.2 Literature review

The public, commercial sector, nonprivate sector, and government should all take water contamination seriously because it is a harmful issue. That is because water makes up 70% of the world and more than 60% of the human body (Taru and Karwanker, 2017). Other than that, 99% of Malaysia's surface water comes from surface sources, and 1% of it comes from the subsurface (Rahman, 2021). Cholera, diarrhea, hepatitis, skin diseases, typhoid, and other illnesses are diseases that can originate from contaminated water and poor hygiene; hence, water—either drinking, domestic use, or industrial use, as well as health—are critical necessities for human survival and well-being (Bull et al., 2020). For instance, due to consumption of contaminated water coming from a farm waterborne diseases erupted in Walkerton, Ontario, Canada in 2000 and impacted 2300 people (Hrudey, Payment, Huck, Gillham, & Hrudey, 2003; Hu, Li, Zeng, & Guo, 2016). It is used as a beverage, for washing, manufacturing, and for irrigation of crops. Producers, consumers, and workers, along with the production ecosystems, are vulnerable to contaminated water; therefore it should be safe and clean. Population growth, industrialization, and urbanization have affected water resources. They are abundant and sufficient, though the quality is wanting due to these factors. The National Water Quality Standards form the basis of the water quality index commonly referred to as WQI, which quantifies the water quality to show the extent of the pollution. Therefore the six indicators used in the determination of the WQI involve dissolved oxygen, total suspended solids, COD, BOD, ammonia (NH_3), and pH level (Nurul-Ruhayu et al., 2015; Razman et al., 2022). Pathogenic microorganisms can be divided into three groups: Viruses (poliovirus), bacteria (*Shigella*, *Vibrio cholera*, and *Salmonella typhi*), and algae. Occasionally, such microbes have been isolated in drinking water, well water, and river water samples (Borchardt, Bertz, Spencer, & Battigelli, 2003; Okoh, Sibanda, & Gusha, 2010). As a result, it becomes pertinent to minimize contact with deadly infections for which there are early warning detections of pathogen presence (EPA, 2013). The representatives of the dangerous microbes are *Cryptosporidium, Legionella, Pseudomonas, Giardia*, and *E. coli*, which are often found in the sources of drinking water . The evaluation also compares previous methods of water quality surveillance between the years 2015 and 2022 with the aim of detecting diverse types of pollutants through various techniques. The IoT, wireless sensor networks, real-time monitoring, optical approaches, and conventional filtration technologies are among the methods of monitoring water quality that are covered. Hence, the principal objective of this analysis is to scrutinize contemporary strategies for water quality surveillance, encompassing real-time IoT monitoring, virtual sensing, and cyber-physical systems (CPS) that are dependent on time, equipment, various water quality parameters, or contaminants. The assurance of performance in safety-critical scenarios and the facilitation of real-time monitoring through CPS can be ensured as indicated by Wang et al. (2015).

9.2.1 Water quality parameter and measurement

Monitoring the water quality parameters is important for protecting human health, maintaining ecological balance, and ensuring the overall safety and usability of water resources. Water quality measurements involve various techniques and instruments to measure the physical, chemical, and biological properties of water. Here are the following common methods and tools used for measuring the water quality parameters:

9.2.1.1 Physical parameters

- **Temperature:** Temperature, which mostly measures in degrees Celsius and is measured in the field using thermometry or a thermistor, has a weak correlation with pH and a strong correlation with electrical conductivity. It significantly affects aquatic life, as well as the amount of dissolved oxygen and gas transfer rates. It may also change the concentration or shape of some components (Ali & Qamar, 2013; Khatoon, Khan, Rehman, & Pathak, 2013; Verma & Singh, 2012).
- **Turbidity:** The quality of water in natural rivers should be regulated because it has an impact not only on organisms residing in the water system and the surroundings but also on the techniques used in raw water intake for industries as well as water treatment. Because turbidity can be employed to determine the quality of the environment (in view of pollution) of water bodies or used indirectly to estimate the total suspended solids in the water bodies. The ability of a sample to transmit light through it in relation to the physical purity of a liquid is known as turbidity. Irrigational water quality high is water that is colorless, whereas irrigational water quality low is water with yellow or foggy coloration. Natural rivers therefore contain dissolved colored materials that reflect and absorb light in addition to organic and inorganic microbes, algae, and colored particles hence affecting its turbidity (Anderson, 2005; Mukundan, Pierson, & Schneiderman, 2013). Turbidity harms biological, chemical, and physical actions in water bodies, and water that is opaque reduces commenced light penetration to nothing (Abirhire, Davies, & Guo, 2020). In the same water body, turgidities can be different with respect to time and space depending on physical conditions and local parent materials and colored pigments. Turbidity was also found to be inversely proportional to the stream flow (Gasim, Toriman, & Rahim, 2006); the mobility of suspended particulate matters also tends to reduce as the freshwater flow (Onabule, Mitchell, & Couceiro, 2020). Fluctuations in temperature impact both the in-lab measurements of the sample that is transported to a laboratory and the natural ecosystem sample. Some of the literature detailing prospective environmental influences include Anderson (2005), Kitchener, Wainwright, and Parsons (2017) . Turbidity has been measured using some common instruments, which are as follows: Secchi disks and turbidimeters. Secchi disk enables one to determine the degree of water clarity in lakes and reservoirs, whereas turbidimeters are used to determine the intensity of light that is scattered by the particles suspended in water.

- **Conductivity:** The presence of conductivity in water indicates the potential for electrical conduction, albeit with no immediate benefits in terms of water quality. The advantage lies in the ionic content of water, which governs characteristics such as hardness, alkalinity, and the proportion of dissolved solids. Conductivity levels are influenced by factors like pH, temperature, turbidity, chlorides, sulfates, dissolved oxygen, total dissolved solids, and chemical oxygen demand. Scholars such as Bhandari and Nayal (2008), Ali and Qamar (2013), Khatoon et al. (2013), and Patel and Vaghani (2015) have established that the measurement of conductivity is achieved with the electrometric method.
- **Total dissolved solid:** The amount of soluble organic and inorganic materials that are still present in the water is what it is. It is strongly associated with salinity, and as it rises, the salinity of the water increases. It has a strong relationship with total hardness, total suspended particles, electrical conductivity, chlorides, turbidity, and chemical oxygen demand. It is measured in mg/L using the gravimetric method, which involves drying it at a specific temperature after filtering (Ali & Qamar, 2013; Patel & Vaghani, 2015; EPA 2001; Bhandari & Nayal, 2008; Khatoon et al., 2013).

9.2.1.2 Chemical parameters

- **pH**: The pH of water indicates its acidic or alkaline nature, with the scale ranging from 0 to 14. Acidic levels fall between 0 and 6, while alkaline levels range from 8 to 14, with the most suitable pH range being 6.5 to 8.5. Measurement of pH is typically carried out using electrometry and pH electrodes. It is significantly associated with electrical conductivity, total hardness, sulfates, and total suspended particles (Ali & Qamar, 2013; Patel & Vaghani, 2015; Verma & Singh, 2012; EPA, 2001; Bhandari & Nayal, 2008). Gives accurate readings of the acidity or alkalinity of the water.
- **DO**: This demonstrates how soluble oxygen is in water. The oxygen content in water stems from photosynthesis or absorption from the atmosphere, playing a crucial role in sustaining aquatic life. It has a significant impact on aquatic life. The most common methods for measuring it are Winkler titration and electrometric meters. According to EPA (2001) and Patel and Vaghani (2015), it has a strong correlation with electrical conductivity, biological oxygen requirement, and sulfates.
- **BOD**: This indicates the quantity of oxygen used up by aquatic life, especially bacteria and protozoa. When BOD levels are high enough to exceed DO, there is not enough oxygen available for other species to survive. It is a highly significant indicator of water quality and has a strong correlation with total hardness, dissolved oxygen, and chloride. It is measured using the incubation procedure and expressed in mg/L (EPA, 2001; Ali & Qamar, 2013; Bhandari & Nayal, 2008; Khatoon et al., 2013; Patel & Vaghani, 2015; Verma & Singh, 2012). Oxygen assessments are made using an oxygen meter or the Winkler method. Due to trash from a growing population contaminating the water, rivers and canals now have higher levels of BOD, which can lead to hypoxia and even mortality in aquatic animals. The quantity of BOD in the water has been known to the locals from their own observations.

Water quality monitoring (WQM) is a critical tool for preventing, controlling, and monitoring the misuse of water resources. The data gathered from this monitoring are used to inform the planning and management of water resources for optimal benefit . Utilizing an incubator to regulate the water's temperature, the Azide Modification technique determines the BOD of a given body of water. Consequently, water supplies will be preserved for future sustainable development. To calculate the BOD, measure the amount of oxygen that bacteria use over a predetermined amount of time, usually 5 days.

- **COD:** This is the amount of oxygen employed in the oxidation of the inorganic material, which is already existent, and in the decomposition of the organic material. Like BOD, it is a good marker for the status of water quality and has good relationships with electrical conductivity, turbidity, chlorides, total hardness, and total dissolved solids. Expressed in milligrams per liter (mg/L); it is analyzed by microdigestion, colorimetry, reflux distillation by means of acid potassium dichromate, and titrimetric analysis (EPA, 2001; Patel & Vaghani, 2015; Verma & Singh, 2012). Calculate the overall equivalent quantity of compounds that have the potential to be oxidized in water to qualify for the terminology of organic pollutants.
- **Nutrients**: Used to measure the amounts of phosphorus compounds and nitrogen (nitrate, nitrite, and ammonia). Ion chromatography: Offers accurate concentration assessments of nutrients. UV-Vis spectroscopy is based on the fact that the pollutant molecules in water absorb the UV-Vis light at a particular wavelength. When electrons go from their normal state to a higher state, there is opacity which occurs when a certain wavelength is taken up by the object. UV-Vis spectroscopy can be used to distinguish water pollutants due to the absorption spectrum and the concentration of the substance as analyzed in the Lambert–Beer law according to Alam, Yamamoto, Umino, Nakahara, and Hiraoka (2020). Several factors are commonly used in water quality analysis including the absorbance at 200–250 nm of NO_3-N and NO_2-N. The results within the range of 380–750 show that turbidity and organic materials are eliminated. The expected concentration of nitrate was estimated by Alam et al. (2020) through the PLS method which delivers the best fit for the adjusted concentration of nitrate and its absorbance. In this work, we used UV spectroscopy as the analytical tool complemented with the partial least squares support vector machine model. Li, Feng, and Chen (2016) suggested a quick quantitative prediction approach to find the nitrate concentration in water, and the model's prediction accuracy was much increased. Since the PLSR model can quickly and precisely calculate the total nitrogen level in water, Li et al. (2016) utilized it to predict the effect very efficiently.

9.2.1.3 Biological parameters

- **Membrane filtration**: Membrane filtering can identify indicator microorganisms such as total coliforms. The Enterobacteriaceae family includes all coliforms, yet the group's definition is more operational than phylogenetic. Fecal bacteria are not the only bacteria that fall under the concept of coliforms. Furthermore, different national and public health agencies may have different definitions of total coliforms. For a bacterium to be classified as "total coliform" in the US (1), it must possess the following traits: Rod that is gram-negative, facultatively anaerobic or aerobic, does not produce spores, and ferments lactose at 35°C for 24 hours while producing acid (if using the membrane filtration technique) or acid and gas within 48 hours at 35°C (for multiple-tube fermentation technique, not described in this protocol). *Escherichia, Enterobacter, Klebsiella, Citrobacter,* and *Serratia* are among the genera of bacteria that might be considered coliforms. Not every whole coliform is dangerous. Inside the digestive tract, feces are the vehicle via which a subset of total coliforms are eliminated and found. These indicator microorganisms are more strongly correlated with the existence of fecal contamination. The ability to ferment lactose for 24 hours at 44.5°C, creating acid (and gas, depending on the procedure utilized), sets this group apart. It is usually called thermotolerant coliforms due to their characteristic ability to grow and multiply more in warmer temperatures. Even though most FCs per se are not pathogenic, some of them may be potentially pathogenic. Fecal coliforms are bacteria a part of the *Enterobacter* and *Escherichia* genera. Different water samples are subjected to membrane filtering, which is referred to as screening using membranes. This process enables one to filter the right amount of the sample through a membrane that has an appropriate pore size of 0.45 mm. In the case of an agar plate, there is incubation of a membrane. Again, cells found within the bacterial (and other) stains will also replicate and these colonies, which can be measured, can be used in estimating bacterial density (Forster and Pinedo, 2015). Bacteriological quality of water is determined using membrane filtration. ASM—the American Society for Microbiology.

- PCR: To validate microarray data (Klein, 2002), test for genetically modified organisms, monitor viral load, and perform other pathogen detection applications (Mackay, Arden, & Nitsche, 2002), real-time qPCR is utilized for gene expression quantification (Bustin, 2000), expression profiling (Hurley, Roberts, Bond, Keys, & Chen, 2012), single nucleotide polymorphism analysis (Solinas et al., 2001), and allele discrimination (Wang and Smith, 2007). For real-time PCR, several procedures are available, such as absolute measurements and relative qPCR.
- Determine the amount of chlorophyll in the water to determine the biomass of algae: These are the spectrum characteristics of the principal light-absorbing pigments that exist in all optical methods of algae detection, though in different concentrations based on algae species. The following ideas formed the premise for pulsed LED fluorescence lidar system design. It was required for observing algal growth and was obtained through other metrics of this kind (Table 9.1).

9.2.2 Automated water quality measurement

Automated water quality management systems leverage advanced technology to continuously monitor and manage water quality in real-time. These systems integrate

Table 9.1: Parameters, measurement techniques, and applications that influence water quality measurement.

Parameters	Measurement techniques	Applications	References
1 Physical parameters			
Temperature	Thermometers	Influences chemical reactions and biological activities in water.	Ali and Qamar (2013), Khatoon et al. (2013), Verma and Singh (2012), EPA (2001)
Turbidity	Turbidimeters	Measures the clarity of water; high turbidity can indicate the presence of suspended particles.	Anderson (2005), Kitchener et al. (2017), Fondriest Environmental (2014)
Conductivity	Conductivity meters	Measures the ability of water to conduct electrical current, indicating the presence of dissolved salts and other ions.	Bhandari and Nayal (2008) Patel and Vaghani (2015)
TSS	TDS meters	Represents the combined content of all inorganic and organic substances in water.	EPA (2001), Patel and Vaghani (2015)
2. Chemical parameters			
pH	pH meter	Indicates the acidity or alkalinity of water; critical for aquatic life and chemical processes.	Bhandari and Nayal (2008), Verma and Singh (2012)
DO	DO meters	Essential for the survival of aquatic organisms; low levels can indicate pollution.	EPA (2001), Patel and Vaghani (2015)
BOD	BOD incubators:	Measures the amount of oxygen required by microorganisms to decompose organic matter in water.	Ali and Qamar (2013), Khatoon et al. (2013), Patel and Vaghani (2015)
COD	COD digesters:	Measures the total amount of chemicals in water that can oxidize; used to gauge organic pollutants.	EPA (2001), Verma and Singh (2012), Patel and Vaghani (2015)
Nutrients	Spectrophotometers	Nutrients: includes nitrogen (nitrate, nitrite, ammonia) and phosphorus compounds; high levels can lead to eutrophication.	Alam et al. (2020)
3 Biological parameters			
Coliform bacteria	Membrane filtration	Indicates the presence of pathogenic microorganisms; total coliforms and fecal coliforms (e.g., *E. coli*) are commonly measured.	Forster and Pinedo (2015)
Microbial pathogens	PCR	Includes bacteria, viruses, and protozoa; monitoring ensures water safety for human consumption and recreation.	Hurley et al. (2012), Mackay et al. (2002)
Algal growth	Chlorophyll fluorometers	Excessive growth can indicate nutrient pollution and lead to harmful algal blooms.	Cadondon, Ong, Vallar, Shiina, and Galvez (2022)

various sensors, data transmission technologies, and software to ensure that water quality parameters are within safe limits and to respond quickly to any anomalies (Adu-Manu, Tapparello, Heinzelman, Katsriku, & Abdulai, 2017; Park, Kim, & Lee, 2020). Here is an overview of key components and technologies involved in automated water quality management. Automated water quality management systems leverage advanced technology to continuously monitor and manage water quality in real-time. Real-time monitoring ensures that water supplied to consumers meets safety standards and automated treatment adjusts chemical dosing and filtration processes based on real-time data (Puig et al., 2017; Wu, Cao, Tong, Finkelstein, & Hoek, 2021). Continuous, real-time monitoring provides more accurate and timely data compared to manual sampling and testing. Some of the key components and technologies involved in automated water quality management are described as follows:

Sensors and Probes: These devices can measure multiple water quality parameters, viz., pH, DO, turbidity, conductivity, and temperature simultaneously (Singh, Rai, Singh, & Mishra, 2022). Specific Ion Sensors: Designed to detect specific ions like nitrate, ammonia, phosphate, and heavy metals (Mahmud et al., 2020). Biological Sensors: For detecting microbial contaminants such as coliform bacteria or pathogens (Kumar et al., 2019). Radiological Sensors: For measuring radioactivity in water. To monitor these sensors some advanced technologies such as IoT are used to connect sensors and control systems to the Internet, enabling real-time data collection, remote monitoring, and control (Kopetz & Steiner, 2022). It also provides integration and management of various IoT devices and data streams. Data acquisition systems include data loggers that collect and store data from sensors for analysis. On the other hand, remote data transmitters use cellular, satellite, or radio networks to transmit data from remote locations to central monitoring stations (Islam & Jin, 2019).

Some wireless communications, such as cellular networks, provide wide-area coverage for data transmission. Long-range wide-area network offers long-range communication with low power consumption, ideal for remote areas (Ikpehai et al., 2018). Satellite communication uses locations where terrestrial communication infrastructure is unavailable. Wastewater treatment, including effluent quality monitoring, ensures that discharged water meets regulatory standards and process optimization enhances the efficiency of biological and chemical treatment processes (Preisner, Neverova-Dziopak, & Kowalewski, 2020). Rapid identification of water quality issues allows for prompt corrective actions, minimizing health risks and environmental impacts.

Real-time monitoring software displays current water quality data, provides alerts for parameter exceeds, and allows remote access and control (Glasgow, Burkholder, Reed, Lewitus, & Kleinman, 2004). Data analytics uses machine learning and statistical methods to analyze data trends, predict potential issues, and optimize water management practices. On the other hand, cloud-based platforms facilitate data storage, analysis, and sharing among stakeholders. Anomaly detection identifies unusual patterns in data that may indicate equipment failure, contamination events, or other issues (Velasquez Rendón, 2023). Surface water quality monitors rivers, lakes, and coastal waters to detect pollution and protect ecosystems. Groundwater monitoring detects contaminants and manages aquifer health. It reduces the need for frequent manual sampling and laboratory analysis, lowering operational costs. Regulatory compliance helps to meet regulatory requirements by providing consistent and documented water quality data and effluent monitoring ensures industrial discharge complies with environmental regulations (Schellenberg, Subramanian, Ganeshan, Tompkins, & Pradeep, 2020). It optimizes the use of water treatment chemicals and energy, leading to more sustainable water management practices. Automated water quality management systems are crucial for ensuring safe, reliable, and sustainable water resources in various applications, from municipal water supplies to environmental conservation (Ikpehai et al., 2018).

Cost-benefit analysis: The cost analysis carried out here is an incremental cost analysis, and the costs estimated pertain to the cost of expanding the coverage of water supply and sanitation services to the uncovered population. International costs refer to all resources that may be needed for the implementation and sustenance of the interventional measures as well as all costs consequent upon an intervention. These are distinguished by investment and recurrent expenditures elaborated in Haller, Staeheli, and Kochs (2007). The incurred costs related to an investment include education, construction, house modifications, planning and supervision of construction, hardware, and the protection of water supplies. As for the estimates of the initial investment cost of the WS&S interventions, the main source of cost data was the Global Water Supply and Sanitation Assessment 2000 Report (World Health Organization, 2020), which reported the investment cost per person covered according to three major world regions: Africa, Latin America and the Caribbean, and Asia/Oceania. All the above report was initially found in Haller et al. (2007). An earlier study was the basis of the cost data pertaining to water purification with chlorine at the point of use. As the paper has pointed out, the current most useful type of economic evaluation, which is employed in the comparison of resource allocation for government-financed activities and also within productive sectors, is cost-benefit analysis, whereas cost-effective analysis has been identified as the most suitable in resource allocation decisions within the health sector (Curry & Weiss, 1993; Hutton, 2000; Layard & Glaister, 1994).

9.3 Initial investment and operational saving costs

Conducted with regards to the new costs that will be incurred when increasing the coverage of sanitation and water supply to individuals who are not within the current coverage, the cost analysis is incremental in this case. Any other costs incurred after the implementation of an intervention, costs, and materials required for the successful running of the interventions are part of incremental costs. These are distinguished by investment and recurrent expenses, as presented in Haller et al. (2007) (Fig. 9.1).

9.3.1 Hardware costs (sensors, data loggers, and communication infrastructure)

- Lambrou, Anastasiou, Panayiotou, and Polycarpou (2014) addressed the development and implementation of a portable, mobile, cost-effective, and reliable water level management system. The sensors were utilized to measure a range of factors such as temperature, pH, turbidity, and dissolved oxygen. The adoption of a wireless system helps to reduce installation costs.
- Prasad, Vaidya, and Anil Kumar (2015) introduced intelligent WQM devices for Fiji that makes use of IoT and remote sensing technologies. To supervise water quality in the Pacific Islands, Fiji must consistently gather, analyze, and upload data to a dedicated website. The authors have leveraged IoT and remote sensing technologies to enhance WQM, with remote sensing offering potential improvements to current measurement practices.
- Omar Faruq, Kim, Padalhin, Lee, and Lee (2017) highlighted the sensitivity of the device to multiple water parameters, including temperature, turbidity, and pH, and its high accuracy. The LCD screen displays the pH value, providing precise readings for each parameter and calculating errors.

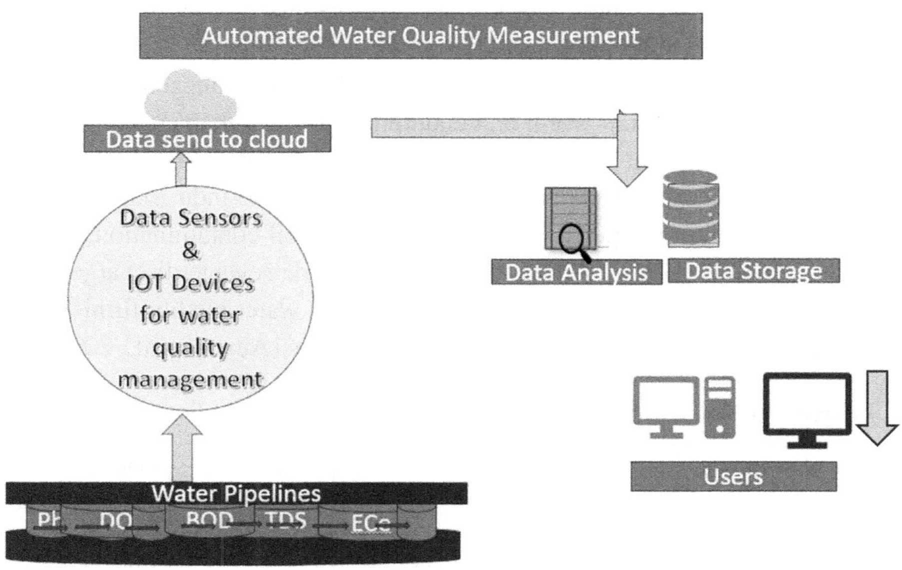

Figure 9.1
Automated water quality management by using hardware and software.

- By utilizing the Wi-Fi data transmission module ESP8266 (NodeMCU) to relay data to the central server, MCU processes, and updates sensor data to the Thing Speak server (Daigavane & Gaikwad, 2017).
- An ultrasonic sensor module consists of an ultrasonic transmitter (Trigger pin), receiver (Eco pin), and control circuit. Emitting a 40 kHz high-frequency sound wave, the sensor measures the time interval between transmission and reception of the echo, determining the distance to an object (Table 9.2).

9.3.2 Software cost

Water quality management software and data analysis and reporting tools are cost-effective and user-friendly due to their endorsement by academic institutions and scientific communities (Buytaert et al., 2014). These tools are consistently updated and accompanied by comprehensive documentation, tutorials, and manuals to ensure a thorough understanding of their functionality. Notably, open-source alternatives like QGIS, SAGA GIS, HEC-RAS, HEC-HMS, and iRIC eliminate cost barriers associated with licensing fees and are compatible with various operating systems (Dambe, 2020). Additionally, freely available software such as MODFLOW, SWAT, MT3DMS, Open FOAM, and Python offer valuable resources to the public.

9.3.3 Strategic advantages

Investing in automated water quality control systems offers strategic advantages, operational efficiencies, and notable cost savings. Despite the initial excessive cost, water management entities should consider the potential substantial return on investment (ROI) and rapid payback period. Embracing automated systems for water quality management brings strategic benefits that can enrich the overall worth and endurance of water management activities. Monitoring water quality aids in pinpointing specific contaminants, individual chemicals, and pollution sources. Water pollution stems from various origins like industrial activities, river and ocean dumping, agricultural pesticide and fertilizer use, oil contamination, port operations, shipping, and oil leaks. Additional sources include sewage discharge and agricultural practices. Regular monitoring and evaluation of water quality furnish data useful for pinpointing prevailing issues and their underlying triggers (Auwal et al., 2023).

Table 9.2: Used sensors and range in automated water quality monitoring.

Sl. No	Name of sensors	Range	References
1	pH sensors	6.5–7.5	Bande and Nandedkar (2016)
2	Turbidity sensors (output both digital and analog 5 V and 0–4.5 V)	100°C–900°C	Shafi et al. (2018)
3	The ultrasonic sensor	2 cm–4 m	Zin Myint et al. (2017)

Uncovering patterns in short- and long-term water quality is crucial. Over time, gathered data reveals trends such as escalating nitrogen pollution levels in inland water bodies like rivers. Subsequently, by utilizing all available data, crucial water quality attributes can be identified (Auwal et al., 2023). Effectively preventing and managing water pollution constitutes a vital component of environmental planning. Essential elements for devising a successful water quality plan encompass data gathering, analysis, and utilization. Conversely, outdated information hinders strategy formulation and limits the ability to influence pollution control measures. Employing digital systems and technologies for data collection and management serves as a remedy to this challenge (Auwal et al., 2023). Studies underscored the necessity of water quality surveillance during emergencies. Notable instances include major oil spills from vessels and flooding triggered by excessive runoff from precipitation. In times of crisis, swift action is imperative, necessitating real-time data access to determine the impact of pollution levels on water quality (Auwal et al., 2023).

9.4 Discussion

By means of responsive control mechanisms and real-time monitoring, automated water quality management systems constitute a significant improvement in guaranteeing the sustainability and safety of water resources. The main elements, advantages, disadvantages, and tactical advantages of these systems are examined in this conversation. Water quality factors including pH, DO, turbidity, and precise ions can be continuously assessed and managed thanks to the integration of sensors, data transmission technology, and real-time monitoring software. These systems allow for proactive adjustments to treatment processes, ensuring that water quality meets regulatory standards and minimizing environmental and health risks through real-time data analysis. Automated systems not only enhance operational efficiency but also deliver environmental benefits and significant cost savings. By optimizing chemical and energy usage in water treatment processes and reducing labor costs associated with manual sampling and laboratory analyzes, these systems contribute to operational sustainability and regulatory compliance. Future research efforts can focus on enhancing the integration and interoperability of IoT devices within automated systems to expedite data collection and management. Exploring the adaptability and scalability of these technologies in diverse socioeconomic contexts can further enhance their utility and accessibility. A cost-benefit analysis demonstrates the financial viability of investing in automated WQM equipment. The substantial annual savings in labor, chemical usage, energy consumption, maintenance, and regulatory compliance justify the initial investment costs. Hardware expenses consist of sensors and data recorders, while software costs cover data processing and reporting functionalities. The payback period and ROI calculations show the systems' potential long-term advantages and financial sustainability in the management of water resources. Strategic adoption of automated technologies enhances water quality management by ensuring compliance with stringent purity standards, improving data accuracy for informed decision-making, and streamlining environmental regulatory adherence. Moreover, these systems promote sustainable practices and foster public trust in business operations by optimizing resource utilization. Future long-term benefits of automated water quality management include longevity of water treatment infrastructure, adaptability to future regulatory changes, and positive public health outcomes. These benefits highlight the transformative impact of technological progress in securing water resources for current and future generations. In short, automatic water quality management systems are a major step forward in modernizing water resource management practices. Using advanced technologies and systematic monitoring methods, these systems not only increase operational efficiency and cost-effectiveness but also maintain environmental protection and public health priorities. Future research and continued innovation in this area are needed to maximize the potential of automated systems to effectively address global water challenges.

9.5 Conclusion

In conclusion, it is essential for society, the public sector, and governmental bodies to treat water contamination with gravity due to its detrimental nature. To effectively tackle this issue, it is crucial that authorities establish a robust and continuous real-time WQM system that can furnish valuable output data for prompt action. Consequently, the objectives of this review encompass an examination of past methodologies in WQM systems, a comparison between modern and conventional approaches, and an assessment of varying strategies adopted by different countries. There is a thorough analysis of various approaches for monitoring water quality, including the CPS approach, optical techniques, virtual sensing systems, IoT approach, and electronics sensing methods. Based on the review, it is deduced that CPS proves to be suitable and effective in WQM systems. This study puts forth a cost-effective IoT framework for real-time WQM, which interfaces seamlessly with multiple sensors through the utilization of Arduino Mega and NodeMCU target boards. In real time, an effective algorithm is intended to monitor the water's quality. The pH levels for groundwater and Hyderabad Metropolitan City supply water are 7−8.5 and 6.5−7.5, respectively. Turbidity measurements for groundwater and water supplied to Hyderabad Metropolitan City span from 600 to 2000 Nephelometric Turbidity Units (NTU). Monitoring of various parameters, including pH, turbidity, tank level, temperature, and humidity in the ambient air, is facilitated through the implementation of an online program named Thing Speak, which operates via a web server.

References

Abirhire, O., Davies, J.-M., & Guo, X. (2020). Understanding the factors associated with long-term reconstructed turbidity in Lake Diefenbaker from Landsat-imagery. *Science of the Total Environment*, *724*. Available from https://doi.org/10.1016/j.scitotenv.138222.

Adu-Manu, K. S., Tapparello, C., Heinzelman, W., Katsriku, F. A., & Abdulai, J. D. (2017). Water quality monitoring using wireless sensor networks: Current trends and future research directions. *ACM Transactions on Sensor Networks (TOSN)*, *13*(1), 1–41.

Alam, F., Yamamoto, T., Umino, T., Nakahara, S., & Hiraoka, K. (2020). Estimating nitrogen and phosphorus cycles in a timber reef deployment area. *Water*, *12*(2515).

Ali, M. & Qamar, A. M. (2013). Data analysis, quality indexing and prediction of water quality for the management of Rawal Watershed in Pakistan. In *Eighth international conference on digital information management (ICDIM 2013)*.

Anderson, C. W. (2005). Chapter A6.7, Turbidity. In Wilde, F. D. (Ed.). National field manual for the collection of water-quality data: U.S. Geological Survey techniques of water-resources investigations. U.S. Geological Survey, book 9, chaps. A1–A10.

AlMetwally, S. A. H., Hassan, M. K., & Mourad, M. H. (2020). Real time internet of things (IoT) based water quality management system. *Procedia CIRP*, *91*, 478–485.

Auwal, R. D., Zainab, A., Suleiman, B. A., Amina, I., Ibrahim, U., & Muhammad, A. B. (2023). Review of the goals and benefits of monitoring water quality. *Iris Journal of Astronomy and Satellite Communication*, *1*(1), 1–5.

Bande, P. N., & Nandedkar, S. J. (2016). Low cost sensor network for real time water quality measurement system. *System*, *5*(12).

Bhandari, N. S., & Nayal, K. (2008). Correlation study on physico-chemical parameters and quality assessment of Kosi river water, Uttarakhand. *Journal of Chemistry*, *5*(2), 342–346.

Borchardt, M. A., Bertz, P. D., Spencer, S., & Battigelli, D. A. (2003). Incidence of enteric viruses in groundwater from household wells in Wisconsin. *Applied and Environmental Microbiology*, *69*, 172–180.

Bull, F. C., Al-Ansari, S. S., Biddle, S., Borodulin, K., Buman, M. P., Cardon, G., ... Dempsey, P. C. (2020). World Health Organization 2020 guidelines on physical activity and sedentary behaviour. *British journal of sports medicine*, *54*(24), 1451–1462.

Bustin, S. A. (2000). Absolute quantification of mRNA using real-time reverse transcription polymerase chain reaction assays. *Journal Molecular Endocrinology*, *25*(2), 169–193.

Buytaert, W., Zulkafli, Z., Grainger, S., Acosta, L., Alemie, T. C., Bastiaensen, J., ... Foggin, M. (2014). Citizen science in hydrology and water resources: Opportunities for knowledge generation, ecosystem service management, and sustainable development. *Frontiers in Earth Science*, *2*, 26.

Cadondon, J. G., Ong, P. M. B., Vallar, E. A., Shiina, T., & Galvez, M. C. D. (2022). Chlorophyll-a pigment measurement of spirulina in algal growth monitoring using portable pulsed LED fluorescence lidar system. *Sensors*, *22*(8), 2940.

Curry, S., & Weiss, J. (1993). *Project analysis in developing countries* (p. 306) New York: St. Martin's Press, pp.

Daigavane, V. V., & Gaikwad, D. M. A. (2017). Water quality monitoring system based on IOT. *Advances in Wireless and Mobile Communications*, *10*(5), 1107–1116.

Dambe, N. (2020). Riverine flooding using GIS and remote sensing.

Devalal, S. & Karthikeyan, A. (2018). LoRa technology—an overview. In *Second international conference on electronics, communication and aerospace technology (ICECA)* (pp. 284–290).

EPA. (2013). *Revised total coliform rule and total coliform rule, drinking water requirements for states and public water systems*. https://www.epa.gov/dwreginfo/revised-total-coliform-rule-and-total-coliform-rule Accessed 27.06.2016.

Faruq, O., Kim, B., Padalhin, A. R., Lee, G. H., & Lee, B. T. (2017). A hybrid composite system of biphasic calcium phosphate granules loaded with hyaluronic acid–gelatin hydrogel for bone regeneration. *Journal of Biomaterials Applications*, *32*(4), 433–445.

Fernandez-Luqueno, F., Lopez-Valdez, F., Gamero-Melo, P., Luna-Suarez, S., Aguilera-Gonzalez, E. N., Martínez, A. I., ... Pérez-Velázquez, I. R. (2013). Heavy metal pollution in drinking water-a global risk for human health: A review. *African Journal of Environmental Science and Technology*, *7*(7), 567–584.

Gasim, M. B., Toriman, M. E., & Rahim, S. A. (2006). Hydrology and water quality and land-use assessment of Tasik Chini's feeder rivers, Pahang Malaysia. *Geografia*, *3*(3), 1–16.

Glasgow, H. B., Burkholder, J. M., Reed, R. E., Lewitus, A. J., & Kleinman, J. E. (2004). Real-time remote monitoring of water quality: A review of current applications, and advancements in sensor, telemetry, and computing technologies. *Journal of Experimental Marine Biology and Ecology*, *300*(1-2), 409–448.

Haller, O., Staeheli, P., & Kochs, G. (2007). Interferon-induced Mx proteins in antiviral host defense. *Biochimie*, *89*(6-7), 812–818.

Hrudey, S., Payment, P., Huck, P., Gillham, R., & Hrudey, E. A. (2003). Fatal waterborne disease epidemic in Walkerton, Ontario: Comparison with other waterborne outbreaks in the developed world. *Water Science and Technology*, *47*, 7–14.

Hu, C., Li, M., Zeng, D., & Guo, S. (2016). A survey on sensor placement for contamination detection in water distribution systems. *Wireless Networks*, *24*, 647–661.

Hurley, J., Roberts, D., Bond, A., Keys, D., & Chen, C. (2012). Methods in Molecular Biology, Vol. 822 (pp. 33–52).

Hutton, J. L. (2000). Number needed to treat: Properties and problems. *Journal of the Royal Statistical Society: Series A (Statistics in Society*, *163*(3), 381–402.

Ikpehai, A., Adebisi, B., Rabie, K. M., Anoh, K., Ande, R. E., Hammoudeh, M., ... Mbanaso, U. M. (2018). Low-power wide area network technologies for Internet-of-Things: A comparative review. *IEEE Internet of Things Journal*, *6*(2), 2225–2240.

Islam, M., & Jin, S. (2019). An overview of research on wireless communication networks. *Networks*, *5*(1), 19–28.

Khatoon, N., Khan, A. H., Rehman, M., & Pathak, V. (2013). Correlation study for the assessment of water quality and its parameters of Ganga River, Kanpur, Uttar Pradesh, India. *IOSR Journal of Applied Chemistry*, *5*(3), 80–90.

Kitchener, B. G. B., Wainwright, J., & Parsons, A. J. (2017). A review of the principles of turbidity measurement. *Progress in Physical Geography*, *41*(5), 620–642.

Klein, D. (2002). Quantification using real-time PCR technology: applications and limitations. *Trends in Molecular Medicine*, *8*(6), 257–260.

Kopetz, H., & Steiner, W. (2022). *Internet of Things. in Real-time systems: Design principles for distributed embedded applications* (pp. 325–341). Cham: Springer International Publishing.

Kumar, S., Nehra, M., Mehta, J., Dilbaghi, N., Marrazza, G., & Kaushik, A. (2019). Point-of-care strategies for detection of waterborne pathogens. *Sensors*, *19*(20), 4476.

Lambrou, T. P., Anastasiou, C. C., Panayiotou, C. G., & Polycarpou, M. M. (2014). A low-cost sensor network for real-time monitoring and contamination detection in drinking water distribution systems. *IEEE Sensors Journal*, *14*(8), 2765–2772.

Layard, R., & Glaister, S. (Eds.), (1994). *Cost-benefit analysis*. Cambridge University Press.

Li, D., Feng, W., & Chen, X. A. (2016). Seawater nitrate online monitoring system based on ultraviolet spectroscopy. *Spectroscopy and Spectral Analysis*.

Liu, L., Johnson, H. L., Cousens, S., Perin, J., Scott, S., Lawn, J. E., ... Mathers, C. (2012). Global, regional, and national causes of child mortality: an updated systematic analysis for 2010 with time trends since 2000. *The lancet*, *379*(9832), 2151–2161.

Mackay, I. M., Arden, K. E., & Nitsche, A. (2002). Real-time PCR in virology. *Nucleic Acids Research*, *30*(6), 1292–1305.

Mahmud, M. P., Ejeian, F., Azadi, S., Myers, M., Pejcic, B., Abbassi, R., ... Asadnia, M. (2020). Recent progress in sensing nitrate, nitrite, phosphate, and ammonium in aquatic environment. *Chemosphere*, *259*, 127492.

Mukundan, R., Pierson, D. C., & Schneiderman, E. M. (2013). Factors affecting storm event turbidity in a New York City water supply stream. *Catena*, *107*, 80–88.

Netinant, P., & Rukhiran, M. (2020). IoT architecture based on information flow diagram for vermiculture smart farming kit. *Tem Journal*, *9*(4), 1330–1337.

Okoh, A. I., Sibanda, T., & Gusha, S. S. (2010). Inadequately treated wastewater as a source of human enteric viruses in the environment. *International Journal of Environmental Research and Public Health, 7*, 2620–2637.

Onabule, O. A., Mitchell, S. B., & Couceiro, F. (2020). The effects of freshwater flow and salinity on turbidity and dissolved oxygen in a shallow Macrotidal estuary: A case study of Portsmouth Harbour. *Ocean & Coastal Management, 191*. Available from https://doi.org/10.1016/j.ocecoaman.2020.105179.

Park, J., Kim, K. T., & Lee, W. H. (2020). Recent advances in information and communications technology (ICT) and sensor technology for monitoring water quality. *Water, 12*(2), 510.

Patel, J. Y., & Vaghani, M. V. (2015). Correlation study for assessment of water quality and its parameters of Par River, Valsad, Gujarat, India. *International Journal of Innovative and Emerging Research in Engineering, 2*, 150–156.

Prasad, K. D. V., Vaidya, R., & Anil Kumar, V. (2015). A study on causes of stress among the employees and its effect on the employee performance at the workplace in an International Agricultural Research Institute, Hyderabad, Telangana, India. *International Journal of Management Research and Business Strategy, 4*(4), 68–82.

Preisner, M., Neverova-Dziopak, E., & Kowalewski, Z. (2020). An analytical review of different approaches to wastewater discharge standards with particular emphasis on nutrients. *Environmental Management, 66*(4), 694–708.

Puig, V., Ocampo-Martínez, C., Pérez, R., Cembrano, G., Quevedo, J., & Escobet, T. (Eds.), (2017). *Real-time monitoring and operational control of drinking-water systems*. Cham: Springer International Publishing.

Rahman, H. A. (2021). Water issues in Malaysia. *International Journal of Academic Research in Business and Social Sciences, 11*, 860–875.

Sartori, S., Latrônico, F., & Campos, L. (2014). Sustainability and sustainable development: a taxonomy in the field of literature. *Ambiente & sociedade, 17*, 01–22.

Schellenberg, T., Subramanian, V., Ganeshan, G., Tompkins, D., & Pradeep, R. (2020). Wastewater discharge standards in the evolving context of urban sustainability—The case of India. *Frontiers in Environmental Science, 8*, 30.

Shafi, U., Mumtaz, R., Anwar, H., Qamar, A. M., & Khurshid, H. 2018, Surface water pollution detection using internet of things. In *2018 15th international conference on smart cities: improving quality of life using ICT & IoT (HONET-ICT) (pp. 92–96)*. IEEE.

Singh, S., Rai, S., Singh, P., & Mishra, V. K. (2022). Real-time water quality monitoring of River Ganga (India) using internet of things. *Ecological Informatics, 71*, 101770.

Solinas, A., Brown, L. J., McKeen, C., Mellor, J. M., Nicol, J., Thelwell, N., & Brown, T. (2001). *Nucleic Acids Research, 29*(20), e96.

Ullo, S. L., & Sinha, G. R. (2020). Advances in smart environment monitoring systems using IoT and sensors. *Sensors, 20*(11), 3113.

Velasquez Rendón, D. (2023). Machine learning based anomaly detection for Industry 4.0 systems.

Verma, A. K., & Singh, T. N. (2012). Prediction of water quality from simple field parameters. *Environmental Earth Sciences, 69*(3), 821–829.

Wang, C.-J., & Smith, B.F. (2007). *American Journal of Veterinary Research, 68*(3), 231–235.

Wang, Z., Song, H., Watkins, D. W., Ong, K. G., Xue, P., Yang, Q., & Shi, X. (2015). Cyber-physical systems for water sustainability: Challenges and opportunities. *IEEE Communications Magazine, 53*, 216–222.

World Health Organization. (2020). Geneva, Switzerland. https://www.who.int/news-room/fact-sheets/detail/drinking-water Accessed 18.07.2020.

Wu, J., Cao, M., Tong, D., Finkelstein, Z., & Hoek, E. M. (2021). A critical review of point-of-use drinking water treatment in the United States. *NPJ Clean Water, 4*(1), 40.

Further reading

Edejer, T. T. T., & Edejer, T. T. T. (Eds.), (2003). *Making choices in health: WHO guide to cost-effectiveness analysis* (1). World Health Organization.

Fondriest Environmental, Inc. 2014, Measuring turbidity, TSS, and water clarity. Fundamentals of environmental measurements. https://www.fondriest.com/environmental-measurements/measurements/measuring-water-quality/turbidity-sensors-meters-and-methods/ Updated 05.09.2014.

Hongve, D., & Akesson, G. (1996). Spectrophotometric determination of water color in hazen units. *Water Research, 30*, 2771e2775. Available from https://doi.org/10.1016/S0043-1354(96)00163-7.

Moparthi, L., & Koch, S. (2019). Wnt signaling in intestinal inflammation. *Differentiation; Research in Biological Diversity, 108*, 24–32.

Myint, Z. W., & Goel, G. (2017). Role of modern immunotherapy in gastrointestinal malignancies: a review of current clinical progress. *Journal of hematology & oncology, 10*, 1–12.

Taru, Y. K., & Karwankar, A. (2017). Water monitoring system using Arduino with LabVIEW. In *Proceedings of the international conference on computing methodologies and communication (ICCMC)*, Erode, India, 18–19 July 2017, pp. 416–419.

Unicef, 2007. *Annual report 2003*. Unicef.

CHAPTER 10

Significance of automation in water treatment processes

Vijaya Ilango[1] and Karthiyayini Sridharan[2]

[1]Department of General Sciences/Chemistry, Birla Institute of Technology & Science, Pilani, Dubai Campus, Dubai, United Arab Emirates, [2]RHS, RoswellNew Mexico, United States

10.1 Introduction

Water treatment is as important as that of water for living beings. It is vital to drink pure water. Hence, production of clean water becomes essential. Water quality is being improved by various mechanical and automatic methods. In this chapter, it is intended to analyze and understand the automation of water treatment processes. Automation in water treatment processes refers to the use of technology and control systems to operate and monitor various aspects of water treatment facilities automatically, without constant human intervention. This automation aims to improve efficiency, accuracy, and reliability in managing water quality and treatment operations.

Some of the main processes in water treatment are flocculation, coagulation, sedimentation, filtration, disinfection, pH adjustment, aeration, activated carbon treatment, chemical addition, desalination, biological treatment, and advanced treatment like nanofiltration, ion exchange, ultrafiltration, and reverse osmosis. These processes may vary depending on the specific requirements and the quality of the raw water being treated. The goal is always to ensure that the treated water meets safety and quality standards set by regulatory bodies.

Automating a water treatment process involves the integration of various technologies and systems to control and monitor the entire treatment procedure with minimal human intervention. It works through a combination of sensors, actuators, controllers, and software systems designed to monitor, control, and optimize various aspects of the treatment processes (Kartunov & Kosev, 2019). Seventy-five percent of the globe is covered with water, which makes it essential for human survival. Freshwater is only available in small quantities, roughly 1%, despite this abundance.

By 2030, the world's water demand is predicted to increase from 4200 Bm^3 in 2015 to 6900 Bm^3 (Ng et al., 2015). Water and wastewater treatment (WWT) facilities, along with

Computational Automation for Water Security.
DOI: https://doi.org/10.1016/B978-0-443-33321-7.00019-6

many other industrial and natural systems that depend on the source, require sustainable and clean water availability. Apart from satisfying customer demands and delivering essential infrastructure improvements to enhance quality of life, treatment facilities also have to cope with intricate regulatory protocols to maintain rising quality benchmarks (Faherty, 2020).

As technology has advanced and treatment criteria have tightened, so has the level and significance of automation in municipal WWT facilities. The fact that many countries still have seriously contaminated rivers, which affects both aquatic and terrestrial animals as well as human life, complicates matter even further. It is evident that as countries continue to industrialize and modernize, these issues are becoming worse (Ebenstein, 2012).

In the last few decades, a great deal of research has been done on water quality prediction approaches in an attempt to enhance early warning systems and produce effective management strategies. The wastewater plant is divided into several phases, with each stage's output serving as the next step's input. Wastewater treatment plants (WWTPs) use automation techniques with a variety of control devices and components, just as other industries do.

Obtaining uncontaminated drinking water is a crucial aspect of the UN Sustainable Development Goals, and it represents the greatest problem of our time (Okafor et al., 2024). Water and wastewater quality have been analyzed and predicted using a variety of techniques, such as membrane filtration, desalination, coagulation/flocculation processes, chemical oxygen demand (COD), and biological oxygen demand (BOD) (Zhan et al., 2021). The United Nations' International Maritime Organization (IMO) has classified the Baltic Sea as a "Particularly Sensitive Sea Area," indicating that it is of particular concern. Annual nutrient reduction targets are set forth in the Baltic Sea Action Plan for each of the nine Baltic coastline countries (Latvia, Estonia, Sweden, Denmark, Germany, Lithuania, Poland, Finland, and Russia) according to the loads and nutrient reduction requirements between 1997 and 2003. Specifically, Finland has been allotted 150 and 1200 tons of nitrogen and phosphorus load reductions annually (Haimi, Mulas, and Vahala, 2010).

Municipalities are the source of a sizable portion of the total nutrient load that human activity causes to water systems. In 2005 the proportions of phosphorus and nitrogen loads from municipalities were 5.0% and 15.1%, respectively, of the total nutrient loads brought on by human activity (Pihlajamäki and Tynkkynen, 2011). Therefore the world's most pressing issue is the necessity to manage water supplies effectively.

Numerous case studies have demonstrated extraordinarily quick payback periods and large operating cost benefits, including energy savings. Automated system utilization is an efficient method of managing the scarce water supplies. For both new and existing facilities, digitization in the water sector enables more sustainable and efficient procedures. It involves two levels: using digital apps and building digital twins to optimize operations. The activities that release wastewater from cities, factories, or farms determine the type of toxins that are present in the wastewater. The four primary categories of contaminants found in water are microorganisms, suspended solids, and organic and inorganic pollutants. These pollutants are still present in the water in suspended, colloidal, or solvated forms. Table 10.1 provides the type of contaminants, their sources, and impacts found in the wastewater.

The physicochemical characteristics, degree of their impact on living things, and environmental persistence of these pollutants vary from one another. Many companies still lack the equipment and infrastructure necessary to properly treat their wastewater, which leads to the release of untreated or insufficiently treated effluent into water bodies.

10.2 Various automation used in water treatment processes

In contemporary water treatment facilities, the integration of automated systems and technology has grown more prevalent. Several parts of water treatment operations, such as data analysis, equipment operation, chemical dosage, and monitoring and control, are optimized and streamlined by automation. To maintain constant water quality and regulatory compliance, automation is involved, and it increases treatment performance and boosts operating efficiency for real-time monitoring and modification. The various steps involved in the automation of water treatment processes are given below (Moldovan & Nuca, 2019).

Table 10.1: Details of the type of contaminants present in wastewater.

Contaminants present	Sources and examples	Impacts	References
1. Organic pollutants	Residential, food processing, urine, and fecal matter. E.g., Acrylamide, tetrachloroethylene, volatile organic compounds	Contamination of water, soil, reduction of dissolved oxygen, peripheral nervous system damage, placental diseases, etc.	Li, Pal, and Kannan (2021); Carwile et al. (2014); Mojska et al. (2015)
2. Inorganic pollutants	Municipal wastewater, nonbiodegradable substances, often stemming from industrial, agricultural, and residential sources. Metals, nitrates, phosphates, and sulfates	Eutrophication of water bodies, toxic to microorganisms, plants and animals, produces toxins.	Wasewar (2021); Somayeh et al. (2018)
3. Pathogens/ microbes	Sewage discharge, slaughterhouse wastewater, fecal contamination. E.g., Fungus, algae, viruses, protozoa, amebas, plankton, etc.	Polio, cholera, hepatitis, and typhoid	Devane et al. (2020)
4. Sediments and suspended solids	Garbage, trash, industrial and storm sewage. E.g., Medical packaging, wet wipes, diapers, etc.	Gastrointestinal problems, negative impact on the food chain, and water pollution.	Lu et al. (2021); Turjja (2022)

(i) **Pretreatment**: Automation of screening, coagulation, and flocculation processes, with sensors adjusting chemical dosing and mixing times.
(ii) **Sedimentation**: Automated control of sludge removal based on sediment levels.
(iii) **Filtration**: Automated monitoring and backwashing of filters.
(iv) **Disinfection**: Automated dosing of disinfectants such as chlorine or adjustments for UV treatment.
(v) **Advanced Treatment**: Automation of membrane filtration, activated carbon treatment, or other advanced processes as needed.

10.2.1 Instrumentation, control, and automation

It has been shown that today's biological WWTPs can remove 10%–30% more nutrients from the water with the help of instrumentation, control, and automation (ICA). In the urban water system, ICA is currently used in WWT facilities, sewage networks, water distribution networks, and water treatment plants (Yuan et al., 2019). The use of ICA in WWT systems has advanced significantly and is currently regarded as a cherished and established field of technology in the industry. This advancement has been made possible by a variety of elements working together (Olsson, 2007).

10.2.2 Data acquisition systems

* **Supervisory Control and Data Acquisition (SCADA)**: A central system that gathers data from all sensors and instruments. A process control system (PCS), often known as SCADA system, is essential for the efficient, safe, and reliable operation of numerous industrial processes.

10.2.3 Parameter measurement

Sensors are strategically placed throughout the treatment facility to monitor essential parameters, including water flow rates, turbidity, pH levels, dissolved oxygen (DO), temperature, and chemical concentrations (such as chlorine and coagulants).

10.2.3.1 Sensors

- Water Quality Sensors: Measure parameters like pH, turbidity, conductivity, DO, and specific ions or compounds.
- Flow Meters: Monitor the water's flow rate as it progresses through the several treatment phases.
- Level Sensors: Water levels in tanks and reservoirs are to be monitored.
- Pressure Sensors: Ensure appropriate pressure in pipes and filters.
- Temperature Sensors: Measure water temperature, which can impact treatment efficiency.

Feedback Loop: These sensors continuously transmit real-time data to the control system, offering an up-to-date overview of current process conditions (Kartunov, Kosev, 2019). Sensors for water purification cites the Waste Water Treatment Plant Gabrovo as an example of an industrial process that operates reliably.

10.2.3.2 Process control system

PCS is extensively used in various sectors, such as infrastructure (including electric power, water, petroleum, and natural gas) and different manufacturing operations. Electronic automation through PCS allows for faster and more coordinated system management compared to manual operation, and in many cases, PCS is indispensable. Water treatment plants must balance providing high-quality water with maintaining low operational costs.

Due to the complex chemical, biological, and physical interfaces, managing water treatment methods can be challenging, and making accurate predictions is often difficult. The development, implementation, and commissioning of an advanced and smart monitoring and control system for a water treatment plant, designed to meet international standards and successfully applied in a large gas refinery, is discussed in the literature (Firoozshahi, 2010).

10.2.3.3 Programmable logic controllers and distributed control systems

Distributed control systems and Programmable Logic Controllers (PLCs) are crucial components of automation infrastructure. These systems receive data from sensors and use algorithms and logic to make decisions and issue commands to actuators. PLCs are industrial digital computers designed for real-time control of processes (Yua, 2015). Ensuring the availability of clean and safe water for various applications necessitates effective water treatment. Recently, the integration of Internet of Things (IoT) technology with water purification systems has demonstrated significant potential to enhance the efficiency and effectiveness of the treatment process. A major challenge in this domain is monitoring sewage disposal in treatment facilities.

To address this, SCADA system incorporating IoT has been proposed to ensure the proper operation of sewer systems, minimizing the risks of overflow and malfunction (Dwarakanath et al., 2023). The impacts of climate change and the rapid expansion of societies often lead to water scarcity and compromised water quality, contributing to numerous diseases. The digitalization of infrastructure and the implementation of digital twins are presented as innovative solutions to optimize resources and essential infrastructure throughout the water cycle (Rodríguez-Alonso et al., 2024).

10.2.3.4 Algorithmic control

Algorithms within the control system analyze incoming data, comparing it against predefined setpoints or desired operating ranges. Based on this analysis, the system determines appropriate actions to maintain process stability and quality (Sardarova and Mustafazade, 2023).

Examples of control systems include:

- **Automated Valves and Pumps**: Controlled by PLCs to regulate water flow, chemical dosing, and other operations.
- **Chemical Dosing Systems**: Automatically add the correct amounts of treatment chemicals based on real-time water quality data.
- **Filtration Systems**: Automated control of backwashing and cleaning cycles to maintain filter efficiency.

The technical issues of automation control systems, aiming to advance the automation industry in terms of management and process effectiveness, are also discussed by researchers (Dubey et al., 2016).

10.2.4 Example workflow of an automated water treatment plant

1. Intake and Pretreatment
   - Raw water is pumped from a source and passes through screens to remove large debris.
   - Flow meters measure intake volume, and water quality sensors determine initial parameters.
2. Coagulation and Flocculation
   - Based on sensor readings, the PLC adjusts the coagulant dosage.
   - Automated mixers ensure proper mixing for floc formation.
3. Sedimentation
   - The flocculated water flows into sedimentation tanks.
   - Sensors monitor sludge levels, and automated valves control sludge removal.
4. Filtration
   - Water passes through filters, with turbidity sensors ensuring water clarity.
   - Backwashing is automatically triggered based on differential pressure readings across filters.
5. Disinfection
   - Disinfection agents are dosed automatically based on real-time microbial data.
   - UV systems adjust intensity based on flow rate and water quality.
6. Posttreatment and Distribution
   - Final water quality checks are performed.
   - Treated water is pumped into storage or distribution systems, with automated valves and pumps controlling flow (Lowe, Qin, and Mao, 2022). Workflow of an automated water treatment plant is given in Fig. 10.1.

10.2.5 Features of automated water treatment systems

- **Real-time Monitoring**: Continuous tracking of characteristics pertaining to water quality parameters.

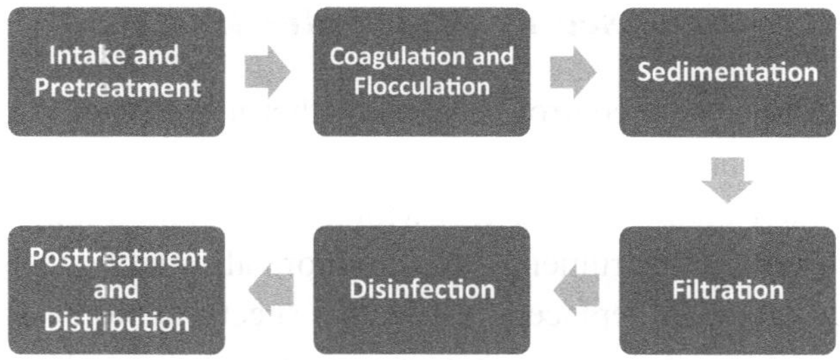

Figure 10.1
Workflow of an automated water treatment plant.

- **Alarms & Alerts**: Automated notifications for any deviations in parameters or equipment malfunctions.
- **Data Logging**: Recording historical data for compliance and performance analysis.
- **Remote Access**: Operators can monitor and control the system remotely through SCADA interfaces (Prajapati, Ashish, and Patel, 2014).

10.2.6 Integration of IoT for water quality monitoring

Sanitation and clean water for all people are prioritized in UN Sustainable Development Goal 6. Despite this, a third of people on the planet do not have access to clean, safe water. When it comes to early monitoring and control of water quality parameters, traditional techniques of filtration are inadequate. This study investigates the development of a real-time water quality monitoring and control system for water treatment plants using the IoT. The World Health Organization (WHO) uses a microcontroller interfaced with sensors to identify deviations from its standards (Chafa, Chirinda, and Matope, 2022).

10.2.7 Addressing water pollution and contaminants of emerging concern

Water pollution, driven by both natural and human activities, is a critical issue affecting global water resources. Pharmaceutically active compounds (PhACs), in particular, are contaminants of emerging concern that pose significant threats to human health and ecosystems. Recent advancements in analytical methods have enabled the identification and detection of a wide range of pollutants. The necessity for a comprehensive understanding of the environmental occurrence, fate, and risks associated with PhACs, as well as effective degradation methods to regulate pharmaceutical exposure in the environment, has been emphasized. It examines the most recent developments, cutting-edge analytical techniques, difficulties in detecting PhACs, and their distribution patterns in various water matrices (Papagiannaki et al., 2022).

10.3 Process automation used in wastewater treatment

The process of manual operation control involves gathering physical samples and testing them in a conventional laboratory. Based on the results, control components are manually changed to keep the water quality parameters within the specified ranges. As online sensors and other direct measurement instruments become more affordable, technology standard laboratory testing methods were replaced by portable spectrophotometers and handheld meters; nonetheless, because of the delay between measurement and adjustment of control elements, optimization remains unattainable (Animireddy and Sharma, 2015). One of the most important design criteria for effective WWT is keeping an eye on the DO levels at different operating cycles. The treated effluent will not be able to continuously fulfill the discharge criteria for BOD, COD, ammoniacal nitrogen, and phosphorus levels in the absence of monitoring equipment (Sanamdikar and Harne, 2012). High-purity water is needed in the Pharma industry for injectables and formulations (purified water). It is necessary to maintain conductivity (<1 mS/cm) and total organic carbon (<500 ppb) as quality standards.

Aside from this, the majority of high-purity water systems in use today favor reverse osmosis-electro deionization ultrafiltration systems and other nonchemical treatment methods. The need that there be no free chlorine content is one of the feed restrictions for RO. An oxidation–reduction potential meter is therefore required. The effective use and verification of sophisticated control strategies created utilizing model-based methodologies in an industrial wastewater treatment facility owned by a major pharmaceutical company has been demonstrated (Demey et al., 2001). One of the best routes to a circular economy is to generate zero waste, which transforms WWT facilities dramatically through commercialization and the addition of value management procedures (Matheri et al., 2022).

Artificial intelligence (AI) signifies the knowledge of endowing algorithms with the capability to achieve tasks and make interpretations that would require an intelligent human in the same position. It is the ability of a computer program to mimic human intelligence to realize autonomous learning, reasoning, judgment, and decision-making. AI has gained a lot of attention and has been applied extensively in a variety of fields in recent years as a strong and quickly evolving technology to solve real-world issues. Many AI models have been used in WWT to optimize processes, forecast efficiency, and assess performance to find more affordable solutions. This is due to the models' robust learning and precise prediction capabilities (Wang et al., 2023).

Numerous studies showed the effective use of various AI techniques for the modeling and process optimization of water treatment, including the elimination of toxins from water (Ghaedi and Vafaei, 2017). AI technologies are mostly used in systems to improve computer capabilities that are relevant to human knowledge, such as knowledge, problem-solving, perception, and cognition. Many researchers have used AI models in the past 10 years because of their ease of use, quick processing speed, and ability to tolerate error without requiring an understanding of physical issues (Safeer et al., 2022).

When combined with conventional methods and IoT architecture, AI models can enhance WWT's intelligent design systems and facilitate their reuse. AI models are useful and effective tools for modeling, forecasting, and streamlining the WWT process. They have been widely used in many different areas of WWT, such as the removal of pigments, heavy metals, organic matter, sediments, microbiological pollutants, drugs, fertilizers, and pesticides from water (Mazloom et al., 2020). The modeling of water quality has changed since the turn of the 20th century. The innovative work of Streeter and Phelps (1958) laid the groundwork for the advancement and development of mathematical models used to address issues with water quality. AI models have been extensively used in many elements of WWT, including the removal of dyes, heavy metals, nutrients, organics, solids, microbiological contaminants, medicines, and pesticides. AI models are a helpful and powerful tool for the modeling, prediction, and optimization of the WWT process (Altowayti et al., 2022).

Using both custom and standard software, SCADA is an advanced industrial measurement and control system that is utilized for remote device and data element monitoring and control, or a central host or master operating in data acquisition and control units in one or more fields. Thus the main tasks carried out by SCADA software are monitoring, control, data gathering, data recording, and storage. SCADA systems are being actively utilized in many different industrial domains these days. The primary ones, which are directly tied to our daily lives, are energy management, natural resource management, the manufacturing and transportation sectors, and clean water and wastewater management. High-tech engineering solutions like SCADA systems are necessary to keep life in the 21 century going normally. Real-time monitoring, remote access and control, alarm and event notification, data collecting and analysis, and energy and resource optimization are just a few of the benefits of utilizing SCADA systems in wastewater automation. Additionally, by facilitating integration between various equipment and processes throughout the facility, SCADA is used in drinking water treatment and wastewater facilities to allow more flexible and compatible operations (Mikrodev, 2023).

In recent years, machine learning (ML) has shown to be a potent technique for streamlining the modeling of WWTP processes. By using ML, machines can learn to carry out tasks that are typically performed by people by taking past experiences into account. Therefore it is possible to identify potential issues in WWTPs and activate the necessary control actions by developing models that are solely based on input–output data and do not incorporate knowledge of physical, chemical, or biological processes within the model structure (Malviya and Jaspal, 2021). Various automations used in wastewater treatment processes are given in Table 10.2.

Table 10.2: Automations used in Waste Water Treatment Processes (WWTPs).

Techniques	Features	Advantages	References
1. Artificial neural network (ANN)	Performance assessment tool for the variables such as BOD, COD, and TDS (total dissolved solids) used by the plant operators to make the appropriate decision.	Efficient analyzing and robust tool in predicting WWTPs performance.	Nasr et al. (2012)
2. Recurrent neural network (RNN)	Prediction of effluent water quality in WWTPs. Aids the operation and process optimization.	Excellent for developing a predictive model for sequential Big Data in WWTPs	Praewa, Park (2023)
3. Big Data analytics	Methods include data collection, data understanding, data preparation, data mining, and evaluation along with the statistical analysis.	Constructing a wastewater treatment plant model	Wongburi and Park (2021)
4. AI techniques	Larger amount of water leak, contamination rates, and water pipe deterioration modeling in WWTPs	Predicting and assessing water distribution network deterioration	Krishnan et al. (2022) Dada et al. (2024)
5. IoT-based wastewater management system (WMS)	The Internet of Things (IoT)-based wastewater management approach is made possible by sensors and actuators in every smart city home and business.	Processing and monitoring various stages of wastewater management	Alzahrani, Chauhdary, and Alshdadi (2023)
6. ML framework based on RF, DNN, VIM, and PDP	Plant operational variables and the factors that influence effluent quality	Potential to improve effluent quality control strategies in WWTPs	Wang et al. (2021)
7. Deep learning (DL) methods	Performance prediction, process automation and control, soft sensing, defect detection, diagnostics, and missing data imputation.	DL methods are evolving as competent tools for forecasting key performance and factors related to effluent quality. WWTP studies use various deep-learning algorithms	Alvi et al. (2023)
8. Digital twins	An online, real-time simulation produced by monitoring systems and sensors that gather information on the operations of WWTPs.	Adapt and optimize the wastewater treatment strategies	Wang et al. (2024)
9. Supervisory control and data acquisition (SCADA)	Optimize the plant operations to save money on maintenance, chemicals, energy, and operations.	Potential deviations from the wet-well level and effluent discharge can be quickly detected and fixed.	Dieu (2001)
10. Support vector machine (SVM)	Modeling effluent quality with cyclic-activated sludge process. Optimization of COD.	Better prediction of effluent quality and good results.	Wang, Chen (2008), Yasmin et al. (2020)

Note: *BOD*, Biological oxygen demand; *COD*, chemical oxygen demand; *WWTPs*, wastewater treatment plants; *AI*, artificial intelligence; *ML*, machine learning; *RF*, Random forest; *DNN*, Deep neural network; *VIM*, Variable importance measures; *PDP*, Partial dependence plot.

10.4 Importance of automation in water treatment plants

Water treatment plant automation has become crucial. It enhances operational conditions, improves plant performance, assists in supervision, minimizes human error, and boosts operational safety. This paper provides a brief overview of conventional chemical control techniques, discusses the challenges in designing control loops, and suggests possible improvements for automating water treatment processes (Al-Sum, Sattar, and Aziz, 1993).

To safeguard the environment and public health, urban water systems must run constantly, delivering consistently high-quality drinking water and WWT that satisfies regulatory requirements. Dynamic operation is now possible due to new financial incentives, developments in sensor and computer technology, and enhanced process knowledge. Traditionally, designs were static and had huge safety margins. With smaller amounts and more sophisticated control and operation, this method can tolerate significant changes, guaranteeing the system's resilience and constant water quality (Olsson and Pernille, 2003).

10.4.1 Optimization and maintenance

Key Features of Automated Water Treatment Systems

- **Predictive Maintenance**: Utilizing data analytics to forecast equipment failures and schedule maintenance accordingly.
- **Process Optimization**: Leveraging data to enhance the efficiency and cost-effectiveness of treatment processes.
- **Energy Management**: Monitoring and optimizing energy consumption throughout the plant (Bencsik et al., 2024).

10.4.2 General model for cost analysis and maintenance

A recent model, supported by specialized software, has been introduced to analyze exploitation costs in multistate renewable systems. This model considers costs related to preventive inspections, repairs, and other reliability-associated expenses, such as the financial impact of reduced system reliability and serviceability. By employing a multistate reliability analysis, the model determines the appropriate reliability state threshold that should trigger system repairs (Blokus-Dziula et al., 2023).

In the context of striving for a circular economy and carbon neutrality, revolutionary solutions for WWT are emerging. These solutions focus on automation and advanced process control using smart technologies, significantly enhancing the sustainability and efficiency of water treatment processes (Cairone et al., 2024). The importance of automation in the wastewater industry is given in Fig. 10.2.

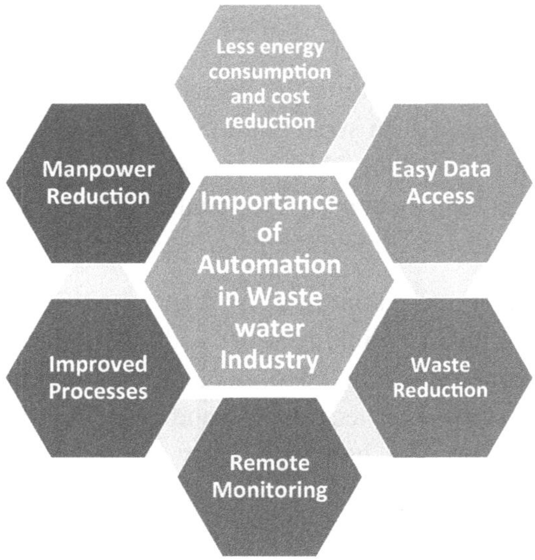

Figure 10.2
Importance of automation in the wastewater industry.

10.5 Discussion

Automating water treatment plants has become authoritative as it enables real-time supervision and curtails the risk of human error, improves operational conditions and plant performance, and increases operational safety. A study demonstrated that the automated drinking water treatment facility has been suggested using an Arduino to remove pollutants from water. The Arduino UNO microcontroller's integration provided a high degree of automation and accurate control. Using this technology in water treatment facilities has the potential to lower the expenses associated with water treatment while improving the quality and safety of drinking water. The system integrates an Arduino Uno microcontroller with various sensors using the IoT technology. This pioneering approach boosts current systems, monitoring, and control.

Key parameters, including TDS, pH, and turbidity, were monitored and controlled according to prescribed standards for the collected samples. A microcontroller and several connected sensors comprised a basic prototype that was used for weekly on-site tests at daily intervals. It was found that although the system is reliable, it depends on human support and is prone to inaccurate data (Hong et al., 2021). The primary obstacles that need to be addressed for a broader application of AI models were found to be data collection and curation. The most explored uses are influent flow prediction and effluent characterization. However, models for anomaly detection and energy consumption optimization, for instance, still have a lot of space for improvement.

In one of the studies multi-criteria decision-making theories were applied to various WWT methods. Due to the massive production of wastewater discharged by urbanization, industrialization, and population growth, these wastewaters require thorough treatment before being released. Consequently, there is a need for WWT systems that are low-cost, low-energy, and easy to maintain. The WWT technologies were examined in this study using a set of fuzzy PROMETHEE criteria (Mukhtar et al., 2020). The fuzzy PROMETHEE decision-making tool has shown to be a useful tool for determining the most desired approach among several WWT systems. According to the results, nanofiltration is the best way to treat wastewater, while activated sludge (AS) is the second most popular approach. However, developing in areas with problematic power supplies may not be practical for AS.

ICA in WWT systems has advanced significantly and is now regarded as a reputable and established field of technology in the industry. Actually, a WWT facility ought to be viewed as a plant that recovers energy and nutrients. There is a significant amount of untapped potential in anaerobic digestion when we take into account its energy potential. One excellent example is the WWTP in Göteborg, Sweden: The factory consumes 41 kWh of electricity per person annually. The facility simultaneously generates 72 kWh of biogas per person per year. Heat pumps handle the heat content of the effluent water. The potential for generation here is 336 kWh/person per year (Olsson, 2007). Actually, the plant produces a significant amount of energy. Information acquisition and process control are considered to be the two main purposes of automation in the water industry. There is a fair amount of automation for the former function. The SCADA systems of treatment plants frequently collect thousands of variables online, and sophisticated data analyses are routine parts of the treatment process.

It is important to evaluate the benefits of automation on water and WWTPs, as water is a critical utility in many industries (power, pharmaceuticals, electronics, etc.). PLC is a small computer used for automation of real-world processes, such as control of machinery on factory assembly lines, and makes use of a microprocessor. The program is often written by engineers and can often control complex sequencing (Burali, 2012). The progress made possible the creation of an intelligent, decentralized network for automation, and it also invigorated the establishment of an information system within an integrated company for management and control. These days, the control sludge in the sedimentation tank is reached by PLC, which is based on the law of deposition and sludge level. The host computer, PLC, suction pump control, fault alarm, and other components make up the PLC control system (Shunzhou YuXiaodong Xu, 2012).

The primary obstacles that need to be addressed for a broader application of AI models were found to be data collection and curation. The most researched uses are influent flow prediction and effluent characterization. However, models for anomaly detection and energy consumption optimization, for instance, still have a lot of space for improvement.

10.5.1 Challenges faced in the automation of water /wastewater treatment processes

One of the most fundamental human needs, not only in the industrialized world but all throughout the world, is access to safe, clean drinking water. At the center of this process is the water and WWT sector, and companies and individuals working to create new technologies to replenish rivers and lakes with clean water have a significant part to play. With the ever-increasing demand for clean water and rigorous effluent quality standards, both water and WWTPs need to upgrade their technology and equipment to progress their processes regularly.

- **Energy consumption:**
  While operating a water treatment facility, the biggest cost associated is energy usage. Throughout the whole year, a number of facilities are in use continuously, with the primary and secondary phases of WWT with the highest volume of use. Up to 30% of the entire cost of operations may be attributed to energy expenses.
- **Treatment Chemical Costs:**
  Both in the water and WWTPs, chemicals are used to separate the contaminants. WWT operations have faced challenges in the recent past due to supply chain difficulties. The need for innovative technology has been exacerbated by price hikes, product shortages, and replacements. To maintain operational goals and objectives, wastewater operators have had to get inventive.
- **Sludge Management:**
  Sludge is the term for the remnants created throughout the various stages of wastewater treatment. The amount of collected sludge that is disposed of in rivers and lakes should be decreased since it harms the ecology and presents a serious environmental problem. Organic stuff found in the sludge can be recycled and applied to the land or used as fertilizer for agriculture. A few technological developments have been shown to lessen the quantity of sludge that WWTP produces.
- **Workforce:**
  To handle technical problems and guarantee the WWTP operates smoothly and efficiently, qualified and experienced staff are needed. Since every step of the WWT process is automated, technicians must be on call around the clock. Operators of WWTPs face difficulties in finding and keeping skilled and experienced personnel. Automating every step of the treatment procedure will minimize the workload for the employees. Improved workforce management is facilitated by automation.
- **Inefficient Plant Management:**
  Many water treatment plants struggle to address the issue of inefficient plant management. Poor wastewater management results from the complex and decentralized nature of the water treatment process, which scatters data and makes it difficult for managers to oversee the entire facility. Furthermore, operators still have to make

important judgments, and if they lack experience, human error is a possibility and might result in additional expenses for correcting unforeseen errors.

- **Data Challenges:**
  The fact that the AI models heavily rely on the chosen data presents another significant problem. A good model is often developed through a learning/training process, and choosing the right dataset is a crucial part of that process. This means that minor variations in natural or therapeutic processes that were not observed during training can render a previously effective model effectively nonfunctional, unable to comprehend newly given data since it was never trained to handle it. The fact that many real-world systems lack the large datasets required to power many of these ML models can make the issue much worse.
  (Lowe, Qin, and Mao, 2022)
- **Integration:**
  Implementing automation successfully also requires integration. It entails integrating several tools, systems, and procedures to function well together toward a single objective. Given that it needs meticulous planning, coordination, and testing, this activity can be challenging. Skilled engineers can be very important in making sure that every part is integrated correctly, reducing errors, and increasing productivity.
- **Cybersecurity:**
  Cybersecurity becomes a major worry as water treatment plants become increasingly networked and dependent on digital technology. To defend against cyberattacks and guarantee the integrity of water treatment processes, strong security measures are required.

WWTPs are up against new difficulties. Operating them is becoming more complicated due to stricter quality requirements for sludge and plant effluent as well as to reduce noise, odors, and greenhouse gas emissions. Efficiency, dependability, and water quality are all significantly improving as a result of automation, which is revolutionizing the water treatment sector. Automation offers improved water quality, cost savings, and improved operational performance to residential, commercial, and industrial users. The long-term benefits of automation outweigh the difficulties posed by upfront costs and skill requirements, making it an investment worth in the water treatment sector. Utilities can obtain recommendations for actions to take, anticipate possible issues, and optimize plant processes by applying mathematical models and algorithms to these data. The trend is to put in place systems that may initiate any necessary actions in the plant autonomously, which leads to resilience, cost savings, and improved operations.

10.6 Conclusion

The use of automation is spreading throughout society. Automation is swiftly permeating every aspect of our lives, from thermostats to home security systems to refrigerators. Automation should be able to assist more broadly if it can be such a benefit in day-to-day living. The development of WWT has been largely shaped by automation, which provides unmatched advantages in terms of efficacy, affordability, and environmental sustainability. WWTPs can address present issues and set themselves up for future breakthroughs in resource management and water treatment by using automation technologies.

A cleaner, healthier environment for future generations can only be achieved by integrating automation. Smart technologies, emerging AI, and ML are bridging a gap in water applications that was previously unmet by conventional methods and approaches. The application of technology in WWT ought to be given higher attention. Investing in cutting-edge technologies will enhance data gathering, facility monitoring in real time, and quick intervention. Since data management offers precise information to support decisions, it is also a critical component of sustainable resource management.

Mechanistic models for the various components of water and wastewater systems are available, but a strong integration with data-driven models is still lacking to strike the best possible balance between the computational power needed and the models' capacity for prediction. Therefore the use of mixed mechanistic and data-driven models should be the main emphasis of future study. This strategy will help WWTPs to become more environmentally sustainable by improving their operational and financial efficiency. To develop better techniques for overcoming the obstacles now facing water management and automation in the water sector, a full assessment of these issues is necessary. These solutions should be assessed according to the costs and labor they can save, making them easily implementable in both industrial and domestic applications.

The development of WWT has been largely shaped by automation, which provides unmatched advantages in terms of efficacy, affordability, and environmental sustainability. WWTPs can address present issues and set themselves up for future breakthroughs in resource management and water treatment by using automation technologies. A cleaner, healthier environment for future generations can only be achieved by integrating automation.

References

Al-Sum, E. A., Sattar, A., & Aziz, M. A. (1993). Automation of water treatment plants and its application in power and desalination plants. *Desalination*, *92*(1), 309–321. Available from https://doi.org/10.1016/0011-9164(93)80087-4.

Altowayti, W. A. H., Shahir, S., Othman, N., Eisa, T. A. E., Yafooz, W. M. S., Al-Dhaqm, A., ... Ali, A. (2022). The role of conventional methods and artificial intelligence in the wastewater treatment: A comprehensive review. *Processes*, *10*(9), 1832. Available from https://doi.org/10.3390/pr10091832.

Alvi, M., Batstone, D., Mbamba, C. K., Keymer, P., French, T., Ward, A., ... Cardell-Oliver, R. (2023). Deep learning in wastewater treatment: A critical review. *Water Research*, *245*, 120518. Available from https://doi.org/10.1016/j.watres.2023.120518.

Alzahrani, A. I. A., Chauhdary, S. H., & Alshdadi, A. A. (2023). Internet of Things (IoT)-based wastewater management in smart cities. *Electronics*, *12*(12), 2590. Available from https://doi.org/10.3390/electronics12122590.

Animireddy, S., & Sharma, M. (2015). Automation of common effluent treatment plant. [online]. *International Journal of Advanced Technology in Engineering and Science*. Available from http://www.ijates.com/images/short_pdf/1424706712_P197-205.pdf, Accessed 26.06.2024.

Bencsik, D., Wadhawan, T., Házi, F., & Karches, T. (2024). Plant-wide models for optimizing the operation and maintenance of BTEX-contaminated wastewater treatment and reuse.. *Environments*, *11*(5), 88. Available from https://doi.org/10.3390/environments11050088, −88.

Blokus-Dziula, A., Dziula, P., Kamedulski, B., & Michalak, P. (2023). Operation and maintenance cost of water management systems: Analysis and optimization. *Water*, *15*(17), 3053. Available from https://doi.org/10.3390/w15173053.

Burali, Y. (2012). PLC based industrial crane automation & monitoring. *Research Inventy: International Journal of Engineering and Science [online]*, *1*(3), 1−4. Available from https://www.researchinventy.com/papers/v1i3/A0130104.pdf, Accessed 28.07.2024.

Cairone, S., Hasan, S. W., Choo, K.-H., Lekkas, D. F., Fortunato, L., Zorpas, A. A., … Naddeo, V. (2024). Revolutionizing wastewater treatment toward circular economy and carbon neutrality goals: Pioneering sustainable and efficient solutions for automation and advanced process control with smart and cutting-edge technologies. *Journal of Water Process Engineering*, *63*, 105486. Available from https://doi.org/10.1016/j.jwpe.2024.105486, 105486.

Carwile, J. L., Mahalingaiah, S., Winter, M. R., & Aschengrau, A. (2014). Prenatal drinking-water exposure to tetrachloroethylene and ischemic placental disease: A retrospective cohort study. *Environmental Health*, *13*(1). Available from https://doi.org/10.1186/1476-069x-13-72.

Chafa, A. T., Chirinda, G. P., & Matope, S. (2022). Design of a real−time water quality monitoring and control system using Internet of Things (IoT). *Cogent Engineering*, *9*(1). Available from https://doi.org/10.1080/23311916.2022.2143054.

Dada, M. A., Majemite, M. T., Obaigbena, A., Daraojimba, O. H., Oliha, J. S., Nwokediegwu, Z. Q. S., … Nwokediegwu, Z. Q. S. (2024). Review of smart water management: IoT and AI in water and wastewater treatment. *World Journal of Advanced Research and Reviews*, *21*(1), 1373−1382. Available from https://doi.org/10.30574/wjarr.2024.21.1.0171.

Demey, D., Vanderhaegen, B., Vanhooren, H., Liessens, J., Van Eyck, L., Hopkins, L., & Vanrolleghem, P. A. (2001). Validation and implementation of model based control strategies at an industrial wastewater treatment plant. *Water Science and Technology*, *44*(2-3), 145−154. Available from https://doi.org/10.2166/wst.2001.0764.

Devane, M. L., Moriarty, E., Weaver, L., Cookson, A., & Gilpin, B. (2020). Fecal indicator bacteria from environmental sources; strategies for identification to improve water quality monitoring. *Water Research*, *185*, 116204. Available from https://doi.org/10.1016/j.watres.2020.116204.

Dieu, B. (2001). Application of the SCADA system in wastewater treatment plants. *ISA Transactions*, *40*(3), 267−281. Available from https://doi.org/10.1016/s0019-0578(00)00053-7.

Dubey, S., Agarwal, M., Gupta, A. B., Dohare, R. K., & Upadhyaya, S. (2016). Automation and control of water treatment plant for defluoridation. *International Journal of Advanced Technology and Engineering Exploration*, *4*(26), 6−11. Available from https://doi.org/10.19101/ijatee.2017.426002.

Dwarakanath, B., Kalpana Devi, P., Kumar, R., Sayed, A., Ghulam Abbas, A., & Thamineni Bheema, L. (2023). Smart IoT-based water treatment with a Supervisory Control and Data Acquisition (SCADA) system process. *Water Reuse*, *13*(3), 411−431. Available from https://doi.org/10.2166/wrd.2023.052.

Ebenstein, A. (2012). The consequences of industrialization: Evidence from water pollution and digestive cancers in China. *Review of Economics and Statistics*, *94*(1), 186−201. Available from https://doi.org/10.1162/rest_a_00150.

Faherty, A. (2020). Tapped out: How Newark, New Jersey's lead drinking water crisis illuminates the inadequacy of the federal drinking water regulatory scheme and fuels environmental injustice throughout the nation. *Environmental Claims Journal*, 1−31. Available from https://doi.org/10.1080/10406026.2020.1848078.

Firoozshahi, A. (2010). Innovative and intelligent industrial automation for water treatment plant in large gas refinery. In *2010 second international conference on advanced computer control*. Available from https://doi.org/10.1109/icacc.2010.5487036.

Ghaedi, A. M., & Vafaei, A. (2017). Applications of artificial neural networks for adsorption removal of dyes from aqueous solution: A review. *Advances in Colloid and Interface Science*, *245*, 20–39. Available from https://doi.org/10.1016/j.cis.2017.04.015.

Haimi, H., Mulas, M., & Vahala, R. (2010). *Official Publication of the European Water Association (EWA)* [online]. https://www.ewa-online.eu/tl_files/_media/content/documents_pdf/Publications/E-WAter/documents/5_Haimi_Mulas_Vahala_Process_Automation_051010.pdf.

Hong, W. J., Shamsuddin, N., Abas, E., Apong, R. A., Masri, Z., Suhaimi, H., ... Noh, M. N. A. (2021). Water quality monitoring with Arduino based sensors. *Environments*, *8*(1), 6. Available from https://doi.org/10.3390/environments8010006.

Kartunov, S., & Kosev, B. (2019). Sensors for water purification using the example of wastewater treatment plant Gabrovo, Bulgaria. *Advances in Science, Technology & Innovation*, 477–480. Available from https://doi.org/10.1007/978-3-030-13068-8_119.

Krishnan, S. R., Nallakaruppan, M. K., Chengoden, R., Koppu, S., Iyapparaja, M., Sadhasivam, J., & Sethuraman, S. (2022). Smart water resource management using artificial intelligence—A review. *Sustainability*, *14*(20), 13384. Available from https://doi.org/10.3390/su142013384.

Li, A. J., Pal, V. K., & Kannan, K. (2021). A review of environmental occurrence, toxicity, biotransformation and biomonitoring of volatile organic compounds. *Environmental Chemistry and Ecotoxicology*, *3*, 91–116. Available from https://doi.org/10.1016/j.enceco.2021.01.001.

Lowe, M., Qin, R., & Mao, X. (2022). A review on machine learning, artificial intelligence, and smart technology in water treatment and monitoring. *Water*, *14*(9), 1384. Available from https://doi.org/10.3390/w14091384, [online.

Lu, Y., Zhang, Y., Zhong, C., Martin, J. W., Alessi, D. S., Goss, G. G., ... He, Y. (2021). Suspended solids-associated toxicity of hydraulic fracturing flowback and produced water on early life stages of zebrafish (Danio rerio). *Environmental Pollution*, *287*, 117614. Available from https://doi.org/10.1016/j.envpol.2021.117614, −117614.

Malviya, A., & Jaspal, D. (2021). Artificial intelligence as an upcoming technology in wastewater treatment: A comprehensive review. *Environmental Technology Reviews*, *10*(1), 177–187. Available from https://doi.org/10.1080/21622515.2021.1913242.

Matheri, A. N., Mohamed, B., Ntuli, F., Nabadda, E., & Ngila, J. C. (2022). Sustainable circularity and intelligent data-driven operations and control of the wastewater treatment plant. *Physics and Chemistry of the Earth, Parts A/B/C*, *126*, 103152. Available from https://doi.org/10.1016/j.pce.2022.103152.

Mazloom, M. S., Rezaei, F., Hemmati-Sarapardeh, A., Husein, M. M., Zendehboudi, S., & Bemani, A. (2020). Artificial intelligence based methods for asphaltenes adsorption by nanocomposites: Application of group method of data handling, least squares support vector machine, and artificial neural networks. *Nanomaterials*, *10*(5), 890. Available from https://doi.org/10.3390/nano10050890.

Mikrodev. (2023). *SCADA systems and their role in automation processes*. Available from https://www.mikrodev.com/scada-systems-and-their-role-in-automation-processes/.

Moldovan, A., & Nuca, I. (2019). Automation of wastewater treatment plant. *International Conference on Electromechanical and Energy Systems (SIELMEN)*. Available from . https://doi.org/10.1109/sielmen.2019.8905867.

Mojska, H., Gielecińska, I., Zielińska, A., Winiarek, J., & Sawicki, W. (2015). Estimation of exposure to dietary acrylamide based on mercapturic acids level in urine of Polish women post partum and an assessment of health risk. *Journal of Exposure Science & Environmental Epidemiology*, *26*(3), 288–295. Available from https://doi.org/10.1038/jes.2015.12.

Nasr, M. S., Moustafa, M. A. E., Seif, H. A. E., & El Kobrosy, G. (2012). Application of artificial neural network (ANN) for the prediction of EL-AGAMY wastewater treatment plant performance-EGYPT. *Alexandria Engineering Journal*, *51*(1), 37–43. Available from https://doi.org/10.1016/j.aej.2012.07.005.

Ng, K. C., Thu, K., Oh, S. J., Ang, L., Shahzad, M. W., & Ismail, A. B. (2015). Recent developments in thermally-driven seawater desalination: Energy efficiency improvement by hybridization of the MED and AD cycles. *Desalination, 356*, 255−270. Available from https://doi.org/10.1016/j.desal.2014.10.025.

Okafor, C. O., Ude, U. I., Okoh, F. N., & Eromonsele, B. O. (2024). *Safe drinking water: The need and challenges in developing countries* [online]. https://www.intechopen.com/chapters/84994..

Olsson, G. (2007). Automation development in water and wastewater systems. *Environmental Engineering Research, 12*(5), 197−200. Available from https://doi.org/10.4491/eer.2007.12.5.197.

Olsson, G., & Pernille, I. (2003). Automation in wastewater treatment plants. *Proceedings of the Water Environment Federation, 2003*(9), 238−254. Available from https://doi.org/10.2175/193864703784639570.

Papagiannaki, D., Belay, M. H., Gonçalves, N. P. F., Robotti, E., Bianco-Prevot, A., Binetti, R., & Calza, P. (2022). From monitoring to treatment, how to improve water quality: The pharmaceuticals case. *Chemical Engineering Journal Advances, 10*, 100245. Available from https://doi.org/10.1016/j.ceja.2022.100245.

Pihlajamäki, M., & Tynkkynen, N. (2011). The challenge of bridging science and policy in the baltic sea eutrophication governance in Finland: The perspective of Science. *Ambio, 40*(2), 191−199. Available from https://doi.org/10.1007/s13280-010-0130-4.

Praewa, W., & Park, J. K. (2023). Prediction of wastewater treatment plant effluent water quality using recurrent neural network (RNN) models. *Water, 15*(19), 3325. Available from https://doi.org/10.3390/w15193325, 3325.

Prajapati, M., Ashish, G., & Patel (2014). *PLC & SCADA based automation of industrial reverse osmosis desalination plants* [online]. https://www.ijert.org/research/plc-scada-based-automation-of-industrial-reverse-osmosis-desalination-plants-IJERTV3IS031425.pdf.

Rodríguez-Alonso, C., Pena-Regueiro, I., & García, Ó. (2024). Digital twin platform for water treatment plants using microservices architecture. *Sensors, 24*(5), 1568. Available from https://doi.org/10.3390/s24051568, 1568.

Safeer, S., Pandey, R. P., Rehman, B., Safdar, T., Ahmad, I., Hasan, S. W., & Ullah, A. (2022). A review of artificial intelligence in water purification and wastewater treatment: Recent advancements. *Journal of Water Process Engineering, 49*, 102974. Available from https://doi.org/10.1016/j.jwpe.2022.102974.

Sanamdikar, S. T., & Harne, K. R. (2012). Advanced method for sewage water treatment.. *International Journal of Advanced Technology in Civil Engineering*, 175−179. Available from https://doi.org/10.47893/ijatce.2012.1026.

Sardarova, I., & Mustafazade, T. (2023). *Control system for wastewater treatment process* [online]. https://papers.ssrn.com/sol3/papers.cfm?abstract_id = 4586708 Accessed 29.06.2024.

Shunzhou, Y., & Xiaodong, X. (2012). An input and output monitoring system for FPGA-based hardware PLC. *International Journal of Engineering and Industries, 3*(1), 34−44. Available from https://doi.org/10.4156/ijei.vol3.issue1.4.

Somayeh Asadi, H., Hamood, A., Ibrahim, Z., & Shafinaz, S. (2018). Arsenic biosorption using pretreated biomass of psychrotolerant Yersinia sp. strain SOM-12D3 isolated from Svalbard, Arctic. *Environmental Science and Pollution Research International, 25*(28), 27959−27970. Available from https://doi.org/10.1007/s11356-018-2799-z.

Streeter, H. W., & Phelps, E. B. (1958). *A study of the pollution and natural purification of the Ohio River* [online]. https://udspace.udel.edu/items/03d5883b-d6e7-4cf6-8a35-e14f9190302f Accessed 27.06.2024.

Turjja, S. (2022). *Controlling the contamination: preventing environmental impacts of combined sewage overflows in NYC*. Student Theses 2015-Present [online]. https://research.library.fordham.edu/environ_2015/127/.

Wang, Y., Cheng, Y., Liu, H., Guo, Q., Dai, C., Zhao, M., & Liu, D. (2023). A review on applications of artificial intelligence in wastewater treatment. *Sustainability, 15*(18), 13557. Available from https://doi.org/10.3390/su151813557, −13557.

Wang, A., Li, H., He, Z., Tao, Y., Wang, H., Yang, M., ... Ren, N. (2024). Digital twins for wastewater treatment: A technical review. *Engineering*. Available from https://doi.org/10.1016/j.eng.2024.04.012.

Wang, D., Thunéll, S., Lindberg, U., Jiang, L., Trygg, J., Tysklind, M., & Souihi, N. (2021). A machine learning framework to improve effluent quality control in wastewater treatment plants. *Science of the Total Environment, 784*, 147138. Available from https://doi.org/10.1016/j.scitotenv.2021.147138.

Wang, L.-J, & Chen, C.-B (2008). *Support vector machine applying in the prediction of effluent quality of sewage treatment plant with cyclic activated sludge system process*. Available from . https://doi.org/10.1109/kamw.2008.4810572.

Wasewar, K. L. (2021). Process intensification in wastewater treatments: basics of process intensification and inorganic pollutants (pp. 313–337). Elsevier eBooks. Available from https://doi.org/10.1016/b978-0-12-824058-8.00011-6.

Wongburi, P., & Park, J. K. (2021). Big Data analytics from a wastewater treatment plant. *Sustainability, 13* (22), 12383. Available from https://doi.org/10.3390/su132212383.

Yahya, M. N., Gökçekuş, H., & Ozsahin, D. U. (2020). Comparative analysis of wastewater treatment technologies. *Jurnal Kejuruteraan, 32*(2), 221–230. Available from https://doi.org/10.17576/jkukm-2020-32(2)-06.

Yasmin, A., Norhaliza Abdul, W., & Aznah Nor, A. (2020). Improved support vector machine using optimization techniques for an aerobic granular sludge. *Bulletin of Electrical Engineering and Informatics, 9*(5), 1835–1843. Available from https://doi.org/10.11591/eei.v9i5.2264.

Yua, H. (2015). *Data acquisition system of water treatment equipment based on GSM*. [online] www.atlantis-press.com. doi: https://doi.org/10.2991/isrme-15.2015.292.

Yuan, Z., Olsson, G., Cardell-Oliver, R., van Schagen, K., Marchi, A., Deletic, A., … Jiang, G. (2019). Sweating the assets—The role of instrumentation, control and automation in urban water systems. *Water Research, 155*, 381–402. Available from https://doi.org/10.1016/j.watres.2019.02.034.

Zhan, M., You, M., Liu, L., Zhang, Y., Yuan, F., Guo, B., … Xu, W. (2021). Numerical simulation of mechanical flocculation in water treatment. *Journal of Environmental Chemical Engineering, 9*(4), 105536. Available from https://doi.org/10.1016/j.jece.2021.105536.

CHAPTER 11

Water sustainability: a review of advances in water quality management technologies

Shama E. Haque, Farhan Sadik Snigdho and Nishat Tasneem
North South University, Bashundhara, Dhaka, Bangladesh

11.1 Introduction

Water is the most important resource on Earth that is fundamental to human survival.Water is considered a renewable resource because it is replenished in the water cycle. However, water is only renewable when managed properly. Water shortage can occur when communities are incapable of meeting their water needs, either owing to inadequate supplies or poor infrastructure. According to the United Nations (2018), by 2050, an estimated 6 billion people will be without access to clean water. This is due to rising water demand, dwindling water resources, and increased water pollution, all of which are being driven by accelerated growth in population, economic development, cropland expansion, increased need for irrigation, and shifts in global climatic patterns.

Water pollution is the introduction of various unwanted and undesirable substances—chemicals and microorganisms—into water bodies. There are two types of water pollution: (1) point source pollution, which occurs when the source of the pollution can be identified, such as a spill or discharge; and (2) nonpoint source contamination, which occurs when the source of pollution cannot be identified or originates from various sources. Pollution from both point and nonpoint sources affects water quality (WQ) and usage (du Plessis, 2022). WQ refers to the biological, chemical, and physical characteristics of water supply system that impact its suitability for a particular use or as an environmental condition (Pule et al., 2017). Both natural and anthropogenic processes can lead to variations in WQ in surface and groundwater resources. Particularly, processes leading to changes in WQ include factors such as climate change, shifts in temperature and precipitation patterns, water flow alternation, soil type, wind deposition, geology, and various natural and anthropogenic activities (Khatri & Tyagi, 2014). Past researchers found that globally, the most widespread WQ issue is associated with nutrient loading, which is linked to pathogens, endocrine-disrupting compounds, heavy metals, pharmaceuticals, and residual drugs along with loading from untreated sewage and wastewater disposal (Wear et al., 2021). Wastewater is

Computational Automation for Water Security.
DOI: https://doi.org/10.1016/B978-0-443-33321-7.00018-4

defined as water that has been degraded in terms of physical, chemical, and biological composition due to human activities (Crini and Lichtfouse2018a, 2018b). The availability of clean and sufficient water sources is directly related to wastewater management practices (Furlong et al., 2019). Increased demand for water supplies has prompted many utilities to consider potable water reuse, which refers to the process of utilizing treated wastewater to produce drinking water (Furlong et al., 2019). Treated wastewater can be considered as a sustainable source for water supply, which can also alleviate pressure on primary water resources. Achieving almost all of the UN SDGs depends on access to clean water, as it affects numerous aspects of human existence, health, and development. Water sustainability necessitates an integrated approach that meets current water requirements while preserving future generations' ability to satisfy their own water needs (EL-Nwsany et al., 2019).

WQ assessment refers to the total process of analyzing the biological, chemical, and physical characteristics of water concerning human consequences as well as intended water applications. This is an essential part of WQM and management. WQ monitoring is critical for water pollution surveillance. For example, if particular activities (e.g., farming, industrial, commercial, etc.) tend to produce specific pollutants, a possible connection may be established, which would necessitate additional inquiry or monitoring (USEPA, 2012). When using treated wastewater, WQ must be strictly monitored and managed, taking into account variables such as the possible accumulation of toxins that may be harmful to human health and the environment (Dickin et al., 2016). As such, WQM may assist in water resource management for clean and safe water supply (Essamlali et al., 2024). The possible improvement in accuracy and efficiency of water resource management through the use of modern technologies in WQM and prediction can ensure the long-term availability, accessibility, and safe use of a consistent quantity and quality of water, as well as proactive responses to environmental concerns. Manual sampling and laboratory analysis are labor-intensive, costly, and logistically problematic in traditional WQM approaches. In recent years, sensor technology advancements have altered WQM, which allow for real-time data collection and take corrective actions instantaneously.

This chapter aims to present a detailed analysis of previously published scholarly and nonscholarly works to identify the available references on water sustainability and advances in WQ management technologies. Specifically, the review covers wastewater treatment, WQM, and the influences of global climate change on WQM. Additionally, this chapter provides a discussion regarding the recent advances in Internet of-Things and machine learning (ML), which can be applied in the development of more effective and low-cost WQM systems with real-time capabilities. There are a large number of scientific studies available on wastewater treatment and WQM; however, the issue of water sustainability in this context has not been thoroughly addressed. As such, this chapter also focuses on water sustainability and UN SDGs concerning the imperative of guaranteeing access to good health and well-being (SDG 3), clean water and sanitation (SDG 6), climate action (SDG 13), and life below water (SDG 14). It is anticipated that the findings will assist policymakers in understanding how technology might support SDG attainment while resolving possible trade-offs.

11.2 Methodology

In this review, Google Scholar, Web of Science, ScienceDirect, and Scopus were searched for relevant scholarly literature. Documents were identified using keywords such as "water quality," "wastewater," "water sustainability," "monitoring," "biosensors," "real-time," "internet of things (IoT)," and "water quality management technologies." More focus was placed on current, highly cited research papers, and the timeline considered was 1998–2024. Briefly Subsections 11.2.1 and 11.2.2 present an overview of wastewater treatment processes and WQM, respectively. Subsection 11.2.3 provides a summary table of recent articles on WQM and wastewater management issues. Section 11.3 covers IoT and ML in WQM, and Section 11.4 presents a discussion on water sustainability, WQM, and their connection to attaining UN SDGs. Section 11.5 discusses the key findings of the review. Finally, Section 11.6 presents the conclusions of the review. Fig. 11.1 presents the systematic literature review process.

11.2.1 Wastewater treatment

Wastewater is water that has already been used. Wastewater streams arise from various domestic sources such as toilets, sinks, showers, and washing machines, as well as agricultural, commercial, and industrial processes (Gazi-Khan & Haque, 2022; Haque, 2022; Haque et al., 2020). Wastewater treatment is essential before it can be released into the environment; the complexity and variety of the contaminants in wastewater necessitates

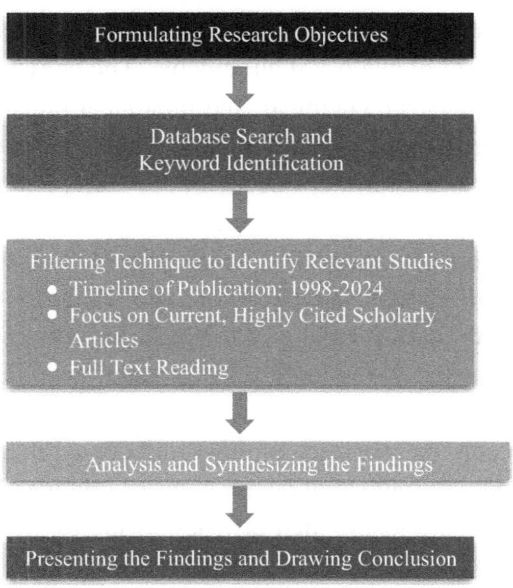

Figure 11.1
Literature review process.

the employment of highly efficient technology to meet the WQ criteria. Wastewater treatment facilities accelerate the natural process of purifying water that is unable to process the amount of waste produced by society (USEPA, 1998). The primary purpose of wastewater treatment plants is to safeguard humans and local ecosystems from toxic chemicals found in wastewater.

Wastewater reuse necessitates treatment and the implementation of adequate wastewater treatment systems. Wastewater treatment is a complex, nonlinear system that involves applying technologies to decontaminate wastewater and change it into an improved or upgraded effluent that can be: (1) reintroduced to the water cycle with a lesser environmental effect or (2) directly reused (Asrav et al., 2023; Nirmala & Prakash, 2023). To remove and eliminate soluble contaminants and insoluble particles from effluents, the conventional treatment method involves gathering wastewater in a wastewater treatment plant and treating it using a blend of physical, chemical, and biological processes (Crini and Lichtfouse2018a, 2018b). Water treatment plants and distribution systems each use their own set of WQM procedures to detect contaminants and determine whether the water is safe to drink. Over the past few decades, analytical research on sensing and WQM has resulted in reliable and efficient techniques with minimal operational costs and energy usage. Additionally, wastewater treatment is regarded as a significant contributor to the circular economy as a useful resource rather than a liability (Neczaj & Grosser, 2018).

11.2.1.1 Waste water processes

Wastewater treatment processes are designed based on the contaminant type and concentration, site location, and regulatory requirements. The conventional wastewater treatment method combines physical, chemical, and biological procedures to reduce and eliminate various undesirable suspended and dissolved solids, inorganic and organic compounds, and nutrients from wastewater (Gedda et al., 2021; Pabbati et al., 2017) to achieve treated wastewater that meets specified requirements. Wastewater treatment plants often use a series of water treatment steps that include the addition of coagulant chemicals, flash mixing, coagulation/flocculation, sedimentation, and filtration (Obaideen et al., 2022; Zaher & Shehata, 2021). Note that over the past few decades, environmental contaminants have emerged considerably owing to the large-scale production and consumption of medications, personal care products, plastic products, antibiotics, and hormones (Haque, 2022; Petrie et al., 2015). As a result, the presence of emerging contaminants in wastewater has become a serious public and environmental health concern (Haque, 2022). Conventional wastewater treatment methods are losing their efficacy owing to the evolving nature of the pollutants. Wastewater treatment plants are developing new strategies to identify contaminants of concern and improve the WQ to meet the safety requirements of the effluent after treatment (Moretti et al., 2024). In recent years, treatment technologies such as advanced oxidation processes, constructed wetlands, bioelectrical systems, and enzymatic treatment have been advanced for the removal of emerging contaminants (Rout et al., 2021).

Worldwide, for reasons of quality and safety, monitoring wastewater quality of both treated and untreated effluent is becoming increasingly important. Advancements in nanotechnology (which may include membrane filtration, ultraviolet irradiation, enhanced oxidation ion exchange, and biological filtration) in water treatment and purification have modernized wastewater treatment procedures (Qu et al., 2013; Shayo et al., 2024). Depending on the type of contaminants and required efficiency of treatment, nanomaterials for wastewater treatment can include carbon, single-walled carbon nanotubes, multiple-walled carbon nanotubes, covalent organic frameworks, metal, and metal oxide-based nanoparticles (Hairom et al., 2021). Efficient eradication of microorganisms is also accomplished through nanotechnology-based strategies (Nishu & Kumar, 2023). Nanotechnology enables the development of low-cost systems, making them accessible to people with lower incomes (Shayo et al., 2024). According to Enuh (2023), the global market for nanosensors is estimated to increase at a compound yearly growth rate of 11.0% between 2019 and 2026, reaching approximately USD 1.3 billion in 2026. The increased need for these sensors in a range of industries, including WQM, is driving the market's growth.

11.2.2 Water quality monitoring

The quality of water bodies can be determined using three major parameters: physical (e.g., turbidity, color, temperature, odor, etc.), chemical (total dissolved solids, organics, heavy metals, etc.), and biological (various waterborne pathogens) features.

11.2.2.1 Conventional WQM techniques

The findings of this literature review indicate that in general, the conventional surface water monitoring method involves collecting water samples on-site where trained field personnel travel to a source of water (Paepae et al., 2021). Subsequently, the collected samples are evaluated chemically, physically, and microbiologically in a laboratory, which is time-consuming and does not provide real-time WQ information. In these techniques, errors resulting from field sampling and mis-calibrated equipment can cast doubt on results (Zainurin et al., 2022). Additionally, it is expensive, requires a considerable amount of time and resources, and lacks integration (Khatri et al., 2019). New technologies were introduced in the latter part of the 2000s to address the problems posed by the limitations of traditional WQM. In particular, new sensor technologies created to monitor WQ characteristics in situ include fiber optics, laser technology, biosensors, and microelectronic mechanical systems (Adu-Manu et al., 2017; Bhardwaj et al., 2015).

11.2.2.2 Advanced WQM techniques

Over the last few decades, there has been significant advancement in research on sensing and monitoring challenges to develop reliable and cost-effective approaches. Remote sensing, data analytics, and the IoT have recently been identified as the primary technologies revolutionizing WQM by continuous monitoring and assessment of WQ parameters in real time (Ahmed et al., 2019; Essamlali et al., 2024; Rajitha et al., 2024). Additionally, innovative and smart technologies applied across diverse water infrastructure components (e.g., pipelines, reservoirs, and treatment plants) can allow for large-scale monitoring, data transmission, and control of water treatment activities, all of which contribute to sustainable water resource management (Iancu, Ciolofan, & Drăgoicea, 2024; Palermo et al., 2022).

Jabbar et al. (2024) created a real-time WQM device with long-range wide-area network technology that is appropriate for rural locations. The device is waterproof, portable, and IoT-enabled and runs on solar energy. The Things Network server receives the collected data and fully integrates it with the ThingSpeak web-based IoT server and the ThingView mobile applications. Thus the users can access critical WQ information online in real time. According to Miller et al. (2023), the most difficult problem in constructing an IoT system for WQM consists of multiple steps. The first step in developing an IoT system for WQM is to determine the specific parameters that must be measured. This relies on the application's precise needs and regulatory requirements. Additionally, when developing an IoT system for WQM, it is critical to ensure data security and privacy. As IoT systems capture and store large amounts of sensitive data, unauthorized individuals may gain access to the data.

Yaroshenko et al. (2020) present the recent advances in real-time WQM, particularly employing chemical sensors that were tested online for their suitability for long-term use using real water samples. The most effective chemical sensing systems, either alone or in conjunction with other techniques, depend on the specificity of a coating material that can identify certain water pollutants with high accuracy; molecularly imprinted polymers offer increased system design flexibility. ML algorithms, on the other hand, analyze huge datasets to reveal trends that humans may miss, anticipating changes in WQ and directing management decisions (Rana et al., 2023). Abbas et al. (2024) studied ML models including XGBoost, Gradient Boosting, Random Forest, and SVM, for the prediction of WQ in Mirpukhas, Pakistan. These researchers report that the applied models demonstrate excellent accuracy for accurate evaluation of WQ. Feature importance analysis emphasizes the vital role of individual variables, underlining the need for targeted monitoring and management efforts. However, the findings indicate that despite the positive results, it is important to recognize the limitations, such as dataset size and variability in scope.

The sustainable use of water resources necessitates planning and control approaches, which requires the incorporation of a wide range of spatial and temporal parameters. A geographic information system (GIS) can be successfully employed to assess WQ, availability, and risk at the local and regional scales. For example, Criollo et al. (2019) used AkvaGIS, a new open-source module included in the FREEWAT plugin for QGIS, to produce input files for the majority of groundwater flow numerical models in all QGIS formats. The researchers state that AkvaGIS is being used in the Walloon Region of Belgium to demonstrate its capabilities. The findings contribute to a better understanding of the hydro-chemical interaction between groundwater resources of the area and can serve as a baseline for future studies. Sharvelle, Dozier, Arabi, & Reichel (2017) developed an Integrated Urban Water Model to estimate urban water demand. It offers choices for evaluating the effects of water reuse and conservation. The application of the model at the building to municipal scales is made easier by its use as an online tool with GIS interfaces. The efficacy of the model at several spatial scales was confirmed by extensive use of water metering data from the City of Fort Collins, USA. Furthermore, the calibrated model produced remarkably accurate demand predictions for the municipal service area as well as for individual block groups.

11.2.2.3 Global climate change and WQM

Climate change has complicatedly affected our planet's water supplies with unpredictable temperature and precipitation patterns, melting ice sheets, increasing sea levels, frequent floods, and severe droughts, which are among its many repercussions (Fig. 11.2; Haque, 2023; Hanak & Lund, 2011; Haque & Nahar, 2023). Additionally, rising

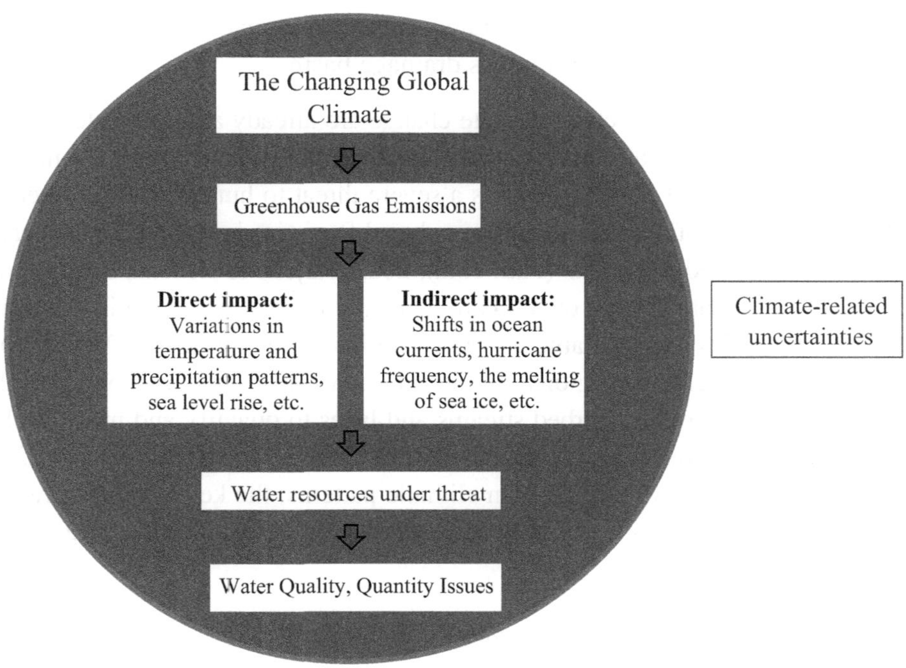

Figure 11.2
Climate change impacts.

sea levels, stronger storm surges, and shifts in precipitation patterns are key drivers of factors enhancing saltwater intrusion, which is likely to contaminate many freshwater resources. In addition, the rising global temperature is expected to impact the physical, chemical, and biological aspects of WQ and biodiversity (Bolan et al., 2024; Molina-Navarro et al., 2014). For example, changing climate may rewet peatland after a drought, increasing dissolved organic carbon levels in water bodies and subsequently, adversely altering WQ (Li et al., 2021; Worrall, Burt, & Adamson, 2004). Moreover, climatic eutrophication is expected to increase the effects of excessive plant and algal growth in fresh and coastal waters through increased diffuse nutrient loading due to severe storms thus altering the entire ecosystem (Haque, 2023; Strecker et al., 2004; Wade et al., 2022). According to Kurniawan et al. (2024), cutting-edge technologies to monitor and understand harmful algal blooms will have long-term benefits for developing nations that are highly susceptible to the influences of changing global climate. Furthermore, the researchers suggest that water security can be attainable in the face of various climate change scenarios if suitable adaptation measures are implemented. Suresh et al. (2023) emphasized the necessity for comprehensive indicators that reflect the complex mechanisms of eutrophication in lake systems, as this is expected to drive the global expansion of WQM networks and aid in the development of integrated eutrophication control strategies. Additionally, these researchers report that the indicators currently used to monitor and evaluate lake eutrophication are limited to components of WQ (such as total phosphorus and total nitrogen), and these may not accurately reflect man-made influences and changes in the global environment that affect a lake's drainage basin.

It is well established that the effects of climate change are already altering water-related hazards and human access to safe drinking water in many parts of the world (Ahmed et al., 2020). Even though a lack of potable water is a severe threat to human health, water scarcity can have long-term implications. For instance, reduced flows in rivers and streams can lead to elevated levels of dissolved pollutants (Patil et al., 2022; Rajesh & Rehana, 2022). Although the consequences of climate change on WQ are generally understood, there is a scarcity of published research articles and available information on related advances in WQM. Currently, the USEPA (2024) is developing regional monitoring networks to collect biological, thermal, and hydrologic data from least-disturbed streams and lakes to quantify and monitor changes in conditions, including climate-related deviations. Barlett and Dedekorkut-Howes (2023) report that in WQM dealing with changes in climatic patterns, the key issue with regulatory systems is a lack of cooperation between government levels and conflicting policies.

11.2.3 Summary table of literature review

Table 11.1 provides a summary of some recent studies conducted between 2015 and 2023 on recent advances in WQM and wastewater treatment.

Table 11.1: A summary of selected recent studies conducted on recent advances in WQM and wastewater management issues (timeline 2015–23).

Author year	Summary
Jamroen et al. (2023)	These researchers developed a standalone photovoltaic/battery energy storage system-powered WQM system. The main components included the photovoltaic/battery energy storage system, monitoring system, and prototype. A photovoltaic module designed with Maximum Power Point Tracking technology provided electricity, whereas a charge controller handled battery operation. The monitoring system received data from WQ and electrical sensors, which were then processed by an Arduino Mega 2560 microcontroller and transmitted to a cloud server via NB-IoT technology. Grafana was used for data visualization that was subsequently appraised using a WQ Suitability Index. A prototype confirmed the system design, and the power balance strategies and techno-economic studies ensured reliability and cost-efficiency.
Palermo et al. (2022)	The research offers an up-to-date and useful overview for academics, consumers, and stakeholders about the implementation of smart and sustainable technologies for water resource management, mainly for building-scale applications. The findings indicate that recent advances in information and communication technology and the internet have created new potential for real-time monitoring and control of city buildings, infrastructure, and services. The findings indicate water scarcity because of a growing human population, which emphasizes the need for effective water management. The research revealed that building management systems connect with data collected via smart water devices to determine WQ consumed by occupants. By adopting this approach, smart buildings provide novel methods to support sustainable urban development and help tackle environmental issues caused by growing populations and the influences of changing global climate.
Li et al. (2021)	To promote the cleaner manufacturing of pulp and paper, researchers developed a hybrid deep-learning CLSTMA model that monitors WQ in a paper industrial wastewater treatment system. The model is built on a sequential fusion convolutional neural network, long short-term memory, and attention mechanism. A paper factory in Dongguan City, China, provided the study's data. Additionally, the model reduces emissions and improves energy conservation. However, previous studies on multivariable complex network regression on wastewater treatment processes could not be directly compared or referenced owing to data scale constraints.
Konde and Deosarkar (2020)	This study uses a field-programmable array design board, sensors, a Zigbee-based wireless communication module, and a personal computer to present a reconfigurable sensor interface device for WQM systems in an IoT environment. Water parameters taken into consideration included temperature, turbidity, humidity, carbon dioxide, pH, and level. Data collection occurred in parallel, real time, and at a high speed from several separate sensor nodes. The proposed system gains fast execution speed and reusable Intellectual Property design by incorporating the field programmable array board. The aquatic ecosystems are expected to be better protected with the help of the suggested system. As a component of environmental management, the smart WQ management system reduces the monetary and time requirements to assess the WQ of a reservoir.
Pasika and Gandla (2020)	The study proposed an efficient and cost-effective system that includes numerous sensors to assess characteristics such as pH, turbidity in the water, tank level, temperature, and humidity in the surrounding atmosphere. The developed system, which includes Arduino Mega and NodeMCU target boards, successfully communicates with many sensors. To monitor WQ in real time, an efficient algorithm is built. In addition, the microcontroller unit is interfaced with these sensors, and further processing is performed on a personal computer. The acquired data are sent to the cloud ThinkSpeak, a web-based application to monitor WQ.

(*Continued*)

Table 11.1: (Continued)

Author year	Summary
Jiang et al. (2020)	This review paper offers guidance on how to effectively design and build WQM networks. The results demonstrate that WQMN quantitative design is now undergoing rapid development, and the common categories include topology, geostatistics, information entropy, multivariate statistics, and single- and multiple-criteria optimization. In addition, surface water networks in the age of smart cities display unique features, including dynamic, varied coupling with other urban monitoring infrastructures such as transit, security, and urban flood control. Furthermore, the study stresses that there is still an urgent need for innovative design techniques, particularly for pollution source identification, emergency monitoring, mobile monitoring, and nonpoint source management.
(Mukta, Islam, Barman, Reza, & Hossain Khan, 2019)	This study highlights a smart WQM system that utilizes the IoT to facilitate continuous monitoring of four physical parameters: temperature, pH, electric conductivity, and turbidity qualities. Four sensors are connected separately to an Arduino Uno to detect the properties of water. The sensors' extracted data are transferred to a desktop application developed on the NET platform, where it is compared to values set by the World Health Organization. The suggested system can effectively evaluate the water characteristics using a swift forest binary classifier based on the observed findings to decide whether or not the test water sample is fit for human consumption.
Japitana and Burce (2019)	This study aims to create remote sensing (RS)-based models using Landsat 8 pictures using regression analysis. The findings revealed a significant connection between in situ WQ readings and RS-based datasets in the Tubay River, Philippines. These models are expected to be helpful in estimating seasonal fluctuations and computing WQ parameters. The promising results in estimating WQ parameters can be utilized to develop an operational RS-based monitoring and assessment system that can be scaled up at the regional and national levels if validated with multitemporal datasets.
Wang and Yang (2019)	The purpose of this review paper is to conduct a thorough analysis of current WQ and RS research in China, identifying present difficulties and future directions. The findings revealed that the use of RS approaches in WQ research has increased markedly, whereas the regional distribution is relatively uneven. Additionally, the ground object spectrometer is the commonly used data source. The frequently studied watercolor indicators in China are chlorophyll a and suspended solids. In addition, the semiempirical technique is the most widely used inversion method. The findings indicate that existing research rarely examined anthropogenic influences, limiting the model's robustness and applicability in anthropogenically impacted water bodies.
Koditala and Pandey (2018)	This research offers a realistic and cost-effective strategy for WQM, particularly in rural regions, without the need for human involvement. The study uses IoT, ML, and cloud computing to replace existing methods of WQM. Using this model, once the turbidity value hits a predefined level, an email alert is sent to the appropriate authorities advising them of the issue, compelling them to take urgent action. For usability, the system based on ML controls water temperature, thus people receive water according to the climatic conditions. This is accomplished by estimating exterior temperature using ML methods.
Singh et al. (2018)	The study describes a system that includes pH, conductivity, salinity, and light intensity sensors connected to a microcontroller (Raspberry Pi) and data transmission modules (Zigbee and GSM). Analog data from the sensors are wirelessly transmitted to the microcontroller via Zigbee, which processes and delivers WQ metrics to a smartphone or PC over GSM. Another approach uses a microcontroller to process sensor data and then send it to a central server over Wi-Fi for approved user access. This real-time system efficiently monitors WQ while using low energy, promoting robust processing capabilities and precise data management.

(*Continued*)

Table 11.1: (Continued)

Author year	Summary
Abdullah et al. (2017)	These researchers conducted a review of chemical and biosensor methods commonly employed for wastewater quality management. The major criteria for selecting sensor technology in wastewater quality control are accuracy, selectivity, overall cost, and usability. The findings indicate that the sensors are cost-effective for large-scale installation and long-term maintenance. Additionally, the review found that chemical sensor systems differ depending on the type of ion to be detected, as well as selectivity and sensitivity. The toxicity of the influent may have an adverse effect on the biological treatment of wastewater. This issue can be solved by utilizing an improvised biosensor.
Riadi (2017)	The paper mentions that hybrid treatment systems such as membrane bioreactors and combined anaerobic–aerobic processes are pivotal in advancing WQ management, particularly in challenging wastewater scenarios. They efficiently eliminate pharmaceuticals and personal care products. Combined anaerobic–aerobic systems reduce sludge production and energy use, and effectively convert organic pollutants into biogas. Biofilm support technologies enhance treatment by facilitating microorganism attachment, improving sludge management, and lowering operational costs. Advanced oxidation processes, such as UV/ozone-hydrogen peroxide and Fenton methods, target micropollutants and wastewater with elevated chemical oxygen demand/biochemical oxygen demand, achieving significant contaminant removal. The findings indicate that hybrid methods promise resource recovery, efficient treatment, and minimizing environmental impacts and are crucial for dealing with complex water treatment challenges.
Saseendran and Nithya (2016)	These researchers propose an automated water usage monitoring system that allows individuals to keep track of their water bills while also assisting the government in managing the water supply and modifying the charging scheme based on water usage. Wireless sensors collect real-time data from water outlets, which is then transmitted to a LabVIEW module for remote access. The system architecture includes sensor modules equipped with sensor circuitry and RF transceivers, central modules with Arduino boards, and server modules for data processing. The system aims to alert users of excessive water usage at each outlet and provides real-time monitoring of water bills. It integrates warning and alert systems to minimize water wastage, with LabVIEW facilitating data visualization and control. Results demonstrate effective water management capabilities, enhancing resource conservation.
Murphy et al. (2015)	The designed optical sensor is durable, deployable, and user-friendly. The sensor consists of a multiwavelength light source and two photodiode detectors. This sensor measures light transmission and side scattering in a detector head equipped with five LEDs of various wavelengths. The sensor can give qualitative data on changes in the optical opacity of water.

Note: *WQM*, Water quality monitoring; *WQ*, water quality; *IoT*, Internet of Things; *ML*, machine learning.

11.3 IoT and ML in WQM

The IoT (such as Wi-Fi, RFID, LpWAN, and cellular networks) has transformed the field of WQ evaluation by enabling researchers to conduct physical and chemical measurements in the lab or on-site with a fraction of the effort compared to depending only on manual data gathering (Essamlali et al., 2024). The IoT has led to design and development of smart products, in which the subcomponents are intelligent. Particularly, WQM systems are getting more intelligent, requiring less electricity, and becoming simpler to use owing to the advancements in IoT technology.

Recent research indicates that IoT-based smart systems can continuously monitor large and local water bodies, raw WQ in treatment plants, water treatment, and the quality of drinking water produced (Lakshmikantha et al., 2021). An IoT-based system for real-time WQM, for instance, is designed by Daigavane & Gaikwad (2017) and consists of a few sensors that measure temperature, turbidity, conductivity, pH, and dissolved oxygen, among other aspects of WQ. The Raspberry Pi serves as the primary controller, and the microcontroller processes the measured values from the sensors before sending the processed values to it over the Zigbee protocol. Finally, cloud computing allows sensor data to be viewed on web browser applications. Gothane (2024) employed several sensors, including temperature, turbidity, and ultrasonic sensors, in the design of a WQM system. Each sensor records its values, such as water flow, temperature, and quality, and transmits the information to Arduino. The Arduino central controller accepts sensor values and processes them via Wi-Fi to the internet. The user can track the WQ by using the Wi-Fi module connected to Arduino to view the sensed data on the phone and LCD module. In a separate study, Miller et al. (2023) indicate that even with the abundance of smart WQM devices, there are still numerous issues that require attention. For example, the IoT technology's integration with current infrastructure can provide challenges. Additionally, it can be expensive to establish connectivity with IoT platforms and the installation of IoT sensors. In addition, there may also be issues concerning privacy and ownership of the generated data. Moreover, there may be difficulties in organizing and processing the large amounts of WQ data collected by the sensors.

Essamlali et al. (2024) state that using IoT devices for WQM alone does not provide improved WQ. To enhance WQM, the data collected from these devices must be analyzed and understood. Integrating ML and IoT for WQM provides faster, more accurate, and less expensive Q analysis (Jayaraman et al., 2024). ML has become a powerful tool for WQM. In WQM, four primary categories of ML techniques are employed: (1) supervised, (2) unsupervised, (3) reinforcement, and (4) semisupervised learning (Lawton, 2023). To categorize WQ data according to particular criteria and find abnormalities or changes, supervised learning is used. Unsupervised learning finds patterns or clusters in the data to find new patterns or recognize anomalous WQ occurrences. By giving control systems feedback, reinforcement learning makes treatment procedures more efficient. Semisupervised learning performs classification or grouping by combining labeled and unlabeled data (Essamlali et al., 2024). These ML methods bring important new understandings of WQ. Additionally, based on the quality and availability of data, GIS can also be integrated with ML techniques (Jayaraman et al., 2024). For ML methods to forecast WQ, temporal dynamics and spatial autocorrelation need to be considered. Incorporating spatial analytic techniques such as geostatistics (within GIS) can allow variation to be identified and increase forecast accuracy (Jayaraman et al., 2024). Addressing the issues will help determine robust ML and GIS-based frameworks for WQ index calculation and assist in effective water management and decision-making processes (Tabassum et al., 2023).

11.4 Water sustainability, WQM, UN SDGs

Declining WQ has become a global concern due to the growing human populations, expanding industrial and agricultural activities, and the possibility of substantial shifts to the hydrological cycle brought on by the changing global climate. Water sustainability is crucial for maintaining ecological balance and human safety. Modern WQM technologies developed to complement the conventional method of sampling and laboratory analysis in WQM are being driven by growing demands and resource constraints. The findings of the review indicate that the modern WQM tools provide both qualitative data and quantitative data, which are founded on novel techniques. In addition to being inexpensive and easy to use on-site, these advanced techniques may be seen as tools for sustainable development (Gagnon et al., 2007). Especially in lower-income parts of the world, the use of these monitoring methods is essential for providing safe water and environmental sanitation for preserving the environment, enhancing health, and reducing poverty.

Sustainable water refers to a country that is capable of meeting its own urban, industrial, and agricultural needs by guaranteeing an adequate supply of water (Russo et al., 2014). It also implies that the availability of water will not alter in response to the effects of climate change. Sustainable water management is directly connected to sustainable development. Appropriate management of water resources is expected to increase food security, lower poverty rates, prevent waterborne diseases and protect the health and biodiversity of aquatic environment (Cosgrove & Loucks, 2015; Forio & Goethals, 2020). Access to safe drinking water and appropriate sanitation are key components of human health and well-being and are fundamental in achieving several of the UN SDGs: access to good health and well-being (SDG 3), clean water and sanitation (SDG 6), climate action (SDG 13), and life below water (SDG 14).

The influences of climate change can impact water supplies, leading to scarcity and flooding in certain locations. Many forms of water pollution, including sediments, pathogens, and pesticides, will worsen owing to the impacts of global warming, along with increased water temperatures and an increase in floods and droughts (Fig. 11.2; Bolan et al., 2024). UN SDG target 6.3 is: "By 2030, improve WQ by reducing pollution, eliminating dumping and minimizing release of hazardous chemicals and materials, halving the proportion of untreated wastewater and substantially increasing recycling and safe reuse globally" (United Nations, n.d.). Specifically, indicator 6.3.2 emphasizes the importance of monitoring freshwater quality (United Nations, 2018, 2023). Implementing such goals necessitates sophisticated WQM programs that are both relevant and measurable.

11.5 Discussion

Understanding the possible risks associated with untreated wastewater disposal in the surrounding environment is vital for both public health and the economy, as it can have adverse environmental effects on receiving water bodies. Potable water remains inaccessible to many people across the globe, specifically in the developing countries. Achieving sustainable water supply depends in large part on WQM and appropriate wastewater treatment. Sustainable water management techniques are also crucial as climate change impacts are exacerbating WQ issues and water scarcity by altering temperature and precipitation patterns as well as the entire water cycle.

A review of past investigations on various conventional and modern methods of WQM and wastewater treatment techniques indicates that the complex nature of emerging pollutants in wastewater necessitates the need for newer and more advanced WQM techniques and wastewater treatment processes. Recent innovations such as nanotechnology and IoT-based systems are essential for addressing these new challenges and improving real-time WQ monitoring capabilities. Pilot-scale, laboratory, and in situ investigations have shown that nanotechnology has a promising future in drinking, wastewater treatment, and environmental remediation. Additionally, the findings reveal that real-time WQ data can be collected from inexpensive portable sensors, enabling prompt action to pollution situations. Furthermore, past studies indicate that conventional WQM techniques are often slow, subject to inaccuracies, and require huge human and material resources, whereas newer and more advanced technologies, despite their benefits can pose issues related to cost, integration, and data security concerns. As such in the past several decades, research efforts have been directed toward enhancing conventional wastewater treatment technology. Specifically, advanced wastewater treatment technologies (e.g., advanced oxidation processes, constructed wetlands, bioelectrical systems, and enzymatic treatment) have been widely explored for the removal of emerging contaminants. It is important to note that WQM and wastewater management techniques vary in utilities across the globe. To ensure sustainable water resource management, utilities in many countries have already integrated sustainability considerations along with advanced WQM and wastewater treatment techniques and technologies into their operating processes. Nonetheless, a lot of utilities continue to search for affordable alternatives to simplify and enhance their current approaches.

The findings of the review highlight the need for further research and development in a number of important areas, including (1) increasing the affordability of nanomaterials and IoT technologies; (2) fostering cross-disciplinary collaborations for better technology integration; and (3) implementing pilot programs to test and refine these technologies in various contexts. In addition, a holistic approach in which technical, environmental, socio-economic, and cultural issues are addressed is necessary to ensure effective water management strategies and sustainable development. Moreover, in the future, research and development of WQM and wastewater treatment techniques should prioritize reducing energy consumption, encouraging responsible raw material consumption, and minimizing water shortages.

11.6 Conclusions

An essential instrument for water management is WQM, which works to curb pollution and its deleterious influences on public health and the environment. The findings of the review indicate that recent advances in IoT, ML, and nanotechnology have made it easier to collect WQ data at a lower cost. The reliable assessment of WQ through advanced WQM systems is crucial for sustainable development. Additionally, water treatment and reuse of treated water are important factors in the achievement of sustainable water provision. The findings of the study are expected to guide future research and assist policy-makers in comprehending how technology can help achieve the relevant SDGs.

References

Abbas, F., Cai, Z., Shoaib, M., Iqbal, J., Ismail, M., Arifullah., . . . Albeshr, M. F. (2024). Machine learning models for water quality prediction: A comprehensive analysis and uncertainty assessment in Mirpurkhas, Sindh, Pakistan. *Water*, *16*(7), 941. Available from https://doi.org/10.3390/w16070941.

Abdullah, N. A., Ramli, S., Mamat, N. H., Khan, S., & Gomes, C. (2017). Chemical and biosensor technologies for wastewater quality management. *International Journal of Advanced Research and Publications*, *1*(6). Available from https://www.researchgate.net/publication/322114992_Chemical_And_Biosensor_Technologies_For_Wastewater_Quality_Management.

Adu-Manu, K. S., Tapparello, C., Heinzelman, W., Katsriku, F. A., & Abdulai, J.-D. (2017). Water quality monitoring using wireless sensor networks. *ACM Transactions on Sensor Networks*, *13*(1), 1−41. Available from https://doi.org/10.1145/3005719.

Ahmed, U., Mumtaz, R., Anwar, H., Mumtaz, S., & Qamar, A. M. (2019). Water quality monitoring: From conventional to emerging technologies. *Water Supply*, *20*(1), 28−45. Available from https://doi.org/10.2166/ws.2019.144.

Ahmed, T., Zounemat-Kermani, M., & Scholz, M. (2020). Climate change, water quality and water-related challenges: A review with focus on Pakistan. *International Journal of Environmental Research and Public Health*, *17*(22). Available from https://doi.org/10.3390/ijerph17228518.

Asrav, T., Koksal, E. S., Esenboga, E. E., Cosgun, A., Kusoglu, G., Aydin, D., & Aydin, E. (2023). Physics-informed neural network based modeling of an industrial wastewater treatment unit. *Computer-aided Chemical Engineering*, *52*, 227−234. Available from https://doi.org/10.1016/b978-0-443-15274-0.50037-8.

Bartlett, J. A., & Dedekorkut-Howes, A. (2023). Adaptation strategies for climate change impacts on water quality: a systematic review of the literature. *Journal of Water and Climate Change*, *14*(3), 651−675. Available from https://doi.org/10.2166/wcc.2022.279.

Bhardwaj, J., Gupta, K. K., & Gupta, R. (2015). A review of emerging trends on water quality measurement sensors. *IEEE Xplore*. Available from https://doi.org/10.1109/ICTSD.2015.7095919.

Bolan, S., Padhye, L. P., Jasemizad, T., Govarthanan, M., Karmegam, N., Wijesekara, H., . . . Bolan, N. (2024). Impacts of climate change on the fate of contaminants through extreme weather events. *Science of the Total Environment*, *909*. Available from https://doi.org/10.1016/j.scitotenv.2023.168388, A.

Cosgrove, W. J., & Loucks, D. P. (2015). Water management: Current and future challenges and research directions. *Water Resources Research*, *51*(6), 4823−4839. Available from https://doi.org/10.1002/2014wr016869.

Crini, G., & Lichtfouse, E. (2018a). Green adsorbents for pollutant removal. *Environmental chemistry for a sustainable world*. Springer International Publishing. Available from https://doi.org/10.1007/978-3-319-92162-4.

Crini, G., & Lichtfouse, E. (2018b). *Wastewater treatment: An overview. Environmental chemistry for a sustainable world.* (pp. 1–21). Springer. Available from https://doi.org/10.1007/978-3-319-92111-2_1.

Criollo, R., Velasco, V., Nardi, A., de Vries, L. M., Riera, C., Scheiber, L., ... Vázquez-Suñé, E. (2019). AkvaGIS: An open source tool for water quantity and quality management. *Computers & Geosciences*, *127*, 123–132. Available from https://doi.org/10.1016/j.cageo.2018.10.012.

Daigavane, V. V., & Gaikwad, M. A. (2017). Water quality monitoring system based on iot, Advances in Wireless and Mobile. *Communications*, *10*(5), 1107–1116.

Dickin, S. K., Schuster-Wallace, C. J., Qadir, M., & Pizzacalla, K. (2016). A review of health risks and pathways for exposure to wastewater use in agriculture. *Environmental Health Perspectives*, *124*(7), 900–909. Available from https://doi.org/10.1289/ehp.1509995.

EL-Nwsany, R. I., Maarouf, I., & Abd el-Aal, W. (2019). Water management as a vital factor for a sustainable school. *Alexandria Engineering Journal*, *58*(1), 303–313. Available from https://doi.org/10.1016/j.aej.2018.12.012.

Enuh, B. M. (2023). Detecting water pollutants with nanosensors. https://www.azonano.com/article.aspx?ArticleID = 6463.

Essamlali, I., Nhaila, H., & Khaili, M. E. (2024). Advances in machine learning and IoT for water quality monitoring: A comprehensive review. *Heliyon*, *10*(6). Available from https://doi.org/10.1016/j.heliyon.2024.e27920.

Forio, M. A. E., & Goethals, P. L. M. (2020). An integrated approach of multi-community monitoring and assessment of aquatic ecosystems to support sustainable development. *Sustainability*, *12*(14), 5603. Available from https://doi.org/10.3390/su12145603.

Furlong, C., Jegatheesan, J., Currell, M., Iyer-Raniga, U., Khan, T., & Ball, A. S. (2019). Is the global public willing to drink recycled water? A review for researchers and practitioners. *Utilities Policy*, *56*, 53–61. Available from https://doi.org/10.1016/j.jup.2018.11.003.

Gagnon, B., Marcoux, G., Leduc, R., Pouet, M.-F., & Thomas, O. (2007). Emerging tools and sustainability of water-quality monitoring. *Trends in Analytical Chemistry*, *26*(4), 308–314. Available from https://doi.org/10.1016/j.trac.2006.11.009.

Gazi-Khan, L., & Haque, S. E. (2022). A review of the current status of water quality and eutrophication in Dhaka's water bodies. *International Journal of Students' Research in Technology & Management, [online]*, *10*(2), 01–09. Available from https://doi.org/10.18510/ijsrtm.2022.1021.

Gedda, G., Balakrishn, K., Devi, R. U., & Shah, K. J. (2021). Introduction to conventional wastewater treatment technologies: Limitations and recent advances. *Advances in wastewater treatment I*, 1–36. Available from https://doi.org/10.21741/9781644901144-1.

Gothan, S. (2024). Water quality monitoring and classification using Internet of Things. *International Journal of Management, Technology and Engineering*, *XIV*(III), 2249–7455.

Hairom, N. H. H., Soon, C. F., Mohamed, R. M. S. R., Morsin, M., Zainal, N., Nayan, N., ... Harun, N. H. (2021). A review of nanotechnological applications to detect and control surface water pollution. *Environmental Technology & Innovation*, *24*. Available from https://doi.org/10.1016/j.eti.2021.102032.

Hanak, E., & Lund, J. R. (2011). Adapting California's water management to climate change. *Climatic Change*, *111*(1), 17–44. Available from https://doi.org/10.1007/s10584-011-0241-3.

Haque, S. E. (2022). *Urban water pollution by heavy metals, microplastics, and organic contaminants. Current directions in water scarcity research (Vol. 6)* (pp. 21–43). Elsevier. Available from https://doi.org/10.1016/b978-0-323-91838-1.00001-4.

Haque, S. E. (2023). The effects of climate variability on Florida's major water resources. *Hazards and Sustainability*, *15*(14). Available from https://doi.org/10.3390/su151411364.

Haque, S. E., & Nahar, N. (2023). *Bangladesh: Climate Change Issues, Mitigation and Adaptation in the Water Sector*. ACS ES&T Water. Available from https://doi.org/10.1021/acsestwater.2c00450.

Haque, S. E., Nahar, N., Chowdhury, S., Sakib, A. S., Saif, A., Hasan, S., ... Nezum, S. T. (2020). Impacts of the partial relocation of hazaribagh tanneries on the environment and human health: focus on children and vulnerable population. *International Journal of Students' Research in Technology & Management*, *8*(4), 1–7. Available from https://doi.org/10.18510/ijsrtm.2020.841.

Iancu, G., Ciolofan, S. N., & Drăgoicea, M. (2024). Real-time IoT architecture for water management in smart cities. *Discover Applied Sciences*, *6*, 191. Available from https://doi.org/10.1007/s42452-024-05855-9.

Jabbar, W. A., Tan, M. T., Hamidun, M. F. I., Che Kamarudin, A. H., Wu, W., Sultan, J., ... Ali, M. A. H. (2024). Development of LoRaWAN-based IoT system for water quality monitoring in rural areas. *Expert Systems with Applications*, *242*, 122862. Available from https://doi.org/10.1016/j.eswa.2023.122862.

Jamroen, C., Yonsiri, N., Odthon, T., Wisitthiwong, N., & Janreung, S. (2023). A standalone photovoltaic/battery energy-powered water quality monitoring system based on narrowband internet of things for aquaculture: Design and implementation. *Smart Agricultural Technology*, *3*. Available from https://doi.org/10.1016/j.atech.2022.100072.

Japitana, M. V., & Burce, M. E. C. (2019). A satellite-based remote sensing technique for surface water quality estimation. *Engineering, Technology & Applied Science Research*, *9*(2), 3965–3970. Available from https://doi.org/10.48084/etasr.2664.

Jayaraman, P., Nagarajan, K. K., Partheeban, P., & Krishnamurthy, V. (2024). Critical review on water quality analysis using IoT and machine learning models. *International Journal of Information Management Data Insights*, *4*(1). Available from https://doi.org/10.1016/j.jjimei.2023.100210.

Jiang, J., Tang, S., Han, D., Fu, G., Solomatine, D., & Zheng, Y. (2020). A comprehensive review on the design and optimization of surface water quality monitoring networks. *Environmental Modelling & Software*, *132*. Available from https://doi.org/10.1016/j.envsoft.2020.104792.

Khatri, P., Gupta, K. K., & Gupta, R. K. (2019). Raspberry Pi-based smart sensing platform for drinking-water quality monitoring system: A Python framework approach. *Drinking Water Engineering and Science*, *12* (1), 31–37. Available from https://doi.org/10.5194/dwes-12-31-2019.

Khatri, N., & Tyagi, S. (2014). Influences of natural and anthropogenic factors on surface and groundwater quality in rural and urban areas. *Frontiers in Life Science*, *8*(1), 23–39. Available from https://doi.org/10.1080/21553769.2014.933716.

Koditala, N. K., & Pandey, P. S. (2018). Water quality monitoring system using IoT and machine learning. In IEEE Xplore. https://doi.org/10.1109/RICE.2018.8509050.

Konde, S., & Deosarkar, S. (2020). *IOT based water quality monitoring system*. SSRN. https://papers.ssrn.com/sol3/papers.cfm?abstract_id = 3645467.

Kurniawan, T. A., Bandala, E. R., Othman, M. H. D., Goh, H. H., Anouzla, A., Chew, K. W., ... Khoir, A. N. (2024). Implications of climate change on water quality and sanitation in climate hotspot locations: a case-study from Indonesia. *Water Supply*, *24*, 517. Available from https://doi.org/10.2166/ws.2024.008.

Lakshmikantha, V., Hiriyannagowda, A., Manjunath, A., Patted, A., Basavaiah, J., & Anthony, A. A. (2021). IoT based smart water quality monitoring system. *Global Transitions Proceedings*, *2*(2), 181–186. Available from . https://doi.org/10.1016/j.gltp.2021.08.062.

Lawton, G. (2023). *4 Types of learning in machine learning explained. Enterprise AI*. https://www.techtarget.com/searchenterpriseai/tip/Types-of-learning-in-machine-learning-explained.

Li, X., Yi, X., Liu, Z., Liu, H., Chen, T., Niu., ... Ying, G. (2021). Application of novel hybrid deep leaning model for cleaner production in a paper industrial wastewater treatment system. *Journal of Cleaner Production*, *294*. Available from https://doi.org/10.1016/j.jclepro.2021.126343.

Miller, M., Kisiel, A., Cembrowska-Lech, D., Durlik, I., & Miller, T. (2023). IoT in water quality monitoring—Are we really here? *Physical Sensors*, *23*(2), 960. Available from https://doi.org/10.3390/s23020960.

Molina-Navarro, E., Trolle, D., Martinez-Perez, S., Merlin, A. S., & Jeppesen, E. (2014). Hydrological and water quality impact assessment of a Mediterranean limno-reservoir under climate change and land use management scenarios. *Journal of Hydrology*, *509*, 354–366. Available from https://doi.org/10.1016/j.jhydrol.2013.11.053.

Moretti, A., Ivan, H. L., & Skvaril, J. (2024). A review of the state-of-the-art wastewater quality characterization and measurement technologies. Is the shift to real-time monitoring nowadays feasible? *Journal of Water Process Engineering*, *60*. Available from https://doi.org/10.1016/j.jwpe.2024.105061.

Mukta, M., Islam, S., Barman, S. D., Reza, A. W., and Hossain Khan, M. S. 2019. Iot based Smart Water Quality MonitoringSystem, 2019 IEEE 4th International Conference on Computer and Communication Systems (ICCCS), Singapore, pp. 669–673, Available from https://doi.org/10.1109/CCOMS.2019.8821742.

Murphy, K., Heery, B., Sullivan, T., Zhang, D., Paludetti, L., Lau, K. T., ... Regan, F. (2015). A low-cost autonomous optical sensor for water quality monitoring. *Talanta*, *132*, 520–527. Available from https://doi.org/10.1016/j.talanta.2014.09.045.

Neczaj, E., & Grosser, A. (2018). Circular Economy in Wastewater Treatment Plant–Challenges and Barriers. *Proceedings*, *2*(11), 614. Available from https://doi.org/10.3390/proceedings2110614.

Nirmala, G. N., & Prakash, S. (2023). Optimizing wastewater treatment performance system and achieving greater efficiency to improve water quality for sustainability — a review. Qeios. Available from https://doi.org/10.32388/8EP4HX.

Nishu., & Kumar, S. (2023). Smart and innovative nanotechnology applications for water purification. *Hybrid Advances*, *3*. Available from https://doi.org/10.1016/j.hybadv.2023.100044.

Obaideen, K., Shehata, N., Sayed, E. T., Abdelkareem, M. A., Mahmoud, M. S., & Olabi, A. G. (2022). The role of wastewater treatment in achieving sustainable development goals (SDGs) and sustainability guideline. *Energy Nexus*, *7*. Available from https://doi.org/10.1016/j.nexus.2022.100112.

Paepae, T., Bokoro, P. N., & Kyamakya, K. (2021). From fully physical to virtual sensing for water quality assessment: A comprehensive review of the relevant state-of-the-art. *Sensor Intelligence through Neurocomputing*, *21*(21). Available from https://doi.org/10.3390/s21216971.

Palermo, S. A., Maiolo, M., Brusco, A. C., Turco, M., Pirouz, B., Greco, E., ... Piro, P. (2022). Smart technologies for water resource management: An overview. *Urban Intelligence at the Edge*, *22*(16), 6225. Available from https://doi.org/10.3390/s22166225.

Pasika, S., & Gandla, S. T. (2020). Smart water quality monitoring system with cost-effective using IoT. *Heliyon*, *6*(7). Available from https://doi.org/10.1016/j.heliyon.2020.e04096.

Patil, R., Wei, Y., Pullar, D., & Shulmeister, J. (2022). Effects of change in streamflow patterns on water quality. *Journal of Environmental Management*, *302*. Available from https://doi.org/10.1016/j.jenvman.2021.113991, Part-A.

Petrie, B., Barden, R., & Kasprzyk-Hordern, B. (2015). A review on emerging contaminants in wastewaters and the environment: Current knowledge, understudied areas and recommendations for future monitoring. *Water Research*, *72*, 3–27. Available from . https://doi.org/10.1016/j.watres.2014.08.053.

du Plessis, A. (2022). Persistent degradation: Global water quality challenges and required actions. *One Earth*, *5* (2), 129–131. Available from . https://doi.org/10.1016/j.oneear.2022.01.005.

Pule, M., Yahya, A., & Chuma, J. (2017). Wireless sensor networks: A survey on monitoring water quality. *Journal of Applied Research and Technology*, *15*(6), 562–570. Available from https://doi.org/10.1016/j.jart.2017.07.004.

Qu, X., Alvarez, P. J. J., & Li, Q. (2013). Applications of nanotechnology in water and wastewater treatment. *Water Research*, *47*(12), 3931–3946. Available from . https://doi.org/10.1016/j.watres.2012.09.058.

Rajesh, M., & Rehana, S. (2022). Impact of climate change on river water temperature and dissolved oxygen: Indian riverine thermal regimes. *Scientific Reports*, *12*(1). Available from . https://doi.org/10.1038/s41598-022-12996-7.

Rajitha, A., Aravinda, K., Nagpal, A., Kalra, R., Maan, P., Kumar, A., & Dalael Saad, A.-Z. (2024). Machine learning and AI-driven water quality monitoring and treatment. *E3S Web of Conferences*, 505. Available from https://doi.org/10.1051/e3sconf/202450503012.

Rana, R., Kalia, A., Boora, A., Alfaisal, F. M., Alharbi, R. S., Berwal, P., ... Qamar, O. (2023). Artificial intelligence for surface water quality evaluation, monitoring and assessment. *Water*, *15*(22), 3919. Available from https://doi.org/10.3390/w15223919.

Riadi, L. (2017). Water sustainability: Emerging trends for water quality management. *KnE Life Sciences*, *3*(5), 118–130. Available from https://doi.org/10.18502/kls.v3i5.984.

Rout, P. R., Zhang, T. C., Bhunia, P., & Surampalli, R. Y. (2021). Treatment technologies for emerging contaminants in wastewater treatment plants: A review. *Science of the Total Environment*, *753*. Available from https://doi.org/10.1016/j.scitotenv.2020.141990.

Russo, T., Alfredo, K., & Fisher, J. (2014). Sustainable water management in urban, agricultural, and natural systems. *Water*, *6*(12), 3934–3956. Available from https://doi.org/10.3390/w6123934.

Saseendran, S., & Nithya, V. (2016). Automated water usage monitoring system. In *2016 International conference on communication and signal processing (ICCSP)* (pp. 0099–0103), Melmaruvathur, India. Available from https://doi.org/10.1109/ICCSP.2016.7754501.

Sharvelle, S., Dozier, A., Arabi, M., & Reichel, B. (2017). A geospatially-enabled web tool for urban water demand forecasting and assessment of alternative urban water management strategies. *Environmental Modelling & Software, 97*, ,, 213–228. Available from https://doi.org/10.1016/j.envsoft.2017.08.009.

Shayo, G. M., Elimbinzi, E., & Shao, G. N. (2024). Water-based technologies for improving water quality at the point of use: A review. *Wiley Interdisciplinary Reviews: Nanomedicine and Nanobiotechnology*, *16*(1). Available from https://doi.org/10.1002/wnan.1940.

Singh, S., Kumar, A., Prasad, A., & Bharadwaj, N. (2018). IOT based water quality monitoring system. *International Journal on Recent & Innovative Trend In Technology*, *4*(3), ISSN: 2454-1400.

Strecker, A. L., Cobb, T. P., & Vinebrooke, R. D. (2004). Effects of experimental greenhouse warming on phytoplankton and zooplankton communities in fishless alpine ponds. *Limnology and Oceanography*, *49*(4), 1182–1190. Available from https://doi.org/10.4319/lo.2004.49.4.1182.

Suresh, K., Tang, T., Viliet, M. T. H. V., Bierkens, M. F. P., Strokal, M., Sorger-Domenigg, F., & Wada, Y. (2023). Recent advancement in water quality indicators for eutrophication in global freshwater lakes. *Environmental Research Letters*, *18*(6). Available from https://doi.org/10.1088/1748-9326/acd071.

Tabassum, S., Kotnala, C. B., Masih, R. K., Shuaib, M., Alam, S., & Alar, T. M. (2023). Performance analysis of machine learning techniques for predicting water quality index using physiochemical parameters. In *International conference on sustainable computing and smart systems (ICSCSS)* (pp. 372–377). Available from https://doi.org/10.1109/icscss57650.2023.10169408.

United Nations. (2018). *Progress on ambient water quality – piloting the monitoring methodology and initial findings for SDG indicator 6.3.2*. [online] UN-Water. https://www.unwater.org/publications/progress-ambient-water-quality-piloting-monitoring-methodology-and-initial-findings Accessed 10.06.2024].

United Nations. (2023). *Progress on ambient water quality (SDG target 6.3)*. [online] UN-Water. https://sdg6data.org/en/indicator/6.3.2.

United Nations. (n.d). *Progress on wastewater treatment (SDG target 6.3)*. https://sdg6data.org/en/indicator/6.3.1.

USEPA. (1998). *How wastewater treatment works the basics*. Office of Water: United States Environmental Protection Agency. https://www3.epa.gov/npdes/pubs/bastre.pdf.

USEPA. (2012). *Chapter 5: Water quality conditions | monitoring & assessment* |United States Environmental Protection Agency. [online]. https://archive.epa.gov/water/archive/web/html/vms50.html.

USEPA. (2024). *Water quality monitoring program*. [online] National Aquatic Resource Surveys. https://www.epa.gov/climate-change-water-sector/water-quality-monitoring-program.

Wade, A. J., Skeffington, R., Couture, R.-M., Lampa, M. E., Groot, S., Halliday, S. J., ... Skuras, D. (2022). Land use change to reduce freshwater nitrogen and phosphorus will be effective even with projected climate change. *Ecological Monitoring and Assessment of Freshwater Ecosystems: New Trends and Future Challenges*, *14*(5), 829. Available from https://doi.org/10.3390/w14050829.

Wang, X., & Yang, W. (2019). Water quality monitoring and evaluation using remote sensing techniques in China: A systematic review. *Ecosystem Health and Sustainability*, *5*(1), 47–56. Available from https://doi.org/10.1080/20964129.2019.1571443.

Wear, S. L., Acuña, V., McDonald, R., & Font, C. (2021). Sewage pollution, declining ecosystem health, and cross-sector collaboration. *Biological Conservation*, *255*, 109010. Available from https://doi.org/10.1016/j.biocon.2021.109010.

Worrall, F., Burt, T., & Adamson, J. (2004). Can climate change explain increases in DOC flux from upland peat catchments? *Science of The Total Environment*, *326*(1–3), 95–112. Available from https://doi.org/10.1016/j.scitotenv.2003.11.022.

Yaroshenko, I., Kirsanov, D., Marjanovic, M., Lieberzeit, P. A., Korostynska, O., Mason, A., ... Legin, A. (2020). Real-time water quality monitoring with chemical sensors. *Advanced Sensors for Real-Time Monitoring Applications*, *20*(12), 3432. Available from https://doi.org/10.3390/s20123432.

Zaher, A., & Shehata, N. (2021). Recent advances and challenges in management of urea wastewater: A mini review. *IOP Conference Series: Materials Science and Engineering*, *1046*(1). Available from https://doi.org/10.1088/1757-899x/1046/1/012021.

Zainurin, S. N., Wan Ismail, W. Z., Mahamud, S. N. I., Ismail, I., Jamaludin, J., Ariffin, K. N. Z., & Wan Ahmad Kamil, W. M. (2022). Advancements in monitoring water quality based on various sensing methods: A systematic review. *International Journal of Environmental Research and Public Health*, *19*(21). Available from https://doi.org/10.3390/ijerph192114080.

CHAPTER 12

Future directions in water quality management: integrating advanced technologies and sustainable practices

Rwitabrata Mallick[1] and Sandeep Poddar[2]

[1]Department of Environmental Science, Amity School of Life Science, Amity University Madhya Pradesh, Gwalior, Madhya Pradesh, India, [2]Lincoln University College, Wisma Lincoln, Petaling Jaya, Selangor Darul Ehsan, Malaysia

12.1 Introduction

The deployment of remote monitoring systems will allow continuous monitoring of water quality indicators such as turbidity, pH, dissolved oxygen, and pollutants. These systems can provide real-time data, enabling quicker response times to potential issues. Networks of sensors can be distributed across water bodies to provide comprehensive coverage. These sensors can detect various parameters, including temperature, chemical composition, and biological indicators. Integrating Internet of Things (IoT) devices with intelligent sensors will enhance data collection and transmission. These devices can communicate wirelessly, providing real-time data access and reducing the need for manual sampling. Artificial intelligence (AI) algorithms can process large volumes of data collected by sensors, identifying patterns and anomalies. This capability enables early detection of water quality issues, potentially preventing larger-scale problems. Advanced oxidation processes (AOPs) can effectively remove contaminants from water, including emerging pollutants like pharmaceuticals and personal care products. These processes use reactive species to break down pollutants into less harmful substances. Using nanomaterials in water treatment can enhance the removal of contaminants at the molecular level. For example, nanofiltration (NF) membranes can filter out small particles and pathogens. Utilizing biological organisms and plants to remove or neutralize contaminants can be a sustainable approach to water treatment. These methods can be particularly effective for organic pollutants and heavy metals (Abdelfattah & El-Shamy, 2024).

Technologies for treating and reusing wastewater will become increasingly important, and advanced treatment processes can make reclaimed water safe for various uses, including

Computational Automation for Water Security.
DOI: https://doi.org/10.1016/B978-0-443-33321-7.00020-2

irrigation and industrial processes. Integrated water management platforms can consolidate data from various sources, including remote monitoring systems, treatment plants, and weather stations. This integration enables a holistic view of water resources and supports better decision-making. Predictive modeling can be done by using historical data and machine learning; predictive models can forecast water quality changes and potential contamination events. This capability can help in proactive management and mitigation efforts. Raising public awareness about water conservation and pollution prevention is crucial. Digital platforms and apps can provide real-time water quality updates to communities and encourage responsible water use. Engaging the public in water quality monitoring through citizen science initiatives can expand data collection efforts and promote environmental stewardship (Adekugbe & Ibeh, 2024).

12.2 Concept of water quality management and present technical scenario

Water quality management refers to the systematic process of monitoring, assessing, and regulating water resources to ensure they meet specific standards for various uses, such as drinking, agriculture, industry, and recreation. The goal is to maintain or improve water quality, protect public health, support ecosystems, and sustain economic activities.

12.2.1 Present technical scenario in water quality management

Deployment of IoT devices and smart sensors is very helpful. These sensors can transmit data wirelessly, enabling continuous monitoring and immediate response to changes. There can be use of satellite imagery and drones to monitor large water bodies, track pollution sources, and assess land-use impacts on water quality. There should be automated systems for water sampling and in situ analysis, reducing the need for manual labor and providing more frequent data collection (Chakraborty et al., 2023).

Collected data are incorporated into integrated platforms for comprehensive analysis. There are applications of machine learning and AI for predictive modeling, anomaly detection, and optimization of treatment processes using advanced data visualization tools to present complex data in intuitive formats for decision-makers and the public. It is important to use advanced membrane filtration processes like reverse osmosis (RO), NF, and ultrafiltration for high-efficiency contaminant removal. Biofilters, constructed wetlands, and other nature-based solutions are used for effective and sustainable wastewater treatment (Geetha et al., 2023).

Portable and modular systems are used to treat water at the point of use, particularly in remote or underserved areas. The implementation of small-scale treatment facilities for localized water supply and sanitation promotes resilience and reduces reliance on centralized infrastructure. Techniques for recovering resources like nutrients, energy, and water from wastewater support the circular economy. It is important to take initiative in developing and implementing strategies to adapt to climate change impacts on water resources, such as increasing resilience to extreme weather events and managing water scarcity. Adoption of adaptive management approaches and flexible regulations can respond to emerging challenges and scientific advancements through increased collaboration and information sharing among countries and regions to address transboundary water quality issues and develop best practices (Giri, 2021).

12.2.2 Advanced water quality monitoring

Advanced water quality monitoring encompasses the use of state-of-the-art technologies and methodologies to accurately and efficiently assess water quality in real-time or near-real-time. These systems are crucial for ensuring safe drinking water, protecting aquatic ecosystems, and managing water resources sustainably. Automated sensor networks involve multiparameter sensors, devices that can measure multiple water quality parameters simultaneously, such as pH, temperature, turbidity, dissolved oxygen, conductivity, chlorophyll, and specific ion concentrations (e.g., nitrate, phosphate). These are installed directly in water bodies or distribution systems to provide continuous data without the need for sample collection. Examples include optical sensors for detecting organic pollutants and biosensors for detecting pathogens. This is used for large-scale monitoring of water bodies and for detecting changes in water color, temperature, and vegetation, which can indicate pollution or eutrophication. Cameras and sensors can capture high-resolution images and data over specific areas, which are useful for monitoring hard-to-reach locations and assessing the impact of land use on water quality (Hamel & Tan, 2022).

Real-time data transmission and communication can be done through integration of sensors with IoT devices, which enable the real-time transmission of data to centralize databases or cloud platforms. This facilitates instant access to water quality data from multiple locations. 4G/5G, LoRaWAN, and other communication technologies are used to transmit data from remote sensors to control centers. Techniques such as UV-visible spectroscopy and fluorescence spectroscopy can identify and quantify a wide range of organic and inorganic contaminants. Coupled with chromatography, MS can detect and quantify trace levels of emerging contaminants, pharmaceuticals, and heavy metals. Lab-on-a-chip devices that enable rapid, on-site analysis of water samples offer portable and low-cost solutions for monitoring (Jan, Min-Allah, & Düştegör, 2021).

Machine learning and AI in water quality management can be initiated through advanced algorithms to analyze large datasets to identify patterns, predict future water quality conditions, and detect anomalies. This helps in early detection of contamination events. AI can process images from cameras or microscopes to identify microorganisms, algae, and other biological indicators of water quality. Hydrodynamic and water quality models can be used to simulate water movement and contaminant dispersion in aquatic systems. These models can predict the impact of pollution events and assess the effectiveness of remediation strategies. Climate and weather models are integrated with hydrological models to understand the influence of climatic factors on water quality. Monitoring devices include portable devices and kits for on-site testing of water quality parameters. These can be used by field technicians or citizen scientists. Mobile applications connected to sensors or portable devices are used for real-time monitoring and data visualization (Jayaraman, Nagarajan, Partheeban, & Krishnamurthy, 2024).

Centralized platforms that collect, store, and analyze data from various sources can provide a comprehensive view of water quality. These systems often include dashboards for visualization and alerts. Decision support systems (DSS) have been integrated with monitoring data to provide actionable insights for water management, regulatory compliance, and emergency response (Jayaraman et al., 2024). This can be achieved by using specific organisms, such as algae, invertebrates, and fish, to assess the ecological health of water bodies. Changes in the presence or behavior of these organisms can indicate water quality issues.

DNA and RNA analysis can be done to identify microbial communities and pathogens, providing detailed information about the biological composition of water. This ensures the accuracy, reliability, and integration of data from diverse monitoring systems remains a challenge. It has also been found that advanced monitoring systems can be expensive, and there is a need for cost-effective solutions that can be widely deployed (Kamyab et al., 2023). Developing standardized methods and protocols for data collection, analysis, and reporting is essential for consistent water quality assessment. Involving communities in monitoring efforts can enhance awareness and support for water quality management initiatives.

12.3 Data analytics and predictive modeling

Data analytics and predictive modeling are increasingly important tools in water quality management. They enable more informed decision-making, enhance the efficiency of water treatment processes, and provide early warnings of potential water quality issues. IoT devices enable seamless data collection and transmission to centralize systems, facilitating real-time analysis and remote monitoring. Historical data are analyzed to understand past trends, identify patterns, and establish baselines for water quality parameters. Data are investigated to determine the causes of water quality issues, such as identifying sources of contamination or malfunctioning equipment. This helps in anticipating issues like algal blooms, pollutant spikes, or system failures. Predictive models can estimate future concentrations of contaminants like nitrates, phosphates, heavy metals, and emerging pollutants, helping water managers plan appropriate treatment strategies (Kamyab et al., 2023). Models can forecast the occurrence and severity of harmful algal blooms by analyzing data on nutrient levels, temperature, and other environmental factors. This enables timely interventions to mitigate their impact. Predicting water demand based on consumption patterns and climatic conditions allows for better resource management and infrastructure planning. Predictive analytics can help identify potential failures in water distribution systems, such as pipe bursts or leaks, by analyzing pressure data and flow rates.

AI algorithms can detect anomalies in water quality data, indicating potential contamination events or system malfunctions. AI can analyze satellite images and remote sensing data to monitor large-scale water bodies, identify pollution sources, and assess land-use impacts on water quality. Combining data analytics, predictive models, and simulation tools into a comprehensive DSS provides water managers with actionable insights. They can simulate various scenarios, such as the impact of different treatment methods or the effects of climate change, to support strategic planning. DSS can offer intuitive interfaces for operators and decision-makers, allowing them to easily interpret complex data and model outputs (Moore et al., 2023).

12.3.1 Benefits of data analytics and predictive modeling

- Moving from reactive to proactive management by anticipating and addressing water quality issues before they become critical.
- Efficient allocation of resources, such as personnel, chemicals, and energy, based on predicted needs, leading to cost savings.
- Ensuring that water quality meets regulatory standards through continuous monitoring and predictive insights.
- Predictive models need regular validation and calibration to maintain accuracy, especially as conditions change.
- Training personnel to use advanced data analytics tools and interpret the results is necessary for effective implementation.

12.4 Decentralized water treatment systems and its significance

Decentralized water treatment systems refer to smaller, often modular facilities that treat water or wastewater close to the point of use or discharge. Unlike centralized systems, which transport water or wastewater to a central location for treatment, decentralized systems provide local treatment, making them particularly valuable in various contexts. Decentralized systems can bring water treatment solutions to rural and remote communities that lack access to centralized infrastructure, ensuring safe water supply and sanitation. In areas with limited infrastructure, decentralized systems can provide essential services without requiring extensive investment in centralized networks (Nwokediegwu, Daraojimba, et al., 2024). Many decentralized systems are modular, allowing them to be scaled up or down based on the population size and water demand. This flexibility is ideal for rapidly growing communities or temporary settlements. They can be adapted to local conditions and specific needs, such as treating particular contaminants or managing seasonal variations in water quality.

Decentralized systems can continue to function independently if parts of the centralized system are damaged, providing critical water services during natural disasters or infrastructure failures. By distributing water treatment capabilities, decentralized systems reduce the risk associated with potential failures or attacks on centralized facilities. Decentralized systems enable localized water quality management, allowing for immediate response to contamination events and more tailored treatment processes. Treating water closer to the source or point of use minimizes the risk of contamination during transportation, ensuring higher water quality. By treating water locally, decentralized systems can reduce the energy required for pumping water over long distances. This is particularly significant for remote or elevated areas. Decentralized wastewater systems can treat and reuse effluent for nonpotable purposes such as irrigation, industrial processes, or even potable reuse in some cases, promoting a circular water economy (Nwokediegwu, Ugwuanyi, Dada, Majemite, & Obaigbena, 2024).

They often require less capital investment compared to building and maintaining extensive centralized pipelines and treatment facilities. Smaller, localized systems can be more cost-effective to operate and maintain, especially in low-density areas. The smaller scale of decentralized systems allows for the adoption of innovative technologies and approaches, such as solar-powered desalination, advanced filtration, and on-site treatment units. Decentralized systems can engage local communities in water management, fostering a sense of ownership and responsibility for maintaining water quality. They can operate under local or regional regulations, which may be more flexible and easier to navigate than national standards. Empowering local authorities or community organizations to manage water resources can lead to more responsive and culturally appropriate solutions. Systems for treating wastewater on-site include septic tanks, biofilters, and membrane bioreactors (MBRs) (Okoli & Kabaso, 2024).

12.5 Emerging contaminant removal technique

Emerging contaminants, including pharmaceuticals, personal care products, endocrine-disrupting chemicals, microplastics, and various industrial chemicals, pose significant challenges to water quality management. These substances often evade traditional water treatment methods due to their complex chemical structures and low concentrations. Advanced techniques are being developed and refined to effectively remove these contaminants from water. AOPs generate highly reactive species like hydroxyl radicals that can oxidize and break down complex organic contaminants into simpler, less harmful compounds. Common AOPs include ozone/hydrogen peroxide (O_3/H_2O_2), ultraviolet light with hydrogen peroxide (UV/H_2O_2), and photocatalysis using titanium dioxide (TiO_2). Pharmaceuticals, personal care products, and other organic micropollutants can be effectively removed (Salam, 2024a).

NF and RO use semipermeable membranes to separate contaminants from water based on size and charge. They are highly effective at removing a wide range of emerging contaminants, including pharmaceuticals, hormones, and nanoparticles. MBRs combine biological treatment with membrane filtration to enhance the removal of organic contaminants and pathogens. Activated carbon adsorbs a wide range of organic contaminants due to its high surface area. Granular activated carbon and powdered activated carbon are commonly used in water treatment. Advanced materials such as carbon nanotubes, graphene oxide, and biochar are being developed to improve adsorption efficiency and specificity (Salam, 2024b).

Constructed wetlands use natural processes involving plants, soil, and associated microorganisms to remove contaminants. They can be effective in removing pharmaceuticals, nutrients, and certain organic compounds. Engineered microorganisms or consortia of microorganisms are used to degrade specific contaminants. Genetic engineering and synthetic biology are exploring new ways to enhance biodegradation capabilities. Electrochemical oxidation uses electrodes to generate oxidizing agents in situ, which can degrade organic contaminants. Electrocoagulation involves the generation of coagulant species through the dissolution of sacrificial anodes. These coagulants can remove contaminants by aggregation and sedimentation (Simpa, Solomon, Adenekan, & Obasi, 2024).

Photocatalysis involves the use of light-activated catalysts, such as TiO_2, to produce reactive oxygen species that can degrade contaminants. It is effective for breaking down a wide range of organic pollutants, including pharmaceuticals, dyes, and pesticides. Engineered magnetic nanoparticles can be functionalized with specific surface coatings to target and adsorb specific contaminants. Once adsorbed, these particles can be magnetically separated from the water. It has high specificity, is easily separable, and has the potential for regeneration and reuse. Biochar is produced from biomass pyrolysis and has a high surface area and porous structure, making it effective for adsorbing organic contaminants, heavy metals, and nutrients. Materials like zeolites, clays, and ion exchange resins can be used to target specific contaminants.

Hybrid systems that combine two or more treatment technologies (e.g., AOPs with membrane filtration or biological treatment with adsorption) can provide synergistic effects and enhance overall removal efficiency. Molecularly imprinted polymers are synthetic polymers with highly specific recognition sites tailored to target contaminants. They can selectively adsorb and remove emerging contaminants from water. The cost-effectiveness and scalability of these technologies vary, with some being more suitable for large-scale municipal treatment plants and others for decentralized systems or point-of-use applications. It is essential to consider the potential by-products or secondary pollution from treatment processes, as some advanced methods can produce toxic intermediates. Emerging contaminants are often not fully regulated, necessitating ongoing monitoring and research to understand their occurrence, behavior, and effects in water systems.

12.6 Resource recovery and circular economy

Resource recovery and the circular economy are transformative approaches to water quality management, focusing on the sustainable use of resources, minimizing waste, and recovering valuable materials from waste streams. Extracting valuable nutrients like nitrogen and phosphorus from wastewater can prevent nutrient pollution in water bodies and provide resources for agricultural fertilizers. Technologies such as struvite precipitation and ammonia stripping are used for nutrient recovery. Wastewater treatment plants can generate energy through anaerobic digestion of organic matter, producing biogas that can be used for heating, electricity generation, or as a renewable fuel source. Products and systems that facilitate the reuse of water and materials have been developed. For example, greywater systems can be designed to reuse water from sinks and showers for toilet flushing or landscape irrigation. We are implementing systems that recycle water and materials within a process. This can reduce water pollution from industrial discharges and promote the use of recyclable materials (Thakur & Devi, 2024).

To recover valuable materials and clean water from waste streams, technologies such as membrane filtration, RO, and nanotechnology are used. Using algae in wastewater treatment removes nutrients and produces biomass, which can be harvested for biofuels, animal feed, or other products. Governments can promote resource recovery and circular economy practices through incentives such as tax credits, subsidies, or grants for companies that invest in sustainable technologies (Fig. 12.1). Regulations can also mandate the recovery of specific resources or the use of recycled materials. To ensure safety, quality, and consumer confidence, standards and certifications for recycled water, recovered materials, and sustainable products are being developed.

Educating the public about the benefits of recycled water and items manufactured from recovered materials can boost acceptability and demand, resulting in market growth. Circular economy ideas are promoted and executed by a wide range of stakeholders, including corporations, governments, and communities. The development and implementation of resource recovery technologies and systems are supported through public institutions and private companies. To advance new technologies and processes for resource recovery and the circular economy, R&D investments were made to ensure continuous improvement and adaptation to changing conditions. To maximize the impact on water quality management, circular economy practices are being used at different scales, ranging from local community projects to global supply chains.

Figure 12.1
Sustainable water management system with the application of IoT. *IoT,* Internet of Things.

12.7 Discussion

As climate change alters weather patterns, intensifies extreme events, and shifts ecological conditions, managing water quality becomes increasingly complex. This involves analyzing the potential impacts on water quality, such as increased pollutant loads from runoff or saltwater intrusion in coastal areas. Developing maps that highlight areas at risk of climate-related water quality issues helps prioritize regions for intervention and resource allocation.

To handle extreme weather events and variations in water quality, such as increased turbidity or temperature, water treatment plants' capacity and resilience should be improved. These approaches can help mitigate the impacts of heavy rainfall and flooding. Designing infrastructure and policies to manage water availability during floods and droughts ensures a stable water supply while also preserving water quality standards. Flexible standards can be adjusted in response to changing conditions and emerging threats. This approach allows for timely updates to regulations and guidelines based on new scientific knowledge. Real-time water quality monitoring systems are being implemented (Fig. 12.2) to detect changes quickly and respond appropriately. This includes using advanced sensors, remote sensing, and IoT technologies to monitor parameters like temperature, pH, dissolved oxygen, and contaminant levels (Wai, Chia, Koo, Huang, & Chong, 2022).

Collaboration among different sectors (e.g., agriculture, industry, urban planning) has been promoted to manage water resources holistically and minimize conflicts over water use,

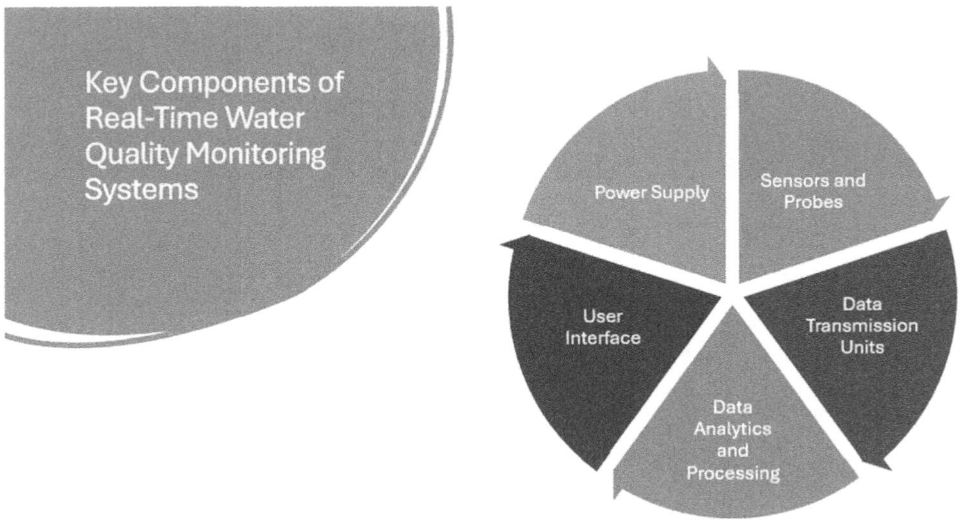

Figure 12.2
Key research and methods in water quality monitoring and modeling.

especially during periods of scarcity. Entire watersheds have been managed to address water quality issues at their source, such as controlling agricultural runoff, protecting wetlands, and managing land use changes. Natural habitats are being restored and protected, as well as ecosystem services like filtration and flood control. Natural processes are used to preserve aquatic and riparian biodiversity, hence enhancing ecosystem resilience and improving water quality.

12.8 Policy and governance innovations

Policy and governance innovations are crucial for effective water quality management, ensuring that water resources are protected, utilized sustainably, and equitably shared. As challenges like pollution, climate change, and population growth intensify, innovative approaches in policy and governance are essential. Policies have been created that can be adjusted in response to new data, technologies, and environmental conditions. This approach allows for timely responses to unexpected challenges, such as climate change impacts. Utilizing scenario-based planning to prepare for various future conditions and uncertainties, ensuring resilience and adaptability.

Decentralized governance and local empowerment are achieved by empowering local communities to manage their water resources while taking into account each area's unique social, cultural, and environmental context. To improve responsiveness and accountability, some governance responsibilities will be shifted to regional or local authorities. Wastewater treatment plants and water purification systems leverage private sector expertise and funding for water infrastructure projects. Green bonds, impact investing, and water tariffs are being investigated as mechanisms for sustainably funding water management projects. Robust legal frameworks that encompass water quality standards, pollution control measures, and water rights are being developed. Regulatory enforcement has been enhanced to ensure compliance with water quality standards and effective penalties for noncompliance (Yan et al., 2024).

Promoting open access to water quality data and allowing greater transparency can help with public awareness and a collaborative problem-solving approach. Utilizing digital platforms and tools for efficient data collection, analysis, and dissemination enables evidence-based policymaking. Climate change considerations should be incorporated into water management policies such as flood management, drought preparedness, and infrastructure resilience. To manage water runoff and improve water quality, promote natural solutions like wetlands restoration, green roofs, and permeable pavements. Agreements and institutions for managing transboundary water resources have been developed to ensure fair allocation and joint management. Engaging in international collaborations encouraged the exchange of knowledge, technology, and resources for water quality improvement. Education campaigns emphasized water conservation, pollution prevention, and the importance of clean water (Zekrifa et al., 2023).

Encouraging public participation in water quality monitoring and data collection fosters a sense of responsibility and stewardship. To promote universal access to safe and affordable drinking water, national and local water policies should be aligned with the United Nations Sustainable Development Goals, particularly SDG 6 (Clean Water and Sanitation). Water quality trading programs have been created, allowing polluters to buy and sell pollution credits, incentivizing reductions in pollutant discharges, and promoting cost-effective solutions. Guidelines and regulations have been created to manage emerging contaminants such as pharmaceuticals, microplastics, and personal care products, which are not yet fully regulated but pose potential risks.

12.9 Conclusion

Small-scale treatment plants serving a group of homes, a small community, or a specific facility like a school or hospital require proper operation and maintenance to prevent failures and ensure consistent water quality. This may require training and capacity building at the local level. Ensuring decentralized systems meeting water quality standards can be challenging, especially with diverse and dispersed installations of water monitoring systems. Regular monitoring and regulation are crucial, and these trends and technologies reflect a multifaceted approach to solving water quality management concerns such as pollution, scarcity, and climate change. To establish long-term solutions for water resource management, sophisticated technology, and data analytics must be seamlessly integrated.

There should be encouragement toward the use of treated wastewater for nonpotable applications such as irrigation and industrial activities to conserve freshwater supplies and minimize pressure on water quality. Modern water treatment technologies, such as desalination and membrane filtration, have been explored to supply safe drinking water in locations where freshwater is scarce. Cooperation and agreements among countries or regions that share water resources have been strengthened to jointly manage water quality and adapt to climate impacts. Securing funding for climate adaptation projects, including grants, loans, and private investments, supports infrastructure upgrades and resilience-building measures. Providing incentives for industries, farmers, and communities encourages them to use water-saving technologies, reduce pollution, and implement sustainable land-use practices.

The future of advanced technologies and sustainable practices in water quality management appears bright as innovations continue to improve, tackling the mounting concerns of water shortages, pollution, and the need for sustainable resource management. The future of water quality management will be determined by improved technologies, sustainable practices, and an increased emphasis on resilience and adaptation in the face of environmental threats. By embracing these technologies and incorporating them into water management methods, we can safeguard and sustain water supplies for future generations.

References

Abdelfattah, I., & El-Shamy, A. M. (2024). Review on the escalating imperative of zero liquid discharge (ZLD) technology for sustainable water management and environmental resilience. *Journal of Environmental Management*, *351*, 119614.

Adekugbe, A. P., & Ibeh, C. V. (2024). Harnessing data insights for crisis management in us public health: lessons learned and future directions. *International Medical Science Research Journal*, *4*(4), 391–405.

Chakraborty, U., Kaur, G., Rubahn, H. G., Kaushik, A., Chaudhary, G. R., & Mishra, Y. K. (2023). Advanced metal oxides nanostructures to recognize and eradicate water pollutants. *Progress in Materials Science*, *139*, 101169.

Geetha, M., Bonthula, S., Al-Maadeed, S., Al-Lohedan, H., Rajabathar, J. R., Arokiyaraj, S., & Sadasivuni, K. K. (2023). Research trends in smart cost-effective water quality monitoring and modeling: special focus on artificial intelligence. *Water*, *15*(18), 3293.

Giri, S. (2021). Water quality prospective in Twenty First Century: Status of water quality in major river basins, contemporary strategies and impediments: A review. *Environmental Pollution*, *271*116332.

Hamel, P., & Tan, L. (2022). Blue–green infrastructure for flood and water quality management in Southeast Asia: evidence and knowledge gaps. *Environmental Management*, *69*(4), 699–718.

Jan, F., Min-Allah, N., & Düştegör, D. (2021). Iot based smart water quality monitoring: Recent techniques, trends and challenges for domestic applications. *Water*, *13*(13), 1729.

Jayaraman, P., Nagarajan, K. K., Partheeban, P., & Krishnamurthy, V. (2024). Critical review on water quality analysis using IoT and machine learning models. *International Journal of Information Management Data Insights*, *4*(1), 100210.

Kamyab, H., Khademi, T., Chelliapan, S., SaberiKamarposhti, M., Rezania, S., Yusuf, M., ... Ahn, Y. (2023). The latest innovative avenues for the utilization of artificial Intelligence and big data analytics in water resource management. *Results in Engineering*, 101566.

Moore, C. C., Corona, J., Griffiths, C., Heberling, M. T., Hewitt, J. A., Keiser, D. A., ... Wheeler, W. (2023). Measuring the social benefits of water quality improvements to support regulatory objectives: Progress and future directions. *Proceedings of the National Academy of Sciences*, *120*(18)e2120247120.

Nwokediegwu, Z. Q. S., Daraojimba, O. H., Oliha, J. S., Obaigbena, A., Dada, M. A., & Majemite, M. T. (2024). Review of emerging contaminants in water: USA and African perspectives. *International Journal of Science and Research Archive*, *11*(1), 350–360.

Nwokediegwu, Z. Q. S., Ugwuanyi, E. D., Dada, M. A., Majemite, M. T., & Obaigbena, A. (2024). Urban water management: a review of sustainable practices in the USA. *Engineering Science & Technology Journal*, *5* (2), 517–530.

Okoli, N. J., & Kabaso, B. (2024). Building a smart water city: iot smart water technologies, applications, and future directions. *Water*, *16*(4), 557.

Salam, A. (2024a). *Internet of things for water sustainability. Internet of Things for sustainable community development: Wireless communications, sensing, and systems* (pp. 113–145). Cham: Springer International Publishing.

Salam, A. (2024b). *Internet of things in water management and treatment. Internet of Things for sustainable community development: Wireless communications, sensing, and systems* (pp. 273–298). Cham: Springer International Publishing.

Simpa, P., Solomon, N. O., Adenekan, O. A., & Obasi, S. C. (2024). Environmental stewardship in the oil and gas sector: Current practices and future directions. *International Journal of Applied Research in Social Sciences*, *6*(5), 903–926.

Thakur, A., & Devi, P. (2024). A comprehensive review on water quality monitoring devices: materials advances, current status, and future perspective. *Critical Reviews in Analytical Chemistry*, *54*(2), 193–218.

Wai, K. P., Chia, M. Y., Koo, C. H., Huang, Y. F., & Chong, W. C. (2022). Applications of deep learning in water quality management: A state-of-the-art review. *Journal of Hydrology*, *613*, 128332.

Yan, X., Xia, Y., Ti, C., Shan, J., Wu, Y., & Yan, X. (2024). Thirty years of experience in water pollution control in Taihu Lake: A review. *Science of The Total Environment*, 169821.

Zekrifa, D. M. S., Kulkarni, M., Bhagyalakshmi, A., Devireddy, N., Gupta, S., & Boopathi, S. (2023). Integrating Machine Learning and AI for Improved Hydrological Modeling and Water Resource Management. *Artificial Intelligence applications in water treatment and water resource management* (pp. 46–70). IGI Global.

CHAPTER 13

Future trends and emerging technologies in water quality management

Vishal Kumar Sandhwar[1], Shivendu Saxena[1], Diksha Saxena[1], Alok Tiwari[1] and Swapnil M. Parikh[2]

[1]Department of Chemical Engineering, Parul Institute of Technology, Parul University, Vadodara, Gujarat, India, [2]Department of Computer Science & Engineering, Parul Institute of Technology, Parul University, Vadodara, Gujarat, India

13.1 Introduction

Water quality management is a cornerstone of environmental protection, public health, and sustainable development. As global populations swell and industrial activities proliferate, the demand for clean, safe water becomes more pressing. At the same time, the challenges posed by pollution, climate change, and resource scarcity accentuate the urgency for innovative solutions in water management. The field of water quality management is experiencing a profound transformation, driven by the rapid development of emerging technologies and evolving trends that address the escalating challenges of water scarcity, pollution, and climate change. As the demand for clean and safe water intensifies globally, innovative solutions are essential to safeguard public health, protect ecosystems, and ensure sustainable development. However, ensuring water quality is becoming increasingly complex due to various factors such as population growth, urbanization, industrial activities, and the impacts of climate change (Akhtar, Syakir Ishak, Bhawani, & Umar, 2021). Pollution from agricultural runoff, industrial discharges, and untreated wastewater poses significant risks to water quality, while climate change exacerbates these challenges by altering precipitation patterns, increasing the frequency of extreme weather events, and raising temperatures. These dynamics necessitate a paradigm shift in water quality management, leveraging cutting-edge technologies and innovative approaches to tackle these multifaceted issues effectively. Water quality management encompasses a wide array of activities to ensure water availability that meets health and safety standards for human consumption, ecosystem support, and industrial use. Traditional methods, while foundational, are increasingly supplemented and sometimes replaced by advanced technologies and novel approaches. These innovations are driven by the necessity to enhance monitoring accuracy, treatment efficiency, and resource sustainability. As we progress through

Computational Automation for Water Security.
DOI: https://doi.org/10.1016/B978-0-443-33321-7.00012-3

the 21st century, integrating smart technologies, data analytics, and biotechnological advancements revolutionizes water quality management, offering promising solutions to long-standing issues. To appreciate the future trends and emerging technologies, it is essential to first understand the current state of water quality management. Presently, water quality is managed through a combination of regulatory frameworks, conventional treatment methods, and monitoring systems (Geissen et al., 2015; Jan, Min-Allah, & Düştegör, 2021). Regulatory bodies establish standards and guidelines to ensure water safety, while water treatment plants employ physical, chemical, and biological processes to purify water. Despite significant progress, several limitations and challenges persist. Many water bodies worldwide are contaminated with pollutants such as heavy metals, pesticides, pharmaceuticals, and pathogens (Morin-Crini et al., 2022). Traditional water treatment methods, although effective to a certain extent, often struggle to address these complex and emerging contaminants. Moreover, the infrastructure for water quality management, particularly in developing regions, is frequently inadequate, leading to disparities in water safety and accessibility. Emerging technologies are set to address many of these challenges, offering innovative solutions that enhance the efficacy and efficiency of water quality management. These technologies span various domains, including real-time monitoring systems, advanced filtration and purification techniques, artificial intelligence (AI) and machine learning applications, and the use of nanotechnology in water treatment. As we navigate the complexities of water quality management in the 21st century, embracing innovation and forward-thinking strategies will be imperative. This chapter serves as a guide to understanding and harnessing the emerging technologies that promise to secure clean and safe water for future generations. It also delves into the cutting-edge advancements and forward-looking trends that are transforming this crucial field. Advanced monitoring and sensing technologies, Big Data analytics, AI, nanotechnology, innovative treatment methods, decentralized systems, and renewable energy-powered solutions are at the forefront of this transformation (Kumar, Gupta, Wang, Kumari, & Korlam, 2023). These advancements hold significant potential to address the escalating challenges of water scarcity, pollution, and climate change, ensuring safe and sustainable water resources for future generations. By synthesizing these trends and technologies, this chapter provides a comprehensive overview of the future of water quality management, offering valuable insights for researchers, policymakers, and practitioners dedicated to achieving this vital goal.

13.2 Innovations and sustainable practices in water treatment technologies

Innovations and sustainable practices in water treatment technologies are crucial for addressing global water challenges, ensuring a clean and safe water supply, and promoting environmental conservation. Key areas of advancement include advanced oxidation processes (AOPs), nanotechnology-based solutions, membrane filtration advancements, electrochemical water treatment methods, nature-based solutions (NbSs), wastewater recycling and reuse, and low-energy and renewable energy technologies, as shown in Table 13.1.

Table 13.1: Water and wastewater treatment technologies and their key innovations.

S. No.	Water and wastewater treatment technologies	Key innovations	References
1	Advanced oxidation processes	i. Visible-light-responsive photocatalysts ii. Use of iron-based MOFs in the Fenton process iii. Ozonation combined with other AOPs iv. Ultrasound-enhanced AOPs	Gomes, Costa, Quinta-Ferreira, and Martins (2022)
2	Nanotechnology-based solutions	i. Use of nanomaterials like CNTs, graphene, and nano-zerovalent iron (nZVI) as adsorbent ii. Use of nanocomposite membranes iii. Low-energy processes such as CDI iv. Development of multifunctional nanomaterials	Park and Srivastava (2015); Mauter et al. (2018)
3	Membrane filtration advancements	i. Development of advanced membrane materials ii. Development of ceramic and nanocomposite membranes iii. Incorporation of FO and MD iv. Use of green and biodegradable materials for membrane fabrication	Yin and Deng (2015); Shah and Malik (2020); Zhou, Yang, and Chen (2019); Kim and Choi (2018)
4	Electrochemical water treatment methods	i. Development of advanced electrode materials ii. Integration of electrochemical methods with other treatment processes iii. Solar-powered electrochemical systems iv. Use of green and recyclable materials in the construction and maintenance of electrochemical systems	Shih and Wang (2018)
5	Nature-based solutions	i. Constructed wetlands ii. Use of genetically modified plants in phytoremediation iii. Floating treatment wetlands (FTWs) iv. Integration of nature-based solutions with conventional water treatment technologies	Masi, Rizzo, and Regelsberger (2018)
6	Wastewater recycling and reuse-based solutions	i. Development of advanced membrane technologies such as RO and FO ii. Implementation of decentralized wastewater treatment systems iii. Recovery of valuable resources from wastewater	Gude (2017); Trussell, Salveson, and Snyder (2016)

(Continued)

Table 13.1: (Continued)

S. No.	Water and wastewater treatment technologies	Key innovations	References
7	Low-energy and renewable energy technologies	i. Solar-powered desalination systems ii. Use of MFCs and MECs iii. Development of high-rate anaerobic reactors and AnMBRs iv. Integration of renewable energy sources, such as wind or hydropower	Logan and Rabaey (2012); Smith, Stadler, and Love (2012)

MOFs, Metal-organic frameworks; AOPs, advanced oxidation processes; CNTs, carbon nanotubes; CDI, capacitive deionization; FO, forward osmosis; MD, membrane distillation; RO, reverse osmosis; MFCs, microbial fuel cells; MECs, microbial electrolysis cells; AnMBRs, anaerobic membrane bioreactors.

13.2.1 Advanced oxidation processes

As the world grapples with the challenge of water scarcity, many water suppliers are turning to once-avoided water sources, such as contaminated surface waters, groundwater, and wastewater, to meet the growing demand. These impaired water supplies can now be considered a valuable part of a sustainable water supply, but they often require advanced treatment with techniques like AOPs to address trace contaminants. AOPs represent a group of chemical treatment procedures designed to remove organic and inorganic contaminants from water and wastewater by oxidation through reactions with hydroxyl radicals ($^{\bullet}OH$). These radicals are highly reactive and can degrade a wide range of pollutants. The AOP is a promising technology that can play a key role in helping industries meet strict treatment standards, either by complementing a biological process or by itself (Bhuta, 2014). This technology offers several advantages, including a small footprint, nonselective destruction of pollutants, and high flexibility, making it a critical component of the future sustainable success of the chemical industry. Here are some key innovations in AOPs such as photocatalysis, Fenton process, ozone-based processes, and electrochemical treatment that have significantly enhanced their efficiency and applicability in wastewater treatment. One of the significant advancements in AOPs for water treatment is the development of visible-light-responsive photocatalysts. Traditional photocatalysts like titanium dioxide (TiO_2) are limited by their activation requirement under ultraviolet (UV) light. To overcome this, researchers have modified TiO_2 and developed new materials such as graphitic carbon nitride (gC_3N_4) that can be activated by visible light. For instance, nitrogen-doped TiO_2 ($NTiO_2$) shows improved photocatalytic activity under visible light, enhancing the degradation of pollutants under natural sunlight conditions, thus making the process more energy-efficient and practical for large-scale applications. Another critical innovation is the advancement of heterogeneous Fenton and photo-Fenton processes. Traditional Fenton reactions, involving hydrogen peroxide (H_2O_2) and ferrous iron (Fe^{2+}), are effective but limited to acidic conditions and involve homogeneous catalysts that are difficult to recover. Recent developments in heterogeneous Fenton catalysts, such as iron-loaded zeolites and iron oxide nanoparticles, have expanded the effective pH range and simplified the recovery and reuse of catalysts. Iron-based metal-organic frameworks, for instance, enhance Fenton-like reactions under visible light, providing a more sustainable and efficient approach for water treatment by reducing the dependence on harsh chemical conditions and facilitating catalyst recovery. Ozonation combined with other AOPs represents another key innovation (Gomes et al., 2022). Ozonation (O_3) alone is effective in generating reactive oxygen species (ROS), but its efficiency can be significantly enhanced when combined with UV irradiation or hydrogen peroxide (O_3/UV, O_3/H_2O_2). These combinations produce synergistic effects, increasing ROS generation and improving the degradation of complex contaminants like pharmaceuticals and personal care products. For example, O_3/UV systems have shown superior performance in breaking down persistent organic pollutants, demonstrating the benefits of integrating multiple AOPs to achieve higher purification efficiencies. Ultrasound-enhanced AOPs also offer significant advancements. Ultrasound irradiation (sonolysis) produces ROS through acoustic cavitation, where the collapse of microbubbles generates localized high temperatures and pressures. This process can be combined with other AOPs, such as photocatalysis (sono-photocatalysis) or ozonation (sono-ozonation), to enhance overall efficiency. For instance, sono-photocatalytic systems using TiO_2 nanoparticles have shown substantial improvements in the degradation rates of organic dyes and other pollutants in water, highlighting the effectiveness of combining ultrasound with other oxidative processes.

13.3 Nanotechnology-based solutions

Nanotechnology has emerged as a transformative approach in water treatment, offering innovative solutions that enhance efficiency, reduce environmental impact, and promote sustainability. The application of nanomaterials in water purification processes leverages their unique properties, such as high surface area, reactivity, and tuneable physical and chemical characteristics, to address various water contamination issues effectively. One significant innovation is the use of nanomaterials like carbon nanotubes, graphene, and nano-zerovalent iron for the removal of contaminants. Carbon nanotubes and graphene have exceptional adsorption capacities due to their high surface area and unique surface properties, making them effective for capturing heavy metals, organic pollutants, and pathogens. These materials can be tailored to target specific contaminants, enhancing their applicability and efficiency in various water treatment processes. Another groundbreaking development is the integration of nanotechnology with membrane filtration systems. Nanocomposite membranes incorporate nanoparticles such as silver, TiO_2, and zinc oxide to improve filtration efficiency and antimicrobial properties. These membranes offer superior resistance to fouling, a common issue in conventional filtration systems, thereby extending their lifespan and reducing maintenance costs. For instance, TiO_2 nanoparticles impart photocatalytic properties to the membranes, enabling them to degrade organic pollutants under light exposure, which not only improves water quality but also reduces the frequency of membrane cleaning (Park and Srivastava, 2015).

Sustainable practices in nanotechnology-based water treatment also focus on energy efficiency and resource recovery (Mauter et al., 2018). For example, the use of magnetic nanoparticles allows for the easy separation and recovery of the nanoparticles after water treatment, minimizing waste and reducing environmental impact. Additionally, advances in nanotechnology have led to the development of low-energy processes such as capacitive deionization, where carbon-based nanomaterials are used to selectively remove salts and other contaminants from water through electrostatic adsorption. This method consumes significantly less energy compared to traditional desalination techniques like reverse osmosis (RO). Furthermore, the development of multifunctional nanomaterials has enhanced the sustainability of water treatment processes. These materials can simultaneously remove various types of contaminants, such as heavy metals, organic pollutants, and microorganisms, thus streamlining the treatment process and reducing the need for multiple stages of treatment. Innovations such as hybrid nanomaterials that combine catalytic, adsorptive, and antimicrobial properties exemplify the potential of nanotechnology to create comprehensive water treatment solutions. These advancements contribute to more efficient use of resources and support the overarching goal of achieving sustainable water management practices.

13.4 Membrane filtration advancements

Membrane filtration has emerged as a pivotal technology in water treatment due to its efficiency in removing contaminants and its adaptability to various scales and types of water sources. One of the major innovations in membrane filtration is the development of advanced membrane materials. Traditional polymeric membranes, while effective, often suffer from limitations such as fouling and limited chemical stability. To address these issues, researchers have developed new materials, including ceramic and nanocomposite membranes. Ceramic membranes, for instance, exhibit superior chemical and thermal stability, making them suitable for treating challenging wastewater streams. Nanocomposite membranes, which incorporate nanoparticles like TiO_2 or silver, enhance the antimicrobial properties and reduce fouling, thus extending the membrane lifespan, and efficiency (Yin and Deng, 2015). Another significant advancement is the incorporation of forward osmosis (FO) and membrane distillation (MD) into water treatment processes. FO uses a natural osmotic pressure gradient to draw water through a semipermeable membrane, requiring less energy compared to RO. This makes FO particularly attractive for applications in resource-limited settings and for treating highly saline or contaminated water sources. MD, on the other hand, leverages thermal gradients to drive water vapor through a hydrophobic membrane, effectively removing nonvolatile contaminants. Both of these processes can be integrated with renewable energy sources, such as solar thermal energy, further enhancing their sustainability by reducing reliance on conventional energy sources. Sustainable practices in membrane filtration also include the use of green and biodegradable materials for membrane fabrication (Shah and Malik, 2020). Recent research has focused on using natural polymers like chitosan and cellulose, which are biodegradable and derived from renewable sources, for membrane production. These materials not only reduce the environmental footprint of membrane production but also offer beneficial properties such as inherent antimicrobial activity and biocompatibility. Furthermore, the recycling and regeneration of membranes are gaining traction as sustainable practices. Techniques such as chemical cleaning, ultrasonic cleaning, and the use of less aggressive cleaning agents can regenerate membranes, reducing waste and operational costs associated with membrane replacement. The integration of smart technologies and real-time monitoring systems represents another innovation in sustainable membrane filtration (Zhou et al., 2019). The application of sensors and Internet of Things (IoT) technologies allows for continuous monitoring of membrane performance, fouling levels, and water quality (Kim and Choi, 2018). This data-driven approach enables predictive maintenance, optimizing membrane lifespan and reducing downtime. Additionally, automated control systems can adjust operational parameters in real time to maintain optimal filtration efficiency and energy use, further contributing to the sustainability of membrane filtration processes. These advancements not only enhance the efficiency and effectiveness of water treatment but also align with global sustainability goals by minimizing resource use and environmental impact.

13.5 Electrochemical water treatment methods

Electrochemical water treatment methods have gained substantial attention due to their high efficiency and versatility in removing a wide range of contaminants. One of the key innovations in this field is the development of advanced electrode materials. Traditional electrodes, such as those made from graphite or platinum, are being enhanced with materials like boron-doped diamond (BDD) and mixed metal oxides. These materials exhibit superior electrocatalytic activity, chemical stability, and durability. BDD electrodes, for instance, generate a higher density of ROS such as hydroxyl radicals, which significantly improve the degradation rates of organic pollutants. This enhancement not only increases the efficiency of contaminant removal but also extends the lifespan of the electrodes, reducing the overall operational costs and environmental impact. Another significant advancement is the integration of electrochemical methods with other treatment processes to create hybrid systems. Combining electrochemical oxidation with techniques such as membrane filtration, advanced oxidation processes, or biological treatments can achieve more comprehensive water purification. For example, coupling electrochemical oxidation with membrane bioreactors (MBRs) enhances the breakdown of complex organic pollutants that are otherwise resistant to biodegradation. This hybrid approach optimizes the strengths of each method, leading to higher purification efficiencies, reduced energy consumption, and minimized production of harmful by-products. Such integrative strategies are pivotal for developing sustainable and effective water treatment solutions, especially in regions with limited resources and complex wastewater streams. Sustainable practices in electrochemical water treatment also focus on energy efficiency and the use of renewable energy sources (Shih and Wang, 2018). One promising area is the development of energy-efficient reactors and the optimization of operational parameters to minimize energy consumption. Research has shown that optimizing the electrode configuration, flow dynamics, and applied current can significantly reduce the energy requirements of electrochemical processes. Furthermore, integrating renewable energy sources, such as solar or wind power, with electrochemical treatment systems can provide a sustainable energy supply. Solar powered electrochemical systems, for instance, utilize photovoltaic panels to convert solar energy into electrical energy, driving the electrochemical reactions. This approach not only reduces the carbon footprint of water treatment but also makes the process more viable in off-grid or remote locations. Moreover, the use of green and recyclable materials in the construction and maintenance of electrochemical systems is gaining traction. Efforts to replace conventional, often toxic, electrode materials with environmentally benign alternatives are underway. For instance, carbon-based materials derived from biomass or waste products are being explored as sustainable electrode materials. These materials are not only renewable but also exhibit good electrical conductivity and catalytic properties. Additionally, recycling and reusing treated water within industrial processes further enhance the sustainability of electrochemical water treatment. Implementing closed-loop water systems where treated water is continuously reused minimizes freshwater consumption and reduces wastewater discharge, aligning with the principles of circular economy and sustainable resource management.

13.6 Nature-based solutions (constructed wetlands, biofilters)

The ever-increasing need for clean water is prompting a shift toward sustainable treatment methods. NbSs are gaining traction, offering eco-friendly alternatives to conventional water treatment plants. These innovative approaches harness natural processes to purify water while providing additional environmental benefits. Constructed wetlands are a prime example of this approach, replicating natural wetlands' ability to filter and cleanse water (Masi et al., 2018). These systems use plants, soil, and associated microbial communities to remove pollutants through physical, chemical, and biological processes. Innovations in the design of constructed wetlands, such as vertical flow and hybrid systems, have significantly improved their efficiency in removing contaminants like nutrients, heavy metals, and pathogens. Vertical flow wetlands, for instance, enhance oxygen transfer and microbial activity, leading to more effective treatment outcomes. Phytoremediation, another NbS, involves using plants to absorb, accumulate, and detoxify pollutants from water and soil. Recent advancements in phytoremediation include the use of genetically modified plants designed to target specific contaminants more effectively. For example, certain transgenic plants have been engineered to express higher levels of enzymes that can breakdown organic pollutants or accumulate heavy metals from contaminated water. This innovation not only improves the efficiency of phytoremediation but also expands its applicability to a broader range of pollutants and environmental conditions. Floating treatment wetlands (FTWs) represent a novel application of wetland technology. These systems consist of floating rafts planted with wetland vegetation, which create a habitat for microbial communities on the raft's roots and underside. FTWs are particularly effective in treating stormwater runoff and other types of surface water pollution. They can be deployed in urban water bodies to mitigate the impacts of nonpoint source pollution by removing nutrients, sediments, and heavy metals. The floating nature of these wetlands allows them to adapt to water level fluctuations, making them versatile and resilient options for water treatment in dynamic environments. The integration of these NbSs with conventional water treatment technologies is also gaining traction as a sustainable practice. By combining the strengths of natural systems with engineered processes, hybrid systems can achieve higher treatment efficiencies and resilience. For instance, coupling constructed wetlands with anaerobic digesters or biofilters can enhance the breakdown of complex organic pollutants and reduce greenhouse gas emissions. These integrated systems not only improve water quality but also promote biodiversity, enhance ecosystem services, and provide additional benefits such as habitat creation and recreational spaces. The adoption of NbSs reflects a shift toward more sustainable and holistic approaches to water management, aligning with global sustainability goals and the principles of the circular economy.

13.7 Wastewater recycling and reuse

The global water crisis has spurred innovation in wastewater recycling and reuse, promoting a more sustainable approach to water management. One of the major advancements is in the development of advanced membrane technologies such as reverse osmosis and forward osmosis. These technologies are highly effective in removing contaminants and recovering high-quality water from wastewater. Recent improvements in membrane materials, such as the use of graphene and other nanomaterials, have enhanced permeability and fouling resistance, making the processes more efficient and cost-effective. Additionally, hybrid systems that combine membrane processes with other treatment methods, like biological treatment or advanced oxidation processes, are being implemented to achieve higher purification levels and resource recovery. Another significant innovation is the implementation of decentralized wastewater treatment systems. Unlike traditional centralized systems, decentralized systems treat wastewater closer to its source, which reduces the infrastructure costs and energy consumption associated with long-distance water transport. These systems often integrate various treatment technologies, such as constructed wetlands, biofilters, and MBRs, tailored to local conditions and water reuse needs. By treating and reusing water locally for irrigation, industrial processes, or even potable uses, decentralized systems can alleviate pressure on freshwater sources and provide a more resilient water supply in the face of growing urban populations and climate variability. Sustainable practices in wastewater recycling also emphasize the recovery of valuable resources from wastewater (Gude, 2017). Technologies like anaerobic digestion and microbial fuel cells (MFCs) not only treat wastewater but also generate biogas or electricity, turning waste into a resource. Anaerobic digestion processes breakdown organic matter in the absence of oxygen, producing methane that can be used as a renewable energy source. MFCs, on the other hand, harness the metabolic activities of microorganisms to generate electricity while treating wastewater. These technologies contribute to a circular economy by recovering energy and other by-products, such as nutrients that can be used as fertilizers, thus reducing the environmental impact of wastewater treatment and promoting resource efficiency.

Moreover, public acceptance and regulatory frameworks play crucial roles in the successful implementation of wastewater recycling and reuse practices. Public education and awareness campaigns are essential to overcoming the “yuck factor” associated with recycled water. Effective communication strategies and demonstrations of successful projects can build trust and acceptance among communities. On the regulatory side, establishing clear guidelines and standards for water quality and reuse applications ensures the safety and reliability of recycled water. Policymakers are increasingly recognizing the importance of supportive regulations that encourage innovation and investment in wastewater recycling infrastructure (Trussell et al., 2016). These comprehensive approaches help to integrate wastewater recycling into mainstream water management practices, enhancing sustainability and resilience.

13.8 Low-energy and renewable energy technologies in water treatment

Innovations in low-energy and renewable energy technologies for water treatment are crucial for developing sustainable solutions that reduce environmental impact and operational costs. One significant advancement is the integration of solar energy with water treatment processes. Solar-powered desalination systems, such as solar stills and photovoltaic-powered reverse osmosis (PV-RO), harness solar energy to purify water, reducing reliance on fossil fuels. Solar stills utilize thermal energy to evaporate and condense water, while PV-RO systems convert solar energy into electricity to power the desalination process. Recent innovations in solar materials and system designs have improved the efficiency and scalability of these technologies, making them more viable for remote and arid regions. Another promising development is the use of MFCs and microbial electrolysis cells (MECs) in wastewater treatment (Logan and Rabaey, 2012). These bioelectrochemical systems leverage the metabolic activities of microorganisms to generate electricity or hydrogen while treating wastewater. MFCs convert organic matter in wastewater into electrical energy through microbial oxidation, offering a dual benefit of energy production and water treatment. MECs, on the other hand, use an external voltage to enhance microbial activity, producing hydrogen gas as a valuable byproduct. Advances in electrode materials and reactor designs have significantly improved the performance and scalability of MFCs and MECs, contributing to more sustainable and energy-efficient wastewater treatment solutions. Anaerobic digestion is another low-energy technology gaining traction in wastewater treatment. This process involves the breakdown of organic matter by anaerobic bacteria in the absence of oxygen, producing biogas (a mixture of methane and carbon dioxide) that can be used as a renewable energy source. Innovations in anaerobic digestion, such as the development of high-rate anaerobic reactors and anaerobic membrane bioreactors (AnMBRs), have enhanced the efficiency and applicability of this technology (Smith et al., 2012). AnMBRs combine anaerobic digestion with membrane filtration, providing effective solid–liquid separation and producing high-quality effluent. These systems not only reduce energy consumption but also enable the recovery of valuable resources from wastewater. Sustainable practices in low-energy water treatment also include optimizing operational parameters and integrating energy recovery mechanisms. For instance, the use of energy efficient pumps and blowers, coupled with advanced process control systems, can significantly reduce the energy footprint of water treatment facilities. Additionally, recovering heat from treated water and using it to preheat incoming water can improve the overall energy efficiency of the treatment process. The integration of renewable energy sources, such as wind or hydropower, with existing water treatment infrastructure further enhances sustainability. These combined approaches not only lower operational costs but also reduce greenhouse gas emissions, contributing to more sustainable water management practices.

13.9 Emerging technologies in water quality monitoring

Water quality monitoring is a critical aspect of environmental management, public health, and sustainable development. With increasing concerns about pollution, climate change, and resource scarcity, the need for innovative and effective water quality monitoring technologies has never been more pressing. Emerging technologies are revolutionizing this field, providing more accurate, efficient, and comprehensive solutions for detecting and addressing water quality issues as represented in Fig. 13.1. This section explores some of the most promising emerging technologies in water quality monitoring, including real-time monitoring systems, remote sensing, biosensors, and advanced data analytics.

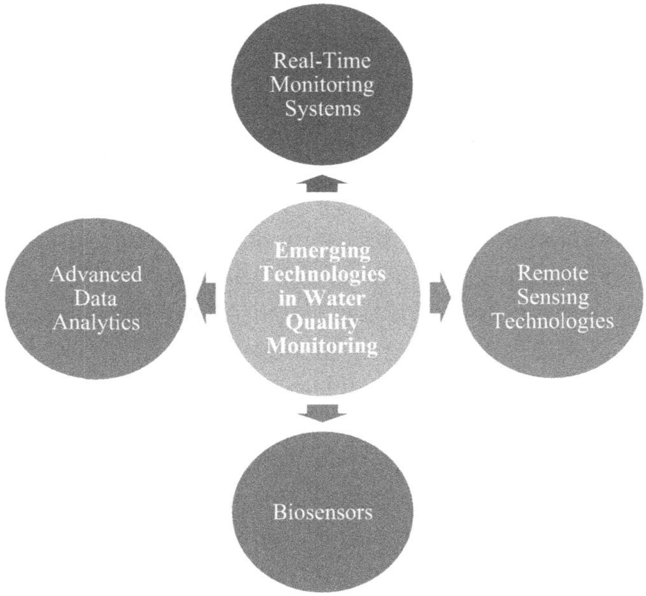

Figure 13.1
Emerging technologies in water quality monitoring.

13.10 Real-time monitoring systems

Real-time monitoring systems represent a significant advancement in water quality management. These systems employ a network of sensors and IoT devices to continuously measure various water quality parameters such as pH, turbidity, dissolved oxygen, temperature, and contaminant levels. The data collected are transmitted in real-time to centralized platforms where it can be analyzed and used to make informed decisions. Sensors can provide rapid data on water quality, allowing for prompt response in the event of pollution (Essamlali, Nhaila, & El Khaili, 2024).

Key components and benefits:

1. Sensors and IoT Devices:
   a. These devices are capable of detecting a wide range of water quality indicators with high precision.
   b. IoT-enabled sensors facilitate the continuous collection and transmission of data.
2. Data Analytics and Cloud Computing:
   a. Integration with cloud computing platforms allows for the storage and real-time analysis of large datasets.
   b. Advanced data analytics can identify trends, patterns, and anomalies, enabling proactive management.

3. Immediate Response:
   a. Real-time data allow for the immediate detection of pollution events, leading to quicker responses and mitigation measures.
   b. This proactive approach can prevent the spread of contaminants and reduce the impact on ecosystems and human health.

13.10.1 Remote sensing technologies

Remote sensing technologies, including satellite and aerial imagery, are transforming the way we monitor large bodies of water such as lakes, rivers, and oceans. These technologies enable the collection of data over vast areas, which is particularly useful for monitoring inaccessible or remote regions.

Key components and benefits:

1. Satellite Imagery:
   a. Satellites equipped with multispectral and hyperspectral sensors can capture detailed images of water bodies.
   b. These images can be analyzed to detect changes in water quality parameters such as algal blooms, sediment load, and temperature variations.
2. Aerial Drones:
   a. Drones equipped with high-resolution cameras and sensors can provide detailed, site-specific data.
   b. They offer flexibility in monitoring specific areas of interest and can be deployed rapidly in response to pollution events.
3. Large-Scale Monitoring:
   a. Remote sensing allows for the monitoring of large and remote areas that are difficult to access through traditional means.
   b. It provides a comprehensive overview of water quality trends and potential sources of pollution.

13.10.2 Biosensors

Biosensors are another promising technology in the field of water quality monitoring. These devices use biological molecules, such as enzymes or antibodies, to detect numerous water quality characteristics, including biological oxygen demand, *Escherichia coli* (*E. coli*), heavy metals, hazardous pollutants, and endocrine-disrupting chemicals (Woldu, 2022). Biosensors are highly sensitive and can provide rapid, on-site analysis of water samples.

Key features and innovations:

1. Specificity and Sensitivity:
   a. Biosensors are designed to target specific contaminants, providing highly accurate results.

   b. They can detect low concentrations of pollutants, making them useful for early warning systems.
2. Rapid Detection:
   a. Biosensors offer quick results, often within minutes, enabling immediate decision-making.
   b. They are portable and can be used in the field for on-site testing.
3. Applications:
   a. Detection of pathogens, such as bacteria and viruses, in drinking water.
   b. Monitoring of chemical pollutants, including pesticides, heavy metals, and pharmaceuticals.

13.10.3 Advanced data analytics

The integration of advanced data analytics, including machine learning and AI, is revolutionizing water quality monitoring. These technologies enable the processing and analysis of vast amounts of data, providing deeper insights and predictive capabilities.

Capabilities and Benefits:

1. Predictive Analytics:
   a. Machine learning algorithms can analyze historical and real-time data to predict future water quality trends.
   b. Predictive models can identify potential risks and inform proactive management strategies.
2. Anomaly Detection:
   a. AI systems can detect anomalies and unusual patterns in water quality data, indicating potential pollution events or system malfunctions.
   b. These systems can differentiate between natural variations and actual contamination, reducing false alarms.
3. Decision Support:
   a. Advanced analytics provide decision-makers with actionable insights, helping to optimize water treatment processes and resource allocation.
   b. Visualization tools and dashboards facilitate the interpretation of complex data, making it accessible to a broader audience.
4. Optical sensors are becoming increasingly essential in water quality monitoring due to their capacity to give real-time data, faster findings, and greater accessibility than older approaches. These sensors use light-based approaches to detect the presence and concentrations of various contaminants, including pollutants, nutrients, dissolved organic matter, algae, and more.

   Key features and innovations:

   - Measure various parameters at the same time: Optical sensors can measure turbidity, chlorophyll-a concentration, dissolved oxygen, temperature, pH, and other parameters in one measurement.

- Excellent accuracy and precision: Automated data collection and analysis reduce the possibility of errors caused by human bias or ineffective sampling strategies.
- Real-time data access: Optical sensors quickly detect changes in water conditions, allowing for rapid response as necessary (Murphy et al., 2015).
- Cost-effectiveness: Automation lowers labor costs involved with manual sampling and reduces equipment maintenance requirements.

13.11 Future trends in water quality management

Water quality management is at a pivotal point, with emerging technologies and innovative practices promising to transform how we monitor, protect, and utilize water resources. The future of water quality management will likely be shaped by trends that emphasize sustainability, resilience, and efficiency. This section explores key future trends as shown in Fig. 13.2 that are poised to redefine the field, including the integration of smart technologies, the adoption of sustainable practices, the rise of decentralized water systems, advancements in bioremediation, and the increasing role of policy and community engagement.

1. Integration of Smart Technologies
   a. Smart Sensors and IoT:
      i. The deployment of smart sensors and IoT devices enables continuous, real-time monitoring of water quality.
      ii. These sensors can detect a wide range of contaminants and water quality parameters with high accuracy.
   b. Data Analytics and AI:
      i. Advanced data analytics, including machine learning and AI, allow for the processing and analysis of large datasets.
      ii. AI algorithms can predict water quality trends, identify potential risks, and optimize treatment processes.
   c. Automation and Remote Control:
      i. Automated systems can adjust treatment processes in real time based on sensor data, ensuring optimal water quality with minimal human intervention.
      ii. Remote control capabilities enable the management of water quality systems from anywhere, increasing operational efficiency.
2. Adoption of Sustainable Practices
   a. Green Infrastructure:
      i. The implementation of green infrastructure, such as constructed wetlands, green roofs, and permeable pavements, helps to manage stormwater, reduce runoff, and improve water quality.
      ii. These natural systems provide ecosystem services, including habitat creation and carbon sequestration.

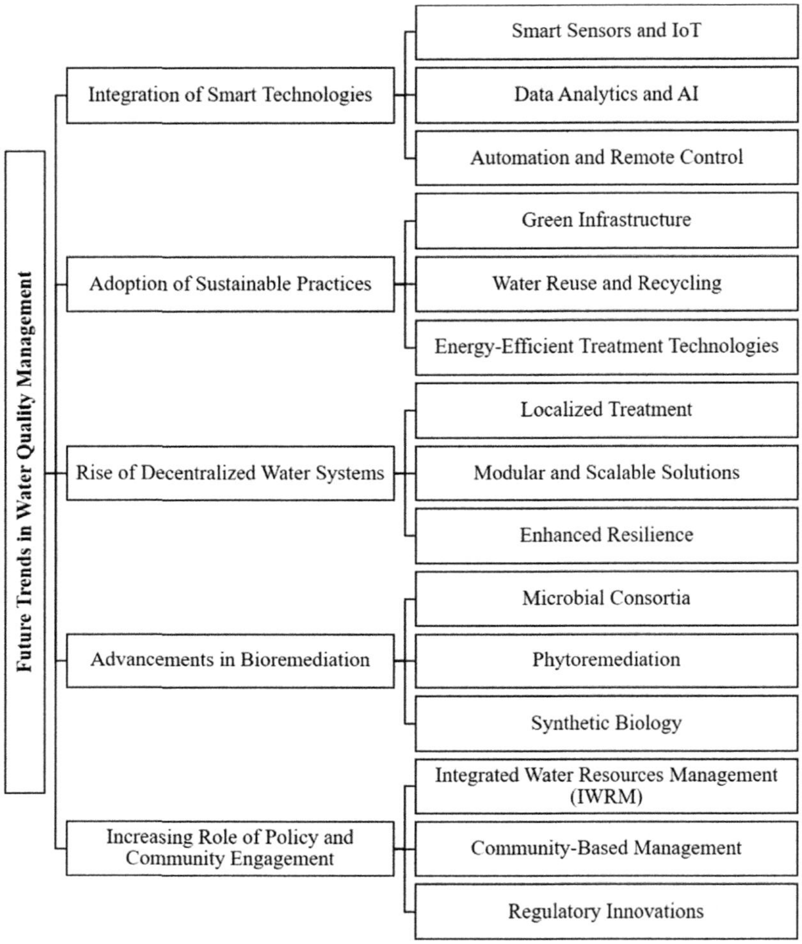

Figure 13.2
Future trends in water quality management.

b. Water Reuse and Recycling:
   i. Advanced treatment technologies are making it feasible to reuse and recycle water for various purposes, including agricultural irrigation, industrial processes, and even potable water supplies.
   ii. Water reuse reduces the demand on freshwater resources and promotes a circular water economy.
c. Energy-efficient Treatment Technologies:
   i. Innovations in water treatment are focusing on reducing energy consumption and utilizing renewable energy sources.
   ii. Techniques such as anaerobic digestion and MFCs are being developed to generate energy from wastewater treatment processes.

3. Rise of Decentralized Water Systems
   a. Localized Treatment:
      i. Decentralized systems treat water close to the point of use, reducing the need for extensive infrastructure and long-distance water transport.
      ii. This approach can be more cost-effective and easier to implement in remote or developing regions.
   b. Modular and Scalable Solutions:
      i. Decentralized systems are often modular and scalable, allowing them to be tailored to the specific needs and capacities of communities.
      ii. They can be expanded or adapted as water demand and quality requirements change.
   c. Enhanced Resilience:
      i. By distributing water treatment across multiple locations, decentralized systems reduce the risk of widespread service disruptions due to infrastructure failures or natural disasters.
      ii. They promote community involvement and ownership in water management practices.
4. Advancements in Bioremediation
   a. Microbial Consortia:
      i. Research is focusing on developing microbial consortia, or communities of microorganisms, that can work together to degrade complex pollutants.
      ii. These consortia can be tailored to target specific contaminants, enhancing treatment efficacy.
5. Phytoremediation:
   i. Phytoremediation involves using plants to absorb, accumulate, and detoxify pollutants from water.
   ii. This method is particularly effective for treating water contaminated with heavy metals and organic compounds.
6. Synthetic Biology:
   i. Advances in synthetic biology are enabling the creation of engineered microorganisms with enhanced capabilities for pollutant degradation.
   ii. These microorganisms can be designed to target specific pollutants and operate under diverse environmental conditions.
7. Increasing Role of Policy and Community Engagement
   a. Integrated Water Resources Management (IWRM):
      i. IWRM promotes a holistic approach to water management that considers the interconnections between water, land, and ecosystems.
      ii. Policies based on IWRM principles encourage coordinated efforts across sectors and stakeholders.

b. Community-based Management:
   i. Involving communities in water quality monitoring and management fosters local stewardship and accountability.
   ii. Community-based approaches can enhance the acceptance and effectiveness of water management initiatives.
c. Regulatory Innovations:
   i. Governments are developing and implementing new regulations that incorporate emerging technologies and address contemporary water quality challenges.
   ii. Policies are also focusing on promoting transparency, data sharing, and public participation in water management decisions.

13.12 Discussion

The exploration of emerging technologies reveals that internet of things, artificial intelligence, advanced treatment technologies, remote sensing, and blockchain are set to transform water quality management. These technologies offer real-time monitoring, predictive analytics, enhanced treatment processes, comprehensive monitoring capabilities, and improved data transparency. The integration of these technologies into water quality management systems represents a significant shift toward more proactive and efficient management practices. Real-time monitoring and predictive analytics enable quicker responses to potential issues, reducing the risk of water contamination and improving public health outcomes. Advanced treatment technologies ensure that even the most persistent contaminants can be effectively removed, providing cleaner and safer water. The adoption of these technologies can lead to substantial improvements in water quality management. Enhanced monitoring and treatment capabilities can help address the challenges posed by climate change, population growth, and industrialization. However, the successful implementation of these technologies requires substantial investment, regulatory support, and the development of a skilled workforce. While the potential of these technologies is significant, there are limitations to consider. The high cost of implementation and maintenance, especially for advanced treatment technologies and IoT infrastructure, can be a barrier for some regions. Additionally, the integration of AI and ML requires high-quality data, which may not be available in all areas. There is also a need for standardized protocols and regulations to ensure the effective use of these technologies. To maximize the benefits of these emerging technologies, several recommendations are proposed: continued investment in R&D is essential to advance these technologies and make them more accessible and cost-effective; training programs should be developed to equip professionals with the necessary skills to implement and manage these technologies; governments should establish supportive policies and regulations to encourage the adoption of innovative water management technologies; and collaboration between the public and private sectors can facilitate the deployment of these technologies on a larger scale. Comparing traditional water quality management methods with emerging technologies highlights the substantial advantages of the latter. Traditional methods often rely on periodic sampling and laboratory analysis, which can be time-consuming and reactive. In contrast, emerging technologies enable continuous, real-time monitoring and predictive analysis, leading to more proactive and effective management. The main challenges in adopting these technologies include high costs, data management issues, and the need for regulatory frameworks. To overcome these challenges, it is suggested to develop cost-effective solutions through innovation and economies of scale; implement robust data management systems to ensure the accuracy and reliability of data; and establish international standards and regulations to guide the use of these technologies. Overall, the adoption of these innovative technologies holds great promise for transforming water quality management, making it more efficient, effective, and sustainable. However, addressing the challenges and limitations through strategic investments, regulatory frameworks, and capacity building will be crucial to realizing their full potential. The comparative analysis underscores the need for a paradigm shift from traditional, reactive approaches to modern, proactive strategies that leverage the power of real-time data, advanced analytics, and cutting-edge treatment methods. This transition will not only improve water quality outcomes but also ensure resilience against future challenges such as climate change, population growth, and increased industrial activities. The recommendations provided aim to facilitate this transition by fostering innovation, building capacity, and encouraging collaborative efforts across sectors. While the journey toward widespread adoption of these technologies may be fraught with challenges, the potential benefits make it a worthy endeavor, promising a future where water quality management is characterized by precision, efficiency, and sustainability.

13.13 Conclusions

The chapter has explored the transformative advancements and innovative practices poised to redefine water quality management. By delving into cutting-edge technologies such as real-time monitoring systems, remote sensing, biosensors, and advanced data analytics, we have highlighted the potential for more precise, efficient, and proactive management of water resources. The integration of sustainable practices, decentralized systems, and bioremediation techniques underscores a shift toward environmentally conscious and resilient water management strategies. As we look to the future, the convergence of technological innovation, robust policy frameworks, and active community engagement will be crucial. Ensuring equitable access to these advancements and fostering global collaboration will help address the disparities and challenges faced by underserved communities. Embracing these trends will enhance our capacity to manage water quality effectively, safeguard public health, and ensure the sustainability of our water resources in the face of climate change and increasing demand. In summary, the future of water quality management is bright, driven by interdisciplinary efforts and a commitment to innovation and sustainability. By leveraging emerging technologies and holistic approaches, we can build resilient and sustainable water management systems that secure clean and safe water for all.

References

Akhtar, N., Syakir Ishak, M. I., Bhawani, S. A., & Umar, K. (2021). Various natural and anthropogenic factors responsible for water quality degradation: A review. *Water*, *13*(19), 2660.

Bhuta, M. (2014). Advanced oxidation processes: Current status and prospects. *Journal of Environmental Management*, *148*, 220–230. Available from https://doi.org/10.1016/j.jenvman.2014.03.015.

Essamlali, I., Nhaila, H., & El Khaili, M. (2024). Advances in machine learning and IoT for water quality monitoring: A comprehensive review. *Heliyon*.

Geissen, V., Mol, H., Klumpp, E., Umlauf, G., Nadal, M., van der Ploeg, M., … Ritsema, C. J. (2015). Emerging pollutants in the environment: A challenge for water resource management. *International Soil and Water Conservation Research*, *3*(1), 57–65.

Gomes, J., Costa, R., Quinta-Ferreira, R. M., & Martins, R. C. (2022). Ozonation and advanced oxidation processes for the treatment of pharmaceutical residues in wastewater. *Chemical Engineering Journal*, *428*, 131019. Available from https://doi.org/10.1016/j.cej.2021.131019.

Gude, V. G. (2017). Desalination and sustainability – An appraisal and current perspective. *Water Research*, *89*, 87–106. Available from https://doi.org/10.1016/j.watres.2015.11.012.

Jan, F., Min-Allah, N., & Düştegör, D. (2021). IoT based smart water quality monitoring: Recent techniques, trends and challenges for domestic applications. *Water*, *13*(13), 1729.

Kim, S., & Choi, Y. (2018). Application of IoT for real-time performance monitoring in membrane water treatment plants. *Sensors*, *18*(12), 4212. Available from https://doi.org/10.3390/s18124212.

Kumar, R., Gupta, S. K., Wang, H. C., Kumari, C. S., & Korlam, S. S. V. P. (2023). From efficiency to sustainability: Exploring the potential of 6G for a greener future. *Sustainability*, *15*(23), 16387.

Logan, B. E., & Rabaey, K. (2012). Conversion of wastes into bioelectricity and chemicals by using microbial electrochemical technologies. *Science (New York, N.Y.)*, *337*(6095), 686–690. Available from https://doi.org/10.1126/science.1217412.

Masi, F., Rizzo, A., & Regelsberger, M. (2018). The role of constructed wetlands in a new circular economy, resource oriented, and ecosystem services paradigm. *Journal of Environmental Management*, *216*, 275–284. Available from https://doi.org/10.1016/j.jenvman.2017.11.086.

Mauter, M. S., Zucker, I., Perreault, F., Werber, J. R., Kim, J. H., & Elimelech, M. (2018). The role of nanotechnology in tackling global water challenges. *Nature Sustainability*, *1*(4), 166–175. Available from https://doi.org/10.1038/s41893-018-0059-8.

Morin-Crini, N., Lichtfouse, E., Liu, G., Balaram, V., Ribeiro, A. R. L., Lu, Z., … Crini, G. (2022). Worldwide cases of water pollution by emerging contaminants: A review. *Environmental Chemistry Letters*, *20*(4), 2311–2338.

Murphy, K., Heery, B., Sullivan, T., Zhang, D., Paludetti, L., Lau, K. T., Diamond, D., Costa, E., & Regan, F. (2015). A low-cost autonomous optical sensor for water quality monitoring. *Talanta*, *132*, 520–527.

Park, N., & Srivastava, R. (2015). Titanium dioxide nanomaterials: Recent advances in photocatalysis and photovoltaics. *Journal of Nanomaterials*, *2015*, 256–276. Available from https://doi.org/10.1155/2015/123456.

Shah, M. A., & Malik, M. A. (2020). Green synthesis of biodegradable polymeric membranes for water purification. *Environmental Chemistry Letters*, *18*(2), 407–423. Available from https://doi.org/10.1007/s10311-020-00960-9.

Shih, Y. H., & Wang, C. K. (2018). Renewable carbon-based materials for sustainable electrochemical water treatment. *Journal of Cleaner Production*, *197*, 1222–1233. Available from https://doi.org/10.1016/j.jclepro.2018.06.302.

Smith, A. L., Stadler, L. B., Love, N. G., et al. (2012). Perspectives on anaerobic membrane bioreactor treatment of domestic wastewater: A critical review. *Bioresource Technology*, *122*, 149–159. Available from https://doi.org/10.1016/j.biortech.2012.04.055.

Trussell, R. S., Salveson, A., & Snyder, S. A. (2016). *Potable reuse: State of the science report and equivalency criteria for treatment trains*. Water Research Foundation.

Woldu, A. (2022). Biosensors and its applications in water quality monitoring. *International Journal of Scientific & Engineering*, *13*, 12–29.

Yin, J., & Deng, B. (2015). Polymer-matrix nanocomposite membranes for water treatment. *Water Research*, *91*, 223–231. Available from https://doi.org/10.1016/j.watres.2015.12.031.

Zhou, S., Yang, H., & Chen, X. (2019). Smart membrane materials for sustainable water treatment. *Journal of Materials Chemistry A*, *7*, 8184–8204. Available from https://doi.org/10.1039/C8TA12446K.

CHAPTER 14

Addressing complex challenges in water quality management: emerging technologies and sustainable strategies

Faheem Ahamad[1], Avnish Chauhan[2], Prabhat K. Chauhan[3], Sudhir K. Upadhyay[3], Anirudh Tomar[4], Nitesh Singh[5] and Teresinha Andrade[6]

[1]Department of Environmental Science, Keral Verma Subharti College of Science (KVSCOS), Swami Vivekanand Subharti University, Meerut, Uttar Pradesh, India, [2]Department of Environmental Sciences, Graphic Era Hill University, Dehradun, Uttarakhand, India, [3]Department of Environmental Science, VBS Purvanchal University, Jaunpur, Uttar Pradesh, India, [4]Department of Physics, Graphic Era Hill University, Dehradun, Uttarakhand, India, [5]Department of Plant Pathology, Faculty of Agriculture Sciences, SGT University, Gurugram, Haryana, India, [6]Nucleus of Applied Research to Sciences (NIAC), Federal Institute of Maranhao, Presidente Dutra, Brazil

14.1 Overview of water quality

The accessibility of clean and drinking water for mankind is a key component of both the Sustainable Development Goals (SDGs) and the Millennium Development Goals (MDGs) (Bain et al., 2018; Ahamad et al., 2023 & 2024). Over 2 billion people lack access to hygienic water, and 844 million people lack even a basic water source, despite significant efforts made by numerous international institutions (Lal, 2019; Makarigakis and Jimenez-Cisneros, 2019; Hayat et al., 2024). Quality drinking water is not available to about 50% of the Indian rural and urban population (Nandi et al., 2017; Wakode et al., 2018; WHO, 2021).

The term "water quality" means the fitness of water in terms of its physicochemical and microbiological assets (Chauhan and Singh, 2010; Omer, 2019). Disturbance in its quality characteristics makes it unsuitable for drinking, bathing, irrigation, and industrial purposes, which impacts the health of human beings, other living beings, vegetation, and other industrial processes (Mekuria et al., 2021; Emodi, 2020; Aqel et al., 2024). Physical characteristics like turbidity, temperature, electrical conductivity, and solid concentrations represent the esthetic beauty of the water and therefore are important. Chemical characteristics like hydrogen ion concentration, chloride, calcium, magnesium, sulfate, nitrate, fluoride, and different heavy metals are all equally important, as slight disturbance

Computational Automation for Water Security.
DOI: https://doi.org/10.1016/B978-0-443-33321-7.00003-2

in these ions directly impacts the healthiness of living (Bhutiani et al., 2021a,b; Khan et al., 2023a,b; Kumar et al., 2023). The notable causes of ionic balance disturbance are anthropogenic and geogenic (Ahamad et al., 2023). Different microbes like bacteria, fungi, algae, and bacterial species work as indicators of the health of aquatic bodies. The presence of *Escherichia coli* (*E. coli*) is a pointer toward fecal adulteration and be harmful to people's health (Amare et al., 2023; Upadhyay et al., 2023). High concentration of sulfate, nitrate, and phosphate (nutrient pollution) in aquatic bodies results in algal blooms, which result in low dissolved oxygen concentrations. Polluted water is responsible for more than 20% of diseases in developing countries like India (Venkatesh et al., 2024). Water may include some radioactive elements, mostly from human activity (nuclear power plants) or from natural sources (radon) (Suresh et al., 2020; Gadewar et al., 2023; Kaur et al., 2024).

Prasad et al. (2019) evaluated the groundwater quality of Uttarakhand state and reported the different radioactive elements and heavy metals in the water samples. Suresh et al. (2020) observed the radon concentration beyond the maximum contaminant level suggested by the United States Environmental Protection Agency (USEPA) in more than 50% of the samples. Balaram et al. (2022) assessed the groundwater and found the uranium in the different parts of geographic regions of India and the world. Sarkar et al. (2022) declared 50% of the samples unfit for drinking collected from the Malda district of Eastern India. Ahamad et al. (2023) assessed the groundwater quality of arid and semiarid areas of India and observed the values of nitrate, arsenic, lead, and fluoride atop the standard limit of the Bureau of Indian Standards (BIS). Similar interpretations were reported by numerous authors from different parts of the India and world (Roy et al., 2023; Lal et al., 2023; Ali et al., 2024; Awasthi et al., 2024; Dey et al., 2024). Degradation of habitats, dwindling fish populations, and biodiversity loss are just a few of the negative consequences that poor water quality can have on ecosystems. It may also harm human endeavors such as fishing, tourism, and agriculture, which could result in financial losses (Akhtar et al., 2021).

Urbanization is creating a lavish life, which results in the depletion of natural resources fast compared to the rural population (Negi and Chandel, 2022 & 2024). Beyond 80% of the water provided to a society returned as effluent. Besides this, India possesses a treatment facility for only about 40% of the effluent generated, and the rest of the effluent is discharged directly or indirectly into surface water bodies and on the ground, making the freshwater polluted (Schellenberg et al., 2020; Tariq and Mushtaq, 2023). Therefore the two problems are growing simultaneously: one is a continuous decrease in the water table due to climate change and other anthropogenic causes, and the second is pollution of the available water resources (Saravanan et al., 2021). Many treatment practices are used to improve the quality of water, such as chemical treatment, filtration, and disinfection (such as chlorination). With the help of these techniques, pollutants and germs can be eliminated to meet drinking water safety regulations and wastewater discharge regulations (Gitis and Hankins, 2018; Pichel et al., 2019; Mazhar et al., 2020; Vaseashta et al., 2022).

This chapter presents an extensive review of research articles available in digital journal publications and book chapters. Our primary focus extended from 2019 to 2024, with a particular emphasis on articles published within the last 5 years to underscore recent advancements in the fascinating field of water quality and their sustainable management approaches. The search was focused on water quality, water-energy nexus, sustainable approaches to water quality management, and advanced sensor technologies. We harnessed 112 unique publications from 12 databases to extract the data and analyzed in the present systematic chapter.

Data were explored for bibliometric analysis as earlier outlined by using keywords within the PubMed database. The VOS viewer 1.6.18 program screened the datasets based on three different matrices, such as (1) strength of total link, (2) number of published articles, and (3) number of citations. Network visualization in Fig. 14.1A and B is illustrated as larger frame size refers to larger dominance, with maximum number of lines showing greater networking power. The more frequently a keyword is used, the bigger the frame size in relation to the co-occurrence of that keyword in maps. Fig. 14.1A and B depicts a network visualization map of the most cited phrases in 112 publications published between 2019 and 2024 (Supplementary source 1). The primary focus of this section is on the water quality and their sustainable management approaches.

14.1.1 Water quality risk

Global water quality is increasingly at risk, posing a significant threat to ecosystems, human health, and economic stability. According to the World Wide Fund for Nature's (WWF) Water Risk Filter 2021, the degradation of water quality is driven by multiple factors, including industrial pollution, agricultural runoff, and inadequate wastewater treatment. These pollutants introduce harmful substances such as heavy metals, pesticides, and pathogens into water bodies, leading to deteriorating water quality.

The report highlights that regions with high industrial activity and intensive agriculture are particularly vulnerable. For example, parts of Asia, Africa, and Latin America face severe water quality challenges due to rapid industrialization and the extensive use of fertilizers and pesticides. The contamination of freshwater resources impacts not only biodiversity but also the availability of clean water for human consumption, agriculture, and other uses.

The WWF Water Risk Filter emphasizes the need for integrated water resource management to address these issues. This includes improving industrial processes to reduce emissions, promoting sustainable agricultural practices, and investing in advanced wastewater treatment technologies. Additionally, policies and regulations need to be strengthened and enforced to ensure compliance and protection of water resources (Fig. 14.2).

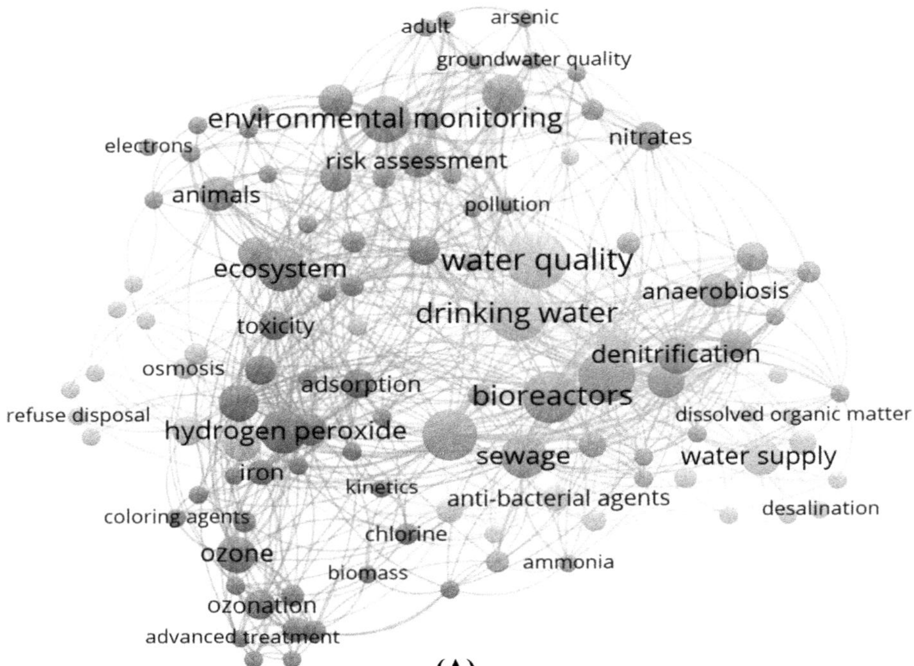

(A)

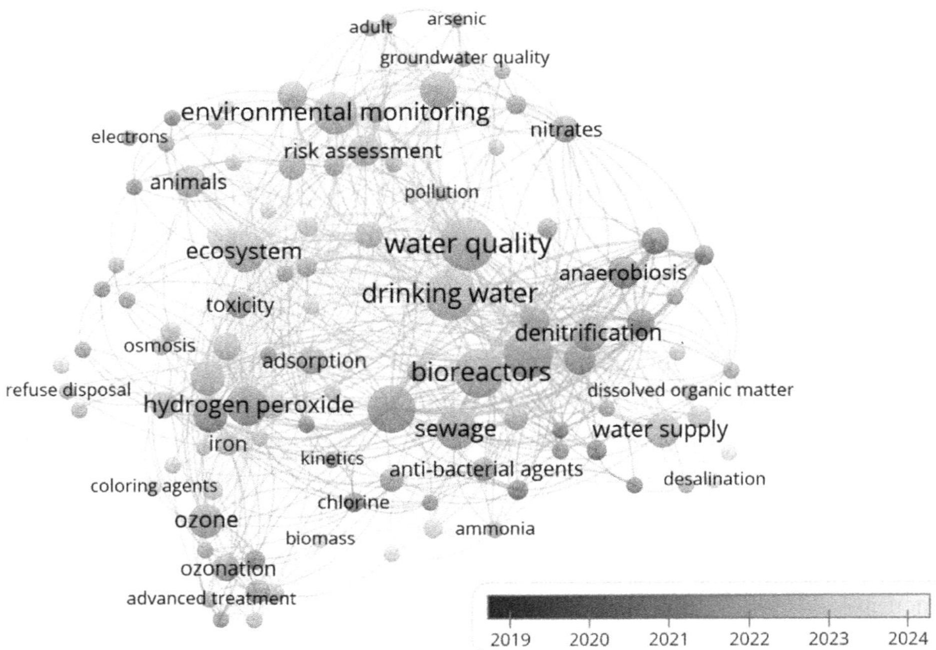

(B)

(Continued)

14.2 The water-energy nexus

The availability, use, and management of water resources directly affect energy generation and vice versa; this is known as the "water-energy nexus," which describes the interdependent relationship between water resources and energy production (Hamiche et al., 2016). This emphasis points to the importance of joint energy and water considerations in policy, infrastructure, and technology development. This source is required for agriculture, industry, and family use, as well as energy. Energy and water can become more competitive resources, especially in regions where they are scarce. Resilience and sustainability can be realized only when these conflicting demands are dealt together via integrated planning and resource management (Khalkhali et al., 2018, Rasheed and R-Moreno, 2024).

Energy-related activities like hydropower generation, cooling at thermal power plants (coal, nuclear, or natural gas plants), and extractive processes like fracking for oil and gas often require large amounts of water. The unavailability of water or its poor quality can reduce energy production and, therefore, its reliability and efficiency (Bortoleto et al., 2021). Meanwhile, water management and treatment processes need energy inputs in terms of pumping, transporting, purifying, and distributing water. For instance, desalination process that produces freshwater from saltwater consumes a very high energy. Energy demand, as well as partner cost and environmental effects, can affect decision-making for water treatment and general water management techniques (Molajou et al., 2021).

The exploitation of water and energy resources can have major environmental effects like territory demolition, water pollution, greenhouse gas radiations, and alterations to hydrological cycles (Prakash, 2021). For example, the use and combustion of fossil fuels contribute to water pollution and climate change, while hydropower dams affect fish populations and river ecosystems. Furthermore, climate change exacerbates the interdependence between water and energy (Singh and Singh, 2017; Bano and Arshad, 2018). It is crucial to address these interconnected issues and implement robust solutions to adapt to the challenges posed by climate change. Rules, regulations, and different management practices must be integrated to recognize the correlation between water and energy demands. Exploring the use of nonconventional energy sources with minimum water requirements and the application of water conservation and recycling technologies are some of the ways. Enhancement in research and development initiatives can also help in stabilizing the water-energy nexus (Gupta et al., 2020; Weerasooriya et al., 2021).

Figure 14.1

Bibliometric analysis based on network clustering of 112 published articles from 2019 to 2024 (Supplementary source 1) that have high match of selected keywords and their citations (A) network visualization and (B) overlay visualization.

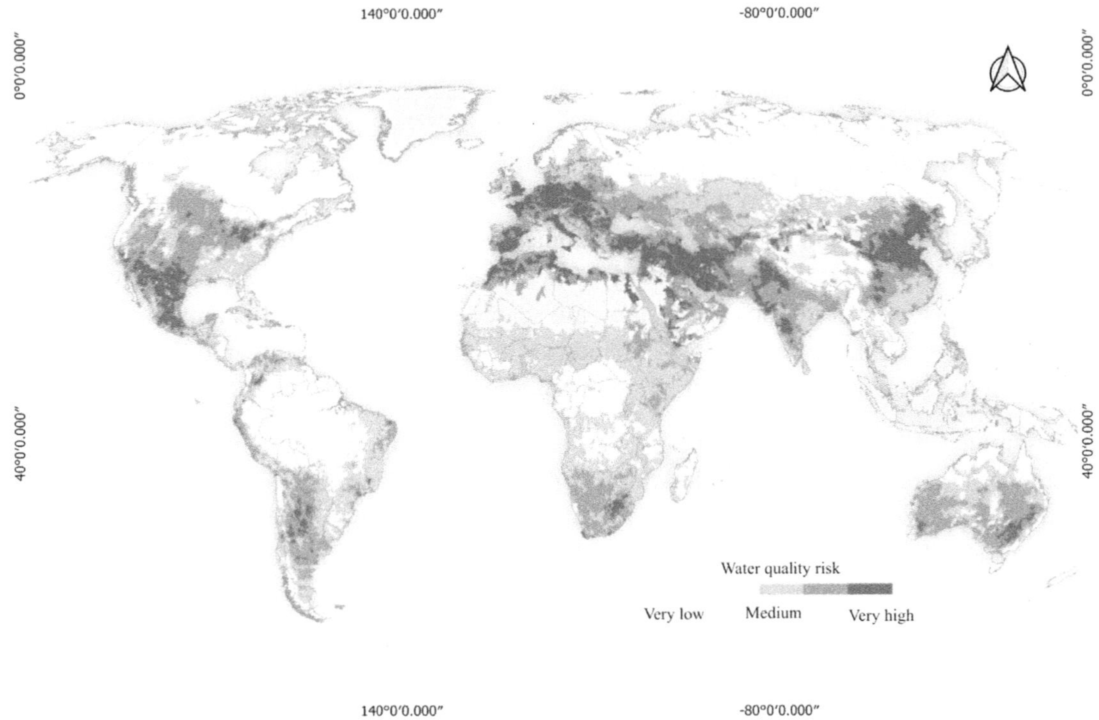

Figure 14.2
Global water quality risk. Map lines delineate study areas and do not necessarily depict accepted national boundaries. Source: *WWF Water Risk Filter 2021.*

14.3 Current challenges in water quality management

The whole world is continuously facing many challenges to make the clean water supplies easily accessible to every person in the world. Due to the lack of awareness, the population is polluting these freshwater resources at a significant speed, thereby generating the wastewater in excess. The discharge of wastewater either generated from domestic or industrial areas along with runoff from agricultural fields both on land and in water bodies is responsible for continuously degrading water quality (Vinayagam et al., 2023). The addition of nutrients from all these sources makes the water bodies eutrophic (Tyagi et al., 2024).

Along with conventional pollutants, nonconventional or emerging pollutants are a hard rock challenge for the efficient management of water quality. The available treatment practices are not able to remove these impurities from the ground or surface water, and at the same time, their health impact chemistry is also not well known (Gupta et al., 2020). Upgradation of the existing treatment technologies and maintenance of distribution systems is the urgent need of the present hour (Bekele et al., 2018). Microbial contaminants are responsible for most of the famine worldwide, and therefore, along with physical and chemical characteristics, microbes and emerging contaminants are also a matter of concern (Ahamad et al., 2023). Uncertainty in the rainfall, extremes of temperature, droughts, and floods are some of the weather events happening due to climate change and putting extra stress on available water resources and management practices (Gupta et al., 2020).

Shortage of water also results in an enhanced level of pollutants and conflicts among the users, which disturb the harmony of the society. The use of alternate water resources is not safe due to lack of advanced treatment systems, and therefore the management of water quality and quantity to meet the needs of the continuously growing population is an ongoing concern (Jan et al., 2021).

The only solutions to all these problems are research and development on existing water treatment technologies, revising existing water use policies, campaigns to enhance the awareness among the society, inclusion of all these concerns in the curriculum at early stages, and involvement of public, private, and nongovernmental organizations (Ameta et al., 2023; Awewomom et al., 2024). Besides these, an eagle watch by the government organizations on the use of corporate social responsibility funds by the companies and making them convince to make use of these funds for the sustainable water resource management can also help in achieving sustainable water management.

14.4 Emerging trends in water quality management

Increasing awareness among the common public, advancements in technologies, and amended environmental laws and policies make small but significant changes in field of water quality management. For example, real-time water monitoring and detection of early pollution incidents become possible due to integration of sensor-based technologies and data analytics (Ahamad et al., 2022; Awewomom et al., 2024). Low-capital treatment devices, onsite water treatment devices, and natural and artificial wetlands are some of the examples of decentralized water treatment systems that are now trending for sustainable and efficient management of water resources (Chavan et al., 2024). Similarly, wetlands and riparian flora are countryside solutions that are more potent in present era of industrialization and urbanization to manage the water sources and control climate change (Pennellini, 2022). Now the removal of emerging pollutants and microbial populations is also possible due to advancements in membrane filtration, electrotreatment, and oxidation processes (Yan et al., 2024). Detection and remediation of microbial contamination is also possible due to advances in molecular and genetic markers, which can result in reduced waterborne diseases soon (Capodaglio, 2021; Tyagi et al., 2024).

Conversion of sludge generated from wastewater treatment plants into energy and compost is due to the advancement of treatment processes and incineration technologies, which helps in circular economy (Boha et al., 2022). Focus on the establishment of green buildings is also helping in the minimization of resource utilization, enhancing the utilization of waste resources, and use of renewable energy resources, which is helping in the control of climate change and conservation of natural resources (Puyol et al., 2017).

Utilization of traditional knowledge of local inhabitants and their involvement in policy-making processes can help in the water quality monitoring programs. This can reduce the burden of government agencies and strengthen the water resource management programs (Garcia and Pargament, 2015). Integrated water resource management practices like centralized water treatment plants, irrigation from centralized borewells, and incentives to farmers for the reduced water use in irrigation practices can help in sustainable management of water resources. Involvement of the National Accreditation Board for Testing and Calibration Laboratories (NABL)-approved laboratories for regional water quality monitoring is also an efficient and cost-saving method (Ahamad et al., 2023).

14.4.1 Hybrid treatment system

Integration of different water treatment processes into a single system is called hybrid treatment system (HTS), which can treat water more efficiently compared to individual ones as it utilizes their pros and minimizes cons (Dhangar and Kumar, 2020). HTS integrates different physicochemical and microbiological treatment processes like simple filtration, adsorption, and disinfection processes to treat the water completely to meet the quality standards (Grover et al., 2011; Badawi and Zaher, 2021). The characteristic property of all HTS is their segmental architecture, which provides flexibility in structure for the required shape, flow rate, and targeted pollutant as per the need. The HTS system may be optimized to meet the required performance as per the goal, nature of water to be treated, and operational circumstances (Dhangar and Kumar, 2020). In HTS, the processes may be combined in such a way that the drawbacks of one treatment system can be eliminated by another. For example, membrane fouling is the common and significant drawback of membrane filtration systems, which can be eradicated by using the advanced oxidation processes (AOPs). The integration of AOPs and membrane filtration may treat the water efficiently and provide the longer life span to the membrane of membrane filtration system along with accepted water quality (Sen et al., 2010).

HTS possesses different barriers, treatment stages, and more retention time to provide stability and continuous water supply in case of equipment failure (Tee et al., 2016). Efficiency of HTS can be enhanced by changing the sequence of processes and utilizing the artificial intelligence and machine learning (AIML) techniques for real-time water monitoring. Optimization of the processes involved and factors affecting the effectiveness of the system can reduce the energy and chemical intake (Nguyen et al., 2014).

The starting investment in applying the HTS is more in comparison to their individual application but can be recovered by reducing the energy and chemical consumption, maintenance cost, and resource recovery, which ultimately help in achieving environmental sustainability (Mahjouri et al., 2017). The environmental sustainability of HTS can also be improved by utilizing the renewable sources of energy, especially solar energy, by applying target-specific treatment technologies that reduce the consumption of chemicals, energy, and human resources (Rodriguez-Narvaez et al., 2017).

14.4.1.1 Membrane bioreactor

Membrane bioreactors (MBRs) are innovative and unified wastewater treatment systems in which membrane filtration technology is coupled with biological treatment procedures for the treatment of domestic and industrial effluent to meet their discharge standards (Iorhemen et al., 2016). In MBRs, the mortification of organic pollutants to transform them into nutrients takes place both in the presence and absence of oxygen. MBRs incorporate both micro- and ultrafiltration membranes to reduce the load of suspended solids and pathogens from the wastewater. The membrane technology holds the wastewater for longer duration compared to traditional activated sludge systems, which accelerates the degradation of organic impurities through biological treatment processes. The well-known drawback of MBRs, i.e., membrane fouling, can be minimized by using backwashing, air scouring, and chemical cleaning, which enhance the system efficiency along with the life span of membranes (Meng et al., 2009; Liu et al., 2024; Table 14.1).

The MBRs require very little space for installation due to combination of both membrane filtration and biological treatment units along with absence of secondary and tertiary clarifiers, which results in their smaller footprint and easy installation in small areas (Santos et al., 2011). It can be applied in water recycling plants, sewage treatment plants, industrial effluent treatment plants (ETPs), etc. (Ruhela et al., 2023). Due to energy and resource recovery operations, MBRs are sustainable and cost-effective (Gukelberger et al., 2019; Rahman et al., 2023).

14.4.1.2 Amalgamated anaerobic—aerobic treatment system

Amalgamated anaerobic—aerobic treatment is a hybrid system for the treatment of complex natural wastewater and managing hydraulic and organic loading rates (Cesaro et al., 2015; Uddin and Wright, 2023). Here in this arrangement, the breakdown of carbon-based matter happens in anaerobic reactors through the processes of hydrolysis, acidogenesis, acetogenesis, and methanogenesis, and thereafter breakdown of remaining organic matter and other pollutants takes place by the bacterial species in aerobic reactor systems like activated sludge processes (ASP) and sequential batch reactors (SBRs) (Liebetrau et al., 2019; Induchoodan et al., 2022), which help in the reduction of nutrient and pathogen loads (Campo et al., 2020; Rajesh Banu et al., 2021). Aerobic reactors efficiently reduce nitrogen

Table 14.1: Selected pollutant removal by the selected hybrid treatment technologies reported in the literature.

S. No.	Hybrid treatment system	Technology used	Type of water	Pollutants removed	References
1		Membrane bioreactor (MBRs)	Domestic wastewater Groundwater Landfill leachate	NH_4-N (100%) As (50%–97%) BOD (94%) COD (92%) TN (74%) TP (73%)	Nguyen et al. (2014) Sen et al. (2010) Saidulu et al. (2021)
			Synthetic wastewater	BOD (94%) COD (97.4%) TN (68.4%) TP (37.5%) TSS (97.8%)	Ashadullah et al. (2021); Abuabdou et al. (2020)
			Textile wastewater	COD (97%) TN (70%)	Xu et al. (2014)
			Domestic wastewater	COD (87%) TSS (100%)	Herrera-Robledo et al. (2010)
2		Biofilm support	Domestic wastewater	COD (89.4%) NH_4-N (83.3%) TN (62.9%) TP (48.7%)	Ji et al. (2015)
			Domestic wastewater	TN (96.0%) TP (91.5%) COD (80.2%)	Yang et al. (2018)
			Synthetic wastewater	BOD (92.5%) P (40%)	Ruiz-Marín et al. (2009)
			Domestic wastewater	NH_3 (99.75%) TN (59.51%) COD (85.06%)	Chen et al. (2018)
			Domestic wastewater	TSS (93.2%) BOD (96.5%) TN (67.4%) TP (36.7%)	Park et al. (2019)
			Domestic and industrial effluent	N (63.6%) P (70.2%)	Katam and Bhattacharyya (2021)
			Domestic wastewater	C (91%) N (70%) P (85%)	Posadas et al. (2013)
3		Combined anaerobic–aerobic treatment system	Domestic wastewater	C (84%) N (71%) P (96%)	Campo et al. (2020)
			Domestic wastewater	BOD (81.7%–92.0%) COD (74.4%–82.3%) TSS (83.7%–91%) VSS (83.4%–92.9%)	Mazhar et al. (2021)

(*Continued*)

Table 14.1: (Continued)

S. No.	Hybrid treatment system	Technology used	Type of water	Pollutants removed	References
			Domestic wastewater	COD (74.2%)	Liu et al. (2020)
			Acid wastewater	COD (99%)	Show et al. (2020)
			Mixed wastewater	COD (88.4%–94.9%)	Wang et al. (2022)
				COD (73%) TDS (90%) COD (69%) Cl (66%) Hardness (77%)	Kathawala et al. (2021)
			Landfill leachate	COD (85.8%) NH_4-N (89.7%) TN (53.8%)	Zhou et al. (2023)
4		Advance oxidation process (AOP)	Landfill leachate	COD (53%)	Deng (2009)
			Landfill leachate	COD (72%) NH_4-N (55%)	Amr et al. (2013)
			Hospital wastewater	Ibuprofen (100%) Carbamazepine (100%) Frusemide (100%)	Khan et al. (2020)
			Industrial wastewater	COD (30.9%–59.8%)	Cai et al. (2020)
			Industrial wastewater	Coliform inactivation (6.7%–20.9%)	Ashrafivala et al. (2022)
			Industrial wastewater	PPCPs ($\geq$ 56.5%)	Pai and Wang (2022)

NH_4-N, Ammoniacal nitrogen; *BOD*, biochemical oxygen demand; *COD*, chemical oxygen demand; *TN*, total nitrogen; *TP*, total phosphate; *TSS*, total suspended, solids; *As*, arsenic; *NH_4*, ammonium; *N*, nitrogen; *P*, phosphorus; *C*, carbon; *VSS*, volatile suspended solids; *Cl*, chloride; *PPCPs*, pharmaceutical and personal care products.

from wastewater, and anaerobic reactors efficiently reduce phosphorus. (Campo et al., 2020). The configuration of this HTS can be rearranged based on the characteristics of wastewater and treatment goals like irrigation, drinking, and discharge on land or in water bodies. The integration of anaerobic–aerobic treatment is helpful in the reduction of cost and environmental impacts (Anukam et al., 2019). The use of biogas produced in anaerobic reactors for electricity production makes this system energy efficient and therefore sustainable as well (Di Maria et al., 2019; Shah et al., 2024), and biosolids as manure in agricultural fields help in resource conservation and circular economy (Wainaina et al., 2020; Awasthi et al., 2022).

14.4.1.3 Biofilm support

The growth of different bacterial colonies sticking on the surface of filtration materials in the form of thin and slimy layers is known as biofilms (Germec et al., 2020; Singh et al., 2023a,b). The "biofilm support" is the material or structure to enhance the formation of biofilms in wastewater treatment processes like biofiltration or bioremediation (Bhutiani et al., 2021a, b). The support materials are available in different forms, like porous media, structured media, and suspended carriers. Small diffuser-like materials made up of plastics, ceramics, and fibers are a part of the porous materials, which create perfect environment and provide high surface area-to-volume ratio (Dutt et al., 2020). Engineered surfaces improve the biofilm attachment and growth due to their special patterns, textures, or coatings, surface roughness, and other microscale characteristics (Oopath et al., 2023). Biofilm carriers composed of natural or synthetic polymers, plastic beads, and granular activated carbon promote the growth of biofilms in wastewater or bioreactors by facilitating biomass retention and improving mixing and oxygen transfer (Jagaba et al., 2021; Table 14.1).

Biofilm support materials are extensively utilized in a variety of wastewater treatment processes, like attached growth processes (such as trickling filters and rotating biological contactors), submerged biofilm reactors (like moving bed biofilm reactors, fixed film activated sludge), and biofiltration units (like biofilters and bioscrubbers). The optimization of support materials entails taking lifetime, fouling resistance, compatibility of the materials, and surface qualities into account (Murshid et al., 2023). For treatment efficacy to be maintained, biofouling to be avoided, and system life span to be extended, regular cleaning and maintenance of biofilm support structures are necessary (Germec et al., 2020).

14.4.1.4 Advanced oxidation processes

AOPs are a class of chemical treatment methods used in water treatment processes that breakdown and eliminate pathogens, refractory chemicals, and emerging contaminants, among other organic and inorganic pollutants (Deng and Zhao, 2015; Tufail et al., 2021). The processes like photolysis (UV light), chemical reactions (ozone or hydrogen peroxide), and sonochemical processes (ultrasound) are used to produce extremely sensitive and stable hydroxyl radicals (•OH) (Sirés et al., 2014). The radicals help in the oxidation of both organic and inorganic contaminants present in wastewater (Miklos et al., 2018) and help in reducing their toxicity. The combination of UV/H_2O_2 helps in the efficient degradation of different agriculture, industrial, and pharmaceutical chemicals. On the other hand, combination of O_3/UV helps in the degradation of organic pollutants and reduction of odor and pathogenic load efficiently. The combination of H_2O_2/Fe^{2+} (Fenton reaction) helps in reducing the organic and persistent chemicals (Bala et al., 2018; Tijani et al., 2014).

Although the AOPs can be applied to wide range of treatment processes, on targeted pollutants, they can work efficiently. Besides this, the process is quick and generates the minimum quantity of by-products (Khan et al., 2023a,b; Table 14.1). The disadvantages of AOPs are high energy, investment, and safety precautions intensiveness (Mahbub and Duke, 2023). Therefore thorough investigations are required regarding design, optimization, and cost-benefit analysis before its implementation to meet the desired characteristics of discharged wastewater (Tufail et al., 2021; Vinayagam et al., 2023).

14.5 Sustainable approaches in water quality management

The basic aim behind the application of sustainable technologies for the management of water quality is to keep in mind the societal, financial, and environmental factors along with the long-term conservation and justified use of water sources. These approaches consist of flexibility, equal opportunities and sharing, and sustainability to execute the intention of water quality management (Sarang et al., 2008). Management of watersheds, adoption of sustainable land use policies, strict application of rules and regulations related to environment, and awareness among the society are some of the sustainable approaches for the better management and conservation of ground and surface water sources (Bekele et al., 2018). Water quality of ground and surface water sources can be maintained by reducing the generation of pollutants or contaminants at the source, like agricultural land, urban and rural storm drains, and industries (Tufail et al., 2021).

Green infrastructure like natural and constructed wetlands, vegetated swales, green roofs, rain gardens, and permeable pavement could be used to control the runoff from surfaces, agricultural, and other plain areas. It enhances the water infiltration rate and thereby helps maintain the water table (Shafique and Kim, 2017; Prudencio and Null, 2018). Nature-based solutions like riparian buffers, replanting, soil protection, and habitat restoration initiatives promote biodiversity, control hydrological cycles, improve water quality, and offer numerous additional advantages to communities and ecosystems (Vasiliev and Greenwood, 2022; Srivastava et al., 2023). Integrated water resources management considers the relationships between water resources, land use, ecosystems, and socioeconomic variables to maintain the sustainability of water systems, balance conflicting water needs, and distribute resources fairly by encouraging the stakeholder for participatory decision-making and adaptive management (Rahaman and Varis, 2005).

14.6 Artificial intelligence and machine learning applications in water quality management

Artificial intelligence (AI) and machine learning (ML) help in the better monitoring of water sources, prediction of future trends, optimization of treatment processes, and based on all these results, better decision-making (Lowe et al., 2022; Krishnan et al., 2022; Alprol et al., 2024). The early detection of trends in water quality helps in the preservation of natural and human resources like animals, humans, and floral species enlisted in the International Union for Conservation of Nature (IUCN) list found in the concerned area. The detection of potential pollution sources exactly helps in the safety of water sources from pollution by controlling the release of these pollutants in the water sources (Ahmed et al., 2019). Through the optimization of physicochemical and biological factors, energy, and chemical consumption, AIML helps in the better and more cost-effective operations of ETPs (Li et al., 2021; Ismail et al., 2021).

The water management department can manage the water demand and supply ratio, different natural and anthropogenic factors, and water resources water supply system efficiently with the help of AI-based decisions (Jan et al., 2021). ML algorithm-based models help in the assessment of hydro-meteorological data and present and future water demands, which results in optimized water distribution systems and reduced frequency of droughts. AIML system can continuously upgrade the developed model by rectifying the existing models based on the feedback, environmental conditions, and other observed problems. With the help of ML algorithm, potential sources of pollution, climate change, and land use changes and their possible health risk assessment can be carried out effectively. AI helps in the efficient management of water sources and optimization of the aspects affecting the effectiveness of wastewater treatment plants, which results in cost reduction (Ismail et al., 2021; Singh et al., 2023a,b; Achite et al., 2023).

14.7 Advanced sensor technologies for monitoring water quality

Advanced sensor technologies helps in the real-time checking of water quality with special reference to physicochemical and biological characteristics (Salunke and Kate, 2017), which results in time and money savings, optimization of the processes, and better decision-making (Venkatesh et al., 2024). On the other hand, multiparameter sensors help in the real-time quantification of the values of multiple parameters simultaneously (Krkljes et al., 2024). Optical sensors help in the quantification of turbidity, dissolved organic matter, chlorophyll-a, and algal pigments using light-based models without causing any harm to the object (Goblirsch et al., 2023). Similarly, fluorescence sensors help in the measurement of above-discussed parameters along with heavy metals (Jung et al., 2016). Electrochemical sensors possess low detection limit of detection of pH, DO, ORP, and ions such as nitrate and ammonium chloride, as well as quick response and low maintenance requirements (Kanoun et al., 2021). Advanced biosensors are customized for specific chemicals, pathogens, or biomarkers, and therefore they help in the detection of target pollutants with high accuracy (Thakur and Kumar, 2022). The depth of the rivers, velocity of flowing water, concentration of sediment, and suspended solids can be monitored accurately with the help of acoustic sensors (Onwuka et al., 2023; Tamang et al., 2022). Spatio-temporal changes in the quality of water can also be studied with remote sensing and geographic information systems (Yang et al., 2022; Ahamad et al., 2022; Román et al., 2022; Bai et al., 2024). Multiple sensors coupled with a networked system can help in real-time water monitoring of large areas and in difficult water infrastructure. Portable shrank sensors are helpful in the water quality monitoring of remote areas (Giles et al., 2024). To facilitate the collection, examination, and comprehension of sensor data, sophisticated sensor technologies are augmented by data integration frameworks, analytical software, and visual facilitates. Different statistical models and ML algorithms are examples of such sophisticated sensor technologies (Thakur and Devi, 2024).

14.8 Discussion

The WWF report on water risk filter (2021) indicates that most of the parts of India are in a medium- to high-risk zone. Besides this, the generation of wastewater due to modern-day lavish life is the highest ever. The ratio of treatment to generation of wastewater is very low, and thus the development of innovative and integrated technologies for the complete treatment of wastewater is one of the focus of modern-day science. Irrespective of this much effort, the whole world is facing water shortages along with the increasing contamination of water resources day by day. In this current chapter, we reviewed the different HTSs of wastewater treatment, like MRBs, biofilm support, AOPs, and combined aerobic and anaerobic treatment systems. One of the biggest challenges that we observed while reviewing the status of wastewater treatment is the selection of treatment methods due to generation and addition of different emerging pollutants. If one methodology is found suitable for a particular pollutant, the same technology is not found suitable for the other reduction of another pollutant. The results of the present review-based study revealed that the integration of existing water treatment technologies results in smart water treatment systems that are more efficient for the reduction of pollutants from the wastewater in comparison to individual ones. Furthermore, the HTS technologies can be modified as per the requirement, i.e., characteristics of water, requirements of the site, and according to targeted pollutants. Due to integration of multiple technologies in HTS, drawbacks of one treatment system are eliminated by the other. A large quantity of water is required for energy production, which puts the extra pressure on water resources along with the significant impacts on the environment. Therefore for the water conservation, the use of renewable energy resources is recommended wherever possible. Although many people are using solar energy, there is still a need for awareness campaigns to make the people aware of the advantages of renewable energy resources.

Due to efforts taken to achieve the goals of SDGs and MDGs, now the common man and governments of the countries are aware of the sustainable management of water resources. The scientific communities are now working on the integration of different technologies, which results in the development of real-time water monitoring networks. Reduction of emerging pollutants becomes possible due to the advancement in membrane treatment systems, electrocoagulation, and oxidation processes. Similarly, detection of microbial contamination becomes possible due to development and advancement of molecular and genetic markers. Incineration technologies help in the minimization of sludge obtained from wastewater treatment plants and domestic and industrial solid waste. In this way, the pollution caused by the dumping of sludge and solid waste can be minimized or eliminated completely.

The complete breakdown of waste is not possible in any aerobic and anaerobic treatment system, but integrated aerobic and anaerobic treatment systems like ASP and SBRs can breakdown or eliminate the physicochemical and microbiological contaminants. The AOPs can reduce the multiple pollutants from the wastewater, but they can work efficiently on single pollutants. Therefore AOPs are not a cost-efficient solution and require further investigations for its optimization. As discussed above in the chapter, membrane filtration, sophisticated oxidation processes, sensor networks, and remote sensing technologies can treat wastewater effectively. Promoting sustainable water quality management techniques requires the implementation of strong legislative frameworks, policies, and incentives. Enforcing environmental legislation, enforcing effective governance processes, and using financial tools like pollution fees or incentives for eco-friendly behavior all work together to promote compliance and drive investments in pollution prevention and control methods. The water quality management process can be optimized with the AIML algorithm, which further helps with time and cost reduction. Real-time water quality monitoring became possible due to advancements in sensor-based technologies. Optical, electrochemical, and fluorescence sensor-based technologies also help in the real-time quantification of physicochemical and heavy metal parameters. The detection of specific pollutants has become possible due to the advancement in biosensors. Similarly, the detection of river depth, flow, etc. becomes possible due to advancements in acoustic sensors.

The literature review suggests that all these integrated technologies are in continuous use for the treatment of domestic, industrial, and hospital wastewater, along with leachate treatment. Among all the technologies, MBR technology was found more efficient for the treatment of any kind of wastewater. Based on the review results, we found the efficiency of the discussed technologies in the given sequence: MBRs > biofilm support > combined anaerobic–aerobic treatment system > AOP. Although these integrated technologies are helping in the sustainable management of wastewater, a thorough investigation is also required for the efficiency optimization and cost reduction.

14.9 Conclusion

The present work was carried out to explore water quality management practices. The complex issues of water pollution, shortage, and environmental degradation will be addressed by a confluence of cutting-edge technologies and emerging trends that will determine the future of water quality management. There is a growing need for creative solutions to protect water resources and guarantee sustainable access to clean water because of population growth, industrialization, and climate change. The trajectory of water quality management is expected to be significantly shaped by a number of important future trends and emerging technologies, including the integration of advanced sensor technologies, adoption of AIML, expansion of remote sensing and satellite monitoring, advancements in water treatment technologies, adoption of nature-based solutions, shift toward decentralized and modular treatment systems, and increased involvement of stakeholders and citizens in the preparation and promotion of policy reform and governance initiatives. Water quality management initiatives can adapt to meet the changing opportunities and difficulties brought about by a fast-changing world by adopting the above-discussed emerging technologies. We may strive toward accomplishing the common objective of protecting water resources for present and future generations by cooperating, being innovative, and being devoted to sustainability.

14.10 Future perspective (recommendation)

To meet the changing opportunities and challenges in protecting water resources and guaranteeing that everybody has access to clean water, it is encouraging to consider the potential applications of developing technology in water quality management. The recommendations listed below can help in improving water quality management:

1. The adoption of sensor networks and the integrated Internet of Things will allow for the creation of networked smart water systems that can collect, monitor, and analyze data in real time.
2. Utilizing cutting-edge nanotechnology-based creative solutions is important since these methods will aid in the abstraction of pollutants from water and wastewater, improve detection skills, and increase treatment efficiency.
3. Acceptance of new blockchain technology since it can improve water quality management's traceability, accountability, and transparency by allowing for safe, decentralized data tracking, verification, and storage. Blockchain-based technologies can help sponsors control the water resources and guarantee adherence to regulations by facilitating data exchange, cooperation, and trust.
4. For water quality management, a wide range of applications of AI and ML-based algorithms will allow more advanced data analytics, predictive modeling, and decision support systems. The AI and ML-based technologies will enable water managers to more precisely and accurately estimate trends in water quality and extract the information needed from large and complicated databases to optimize treatment procedures.
5. Enhanced research and development in nature-based solutions to promote resilience, biodiversity, and sustainability in water management practices.
6. More freshwater production from brackish and saltwater sources will be possible because of developments in desalination and water reuse technologies, which will become more sustainable, economical, and energy efficient.
7. Adopting and advancing the concepts of the circular economy in water management can prompt innovation in waste valorization, resource recovery, and environmentally friendly water usage techniques. The development of technologies for reclaiming materials, energy, and nutrients from wastewater streams will support economic expansion, environmental preservation, and resource conservation.

References

Abuabdou, S. M., Ahmad, W., Aun, N. C., & B'ashir, M. J. (2020). A review of anaerobic membrane bioreactors (AnMBR) for the treatment of highly contaminated landfill leachate and biogas production: Effectiveness, limitations and future perspectives. *Journal of Cleaner Production*, *255*, 120215.

Achite, M., Samadianfard, S., Elshaboury, N., & Sharafi, M. (2023). Modeling and optimization of coagulant dosage in water treatment plants using hybridized random forest model with genetic algorithm optimization. *Environment, Development and Sustainability*, *25*(10), 11189–11207.

Ahamad, F., Bhutiani, R., & Ruhela, M. (2022). Environmental quality monitoring using environmental quality indices (EQI) geographic information system (GIS) and remote sensing: A review. *GIScience for the sustainable management of water resources* (pp. 331–348).

Ahamad, F., Tyagi, I., Bhutiani, R., Kumar, V., Tyagi, K., Chandniha, S. K., & Sen, N. (2024). Evaluation of the health risk of fluoride and nitrate-rich groundwater in arid and semi-arid parts of India: An overview. *Environmental metagenomics, water quality and suggested remediation measures of polluted waters: a combined approach.* (pp. 89–105).

Ahamad, F., Tyagi, S. K., Singh, M., & Sharma, A. K. (2023). Groundwater in arid and semi-arid regions of India: A review on the quality, management and challenges. *Groundwater in arid and semi-arid areas: Monitoring assessment, modelling, and management* (pp. 11–52).

Ahmed, A. N., Othman, F. B., Afan, H. A., Ibrahim, R. K., Fai, C. M., Hossain, M. S., & Elshafie, A. (2019). Machine learning methods for better water quality prediction. *Journal of Hydrology*, *578*, 124084.

Akhtar, N., Syakir Ishak, M. I., Bhawani, S. A., & Umar, K. (2021). Various natural and anthropogenic factors responsible for water quality degradation: A review. *Water*, *13*(19), 2660.

Ali, S., Verma, S., Agarwal, M. B., Islam, R., Mehrotra, M., Deolia, R. K., & Fattahi, M. (2024). Groundwater quality assessment using water quality index and principal component analysis in the Achnera block, Agra district, Uttar Pradesh, Northern India. *Scientific Reports*, *14*(1), 5381.

Alprol, A. E., Mansour, A. T., Ibrahim, M. E. E. D., & Ashour, M. (2024). Artificial intelligence technologies revolutionizing wastewater treatment: Current trends and future prospective. *Water*, *16*(2), 314.

Amare, G., Dobo, B., & Haile, E. (2023). The effect of wet coffee processing plant effluent on physicochemical and bacteriological quality of receiving rivers used by local community: Case of aroresa District, sidama, Ethiopia. *Environmental Health Insights*, *17*, 11786302231165186.

Ameta, S. K., Kamaal, M., & Ahamad, F. (2023). Impact of domestic and industrial effluent disposal on physicochemical characteristics of River Malin at Najibabad City, India. *AgroEnvironmental Sustainability*, *1*(3), 246–256.

Amr, S. S. A., Aziz, H. A., & Adlan, M. N. (2013). Optimization of stabilized leachate treatment using ozone/persulfate in the advanced oxidation process. *Waste Management*, *33*(6), 1434–1441.

Anukam, A., Mohammadi, A., Naqvi, M., & Granström, K. (2019). A review of the chemistry of anaerobic digestion: Methods of accelerating and optimizing process efficiency. *Processes*, *7*(8), 504.

Aqel, H., Sannan, N., Al-Hunaiti, A., & Fodah, R. (2024). Integrated water quality dynamics in Wadi Hanifah: Physical, chemical, and biological perspectives. *PLoS One*, *19*(2), e0298200.

Ashadullah, A. K. M., Shafiquzzaman, M., Haider, H., Alresheedi, M., Azam, M. S., & Ghumman, A. R. (2021). Wastewater treatment by microalgal membrane bioreactor: Evaluating the effect of organic loading rate and hydraulic residence time. *Journal of Environmental Management*, *278*, 111548.

Ashrafivala, M., Mousavi, S. B., Heris, S. Z., Heidari, M., Mohammadpourfard, M., & Aslani, H. (2022). Investigation of H_2O_2/UV advanced oxidation process on the removal rate of coliforms from the industrial effluent: A pilot-scale study. *International Journal of Hydrogen Energy*, *47*(78), 33530–33540.

Awasthi, A., Rishi, M. S., & Panjgotra, S. (2024). Groundwater quality assessment for drinking and industrial purposes in transboundary aquifers of Gurdaspur district, Punjab, India. *International Journal of Environmental Analytical Chemistry*, *104*(2), 389–403.

Awasthi, M. K., Singh, E., Binod, P., Sindhu, R., Sarsaiya, S., Kumar, A., & Zhang, Z. (2022). Biotechnological strategies for bio-transforming biosolid into resources toward circular bio-economy: A review. *Renewable and Sustainable Energy Reviews*, *156*, 111987.

Awewomom, J., Dzeble, F., Takyi, Y. D., Ashie, W. B., Ettey, E. N. Y. O., Afua, P. E., & Akoto, O. (2024). Addressing global environmental pollution using environmental control techniques: A focus on environmental policy and preventive environmental management. *Discover Environment*, *2*(1), 8.

Badawi, A. K., & Zaher, K. (2021). Hybrid treatment system for real textile wastewater remediation based on coagulation/flocculation, adsorption and filtration processes: Performance and economic evaluation. *Journal of Water Process Engineering*, *40*, 101963.

Bai, X., Wang, J., Chen, R., Kang, Y., Ding, Y., Lv, Z., & Feng, H. (2024). Research progress of inland river water quality monitoring technology based on unmanned aerial vehicle hyperspectral imaging technology. *Environmental Research*, 119254.

Bain, R., Johnston, R., Mitis, F., Chatterley, C., & Slaymaker, T. (2018). Establishing sustainable development goal baselines for household drinking water, sanitation and hygiene services. *Water*, *10*(12), 1711.

Bala, M. S., Azoddein, A. A. B. M., Ahdash, M. I., & Alshwal, A. B. (2018). Comparative study for activation of hydrogen peroxide by chemical reagents (Fe^{2+}, Al^{3+}) to reduce chemical oxygen demandin petrochemical wastewater. *Journal of Engineering Technology*, *2*(2), 12.

Balaram, V., Rani, A., & Rathore, D. P. S. (2022). Uranium in groundwater in parts of India and world: A comprehensive review of sources, impact to the environment and human health, analytical techniques, and mitigation technologies. *Geosystems and Geoenvironment*, *1*(2), 100043.

Bano, I., & Arshad, M. (2018). Climatic changes impact on water availability. *Perspectives on water usage for biofuels production: Aquatic contamination and climate change* (pp. 39–54).

Bekele, E., Page, D., Vanderzalm, J., Kaksonen, A., & Gonzalez, D. (2018). Water recycling via aquifers for sustainable urban water quality management: Current status, challenges and opportunities. *Water*, *10*(4), 457.

Bhutiani, R., Ahamad, F., & Ram, K. (2021a). Quality assessment of groundwater at laksar block, Haridwar in Uttarakhand, India using water quality index: A case study. *Journal of Applied and Natural Science*, *13*(1), 197–203.

Bhutiani, R., Ahamad, F., & Ruhela, M. (2021b). Effect of composition and depth of filter-bed on the efficiency of sand-intermittent-filter treating the Industrial wastewater at Haridwar, India. *Journal of Applied and Natural Science*, *13*(1), 88–94.

Bohra, V., Ahamad, K. U., Kela, A., Vaghela, G., Sharma, A., & Deka, B. J. (2022). Energy and resources recovery from wastewater treatment systems. Clean energy and resource recovery (pp. 17–36). Elsevier.

Bortoleto, A. P., Barbosa, P. S. F., Maniero, M. G., Guimarães, J. R., & Junior, L. C. M. V. (2021). A water-energy nexus analysis to a sustainable transition path for Sao Paulo State, Brazil. *Journal of Cleaner Production*, *319*, 128697.

Cai, Q. Q., Wu, M. Y., Li, R., Deng, S. H., Lee, B. C. Y., Ong, S. L., & Hu, J. Y. (2020). Potential of combined advanced oxidation–Biological process for cost-effective organic matters removal in reverse osmosis concentrate produced from industrial wastewater reclamation: Screening of AOP pre-treatment technologies. *Chemical Engineering Journal*, *389*, 123419.

Campo, R., Sguanci, S., Caffaz, S., Mazzoli, L., Ramazzotti, M., Lubello, C., & Lotti, T. (2020). Efficient carbon, nitrogen and phosphorus removal from low C/N real domestic wastewater with aerobic granular sludge. *Bioresource Technology*, *305*, 122961.

Capodaglio, A. G. (2021). Fit-for-purpose urban wastewater reuse: Analysis of issues and available technologies for sustainable multiple barrier approaches. *Critical Reviews in Environmental Science and Technology*, *51* (15), 1619–1666.

Cesaro, A., Russo, L., & Belgiorno, V. (2015). Combined anaerobic/aerobic treatment of OFMSW: Performance evaluation using mass balances. *Chemical Engineering Journal*, *267*, 16–24.

Chauhan, A., & Singh, S. (2010). Evaluation of Ganga water for drinking purpose by water quality index at Rishikesh, Uttarakhand, India. *Report and Opinion*, *2*(9), 53–61.

Chavan, S., Atmakuri, A., Kumar, S., Yadav, B., Tyagi, R. D., Drogui, P., & Wong, J. W. (2024). *Introduction to decentralized treatment. Decentralized sanitation and water treatment.* (pp. 3–18). CRC Press.

Chen, H., Li, A., Wang, Q., Cui, D., Cui, C., & Ma, F. (2018). Nitrogen removal performance and microbial community of an enhanced multistage A/O biofilm reactor treating low-strength domestic wastewater. *Biodegradation*, *29*, 285–299.

Deng, Y. (2009). Advanced oxidation processes (AOPs) for reduction of organic pollutants in landfill leachate: A review. *International Journal of Environment and Waste Management*, *4*(3-4), 366–384.

Deng, Y., & Zhao, R. (2015). Advanced oxidation processes (AOPs) in wastewater treatment. *Current Pollution Reports*, *1*, 167–176.

Dey, S., Singh, S., Raju, N. J., & Mall, R. K. (2024). Hydrogeochemical characterization for groundwater quality and risk assessment in part of central gangetic alluvium, India. *Groundwater for Sustainable Development*, 101108.

Dhangar, K., & Kumar, M. (2020). Tricks and tracks in removal of emerging contaminants from the wastewater through hybrid treatment systems: A review. *Science of the Total Environment*, *738*, 140320.

Di Maria, F., Sisani, F., Norouzi, O., & Mersky, R. L. (2019). The effectiveness of anaerobic digestion of bio-waste in replacing primary energies: An EU28 case study. *Renewable and Sustainable Energy Reviews*, *108*, 347–354.

Dutt, M. A., Hanif, M. A., Nadeem, F., & Bhatti, H. N. (2020). A review of advances in engineered composite materials popular for wastewater treatment. *Journal of Environmental Chemical Engineering*, *8*(5), 104073.

Emodi, E. E. P. (2020). Analysis of industrial effluents as they affect the quality of surface water in Enugu Nigeria. *Iconic Research and Engineering Journals*, *4*, 6.

Gadewar, M., Prashanth, G. K., Babu, M. R., Dileep, M. S., Prashanth, P. A., Rao, S., & Orhan, E. (2023). Unlocking nature's potential: Green synthesis of ZnO nanoparticles and their multifaceted applications-A concise overview. *Journal of Saudi Chemical Society*, 101774.

Garcia, X., & Pargament, D. (2015). Reusing wastewater to cope with water scarcity: Economic, social and environmental considerations for decision-making. *Resources, Conservation and Recycling*, *101*, 154–166.

Germec, M., Demirci, A., & Turhan, I. (2020). Biofilm reactors for value-added products production: An in-depth review. *Biocatalysis and Agricultural Biotechnology*, *27*, 101662.

Giles, A. B., Correa, R. E., Santos, I. R., & Kelaher, B. (2024). Using multispectral drones to predict water quality in a subtropical estuary. *Environmental Technology*, *45*(7), 1300–1312.

Gitis, V., & Hankins, N. (2018). Water treatment chemicals: Trends and challenges. *Journal of Water Process Engineering*, *25*, 34–38.

Goblirsch, T., Mayer, T., Penzel, S., Rudolph, M., & Borsdorf, H. (2023). In situ water quality monitoring using an optical multiparameter sensor probe. *Sensors*, *23*(23), 9545.

Grover, D. P., Zhou, J. L., Frickers, P. E., & Readman, J. W. (2011). Improved removal of estrogenic and pharmaceutical compounds in sewage effluent by full scale granular activated carbon: Impact on receiving river water. *Journal of Hazardous Materials*, *185*(2-3), 1005–1011.

Gukelberger, E., Gabriele, B., Hoinkis, J., & Figoli, A. (2019). MBR and integration with renewable energy toward suitable autonomous wastewater treatment. *Current trends and future developments on (bio-) membranes* (pp. 355–384). Elsevier.

Gupta, A. D., Pandey, P., Feijóo, A., Yaseen, Z. M., & Bokde, N. D. (2020). Smart water technology for efficient water resource management: A review. *Energies*, *13*(23), 6268.

Hamiche, A. M., Stambouli, A. B., & Flazi, S. (2016). A review of the water-energy nexus. *Renewable and Sustainable Energy Reviews*, *65*, 319–331.

Hayat, S., Jamali, H., Jan, M. Z., & Ihtisham, Y. (2024). Who gets water? governance failures & persistent drinking water and sanitation issues. *Qlantic Journal of Social Sciences*, *5*(2), 201–214.

Herrera-Robledo, M., Morgan-Sagastume, J. M., & Noyola, A. (2010). Biofouling and pollutant removal during long-term operation of an anaerobic membrane bioreactor treating municipal wastewater. *Biofouling*, *26*(1), 23–30.

Induchoodan, T. G., Haq, I., & Kalamdhad, A. S. (2022). Factors affecting anaerobic digestion for biogas production: A review. *Advanced organic waste management* (pp. 223–233).

Iorhemen, O. T., Hamza, R. A., & Tay, J. H. (2016). Membrane bioreactor (MBR) technology for wastewater treatment and reclamation: Membrane fouling. *Membranes*, *6*(2), 33.

Ismail, W., Niknejad, N., Bahari, M., Hendradi, R., Zaizi, N. J. M., & Zulkifli, M. Z. (2021). Water treatment and artificial intelligence techniques: A systematic literature review research. *Environmental science and pollution research* (pp. 1–19).

Jagaba, A. H., Kutty, S. R. M., Noor, A., Birniwa, A. H., Affam, A. C., Lawal, I. M., & Kilaco, A. U. (2021). A systematic literature review of biocarriers: Central elements for biofilm formation, organic and nutrients removal in sequencing batch biofilm reactor. *Journal of Water Process Engineering*, *42*, 102178.

Jan, F., Min-Allah, N., & Düştegör, D. (2021). IoT based smart water quality monitoring: Recent techniques, trends and challenges for domestic applications. *Water*, *13*(13), 1729.

Ji, B., Wei, L., Chen, D., Wang, H., Li, Z., & Yang, K. (2015). Domestic wastewater treatment in a novel sequencing batch biofilm filter. *Applied Microbiology and Biotechnology*, *99*, 5731–5738.

Jung, H. S., Verwilst, P., Kim, W. Y., & Kim, J. S. (2016). Fluorescent and colorimetric sensors for the detection of humidity or water content. *Chemical Society Reviews*, *45*(5), 1242–1256.

Kanoun, O., Lazarevi-Pašti, T., Pašti, I., Nasraoui, S., Talbi, M., Brahem, A., & Al-Hamry, A. (2021). A review of nanocomposite-modified electrochemical sensors for water quality monitoring. *Sensors*, *21*(12), 4131.

Katam, K., & Bhattacharyya, D. (2021). Improving the performance of activated sludge process with integrated algal biofilm for domestic wastewater treatment: system behavior during the start-up phase. *Bioresource Technology Reports*, *13*, 100618.

Kathawala, T. M., Gayathri, K. V., & Senthil Kumar, P. (2021). A performance comparison of anaerobic and an integrated anaerobic-aerobic biological reactor system for the effective treatment of textile wastewater. *International Journal of Chemical Engineering*, *2021*(1), 8894332.

Kaur, N., Kumar, R., Alhan, S., Sharma, H., Singh, N., Yogi, R., & Kumar, A. (2024). Lycium shawii mediated green synthesis of silver nanoparticles, characterization and assessments of their phytochemical, antioxidant, antimicrobial properties. *Inorganic Chemistry Communications*, *159*, 111735.

Khalkhali, M., Westphal, K., & Mo, W. (2018). The water-energy nexus at water supply and its implications on the integrated water and energy management. *Science of the Total Environment*, *636*, 1257–1267.

Khan, A. H., Khan, N. A., Ahmed, S., Dhingra, A., Singh, C. P., Khan, S. U., & Ali, I. (2020). Application of advanced oxidation processes followed by different treatment technologies for hospital wastewater treatment. *Journal of Cleaner Production*, *269*, 122411.

Khan, W. U., Ahmed, S., Dhoble, Y., & Madhav, S. (2023a). A critical review of hazardous waste generation from textile industries and associated ecological impacts. *Journal of the Indian Chemical Society*, *100*(1), 100829.

Khan, Z. U. H., Gul, N. S., Sabahat, S., Sun, J., Tahir, K., Shah, N. S., & Wu, J. (2023b). Removal of organic pollutants through hydroxyl radical-based advanced oxidation processes. *Ecotoxicology and Environmental Safety*, *267*, 115564.

Krishnan, S. R., Nallakaruppan, M. K., Chengoden, R., Koppu, S., Iyapparaja, M., Sadhasivam, J., & Sethuraman, S. (2022). Smart water resource management using artificial intelligence—A review. *Sustainability*, *14*(20), 13384.

Krklješ, D. B., Kiti, G. V., Petes, C. M., Birgermajer, S. S., Stanojev, J. D., Bajac, B. M., & Matovi, J. B. (2024). Multiparameter water quality monitoring system for continuous monitoring of fresh waters. *IEEE Sensors Journal*.

Kumar, M., Mehra, R., Yogi, R., Singh, N., Salar, R. K., Saxena, G., & Rustagi, S. (2023). A novel tannase from *Klebsiella pneumoniae* KP715242 reduces haze and improves the quality of fruit juice and beverages through detannification. *Frontiers in Sustainable Food Systems*, *7*, 1173611.

Lal, B. S. (2019). Water for life: Issues and challenges. *International Journal of Scientific Research*, *8*, 1949–1957.

Lal, B., Sengar, S. S., Singh, R., Jhariya, M. K., & Raj, A. (2023). Hydrogeochemistry and groundwater quality assessment in Ambagarh Chowki, Chhattisgarh, India. *Environmental Monitoring and Assessment*, *195*(1), 43.

Li, L., Rong, S., Wang, R., & Yu, S. (2021). Recent advances in artificial intelligence and machine learning for nonlinear relationship analysis and process control in drinking water treatment: A review. *Chemical Engineering Journal*, *405*, 126673.

Liebetrau, J., Sträuber, H., Kretzschmar, J., Denysenko, V., & Nelles, M. (2019). Anaerobic digestion. *Biorefineries*, 281–299.

Liu, J., Liu, X., Gao, L., Xu, S., Chen, X., Tian, H., & Kang, X. (2020). Performance and microbial community of a novel combined anaerobic bioreactor integrating anaerobic baffling and anaerobic filtration process for low-strength rural wastewater treatment. *Environmental Science and Pollution Research*, *27*, 18743–18756.

Liu, L., Guo, Z., Wang, Y., Yin, L., Zuo, W., Tian, Y., & Zhang, J. (2024). Low Energy-consumption Oriented Membrane Fouling Control Strategy In Anaerobic Fluidized Membrane Bioreactor. *Chemosphere*, 142254.

Lowe, M., Qin, R., & Mao, X. (2022). A review on machine learning, artificial intelligence, and smart technology in water treatment and monitoring. *Water*, *14*(9), 1384.

Mahbub, P., & Duke, M. (2023). Scalability of advanced oxidation processes (AOPs) in industrial applications: A review. *Journal of Environmental Management*, *345*, 118861.

Mahjouri, M., Ishak, M. B., Torabian, A., Abd Manaf, L., & Halimoon, N. (2017). The application of a hybrid model for identifying and ranking indicators for assessing the sustainability of wastewater treatment systems. *Sustainable Production and Consumption*, *10*, 21–37.

Makarigakis, A. K., & Jimenez-Cisneros, B. E. (2019). UNESCO's contribution to face global water challenges. *Water*, *11*(2), 388.

Mazhar, M. A., Khan, N. A., Ahmed, S., Khan, A. H., Hussain, A., Changani, F., & Vambol, V. (2020). Chlorination disinfection by-products in municipal drinking water—A review. *Journal of Cleaner Production*, *273*, 123159.

Mazhar, M. A., Khan, N. A., Khan, A. H., Ahmed, S., Siddiqui, A. A., Husain, A., & Radwan, N. (2021). Upgrading combined anaerobic-aerobic UASB-FPU to UASB-DHS system: Cost comparison and performance perspective for developing countries. *Journal of Cleaner Production*, *284*, 124723.

Mekuria, D. M., Kassegne, A. B., & Asfaw, S. L. (2021). Assessing pollution profiles along Little Akaki River receiving municipal and industrial wastewaters, Central Ethiopia: Implications for environmental and public health safety. *Heliyon*, *7*(7).

Meng, F., Chae, S. R., Drews, A., Kraume, M., Shin, H. S., & Yang, F. (2009). Recent advances in membrane bioreactors (MBRs): Membrane fouling and membrane material. *Water Research*, *43*(6), 1489–1512.

Miklos, D. B., Remy, C., Jekel, M., Linden, K. G., Drewes, J. E., & Hübner, U. (2018). Evaluation of advanced oxidation processes for water and wastewater treatment—A critical review. *Water Research*, *139*, 118–131.

Molajou, A., Afshar, A., Khosravi, M., Soleimanian, E., Vahabzadeh, M., & Variani, H. A. (2021). A new paradigm of water, food, and energy nexus. *Environmental Science and Pollution Research*, 1–11.

Murshid, S., Antonysamy, A., Dhakshinamoorthy, G., Jayaseelan, A., & Arivalagan, P. (2023). A review on biofilm-based reactors for wastewater treatment: Recent advancements in biofilm carriers, kinetics, reactors, economics, and future perspectives. *Science of the Total Environment*, 164796.

Nandi, A., Megiddo, I., Ashok, A., Verma, A., & Laxminarayan, R. (2017). Reduced burden of childhood diarrheal diseases through increased access to water and sanitation in India: A modeling analysis. *Social Science & Medicine*, *180*, 181–192.

Negi, R., & Chandel, M. K. (2022). Assessment on embodied energy and greenhouse gas emissions in urban water system from life cycle perspective: A typical case of India. *Sustainable Cities and Society*, *86*, 104152.

Negi, R., & Chandel, M. K. (2024). Embodied energy and greenhouse gas emissions from wastewater reuse strategies in Indian Himalayan region. *Renewable and Sustainable Energy Reviews*, *192*, 114269.

Nguyen, D. D., Ngo, H. H., & Yoon, Y. S. (2014). A new hybrid treatment system of bioreactors and electrocoagulation for superior removal of organic and nutrient pollutants from municipal wastewater. *Bioresource Technology*, *153*, 116–125.

Omer, N. H. (2019). *Water quality parameters. Water quality-science, assessments and policy (Vol. 81)* (pp. 1–34).

Onwuka, I. S., Scinto, L. J., & Fugate, D. C. (2023). High-Resolution Estimation Of Suspended Solids And Particulate Phosphorus Using Acoustic Devices In A Hydrologically Managed Canal In South Florida, USA. *Sensors*, *23*(4), 2281.

Oopath, S. V., Baji, A., Abtahi, M., Luu, T. Q., Vasilev, K., & Truong, V. K. (2023). Nature-inspired biomimetic surfaces for controlling bacterial attachment and biofilm development. *Advanced Materials Interfaces*, *10*(4), 2201425.

Pai, C. W., & Wang, G. S. (2022). Treatment of PPCPs and disinfection by-product formation in drinking water through advanced oxidation processes: Comparison of UV, UV/Chlorine, and UV/H_2O_2. *Chemosphere*, *287*, 132171.

Park, J. B., Sukias, J. P., & Tanner, C. C. (2019). Floating treatment wetlands supplemented with aeration and biofilm attachment surfaces for efficient domestic wastewater treatment. *Ecological Engineering*, *139*, 105582.

Pennellini, S. (2022). Appropriate technologies and natural systems for wastewater treatment in low-and middle-income countries.

Pichel, N., Vivar, M., & Fuentes, M. (2019). The problem of drinking water access: A review of disinfection technologies with an emphasis on solar treatment methods. *Chemosphere*, *218*, 1014–1030.

Posadas, E., García-Encina, P. A., Soltau, A., Domínguez, A., Díaz, I., & Muñoz, R. (2013). Carbon and nutrient removal from centrates and domestic wastewater using algal–bacterial biofilm bioreactors. *Bioresource Technology*, *139*, 50–58.

Prakash, S. (2021). Impact of climate change on aquatic ecosystem and its biodiversity: An overview. *International Journal of Biological Innovations*, *3*(2).

Prasad, M., Kumar, G. A., Sahoo, S. K., & Ramola, R. C. (2019). Health risks associated with the exposure to uranium and heavy metals through potable groundwater in Uttarakhand state of India. *Journal of Radioanalytical and Nuclear Chemistry*, *319*, 13–21.

Prudencio, L., & Null, S. E. (2018). Stormwater management and ecosystem services: A review. *Environmental Research Letters*, *13*(3), 033002.

Puyol, D., Batstone, D. J., Hülsen, T., Astals, S., Peces, M., & Krömer, J. O. (2017). Resource recovery from wastewater by biological technologies: Opportunities, challenges, and prospects. *Frontiers in Microbiology*, *7*, 2106.

Rahaman, M. M., & Varis, O. (2005). Integrated water resources management: Evolution, prospects and future challenges. *Sustainability: Science, Practice and Policy*, *1*(1), 15–21.

Rahman, T. U., Roy, H., Islam, M. R., Tahmid, M., Fariha, A., Mazumder, A., & Islam, M. S. (2023). The advancement in membrane bioreactor (MBR) technology toward sustainable industrial wastewater management. *Membranes*, *13*(2), 181.

Rajesh Banu, J., Merrylin, J., Kavitha, S., Yukesh Kannah, R., Selvakumar, P., Gopikumar, S., & Kumar, G. (2021). Trends in biological nutrient removal for the treatment of low strength organic wastewaters. *Current Pollution Reports*, *7*(1), 1–30.

Rasheed, M. B., & R-Moreno, M. D. (2024). An integrated model with interdependent water storage for optimal resource management in Energy-Water-Food Nexus. *Journal of Cleaner Production*, 142648.

Rodriguez-Narvaez, O. M., Peralta-Hernandez, J. M., Goonetilleke, A., & Bandala, E. R. (2017). Treatment technologies for emerging contaminants in water: A review. *Chemical Engineering Journal*, *323*, 361–380.

Román, A., Tovar-Sánchez, A., Gauci, A., Deidun, A., Caballero, I., Colica, E., & Navarro, G. (2022). Water-quality monitoring with a UAV-mounted multispectral camera in coastal waters. *Remote Sensing*, *15*(1), 237.

Roy, G. K., Kumar, S., Alam, F., Kujur, A., Anand, S., & Srivastava, S. K. (2023). Groundwater quality assessment using water quality index in Ranchi Urban area, Jharkhand (India): In parts of Subarnrekha River Basin. *International Journal of Environmental Analytical Chemistry*, *103*(17), 4970–4983.

Ruhela, M., Ahamad, F., Bhardwaj, S., & Gupta, U. (2023). Efficiency assessment of 3.2 MLD MBR based sewage treatment plant of IFFCO township Aonla, Bareilly, Uttar Pradesh, India. *Environment Conservation Journal*, *24*(2), 347–354.

Ruiz-Marín, A., Campos-Garcia, S., Zavala-Loría, J., Solana, F., & Canedo-López, Y. (2009). Assessment of Luffa cylindrica as support in biofilms reactors for the biological treatment of domestic wastewater. *Water, Air, and Soil Pollution*, *199*(1), 13–21.

Saidulu, D., Majumder, A., & Gupta, A. K. (2021). A systematic review of moving bed biofilm reactor, membrane bioreactor, and moving bed membrane bioreactor for wastewater treatment: Comparison of research trends, removal mechanisms, and performance. *Journal of Environmental Chemical Engineering*, *9*(5), 106112.

Salunke, P., & Kate, J. (2017). Advanced smart sensor interface in internet of things for water quality monitoring. In *2017 International Conference on Data Management, Analytics and Innovation (ICDMAI)* (pp. 298–302). IEEE.

Santos, A., Ma, W., & Judd, S. J. (2011). Membrane bioreactors: Two decades of research and implementation. *Desalination*, *273*(1), 148–154.

Sarang, A., Vahedi, A., & Shamsai, A. (2008). How to quantify sustainable development: A risk-based approach to water quality management. *Environmental Management*, *41*, 200–220.

Saravanan, A., Kumar, P. S., Jeevanantham, S., Karishma, S., Tajsabreen, B., Yaashikaa, P. R., & Reshma, B. (2021). Effective water/wastewater treatment methodologies for toxic pollutants removal: Processes and applications towards sustainable development. *Chemosphere*, *280*, 130595.

Sarkar, M., Pal, S. C., & Islam, A. R. M. T. (2022). Groundwater quality assessment for safe drinking water and irrigation purposes in Malda district, Eastern India. *Environmental Earth Sciences*, *81*(2), 52.

Schellenberg, T., Subramanian, V., Ganeshan, G., Tompkins, D., & Pradeep, R. (2020). Wastewater discharge standards in the evolving context of urban sustainability—The case of India. *Frontiers in Environmental Science*, *8*, 30.

Sen, M., Manna, A., & Pal, P. (2010). Removal of arsenic from contaminated groundwater by membrane-integrated hybrid treatment system. *Journal of Membrane Science*, *354*(1-2), 108–113.

Shafique, M., & Kim, R. (2017). Green stormwater infrastructure with low impact development concept: A review of current research. *Desalination and Water Treatment*, *83*(7), 16–29.

Shah, S. V., Lamba, B. Y., Tiwari, A. K., & Chen, W. H. (2024). Sustainable biogas production via anaerobic digestion with focus on CSTR technology: A review. *Journal of the Taiwan Institute of Chemical Engineers*, *162*, 105575.

Show, K. Y., Yan, Y. G., Zhao, J., Shen, J., Han, Z. X., Yao, H. Y., & Lee, D. J. (2020). Laboratory trial and full-scale implementation of integrated anaerobic-aerobic treatment for high strength acrylic acid wastewater. *Science of the Total Environment*, *738*, 140323.

Singh, N., Jiwani, G., Rocha, L. S., & Mazaheri, R. (2023a). Bioagents and volatile organic compounds: An emerging control measures for rice bacterial diseases. *Bacterial diseases of rice and their management* (pp. 255–274). Apple Academic Press.

Singh, P., Tiwari, R. C., Bhutiani, R., & Ahamad, F. (2023b). Efficiency of reactors composed of plant based absorbents in combination with sand and gravel for physicochemical parameters of different category water. *Environment Conservation Journal*, *24*(2), 148–161.

Singh, R. L., & Singh, P. K. (2017). Global environmental problems. *Principles and applications of environmental biotechnology for a sustainable future* (pp. 13–41). .

Sirés, I., Brillas, E., Oturan, M. A., Rodrigo, M. A., & Panizza, M. (2014). Electrochemical advanced oxidation processes: today and tomorrow. A review. *Environmental Science and Pollution Research*, *21*, 8336–8367.

Srivastava, R. P., Kumar, S., Singh, L., Madhukar, M., Singh, N., Saxena, G., & Rustagi, S. (2023). Major phenolic compounds, antioxidant, antimicrobial, and cytotoxic activities of Selinum carvifolia (L.) collected from different altitudes in India. *Frontiers in Nutrition*, *10*, 1180225.

Suresh, S., Rangaswamy, D. R., Srinivasa, E., & Sannappa, J. (2020). Measurement of radon concentration in drinking water and natural radioactivity in soil and their radiological hazards. *Journal of Radiation Research and Applied Sciences*, *13*(1), 12–26.

Tamang, A. M., Singh, N., Chandraker, S. K., & Ghosh, M. K. (2022). Solvent impregnated resin a potential alternative material for separation dyes, metal and phenolic compounds: A review. *Current Research in Green and Sustainable Chemistry*, *5*, 100232.

Tariq, A., & Mushtaq, A. (2023). Untreated wastewater reasons and causes: A review of most affected areas and cities. *International Journal of Chemical and Biochemical Science*, *23*, 121–143.

Tee, P. F., Abdullah, M. O., Tan, I. A. W., Rashid, N. K. A., Amin, M. A. M., Nolasco-Hipolito, C., & Bujang, K. (2016). Review on hybrid energy systems for wastewater treatment and bio-energy production. *Renewable and Sustainable Energy Reviews*, *54*, 235–246.

Thakur, A., & Devi, P. (2024). A comprehensive review on water quality monitoring devices: Materials advances, current status, and future perspective. *Critical Reviews in Analytical Chemistry*, *54*(2), 193–218.

Thakur, A., & Kumar, A. (2022). Recent advances on rapid detection and remediation of environmental pollutants utilizing nanomaterials-based (bio) sensors. *Science of the Total Environment*, *834*, 155219.

Tijani, J. O., Fatoba, O. O., Madzivire, G., & Petrik, L. F. (2014). A review of combined advanced oxidation technologies for the removal of organic pollutants from water. *Water, Air, & Soil Pollution*, *225*, 1–30.

Tufail, A., Price, W. E., Mohseni, M., Pramanik, B. K., & Hai, F. I. (2021). A critical review of advanced oxidation processes for emerging trace organic contaminant degradation: Mechanisms, factors, degradation products, and effluent toxicity. *Journal of Water Process Engineering*, *40*, 101778.

Tyagi, I., Tyagi, K., Ahamad, F., Kothari, R., & Kumar, V. (2024). Aqueous ecosystem: Environmental metagenomics, water quality, and possible remediation measures. *Environmental Metagenomics, Water Quality And Suggested Remediation Measures of Polluted Waters: A Combined Approach* (pp. 1–13).

Uddin, M. M., & Wright, M. M. (2023). Anaerobic digestion fundamentals, challenges, and technological advances. *Physical Sciences Reviews*, *8*(9), 2819–2837.

Upadhyay, A. K., Mallick, S., Singh, R., Singh, L., Singh, N., Mandotra, S. K., & Saxena, G. (2023). Novel cost-effective design for bio-volatilization studies in photosynthetic microalgae exposed to arsenic with emphasis on growth and glutathione modulation. *Frontiers in Microbiology*, *14*, 1170740.

Vaseashta, A., Duca, G., Covaliova, O., & Romanciuc, L. (2022). Water safety, security, and sustainability: Emerging trends and future pathwarys. *Handbook of Research On Water Sciences And Society* (pp. 1–36).

Vasiliev, D., & Greenwood, S. (2022). Making green pledges support biodiversity: Nature-based solution design can be informed by landscape ecology principles. *Land Use Policy*, *117*, 106129.

Venkatesh, J., Partheeban, P., Baskaran, A., Krishnan, D., & Sridhar, M. (2024). Wireless sensor network technology and geospatial technology for groundwater quality monitoring. *Journal of Industrial Information. Journal of Industrial Information Integration*, *38*, 100569.

Vinayagam, V., Palani, K. N., Ganesh, S., Rajesh, S., Akula, V. V., Avoodaiappan, R., & Pugazhendhi, A. (2023). Recent developments on advanced oxidation processes for degradation of pollutants from wastewater with focus on antibiotics and organic dyes. *Environmental Research*, 117500.

Wainaina, S., Awasthi, M. K., Sarsaiya, S., Chen, H., Singh, E., Kumar, A., & Taherzadeh, M. J. (2020). Resource recovery and circular economy from organic solid waste using aerobic and anaerobic digestion technologies. *Bioresource Technology*, *301*, 122778.

Wakode, H. B., Baier, K., Jha, R., & Azzam, R. (2018). Impact of urbanization on groundwater recharge and urban water balance for the city of Hyderabad, India. *International Soil and Water Conservation Research*, *6*(1), 51–62.

Wang, Y., Liang, B., Kang, F., Wang, Y., Zhao, C., Lyu, Z., & Zhang, Z. (2022). An efficient anoxic/aerobic/aerobic/anoxic process for domestic sewage treatment: From feasibility to application. *Frontiers in Microbiology*, *13*, 970548.

Weerasooriya, R. R., Liyanage, L. P. K., Rathnappriya, R. H. K., Bandara, W. B. M. A. C., Perera, T. A. N. T., Gunarathna, M. H. J. P., & Jayasinghe, G. Y. (2021). Industrial water conservation by water footprint and sustainable development goals: A review. *Environment, Development and Sustainability*, *23*(9), 12661–12709.

World Health Organization, & United Nations Children's Fund. (2021). *Progress on household drinking water, sanitation and hygiene 2000-2020: Five years into the SDGs*. World Health Organization.

Xu, M., Bernards, M., & Hu, Z. (2014). Algae-facilitated chemical phosphorus removal during high-density *Chlorella emersonii* cultivation in a membrane bioreactor. *Bioresource Technology*, *153*, 383–387.

Yan, S., Drogui, P., Tyagi, R. D., & Wong, J. W. (2024). Development of Ddecentralized sanitation and water treatment. *Decentralized drinking water treatment plant* (pp. 139–153).

Yang, H., Kong, J., Hu, H., Du, Y., Gao, M., & Chen, F. (2022). A review of remote sensing for water quality retrieval: Progress and challenges. *Remote Sensing*, *14*(8), 1770.

Yang, Z., Pei, H., Hou, Q., Jiang, L., Zhang, L., & Nie, C. (2018). Algal biofilm-assisted microbial fuel cell to enhance domestic wastewater treatment: Nutrient, organics removal and bioenergy production. *Chemical Engineering Journal*, *332*, 277–285.

Zhou, Y., Zhou, J., Yu, J., Huang, X., Niu, X., & Tsang, Y. F. (2023). Improved integrated anaerobic–anoxic–oxic system for landfill leachate treatment using domestic wastewater as carbon source: Performance study and optimization. *Process Safety and Environmental Protection*, *174*, 997–1002.

CHAPTER 15

Introduction to computational automation in water quality management

Nuha Awang[1], Azyyati Johari[1], Sairul Izwan Safie[1], Mohd Al-Fatihhi Mohd Szali Januddi[1], Aliff Radzuan Mohamad Radzi[2], Mohd Anuar Ismail[1], Che Hafizan Che Hassan[1], Ahmad Firdaus Abdul Jalil[3], Syahidah Nurani Zulkifli[4] and Nur Naqiuddin Mdd Nordin[5]

[1]Plant Engineering Technology (PETech), Universiti Kuala Lumpur - Malaysian Institute of Industrial Technology (UniKL MITEC), Jln Persiaran Sinaran Ilmu, Bandar Seri Alam, Masai, Jahore Malaysia, [2]Universiti Kuala Lumpur Branch Campus — Malaysian Institute of Chemical and Bioengineering Technology (UniKL MICET), Lot 1988 Vendor City, Taboh Naning, Alor Gajah, Melaka, Malaysia, [3]Indah Water Konsortium Sdn. Bhd, Kuala Lumpur, Malaysia, [4]School of Electrical Engineering, Faculty of Engineering, Univeristi Teknologi Malaysia, Skudai, Jahore, Malaysia, [5]Langsat-OSC Sdn. Bhd, PLO 209, Jalan Rasau, Kompleks, Kawasan Perindustrian Tanjung Langsat, Pasir Gudang, Jahore, Malaysia

Abbreviations

Artificial intelligence	AI
Constrained application protocol	CoAP
Decision support systems	DSS
Dissolved oxygen	DO
Electrical conductivity	EC
Geographic information systems	GIS
Global system for mobile communication	GSM
Hypertext transfer protocol	HTTP
Internet of Things	IoT
Machine learning	ML
Message queuing telemetry transport	MQTT
Programmable logic controllers	PLCs
Remote terminal units	RTUs
Robotic process automation	RPA
Supervisory control and data acquisition	SCADA
Total dissolved solids	TDS
Total	TSS

Computational Automation for Water Security.
DOI: https://doi.org/10.1016/B978-0-443-33321-7.00001-9

15.1 Introduction

Water quality management is of utmost importance for various reasons, including its impact on human health, the environment, and industrial processes. Traditional methods of water quality management typically involve manual sampling and laboratory testing, which can be time-consuming and costly (Kumawat, Singh, Jain, & Sharma, 2021). The emergence of computational automation has revolutionized water quality management by providing more efficient and accurate monitoring and analysis capabilities.

Computational automation, defined as the use of computer algorithms and digital technologies to automate processes and tasks, has played a crucial role in modern water quality management by streamlining data collection, analysis, and decision-making processes (Rietveld, van der Helm, van Schagen, & Aa, 2010). The benefits of computational automation in water quality management are numerous. It allows for real-time monitoring of water quality parameters, which enables prompt identification of potential contamination events or changes in water quality.

Furthermore, computational automation can provide insights and predictions through advanced data analytics and machine learning algorithms. These insights can help water quality managers make informed decisions and take proactive measures to protect and improve water quality (Akomea-Frimpong et al., 2023). Historically, water quality monitoring relied heavily on manual sampling and laboratory testing. This approach had several limitations and challenges, including slow turnaround times for results, limited spatial coverage, and the inability to capture real-time changes in water quality.

15.2 Historical background

Proper sampling has always been a crucial condition for the accurate measurement of water quality parameters. Aquatic ecosystems play an essential role in maintaining water quality and serve as valuable indicators of water suitability for various uses. Historically, water quality monitoring relied heavily on manual sampling and laboratory analyzes. This traditional approach, although fundamental, presented several limitations (Rietveld et al., 2010). Manual sampling involves the collection of water samples using various containers from surface water bodies such as rivers, lakes, and reservoirs. The collected samples required meticulous handling to avoid contamination and to ensure the integrity of the samples (Htun, Shamsuzzoha, & Ahamed, 2023). Subsequently, these samples were transported to laboratories under controlled conditions to prevent any chemical or biological alterations (Kumawat et al., 2021).

In the laboratory, water samples underwent various analytical tests to measure parameters like pH, turbidity, dissolved oxygen (DO), and concentrations of nutrients and pollutants. Techniques employed included titration for alkalinity, colorimetric analysis for nutrient detection, and chromatographic methods for identifying organic contaminants. Despite the accuracy of these methods, they were labor-intensive and time-consuming, often taking several hours to days to yield results (Htun et al., 2023). This delay hindered prompt responses to pollution events or sudden changes in water quality (Akomea-Frimpong et al., 2023).

The limitations of traditional methods extended beyond slow turnaround times. The geographic scope of manual sampling was inherently restricted due to constraints in manpower and resources. As a result, large water bodies and extensive river systems frequently remained under-monitored, leading to significant data gaps (Infrastructure for climate action, 2022). Furthermore, the inability of traditional methods to provide continuous monitoring meant that transient events, such as spills, runoff after rainfall, or daily fluctuations in water quality, were often undetected (Htun et al., 2023). These limitations significantly reduced the effectiveness of monitoring programs in detecting and managing episodic pollution incidents (Akomea-Frimpong et al., 2023).

Studies from the late 20th century highlighted these challenges. For instance, research on the Great Lakes in the 1980s and 1990s demonstrated significant delays in detecting and addressing pollution due to reliance on manual methods (Kumawat et al., 2021). According to quantitative studies, it took an average of 24–48 hours for manual methods to give results after sample collection. Moreover, spatial analysis indicated that traditional methods covered only about 10%–15% of large water bodies annually, emphasizing the substantial gaps in data collection and monitoring efforts (Infrastructure for Climate Action, 2022).

15.3 Role of computational automation in modern water quality management

Water quality management is a critical aspect of ensuring the availability of clean and safe water resources for various applications, including domestic, agricultural, and industrial use (Htun, Shamsuzzoha, & Ahamed, 2023; Kumawat et al., 2021; Rietveld et al., 2010; Akomea-Frimpong et al., 2023; Infrastructure for climate action, 2022). In recent years, the role of computational automation has become increasingly important in improving the efficiency and accuracy of water quality management.

Artificial intelligence (AI) techniques have been explored for groundwater quality assessment, which is a crucial component of water quality management. Water quality indices, which aggregate multiple water quality parameters into a single metric, have been widely used to indicate the suitability of water samples for different purposes. Machine learning models have been shown to outperform traditional statistical approaches in predicting water quality indices (Htun et al., 2023; Klompenburg, Kassahun, & Çatal, 2020).

An example of the application of computational automation in water quality management is the development of a web crawler system for water quality monitoring data and data visualization. Such systems can automatically gather data from various sources, process it, and present it in a user-friendly format, enabling water quality managers to make informed decisions.

Another key aspect of computational automation in water quality management is the use of intelligent agent-based systems (Akomea-Frimpong et al., 2023). These systems can continuously monitor water quality parameters, detect anomalies, and trigger appropriate responses, such as adjusting treatment processes or alerting relevant authorities. This can lead to more efficient and responsive water quality management, reducing the risk of water-borne diseases and environmental damage (Lim and Spanger-Siegfried, n.d.).

The integration of cyber-physical systems with water infrastructure has also been a subject of recent research. These systems can gather and analyze massive amounts of data from various sensors, enabling real-time monitoring and control of water treatment and distribution processes.

The role of computational automation in modern water quality management is becoming increasingly important. Techniques such as AI, intelligent agent-based systems, and cyber-physical integration can enhance the efficiency, accuracy, and responsiveness of water quality management, leading to a more sustainable and resilient water infrastructure.

15.4 Emergence of automation in water quality

The emergence of automation in water quality management represents a transformative development in the field, driven by the increasing integration of digital technologies and the growing need for efficient and reliable monitoring and control systems (Kumawat et al., 2021; Rietveld et al., 2010). The early automated systems were primarily focused on the operation and control of drinking water treatment plants, aiming to mimic human operators and optimize the treatment process. These initial systems utilized basic sensors and control algorithms to regulate parameters such as chlorine dosage and filtration rates, thus ensuring consistent water quality output.

Over time, the focus of automation in water quality management has broadened to encompass a more holistic approach. Modern automated systems integrate advanced digital technologies, such as high-precision sensors, data analytics, and sophisticated control systems, to provide comprehensive solutions for water quality management (Rietveld et al., 2010). These integrated systems enable continuous monitoring and real-time data acquisition, offering a detailed and dynamic understanding of water quality across various points in the treatment and distribution network.

The automation of water treatment plants has yielded numerous benefits, including significant energy savings. For instance, the use of automated controls in the energy-intensive processes of aeration and chemical dosing has been shown to reduce energy consumption by up to 30% (Akomea-Frimpong et al., 2023). This is achieved by precisely adjusting operational parameters to match real-time water quality conditions, thereby avoiding overuse of energy and chemicals.

Automation has also greatly improved the ability to detect raw water quality changes early and to manage leaks and bursts in distribution systems. Advanced sensors deployed in real-time monitoring networks can detect minute changes in water quality parameters, allowing for the early identification of potential contamination sources. For example, the City of Toronto implemented a smart water network (SWN) that uses continuous monitoring and advanced analytics to detect and localize leaks within hours instead of days, significantly reducing water loss and maintenance costs (Htun et al., 2023).

Recent advancements in AI and machine learning (ML) have further enhanced the capabilities of automated water quality management systems. These technologies enable the development of intelligent control systems that can adaptively adjust treatment parameters based on real-time data, leading to improved efficiency and reduced operational costs. For example, the Dutch water utility Vitens has integrated AI to predict and prevent water quality issues. By analyzing vast datasets, AI algorithms can predict problems like pipe bursts or contaminant intrusions, allowing for proactive interventions (Infrastructure for Climate Action, 2022).

The integration of automated systems with digital technologies has also facilitated the development of remote monitoring and control capabilities. This advancement allows water utilities to monitor and manage their systems from a central location, even in areas with limited access or challenging terrain. In rural Australia, for instance, the integration of remote sensing and control systems has enabled effective water quality management in extensive and remote agricultural regions. These systems provide real-time data and control capabilities, significantly enhancing the ability to manage water resources efficiently and respond to issues promptly (Klompenburg et al., 2020).

One illustrative example is the deployment of an advanced water quality monitoring system in Singapore's Marina Barrage. This system combines high-frequency sensors, real-time data analytics, and automated control systems to manage the quality of the reservoir's water. The automated system continuously monitors parameters such as salinity, turbidity, and microbial levels and uses AI-driven analytics to predict water quality trends and adjust treatment processes accordingly. This approach has significantly improved water quality management efficiency, reduced operational costs, and ensured a reliable water supply (Lim and Spanger-Siegfried, n.d.).

Another notable example is the use of automated monitoring systems in the Chesapeake Bay, USA. This initiative involves a network of automated sensors and satellite data to monitor nutrient levels, temperature, and other critical water quality indicators. The real-time data collected are used by ML models to forecast harmful algal blooms, providing early warnings and enabling timely mitigation actions. This proactive approach has enhanced the bay's ecosystem health and improved water quality for recreational and commercial use (The Sustainable Development Goals Report, 2023).

The emergence of automation in water quality management has revolutionized the field by integrating advanced technologies to enhance monitoring, control, and decision-making processes. This evolution has led to more efficient, reliable, and scalable water quality management solutions, ultimately contributing to better public health outcomes and more sustainable water resource management.

15.5 Technologies in computational automation

Computational automation is a technology-driven approach that seeks to streamline processes and operations and improve efficiency in water quality management systems. At its foundation, computational automation combines numerous digital technology tools and advanced management computing systems to execute tasks with minimal or no human intervention. Computational automations are becoming increasingly important as the paradigm evolves toward innovative, sustainable, and inclusive water quality surveillance (Kumawat et al., 2021).

Water quality monitoring has been revolutionized through advancements in smart digital technologies such as remote sensing, cloud computing, data acquisition techniques, predictive analytics, and Internet of Things (IoT) devices, resulting in a more refined management system (Rietveld et al., 2010). Recognizing the behavior of water quality in large-scale water systems necessitates a big dataset, which may be collected by continuous observations and property assessments (Akomea-Frimpong et al., 2023). Therefore the adoption of computational automation in water quality management applications is valuable to several real-world scenarios that are deeply influenced by various environmental factors (Htun et al., 2023).

15.5.1 Sensors and IoT technology

The development of cutting-edge sensors and advanced IoT technologies has had a significant impact on improving water quality management. The core formation of IoT systems is driven by sensors and IoT devices that are often embedded to create opportunities for computational automation in water quality management. An IoT system refers to the network of interconnected sensor devices that allows for real-time data acquisition procedures over wireless connections

(Ahmed et al., 2019). The expansion of IoT system architecture comprises several key components, which are sensors, network communications, and application layers (Alshehri & Rahman, 2023), as illustrated in Fig. 15.1.

In other words, the IoT architecture is a unified framework that combines physical components and network configurations to provide functional operational principles (Liu, 2018). The feasibility of computational automation in water management systems relies on the quality of sensors and the effectiveness of integration between both components. The critical aspect of IoT system architecture depends on the dynamic network infrastructure, which comprises device correlation configuration, diverse communication capabilities, and data synchronizations (Power, 2008).

Correlation is the ability to identify numerous occurrences across all networked devices, and device correlation is the capability to maximize device properties or features in data processing. IoT communication protocols are data transmission regulations essential to ensuring IoT applications' foundational interoperability. Several common IoT communication protocols are hypertext transfer protocol (HTTP), message queuing telemetry transport (MQTT), and constrained application protocol (CoAP) (Al-Mansori et al., 2020; Balcerzak, 2000). Data synchronization ensures consistency between multiple devices, particularly the ability to automatically update the system periodically during the data exchange. Establishing synchronization is beneficial for obtaining significant data and solving data acquisition challenges. Data synchronization in IoT systems is crucial because it enables system efficiency by reducing the risk of data redundancy and potential errors.

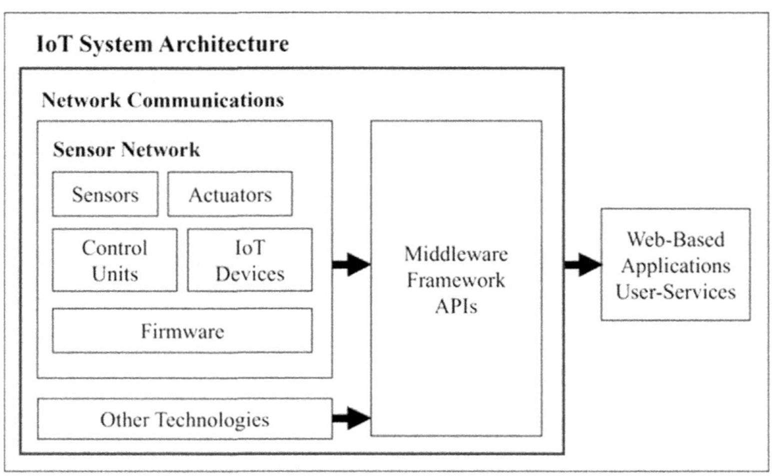

Figure 15.1
Overview of IoT system architecture, including associations between sensor networks and IoT devices. *IoT,* Internet of Things.

The advancement of wireless technologies contributes significantly to the expansion of intelligent water quality management, fostering real-time quality monitoring developments. Modernizing these monitoring systems improves the precision and accuracy of water-based quality properties supervision, particularly in three significant aspects, which are physical, chemical, and biological monitoring (Duraisamy, Khan, & Seetharaman, 2018). These water quality parameters must follow standard limits and regulations provided by national regulatory agencies that the United States Environmental Protection Agency considers acceptable for varying applications, such as agriculture, industrial manufacturing, public drinking water, and potable water (Geographic Information System GIS, n.d.; Masoom et al., 2021; OCSAN Orange County Sanitation District, n.d.). Generally, these three aspects are monitored simultaneously using multiarray quality sensors since all parameters correlate with one another. The integration of various sensors offers optimal quality measurements, reduces operational costs, is time efficient, and offers on-site friendly applications.

15.5.1.1 Physical monitoring

Physical monitoring in water quality management denotes the assessment of physical features in water bodies through sensor observations. Physical water quality parameters are presented in eight principal indicators, which are electrical conductivity (EC), total dissolved solids (TDS), salinity, turbidity, temperature, color, and taste (Zhu et al., 2017). EC measures water capable of conducting electrical current, in unit Siemens per distance (S/m), often correlating to salinity and TDS. Electrical currents are related to ion concentrations in water, and conductivity increases as the number of ions increases (Techet, 2005). Electrolytes are substances present in water that dissolve during conductivity measurements, resulting in the release of ions into water.

EC sensors enable testing and monitoring of the EC of water samples that could indicate the presence of disturbance or contaminants discharge related to the conditions of the water body. A conventional laboratory EC meter utilizes the principle of the potentiometric technique by using four platinum metal-based electrodes. Current industrial conductivity probes are driven to a portability, corrosion-resistant, and noninvasive approach by utilizing two inductively coupled coils able to produce a magnetic field. Subsequently, induced current conductivity measurements are provided (Ji, 2007). Urban and human activities are one of the triggering factors of increasing dissolved solids entering water resources, resulting in a high level of conductivity.

Multiple applications have used EC sensors embedded in an IoT system, which are commonly used in monitoring freshwater, agriculture, and groundwater systems (Ottaviano, 2020). An IoT-based EC monitoring system was introduced to control the shallow stream water level and EC values for plantation nutrient solutions (Eze et al., n.d.). The system comprises several other water quality sensors along with the implementation of web-based applications, resulting in real-time monitoring of water resources for varying plant growth. A global system for mobile communication (GSM) cloud-based river monitoring system was established to observe EC value fluctuations within a specific timeframe (Al-Mansori et al., 2020). EC levels in aquaculture were also demonstrated by combining conductivity sensing capabilities with IoT-based water quality monitoring systems (Boudreau et al., 2015).

TDS is one of the physical parameters of water quality that measures the combination of total organic and inorganic substances, generally expressed in unit milligrams per liter (mg/L) or parts per million (ppm). A TDS meter or sensor is commonly used to monitor TDS level by measuring the resistance through electrical current flow in water, subsequently covert to TDS values. The presence of dissolved solids in water can be determined using TDS indicators if values are not within the safe limits. TDS sensors only provide identification of unwanted substances deemed acceptable in water ecology but not the concentration of specific contaminants, making them suitable for prescreening steps in water quality management (Fen et al., 2010). Surface water quality assessments show that a high concentration of TDS measurements over the TDS range of 2000 mg/L could potentially risk human health and the environment (Golmohammadi et al., 2016).

Turbidity is a measurement of reduction of water clarity or transparency. Cloudy or murky water correlates to suspended particles such as sewage, algae, and plankton, resulting in a higher turbidity reading. Monitoring turbidity levels in water is based on the light scattering technique, commonly used to detect the amount of light scattered through clear or cloudy water. The measurement unit of turbidity is the nephelometric turbidity unit, which complies with optical properties related to the reflectance of light passing through water samples. Turbidity in natural water resources or human-made sources of water could be caused by several environmental factors, such as storm runoff, wind or coastal erosion, dredging, sewer discharge, animals, and algae (Alshehri & Rahman, 2023).

Remote sensing turbidity retrieval is widely applicable due to its ease of use and portability, a direct measurement for on-site monitoring (Lee, Kim, & Kim, 2016). Turbidity distribution in water quality management has been utilized in various aquatic environments, aided with on-site remote sensing technology. Natural sediments and the presence of microbiological contaminants in lakes were monitored through remote sensing turbidity data concerning the normalized difference turbidity index derived from reflectance values of red and green wave bands (Balcerzak, 2000; Skaggs, 1981). Evaluation of turbidity using four remote sensing techniques to detect anomalies in hydroelectric reservoirs during seasonal fluctuations is able to effectively identify anomalous areas (Al Kindi et al., 2021). Remote and low-cost turbidity sensing have been established elsewhere, including water reservoirs and municipal waste treatment plants, alternatively increasing data efficiencies to ease water management decisions (Becker et al., 2023).

15.5.1.2 Chemical monitoring

Chemical parameters in water quality monitoring denote the measurements of chemical properties that reflect water compositions, ensuring its suitability for human and environmental use. Several chemical monitoring sensors are used to assess water health and status. The commonly used chemical water parameters are pH, DO, total suspended solids (TSS), and chlorine (Al Kindi et al., 2021). These parameters are essential to monitor due to their high impact on biological activities and aquatic life. Acidity and alkalinity in water resulting from dynamic chemical equilibrium based on Le Chatelier's principle are measured using a pH meter to understand water conditions (Al-Mansori et al., 2020; Balcerzak, 2000).

An acceptable value of pH varies based on water resources. However, national guidelines recommend that the pH values for water quality regulations fall between 6.5 and 9.0, which is the absence of hazardous pollutants (Becker et al., 2023). DO measures the concentration of oxygen level in water to sustain aquatic life. Low DO values indicate the presence of algal growth in the water ecosystem; on the contrary, high DO represents an increasing level of waste decay (Boudreau et al., 2015; Boyce, 2011). Electrochemical and optical sensors are two commonly used measuring tools for DO levels in water, expressed in unit mg/L or saturation percentages (DO%) (Bürgler et al., 2023; Chapra, 2008). Electrochemical sensors, such as polarographic, pulsed polarographic, and galvanic sensors, are similar to the principle of TDS sensors, which utilize modification of electrolyte solutions and electrode surfaces (Childs, 2010).

The DO percentage depends on the oxygen concentration in water as it passes through the sensor electrode membrane and the oxygen chemical reaction to the electrolyte solution (DHI, 1998). Therefore an electrochemical sensor is suitable for monitoring water quality in rivers, lakes, water municipals, and water distribution systems, as it requires constant mobility in water (Ejigu, 2021). An optical DO sensor measures the dynamic interaction between the concentration of oxygen and luminescent dyes, which emit lights, also known as luminescent DO sensors (El kadi Abderrezzak & Paquier, 2009). The differences between both DO sensors are apparent in the type of water samples being examined, as electrochemical sensors are unsuitable for still water, while optical DO sensors require additional dye sensing elements that complicate on-site measurements (Fen et al., 2010). The presence of anthropogenic activities, sedimentations, erosion, and industrial waste residuals could also be identified using increments of the TSS level in the water system (Glock et al., 2019).

Generally, TSS is measured using filtration, observing any residues or particles that are not dissolved in water samples. A higher value of TSS results in higher turbidity measurements, in which suspended solids decrease water clarity (Golmohammadi et al., 2016). Realizing the effectiveness and practicality of remote sensing in water quality monitoring motivates the development of a multisensor processor, which allows the assessment of various quality parameters to be conducted simultaneously (Hoover, 1991).

A system with several quality sensors and IoT technology could provide information on pH levels and DO. The system utilizes the STM32L151C8 microcontroller unit (MCU) for real-time acquisition via a cloud computing platform through HTTP protocols. A solar-powered, waterproof, and compact array of sensors for lake quality monitoring system able to capture water parameters, such as pH, TDS, turbidity, and temperature, was also developed using an Arduino microcontroller board embedded with long-range wide-area network gateway (Ji, 2007). Web-based IoT platforms, namely, the ThingSpeak website and the ThingView mobile application, were used for data visualization and storage of real-time measurements.

The IoT-based ThinkSpeak application was also used to monitor water tank quality based on critical parameters like turbidity, pH value, and water level, integrated into a MCU (Johnes, 1996). A smart water quality monitoring system was introduced as a surrogate approach in various sensor node locations in distribution systems and rivers, utilizing reconfigurable multisensing devices for pH, turbidity, and oxygen levels using Zigbee WSN IoT-based wireless modules (Joshi, 2024; Kafle & Shakya, 2018).

Another IoT cloud-based platform was established to monitor quality parameters in several water bodies persistently integrated with IoT devices and NodeMCU, consisting of built-in Wi-Fi technology (Liu, 2018). It was also successfully demonstrated that a data logger with an NB-IoT Arduino module uses physiochemical quality sensors, such as pH, temperature, DO, and electroconductivity, to locate areas of water pollution utilizing GPS and Bluetooth communication for remote sensing diagnostics (Negm et al., 2017). IoT-based remote sensing offers computational automation in water quality management that enables cost-effective, reliable, sustainable, and rapid detection of pollution events, including mitigating early warning systems (Niazi et al., 2014).

15.5.1.3 Industrial case studies

With the increasing possibilities for future sustainable computational automation in water quality management, millions of IoT sensing devices are deployed in a broader range of industrial applications. Ericsson Research Center is actively evolving intelligent quality systems, utilizing IoT and AI capabilities as a comprehensive solution to real-world monitoring challenges. An IoT-based technology was embedded in water monitoring networks for water pipeline monitoring, namely, iWater, as a joint project with several other research institutes (What are the main components of a SCADA system?, n.d.). An automated sensing base station was deployed in water treatment plants near Lövon Lake Stockholm to retrieve insights regarding water quality in pipeline systems near the pipeline. Raw quality data were refined using varying ML algorithms, and an online analysis was conducted systematically and in a multidimensional, real-time manner. This project aims to serve as a surrogate approach for obtaining early warnings of potential quality deteriorations in the pipeline system. This project allows for fine-tuning water quality data, providing control over varying information to access authorization.

Another industrial case study deploying an automated IoT cloud-based quality monitoring system in a wastewater treatment plant (WWTP) is the microTERRA project, establishing collaboration with ClearScale to monitor microalgae growth while enhancing fish and livestock framings (Balasundaram & Venkatagiri, 2020). The system monitors water algae-based pollution on-site and purifies the water downstream for farming. A scalable and cost-efficient frontend cloud architecture using Amazon Web Services IoT as the core system was designed, enabling real-time data acquisition and building reliable data analytics in water quality.

A wastewater treatment facility in rural areas of China district has established an intelligent water quality management system employing IoT and Big Data analytical solutions, known as Project NEWATER IDEA (Akomea-Frimpong et al., 2023; Ottaviano, 2020). The automated water monitoring system is placed in over 30 village treatment plants around Shaanxi, Shanxi, and other regions. NEWATER IDEA instrumentation resulted in an 80% reduction in operational and maintenance costs, including an improvement of power consumption by about 35%. This has proven to be a paradigm shift in the evolution of computational automation for water quality management, where IoT-based monitoring systems effectively enhance sustainability and system feasibility.

15.5.2 Data acquisition and transmission

Computational automation in water quality management comprises several automated process systems, including collecting and transmitting real-time quality-related data from various water resources to the central base location. Data acquisition (DAQ) involves acquiring and sampling data comprising multiple sensing mechanisms, DAQ measurement hardware, and programmable software.

Multiple sensing mechanisms may include thermocouples, quality parameter sensors, ultrasonic sensors, and other sensing systems used in real-world scenarios. The DAQ measurement hardware consists of plug-in boards or standalone modules containing critical conversion processes from analog to digital signals, microprocessors, amplification, and filtration circuitry. Programmable software typically processes and analyzes the converted data signals using ready-to-run DAQ software, such as LabView, DewesoftX, DIAdem, Matlab®, FlexPro, and enDAQ LAB. The DAQ analyzer software also includes a data logger for storing, data management, and data visualization. Data transmission refers to sending the collected data to a centralized data processing and analysis system through a communication medium.

15.5.2.1 Real-time data collection methods

As the computational automation system becomes more sophisticated, so do systematic real-time data collection methods. Sensor and IoT-based methods are most commonly used in computational automation systems, embedded in instruments, machinery, infrastructure, and wearable devices for continuous data collection (Balasundaram & Venkatagiri, 2020). For instance, structural health monitoring systems prevent machinery failures by monitoring vibration, force, and strain indicators at regular intervals (Basuli & Roy, 2023). IoT technologies facilitate real-time monitoring of water consumption patterns and water quality deterioration in water management through remote data collection (Choi, R'bigui, & Cho, 2022). Real-time data from smart city infrastructures are collected and stored in plug-and-play online platforms using WSN communication (Ottaviano, 2020).

Real-time data collection in short-range and domestic water environments typically utilizes lightweight quality sensors integrated with LED transmitters, receivers, and PIC microprocessors (Eze et al., n.d.). However, the extensive data generated by IoT-based methods necessitate complex computation and cloud storage, leading to additional costs for storage and processing upgrades (Al Kindi et al., 2021). Security concerns about data leaks and privacy breaches in easily accessible information technology environments are also significant (Al-Mansori et al., 2020).

A network-based system automatically collects real-time data from various online platforms, such as web tracking, web analytics, or web scraping. A nutrient quality tracking system for nitrogen, phosphorus, and sediments uses a web-based tracking program linked to the Agricultural Policy Environmental eXtender model (Balcerzak, 2000). Another system, the Web-geographic information system decision support system, analyzes meteorological data, inflow water quality, reservoirs, geographical water transfer, and water demands based on the water resources vulnerability index measurements (Becker et al., 2023). Real-time databases like MongoDB, Redis, Amazon Dynamo DB, Azure Cosmos DB, and Firebase Database are commonly used for data acquisition in automated monitoring and management systems (Boudreau et al., 2015).

A supervisory control and data acquisition (SCADA) system in IoT captures real-time quality data and monitors pipeline leakages using Arduino Atmega 368 microcontrollers via GSM communication (Boyce, 2011). An industrial SCADA-based system connected to a central sensor-based station extracts quality data parameters from river water resources (Bürgler et al., 2023).

15.5.2.2 Communication protocols and data transmission technologies

Communication protocols ensure successful data transmission across various network mediums. IoT communication protocols include MQTT, data distribution service (DDS), transmission control protocol (TCP), CoAP, and extensible messaging and presence protocol (XMPP) (Chapra, 2008).

MQTT is a low-bandwidth protocol establishing connections between remote devices and clients via MQTT brokers. It is widely used in industrial applications for its simplicity and lightweight nature (Childs, 2010). MQTT protocols are applied in water quality management to transmit data such as water levels and flow measurements using ultrasonic sensors via mobile applications (DHI, 1998). Raspberry Pi 3 modules are used for data collection and processing of water quality parameters through MQTT protocols (Ejigu, 2021).

DDS operates under a publish/subscribe model similar to MQTT, coordinating messages between brokers and clients. It supports scalable solutions for specific-purpose applications, enabling efficient data exchange and real-time constraints (El kadi Abderrezzak & Paquier, 2009). A SWN uses DDS protocol to monitor reliability in physical layers of water management systems (Fen et al., 2010).

TCP and HTTP are standard communication protocols in IoT devices like ESP32 microcontrollers and Arduino NodeMCU. TCP ensures reliable transmission with error detection, retransmission, and data sequencing, suitable for web server communication (Glock et al., 2019). TCP protocol is commonly deployed in IoT-based water quality management systems for its reliability in data transmission (Golmohammadi et al., 2016).

CoAP is a web-based protocol for constrained machine-to-machine applications. CoAP uses RESTful architecture with standard communication commands (GET, PUT, POST, DELETE) to provide minimal latency and power consumption, making it ideal for IoT applications (Hoover, 1991). CoAP is favored in IoT-based resource management for waste reduction and pollution control due to its mobility, data security, and portability features (Ji, 2007).

XMPP is an XML-based protocol facilitating near-real-time data exchange through IM, video and voice chat, and collaboration. XMPP's decentralized architecture and robust security make it suitable for IoT systems (Johnes, 1996). XMPP technology is widely adopted in authorization and access management systems by major internet servers like Google and Github (Joshi, 2024).

Data transmission technologies include Wi-Fi, Bluetooth, near field communication, low power wide-area network (LPWAN), cellular networks, and satellites. Wi-Fi and Bluetooth are commonly used in scalable IoT systems, with Bluetooth operating at low power consumption over short distances and Wi-Fi offering broader coverage with high-speed wireless technology (Kafle & Shakya, 2018). Bluetooth modules like BT04-A are used for data transmission from water quality sensors to mobile applications in rural areas (Liu, 2018).

LPWAN protocols like ZigBee, IPv6 over Low Power Wireless Personal Area Networks (6LoWPAN), SigFox, and Cellular LPWAN enable wireless communication for connecting sensors to centralized networks. LPWAN protocols operate at low bit rates with low power consumption, making them ideal for rural and underground applications (Negm et al., 2017). SigFox, LoRA, and NB-IoT are long-range protocols suitable for remote monitoring, with applications in water quality management for lakes and water resources (Niazi et al., 2014).

Satellite technology addresses coverage challenges in remote areas lacking communication infrastructure. Low-earth-orbit satellites provide Earth observation capabilities for wide-area coverage and environmental monitoring, enhancing early detection and mitigation strategies for water quality management (Papanicolaou et al., 2008). Satellite imagery and remote sensing technologies improve environmental data resolution for decision-making in water resource management (Rao et al., 2022).

15.5.3 Data processing and analysis

Data processing in computational automation involves storage, processing, analysis, and visualization of data. Big Data analytics (BDA) and AI-based algorithms enable analysis of vast datasets for insights into patterns, trends, correlations, and decision-making in water management (Skaggs, 1981). ML models like neural networks and support vector machines (SVMs) predict water quality parameters and optimize wastewater treatment processes (Techet, 2005). AI models like random forests and recurrent neural networks enhance energy consumption management in WWTPs (Thu Minh et al., 2022).

AI-based anomaly detection systems improve predictive maintenance in industrial plants, reducing operational costs and downtime risks. Siemens IoT solutions for Scottish Water integrate AI analytics for predictive maintenance, enhancing equipment lifespan and operational efficiency in wastewater treatment (Tsakiris & Alexakis, 2012). Toshiba Corporation's AI-based anomaly detection system at the Mikawa power plant uses deep learning models to detect and predict equipment anomalies, improving plant performance and maintenance effectiveness (Veerapaga et al., 2019).

15.5.3.1 Big Data analytics and machine learning algorithms

BDA and AI-based algorithms utilize advanced computing technologies and methods to analyze enormous real-time datasets, discovering valuable insights related to patterns, trends, correlations, and preference criteria used to make effective data-driven decision-making in water management practices (Kumawat et al., 2021). Enhancing water quality management through a combination of BDA and ML enables a game-changing strategy for optimizing water quality, conservation, and preservation.

Forecasting water demands using a hybrid model combining extended short-term memory networks, neural networks, and kernel density estimation with an optimization tool using particle swarm optimization model has been used in previous studies (Rietveld et al., 2010).

Water pollution monitoring for freshwater and estuarine reservoirs was developed using two machine learning models, an artificial neural network (ANN) and SVM, to predict the concentration of chlorophyll-a (Chl-a) (Akomea-Frimpong et al., 2023). A 7-year historical dataset related to the detection of algal blooms was used as the input feature for ML models.

Managing the energy consumption in WWTPs employing a random forest algorithm was developed by Htun et al. (2023). Compared to the conventional multiple linear regression model, results show that the proposed model's outstanding performance quantifies input features for obtaining higher energy efficiency in WWTP. Peracetic acid disinfection analysis in secondary municipal wastewater treatment was also analyzed using ANN and RNN prediction models by observing *Escherichia coli* inactivation during the pre/post-disinfectant process (Thu Minh et al., 2022).

ANN is among the most powerful machine-learning tools used for predictive and prescriptive analytics (Zainalfikry et al., 2019). Leveraging historical data, ANN allows automated prediction processes in real-time, self-trains through the adoption of new information, and provides proactive measures or hypotheses to attain the desired outcome. For instance, a process of Fenton reaction (FBR-Fenton) in industrial reverse osmosis requires optimization in distinguishing TOC and minimizing sludge production by controlling the dosage of Fe^{2+} and H_2O_2 concentrations. Therefore a study demonstrates the use of a predictive ANN model trained to monitor and sustain dose inputs, resulting in high TOC removal and optimal performance (Zhang et al., 2023).

Developing a robust AI-BDA model requires enormous datasets and substantial sample sizes, which is one of the critical prerequisites (The Sustainable Development Goals Report, 2023). Despite years of data collected by governmental bodies and the production of historical data, some may concurrently be outdated over time and must be representative of current data in use. This is the main challenge motivating the necessity of real-time computational automation to improve the water-processing process and management. Increasing the quality of data by employing time series water flow data or extracting real-time quality data directly through the analysis of AI-Big Data analytical models could improve prediction outcomes.

15.5.3.2 Industrial applications

Predictive maintenance for WWTPs to reduce operational cost and risk while increasing monitoring effectiveness are the primary goals of Scottish Water collaborating with Siemens. Scottish Water offers billions of liters of water and wastewater services throughout Scotland, and their main goal is effective wastewater treatments to achieve zero carbon footprint by 2040 (Zhang et al., 2023). Siemens provides data-driven digitalization infrastructure solutions installed in more than 300 machineries in 17 wastewater treatment sites to reduce operational expenditure, risk of downtime, energy consumption, and zero emissions. Leveraging the Siemens IoT solution, Scottish Water was able to reduce the cost of risk failures by 10% and extend assets lifespan by 15% by utilizing SIMOTICS CONNECT 400 with a cloud-based analytics application.

The SIMOTICS CONNECT 400 utilizes a sensor module embedded in the aerator's motor, which undergoes a data transmission process through SIDRIVE IQ Fleet cloud apps for monitoring real-time low-voltage motor-driven loads. Data are accessible throughout the plant via wireless LTE/Wi-Fi router access. These data extracted from sensors provide three real-time measurements: magnetic field, temperature, and vibration. AI-based analyzes were calculated using these measurements to predict and estimate several health conditions of motor status, including bearing, unbalanced, and misalignments.

Scottish Water also implemented the SIPLUS CMS1200 condition monitoring system for screw pumps in WWTPs, ensuring the reduction of risk failures or operational errors. For instance, an increment of rotational frequency measurement captured by the sensor indicates alignment issues in the motor gearbox, which can be immediately addressed and identified. Edge Analytics processes these real-time measurements in the Industrial Cloud application. This intelligent condition monitoring system offers better preventive action, higher equipment status transparency, fast responsive measures, and a secure, interoperable solution.

There are other several industrial companies that leverage digitization at operational level, such as Arlanxeo, Sitech, and Covestro, which utilize TrendMiner OSIsoft PI software for optimum maintenance surveillance. The self-service data analytics improves production control, reduces operational costs, increases system efficiencies, and enhances information availability. CURRENTA, one of the largest industrial maintenance services, has been collaborating with TrendMiner in February 2022, establishing a smart monitoring system for wastewater leakages at Leverkusen, Germany. Beforehand, operations control system lacks ways to detect leak occurrences from stacking tank to the treatment plants. By using TrendMiner, the source of leaks was identified, and correctional measures were immediately taken.

Toshiba Corporation has established a novel AI-based anomaly detection system, scalable for large industrial plants to identify varying anomalous signals among sensor values (Zhu et al., 2017). The monitoring system is the first of its kind to obtain high accuracy in anomaly detection under dynamic interaction between operational plants and sensing measurements. The deep learning model predicts sensor values in normal operating conditions and detects anomalies by capturing massive data deviations from forecast values.

Implementing this intelligent system at the Mikawa power plant, operated by SIGMA POWER Ariake Corporation and subsidized by Toshiba Energy Systems & Solutions

Corporation, has resulted in 12 increase in anomaly detection accuracy over other existing monitoring systems. Demonstration of AI-based cloud solution recognized anomalous signals within 6.8 days, allowing proactive measures to be taken by plant operators. The benefits of implementing AI-BDA for maintenance and anomaly monitoring substantially allow time-series data measurements related to condition-based maintenance as well as contributing to the effectiveness of plant operations.

15.6 Software and systems for water quality management

15.6.1 Supervisory control and data acquisition

SCADA systems play a crucial role in industrial automation, integrating hardware and software for data collection, transmission, monitoring, and control (Zhang et al., 2023). SCADA systems are widely adopted in industrial operations for efficient and reliable management of processes (Zhu et al., 2017).

Sensors and instruments deployed within a facility will give out data from real-time measurements like temperature, pressure, flow rate, and equipment status. These data will then be transmitted to a central location, where the data will be processed and visualized. Human operators will then be able to give interpretation of these data and decide to provide control functionalities in response to the data. The operators can make adjustments to the settings remotely, initiate commands, and do optimization on the related processes. Real-time insights and centralized control provided by SCADA system have empowered operators to make informed decisions, ensured smooth operation, and helped in identifying potential issues before they could have caused costly downtime (Rietveld et al., 2010).

15.6.1.1 The SCADA ecosystems: a layered approach

SCADA system can be better visualized as a layered architecture, with each layer performing specific and defined function (Kumawat et al., 2021). As shown in Fig. 15.2, the key components that make up the layers include the following:

1. Field level

   This level usually consists of, but is not limited to, the sensors and instruments as well as the PLCs or remote terminal units (RTUs). Sensors and instruments are like eyes and ears of the system, such as the pressure gauges, thermocouples, and flow meters. They collect real-time data on various defined parameters. These data are then transferred to the intermediates, the PLCs and RTUs, to be preprocessed before transmission to the supervisory level (Kumawat et al., 2021; Akomea-Frimpong et al., 2023).

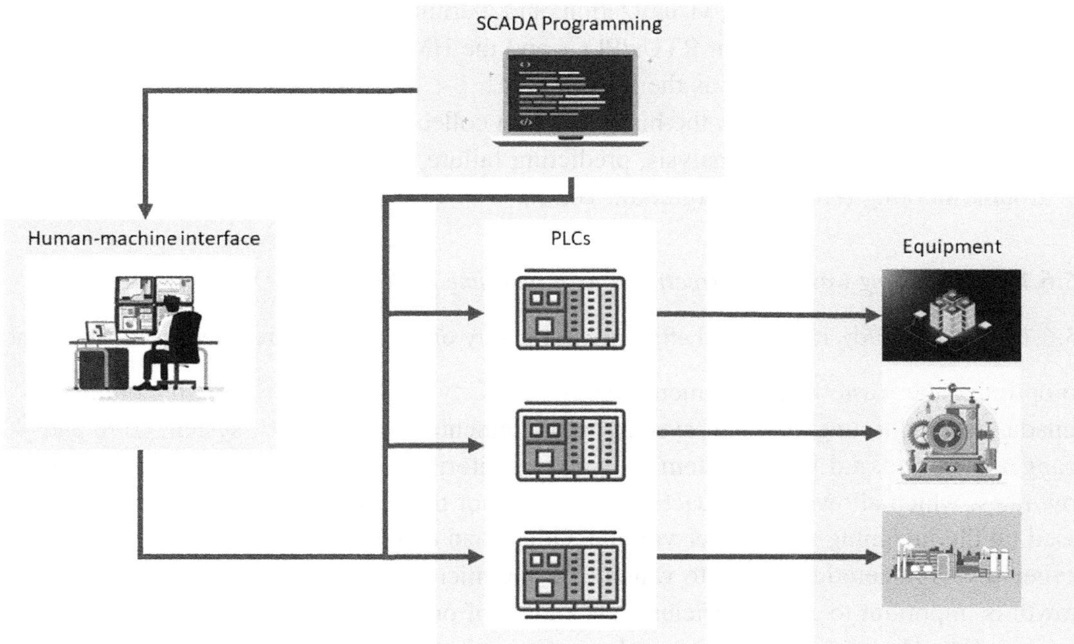

Figure 15.2
Network of SCADA system consisting of the system's program, human–machine interface, programmable logic controllers (PLCs), and the interconnected equipment. *SCADA*, Supervisory control and data acquisition.

2. Communication network

   Data transmission is highly dependent on this level, where the RTUs/PLCs are allowed to send gathered data to the supervisory level and receive control commands from the operators. The commonly used communication protocols include Modbus, DNP3, and IEC 60870-5 (Htun et al., 2023).
3. Supervisory level

   This level usually consists of three main components:
   a. The central processing units are known as the *supervisory computers*. Here, the data from the field level are received, processed, and presented to the operators (Kumawat et al., 2021).
   b. The received and processed data are presented to the operators through the *human–machine interface (HMI)*. It is a window showing real-time data in a user-friendly format, usually on graphical representations like mimic diagrams and trends. Through HMI, operators are allowed to interact with the system to issue control commands and make adjustments to settings (Infrastructure for Climate Action, 2022).
   c. All these processes are controlled and comprehended by the *SCADA software*, which acts like a brain of this level. The software not only manages data

acquisition, processing, visualization, and alarming but also facilitates the communication between RTUs/PLCs and the HMI (Kumawat et al., 2021).

4. Data storage is also known as the historian.

   This component archives the historical data collected by the SCADA system. These data help in performance analysis, predicting failure, trend identification, and troubleshooting (Klompenburg et al., 2020).

15.6.1.2 Optimizing wastewater treatment with real-time control of SCADA system

15.6.1.2.1 Case study 1: enhanced efficiency at the City of Kelowna Wastewater Treatment Plant

To optimize the wastewater treatment process, the City of Kelowna in British Columbia, Canada, decided to upgrade their system by implementing the SCADA system (Lim and Spanger-Siegfried, n.d.). The system provided monitoring and control over their influent flow rates, which allowed for proactive adjustment of the biological treatment processes based on the incoming wastewater volume. Other than that, the systems also monitored the oxygen levels in aeration basins to watch over the microbial growth in real time. Optimized growth is important to ensure efficient breakdown of organic matter. The system also monitored the sludge level, which enables optimization of sludge removal, preventing overflows and improving effluent quality.

Through SCADA upgrade, energy consumption was reduced as the blowers' usage was reduced. This is due to the efficient aeration processes. Consistent compliance with discharge regulations was also observed with constant monitoring and control by the SCADA system. Overall, resources were efficiently utilized and process disturbances were minimized, resulting in lower operational costs.

15.6.1.2.2 Case study 2: automating disinfection at the Orange County Sanitation District

Fluctuating influent characteristics have led the Orange County Sanitation District in California, USA, to have poor maintenance over the chlorine disinfection levels. To mitigate the issues, a SCADA system with chlorine residual analyzer integration was implemented for control automation and monitoring (The Sustainable Development Goals Report, 2023). The characteristics of wastewater influent in terms of the organic load and flow rate were continuously monitored through this system. The chlorine concentration in the effluent was also monitored in real time to ensure proper disinfection.

The system has been reported to adjust chlorine dosing rate automatically in response to variations of influent. Thus the effluent quality was improved and in compliance with the regulatory requirements. As the dosing requirement was optimized, overconsumption of the chemicals was minimized. Operators were freed from manual control tasks, allowing them to focus on process optimization and maintenance efficiently.

15.6.1.2.3 Case study 3: early leak detection at the Thames Water Wastewater Network

SCADA-flow meters integrated system (LeakNavigator, Ovarro, UK) was implemented by Thames Water throughout their wastewater network in England and saved them millions of liters (Tunji-Olayeni, Omuh, Afolabi, Ojelabi, & Eshofonie, 2019). The systems gave constant monitoring of flow rates at various network points to recognize deviations in flow patterns. The deviations, if found, then trigger the alarm in the SCADA system.

Early detection and repair of leaks have minimized discharge of untreated wastewater into the environment. This has a direct effect on the cost savings, as the leak repairs are prompted earlier before they get worse, and subsequent infrastructure and environmental damage can be avoided. Water disruptions were also reduced, thus improving customer's satisfaction.

15.6.2 Geographic information systems

Geographic information systems (GISs) have diversified their technological competency to be implemented in water quality management in terms of visualization, analyzes, and spatial pattern recognition. This integration system benefits significantly for the SCADA system in improving efficiencies of monitoring, assessment, and decision-making for water quality management.

GIS is able to capture, store, analyze, and display geographical information (SCADA, n.d.) beyond simple maps, in which the layers of data with spatial coordinates are incorporated. These layers give representations of physical features on maps such as rivers, lakes, and groundwater aquifers (Mehta & Reddy, 2015). In GIS, we can also recognize the land use patterns, including agriculture, urban areas, and forests. By analyzing spatial patterns of pollutants in water bodies alongside land-use data, GIS can help pinpoint potential sources of contamination, such as industrial facilities or agricultural runoff. Geotagging also enables us to point to locations of water quality sampling sites.

Integration of these various datasets allows users to visualize and analyze the spatial relationships between water quality and other environmental factors. Depending on the water quality parameters, GIS can help in identification of the areas through spatial trend analysis (Wright, 2008). Such identification can prioritize the monitoring effort and targets for areas that require intervention. For example, identifying the potential affected streams in agricultural runoff zones. Furthermore, with spatial information such as topography and rainfall patterns, the water quality data can be mapped into predictive models for water quality issues forecasting and guide for preventive measures (Introduction to Industrial Automation, n.d.). GIS can be used to track how changes in precipitation patterns or rising sea levels affect water quality in various regions. These empowerments of GIS can provide comprehensive understanding and visualization for informed decision-making. Stakeholders can identify areas needing remediation efforts, plan resource allocation effectively, and track progress over time. Interactive online maps with water quality data can raise public awareness about water quality issues and empower citizens to participate in conservation efforts.

15.6.2.1 Delineating watershed boundaries and prioritizing conservation efforts

Some land areas drain into rivers, lakes, or estuaries. These areas are known as watersheds. To effectively manage watersheds, understanding their boundaries is important. Utilization of digital elevation models in GIS, together with the stream network data, has been used to accurately delineate watershed boundaries (Human-machine interface HMI SCADA, n.d.). GIS allows for dataset overlays such as land cover types, soil properties, and potential pollution sources within the area once the watershed boundaries are established (What are the main components of a SCADA system?, n.d.). This spatial relationship is important in identifying conservation of critical areas. For example, the areas with high erosion potential due to deforestation or unsustainable agricultural practices can be pinpointed on GIS. So prioritized efforts can be made for implementation of reforestation initiatives or promotion of sustainable farming practices.

15.6.2.2 Modeling and monitoring water flow and quality

Water flow patterns within the watershed can be simulated using the GIS and hydrological modeling integration. This method has been used in predicting the impact of land-use changes or extreme weather events (OCSAN Orange County Sanitation District, n.d.). The land-use data were altered by simulating increased urbanization. In turn, GIS predicted the changes in water flow patterns, which showed increment in surface runoff and decrement in groundwater recharge.

In 2015 a case study in Wadi Araba, Jordan, was conducted to identify the flood-prone areas. GIS was used alongside analyzed historical precipitation data and soil properties to model and identify flood susceptibility. This method can be used to identify the area with higher drought risk too.

Furthermore, watershed water quality data can be better managed with GIS. The spatial patterns of pollutants like nitrates or sediments can be visualized over time at various monitoring points within the watershed area (LeakNavigator: Water conservation and leak detection, n.d.). Here, the trends can be tracked, and the potential sources of contamination can be identified. This information is important for pollution control measures or in improving the wastewater treatment infrastructure.

15.6.3 Decision support systems

Vast amounts of data, diverse perspectives, and uncertain situations complicate the decision-making process. Thus a decision support system (DSS) has emerged as a powerful tool in assisting well-informed choices for decision-making (Geographic Information System GIS, n.d.). This system incorporates an interactive computer-based system with data integration, analysis, and modeling tools. This system allows for analysis and gathering of relevant data, development and assessment of alternative solutions, evaluation of trade-offs, and identification of optimal solutions.

DSS is often equipped with a user-friendly interface that allows users to interact with the system, navigate the datasets, and explore various functionalities. The system allows for access, storage, and management of data from various sources, ensuring the accuracy, consistency, and accessibility for analysis. Some of the tools included are modeling and simulation, risk assessment, and optimization techniques. These tools help in conducting analysis and support decision-making. This process is supported by a knowledge base, in which relevant information, best practices, and historical data are provided. Fig. 15.3 shows the key features of the DSS.

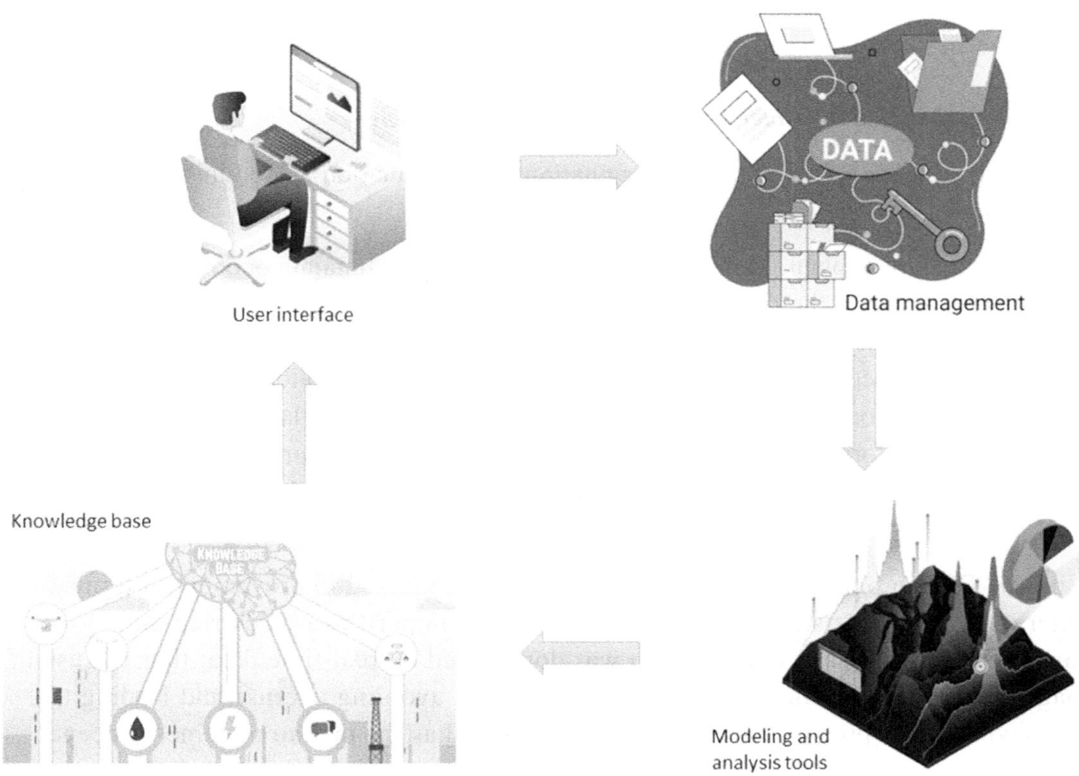

Figure 15.3
The four key features of the DSS. *DSS*, Decision support systems.

DSS can be found in three common types:

1. Data-driven DSS
   This system emphasizes data analysis and visualization tools to help users in identification of trends, patterns, and relationships within the data.
2. Model-driven DSS
   This system utilizes models to simulate potential outcomes of different decision options.
3. Knowledge-driven DSS

This system incorporates expert knowledge and best practices to guide users toward informed decisions.

By providing comprehensive data, analytical tools, and simulation, DSS can improve decision-making process. The time and resources for data analysis and complex calculations can be made automatically; hence, the efficiency of decision-making process is guaranteed. The use of data and models within a DSS promotes transparency and allows for better communication and justification of decisions. Other than that, DSS can facilitate collaboration between stakeholders with diverse perspectives by providing a shared platform for data analysis and decision support.

15.6.3.1 DSS optimizing water treatment: industrial applications

15.6.3.1.1 Case study 1: optimizing coagulation and flocculation at a municipal water treatment plant

The water treatment plant was facing a huge challenge in maintaining consistent water quality. This is due to the fluctuating raw water characteristics (Masoom et al., 2021). DSS was implemented to specifically optimize coagulation and flocculation processes. The system was designed and integrated with real-time sensors to monitor the raw water turbidity, pH, and organic matter content continuously. Historical data on raw water quality and treatment performance were entered into the system as basis for coagulation and flocculation models. These models were used to predict the optimal dosage of coagulants and flocculants.

With the DSS in place, the water treatment plant recognized improvement in terms of water quality as consistent and efficient removal of impurities was carried out. Furthermore, the dosage optimization was done based on real-time data; thus the use of coagulants and flocculants was the strongest match, avoiding overuse and leading to cost savings. The process control was also enhanced as the operators gained a deeper understanding of the relationship between raw water characteristics and optimal treatment parameters.

15.6.3.1.2 Case study 2: predicting and mitigating membrane fouling at a desalination plant

Membrane fouling issues impacted a desalination plant, which compromised its production efficiency and required frequent cleaning cycles (Duraisamy et al., 2018). DSS was implemented to mitigate the issues with integration of sensors to monitor the feed water quality parameters such as turbidity and salinity. The sensors also oversaw the membrane performance metrics, such as the pressure drop and the permeate flow rate. Built-in models were utilized in the system to predict the risk of membrane fouling. This process was assisted by the real-time data from the sensors and the historical data trends. The system was also equipped with cleaning optimization algorithms, which allowed for recommendation of optimal cleaning protocols and schedules based on predicted fouling severity.

DSS proactively predict and address potential fouling events, thus membrane downtime can be minimized and its lifetime can be extended. The production efficiency was also improved as the cleaning frequency was reduced, increasing plant uptime and higher production volumes. Membrane cleaning was efficiently done, thus the use of cleaning chemicals was minimized and requirement for maintenance was reduced. This has led to lower operational costs.

15.7 Automation techniques and methodologies

15.7.1 Machine learning and artificial intelligence

Water quality monitoring and prediction are vital for the sustainability and safety of water resources. Continuous monitoring is essential due to threats from pollution and irrational use, which can hinder sustainable development (Zhang & Fu, 2023). Advanced technologies like AI and ML are increasingly employed to predict water quality indicators, enabling proactive measures to prevent ecosystem damage and meet environmental standards (Zhang et al., 2023). AI refers to the development of machines or systems capable of tasks requiring human intelligence, such as data processing, information interpretation, and knowledge extraction (Zhu et al., 2017). ML, a subset of AI, involves training machines to learn from data and improve performance without explicit programming (Zhu et al., 2017). AI-based prediction models are popular in water quality monitoring for their ability to handle large datasets, predict complex phenomena, and provide accurate results despite missing data (Zhang & Fu, 2023).

The application of AI and ML in water quality analysis has broadened to include predictive water infrastructure management, water demand forecasting, reservoir monitoring, and disaster prediction, demonstrating their diverse uses in environmental management (Zhang et al., 2023).

AI-powered algorithms help address challenges in water resource management, such as predicting river water quality and optimizing variables for sustainability (Zhang et al., 2023).

Supervised ML algorithms, such as SVMs and random forest, have been extensively applied in water quality monitoring. SVM has been used to estimate the water quality index and categorize water quality accordingly (Niazi et al., 2014). It has also been effective in predicting the surface quality index of irrigation systems and river water quality (Rao et al., 2022). Additionally, SVM and other ML methods have been crucial in estimating various water quality parameters. For example, hyperspectral data and ML techniques have been used to monitor parameters like CDOM, Chl-a, diatoms, green algae, and turbidity (Skaggs, 1981).

Random forest has also been widely used in water quality monitoring. It has shown higher accuracy in predicting the presence of Salmonella in agricultural surface waters, particularly for turbidity, compared to other physicochemical attributes (Techet, 2005). Furthermore, the integration of random forest with other techniques, such as correlation analysis and multiple regression analysis, has enhanced the automatic monitoring of water quality indicators, highlighting its effectiveness and versatility (Thu Minh et al., 2022).

Unsupervised learning algorithms play a crucial role in water quality monitoring by providing insights into complex datasets without needing labeled classifications. The K-means clustering algorithm, for example, has been effectively used to identify patterns and group water quality data points based on similarities (Tsakiris & Alexakis, 2012). This application allows for efficient and reliable analysis, aiding in the evaluation and management of water quality across various environments.

Unsupervised clustering models based on graph embedding have also shown potential in water quality monitoring. These models perform spatiotemporal analysis of water quality data, creating clusters with similar characteristics and determining the periodicity of water quality changes across different monitoring points (Veerapaga et al., 2019). Another commonly used unsupervised learning algorithm in water quality monitoring is the expectation-maximization (EM) algorithm. The EM algorithm clusters water quality data to uncover underlying patterns and structures by iteratively estimating the parameters of a statistical model (Yi et al., 2017).

Deep learning techniques, particularly long short-term memory (LSTM) neural networks, have become increasingly important in water quality monitoring due to their ability to enhance prediction accuracy and efficiency (Zainalfikry et al., 2019). Studies have demonstrated the effectiveness of LSTM networks in predicting water quality parameters and indices across various settings (Fen et al., 2010). Additionally, LSTM networks have been pivotal in predicting surface water quality indices and sorting water quality indicators, highlighting their significance in accurate forecasting and effective water quality management (Glock et al., 2019).

Convolutional neural networks (CNNs) are another deep learning technique employed in water quality monitoring. CNNs have been used to enhance water quality prediction and monitoring by extracting local features from water quality parameters (Golmohammadi et al., 2016). For example, a study utilized a hybrid model combining a one-dimensional residual CNN and bi-directional gated recurrent units to predict water quality in the Luan River, demonstrating CNN's effectiveness (Hoover, 1991).

15.7.2 Robotic process automation

Robotic process automation (RPA) enhances process efficiency and reduces costs by automating rule-based, routine tasks (Techet, 2005). RPA involves creating software robots that replicate human interactions to perform tasks, freeing employees from repetitive work (Thu Minh et al., 2022). It mimics high-volume, manual, repeatable, and monotonous tasks typically done by humans (Tsakiris & Alexakis, 2012).

Traditional water quality monitoring faces challenges like manual data collection, reporting, and the potential for human error (Veerapaga et al., 2019). Manual processes are often impractical, labor-intensive, and prone to inaccuracies (Yi et al., 2017). Periodic manual lab sampling for water quality checks can also be cumbersome. RPA addresses these challenges by automating repetitive tasks in water quality monitoring. It streamlines data collection by automating the retrieval of water quality parameters, reducing manual data entry (Zainalfikry et al., 2019). By automating tasks such as data collection and reporting, RPA improves the efficiency of monitoring processes, prevents errors, and ensures regulatory compliance through consistent and accurate performance (Zhang & Fu, 2023). Additionally, RPA enhances productivity and cost savings by freeing human resources for more strategic activities (Zhang et al., 2023).

RPA can effectively automate the structured data collection from various sources, including sensor readings from water quality monitoring equipment, laboratory analysis reports, and external databases such as weather data (Zhu et al., 2017). RPA leverages software robots to replicate human interactions, automating the extraction, processing, and analysis of data from these diverse sources.

For sensor readings from water quality monitoring equipment, RPA can be programmed to collect data at scheduled intervals, input the data into designated systems, and trigger alerts based on predefined thresholds. This automation streamlines the data collection process, reduces manual errors, and ensures real-time monitoring of water quality parameters. Regarding laboratory analysis reports, RPA can extract relevant data, populate databases or spreadsheets, and generate automated summaries or notifications based on the analysis results. By automating the extraction and processing of laboratory data, RPA enhances the speed and accuracy of data handling, facilitating timely decision-making.

Furthermore, when accessing external databases such as weather data, RPA can automate the retrieval of specific information, integrate it with internal systems, and generate reports or alerts based on the analyzed data. This automation ensures that organizations have access to up-to-date external data sources, promoting informed decision-making without manual intervention.

Implementing RPA in water quality monitoring presents several challenges. Water quality data can exhibit significant variability due to environmental factors, seasonal changes, and sensor inaccuracies, necessitating robust mechanisms within RPA to handle different data formats and ensure accurate processing (Skaggs, 1981). Integrating data from various sources, such as sensor readings, laboratory reports, and external databases, into the RPA system is complex, requiring seamless data flow and compatibility between different data formats and systems for effective automation (Techet, 2005). Maintaining data accuracy and quality is also essential, as RPA must include mechanisms to validate and verify data to prevent errors that could impact decision-making processes.

Additionally, sensors require regular calibration and maintenance, and RPA systems need to account for scenarios where sensor drift or malfunction impacts data accuracy. The complexity of integrating RPA with existing monitoring systems, databases, and reporting tools adds another layer of difficulty, necessitating seamless data flow and compatibility for effective implementation (Thu Minh et al., 2022). Furthermore, RPA must process data in near-real-time to enable timely alerts and compliance reporting, as any delays could negatively impact decision-making and regulatory adherence.

Adhering to regulatory standards and requirements in water quality monitoring is critical. Ensuring that the RPA system complies with relevant regulations and guidelines and accurately captures and reports data for compliance purposes is a significant challenge (Tsakiris & Alexakis, 2012). Protecting sensitive water quality data is also essential, requiring RPA systems to have robust security measures to prevent unauthorized access and ensure data integrity (Veerapaga et al., 2019). Resistance from employees who may fear job displacement or changes in their roles due to automation can hinder RPA implementation. Addressing employee concerns through training, communication, and involvement in the implementation process is essential (Yi et al., 2017). Ensuring the ongoing maintenance, governance, and continuous improvement of the RPA system is vital for long-term functionality and error prevention (Zainalfikry et al., 2019). Establishing a robust maintenance strategy and governance framework is crucial for sustained success (Zhang & Fu, 2023).

Scaling RPA implementation to handle increasing data volumes and complexity over time can be challenging. Ensuring that the system can scale effectively to meet growing monitoring needs without compromising performance is essential (Zhang et al., 2023). Additionally, ensuring interoperability with existing systems and technologies within the water quality monitoring infrastructure is crucial. RPA implementation should seamlessly integrate with other tools and platforms to optimize overall monitoring processes (Rao et al., 2022).

15.7.3 Simulation and modeling

15.7.3.1 Introduction to simulation and modeling

Inland surface water systems, including lakes, reservoirs, marshes, rivers, streams, and creeks, are essential for providing a number of essential functions. The maintenance of natural ecosystem structures, ecological processes, and regional ecological habitats is heavily dependent on the water ecosystem. These roles are crucial for the advancement of human civilization in addition to providing resources that are necessary for human life and productivity (Glock et al., 2019). Analyzing the current relationships and overlaps between all the study factors is made possible by studying the behavior and function of river basins (Golmohammadi et al., 2016). For the purpose of making informed management decisions, water system administrators need to understand how these components interact. The main focus of this chapter is on hydrodynamics and water quality, two of the most important factors in river basin models.

Water dynamics was formerly referred to as hydrodynamics. The principles of hydrodynamics and molecular dynamics are similar in that they both involve the conservation of mass and energy as well as Newton's equation of motion for momentum flow (Hoover, 1991). The hydrodynamic properties of the water, such as flow velocity, water depth, and surface elevation, are generally examined in these two analyzes. In contrast, the physical, chemical, and biological characteristics of the water are evaluated in water quality, including temperature, color, odor, sediments, DO, salinity, organic matter and heavy metal content, nutrients, bacteria, viruses, and protozoa (Ji, 2007). Therefore the basic ideas and principles of surface water analysis are covered in this chapter, along with the use of a model that combines hydrodynamics and water quality. The goals are to improve knowledge of the many hydrodynamic processes at play, which will help with surface water system decision-making in the actual world. This chapter also concludes with a review of practical applications of hydrodynamic and water quality modeling to demonstrate how these theoretical predictions compare to the hydrodynamic and water quality properties found in real bodies of water.

15.7.3.2 Hydrodynamic models

The idea behind hydrodynamic modeling is that it deals with the flow of water and the forces acting on it. The main focus of the study is on determining the physical processes that drive flow and how many of them there are. To create a hydrodynamic model, one needs a comprehensive set of governing equations that describe the physical processes involved, the corresponding initial and boundary conditions required to solve the equations numerically, and the numerous parameters that need to be entered to run the model (Johnes, 1996). A multitude of data sources are needed as input or to determine the regulating parameters while running a hydrodynamic model. As an example, the model's surface water reservoir segment was constructed first, taking into consideration the pertinent surface reservoir

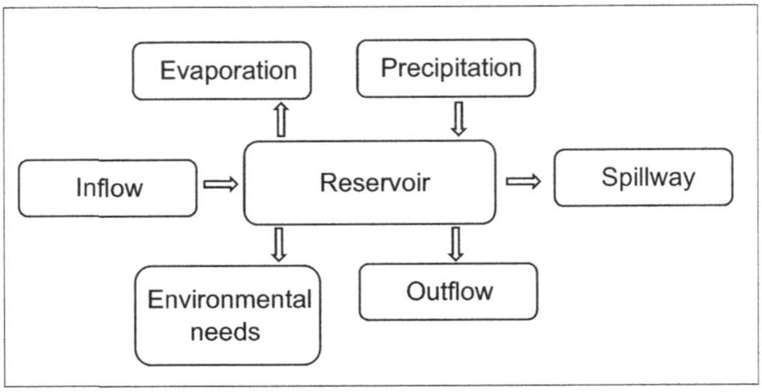

Figure 15.4
Model of the surface water reservoir segment using system dynamics (Niazi et al., 2014).

components (Fig. 15.4) (Joshi, 2024). A few of the variables that were utilized as direct inputs were the geographic coordinates of the coastline, bathymetry, flow rate, water surface elevation, and meteorological data on wind direction, speed, radiation intensity, air pressure, precipitation, and evaporation rate. A water body is susceptible to the following forces in addition to other external forces like wind and the Coriolis force: gravity, pressure, and viscous force. The governing equations for both water flow and pollution movement are based on the conservation laws of mass, momentum, and energy (Kafle & Shakya, 2018).

A control volume technique can be used to derive the rules of fluid motion. This is comparable to a "fluidic black box," where the only information available about the flow is the mass, momentum, and energy entering and leaving the control volume (Liu, 2018). The momentum equation (also known as Newton's second law) and the continuity equation (also known as the mass balance of the water in the flow) are the two main governing equations for hydrodynamic models. The momentum equation shows the relationship between the driving forces and the water's acceleration of motion. Since all surface water systems are, in principle, three-dimensional situations, a fully three-dimensional model would provide the most accurate depiction of their flow characteristics. In most situations, using a 1D, 2D, or quasi-3D model makes sense since it provides sufficient precision to handle nearly all real-world problems (Negm et al., 2017).

Based on application requirements, geographical scale andrepresentation of water bodies, and other factors, researchers have developed a wide range of dimensional models. In general, more information on the water surface elevation, velocities, flow rate, and depth at locations other than the boundaries is needed to calibrate the model, determine hard-to-measure parameters like the horizontal momentum diffusion coefficient and the roughness of the bottom and make sure the model accurately reflects the real scenario. The data measured in the lab or in the field are compared with the modeling results to calibrate and validate the model (Niazi et al., 2014).

15.7.3.2.1 Zero-dimensional models

Zero-dimensional models are based on a sequence of physical parameter measurements rather than hydraulic modeling (Papanicolaou et al., 2008). They are mathematical or physical models without any spatial extent or dimensions. Put otherwise, it symbolizes a point or an item that is not a part of any actual place. Instead of using hydraulic modeling, zero-dimensional models rely on a number of physical parameter measurements. These models calculate the physical habitat conditions for various discharge levels by using interpolation. Nevertheless, these models' forecasting capability is limited to the range of flows for which observational data were gathered (Rao et al., 2022). Furthermore, by disregarding spatial distinctions and accounting only for temporal variations, the zero-dimensional hydrodynamic model of the basin distills the physical system down to its most basic elements. The usual application of these models is to simulate basin-wide phenomena such as water balance and nutrient dynamics. In certain case studies, however, there is insufficient data to run a sophisticated water quality model, regardless of the kind of water system—river or reservoir.

An alternative to understanding the water quality behavior of the water system in the context of limited data is to seek a basic water quality model. A sophisticated water model reservoir or river cannot be run in some case studies due to a lack of data. Consequently, using a simple model is an alternative to comprehending the water behavior of the water system. One of the most basic example models for zero dimensions that can be used to mimic a natural body of water is the continuously stirred tank reactor (Skaggs, 1981). Eq. (15.1) provides an illustration of zero-dimension models based on the sedimentation and reaction processes in river basins, where the accumulation illustrates the change in the system's mass over time (t) in the absence of spatial data.

$$\text{Accumulation} = \text{Inflow} - \text{Outflow} - \text{reaction} - \text{settling} \tag{15.1}$$

One-dimensional model The numerical techniques used to solve the governing equations for water and sediment are more stable and increase computational efficiency by an order of magnitude compared to two- and three-dimensional models. One-dimensional models, like zero-dimensional models, also require the least amount of field data. Zero-dimension models are based on a series of physical parameter measurements rather than hydraulic modeling. Through interpolation, the physical habitat conditions for various discharges are derived (Techet, 2005; Thu Minh et al., 2022). Numerous one-dimensional numerical models with different capabilities and limitations have been proposed in the literature. The development of these models has mostly focused on numerical techniques, grain sorting, and flow resistance relations (Tsakiris & Alexakis, 2012). The governing formulas of the one-dimensional dynamic wave model for open-channel flows are known as the Saint Venant equations or shallow water equations. Eqs. (15.2) and (15.3) illustrate the two primary equations: the continuity equation and the momentum equation (Veerapaga et al., 2019).

$$\text{Continuity equation } \frac{\partial A}{\partial t} + \frac{\partial Q}{\partial x} = 0 \tag{15.2}$$

$$\text{Momentum equation } \frac{\partial Q}{\partial t} + \frac{\partial}{\partial x\left(\frac{Q^2}{A}\right)} + gA\frac{\partial h}{\partial x} + gA\left(S_f - S_o\right) = 0 \tag{15.3}$$

where A is the wetted area, Q is the flow discharge, g is the gravitational acceleration, S_F is the frictional slope, S_O is the slope, h is the depth, and x is the spatial coordinates.

The continuity equation, which emphasizes the concept of mass conservation in flow and states that the change in water volume within a segment equals the difference between inflow and outflow at a spatial coordinate xxx, provides mass conservation unlike the zero-dimensional hydrodynamic model (Yi et al., 2017). According to the momentum equation, the body and surface forces acting on a fluid equal the rate of change of linear momentum of the volume moving with the fluid (Zainalfikry et al., 2019). Table 15.1 illustrates several applications of one-dimensional hydrodynamic models.

Two-dimensional hydrodynamic models are often categorized based on how they handle spatial dimensions, namely, horizontal and vertical. Horizontal and vertical two-dimensional models differ significantly in how they represent the real world and the applications they are suitable for. More specific continuity and momentum equations are used for both horizontal and vertical two-dimensional models. Eqs. (15.4) and (15.5) below illustrate the momentum equation and continuity equation for two-dimensional (vertical) models (Zhang & Fu, 2023). Comparatively, the two types of two-dimensional models—horizontal and vertical—employ different axes for specific objectives.

$$\text{Continuity equation } \frac{\partial h}{\partial t} + u\frac{\partial (h)}{\partial x} + v\frac{\partial (h)}{\partial y} + h\left(\frac{\partial u}{\partial x} + u\frac{\partial u}{\partial x}\right) = 0 \tag{15.4}$$

$$\text{Momentum equation } h\frac{\partial u}{\partial t} + hu\frac{\partial u}{\partial x} + hv\frac{\partial u}{\partial y} - fvh = \frac{1}{\rho}\left[\frac{\partial}{\partial x}\left(\in_{xx} h\frac{\partial u}{\partial x}\right) + \frac{\partial}{\partial y}\left(\in_{xy} h\frac{\partial u}{\partial y}\right)\right.$$

$$-gh\left(\frac{\partial a}{\partial x} + \frac{\partial h}{\partial y}\right) - \left(ugn^2\sqrt{\left(\left(u^2 + v^2\right)\right)}\right)/h^{(1/3)} + \zeta W2\sin\Psi \tag{15.5}$$

where u, v is the depth-averaged velocities in the x,y directions, W is the wind velocity, Ψ is the wind direction, ζ_i is the empirical wind coefficient, f is the Coriolis parameter, g is the gravity acceleration, n is the Manning's roughness coefficient, and $\in$ is the depth-averaged Eddy viscosity.

Table 15.1: List of models' examples along with their uses (Liu, 2018).

Model dimensions	Characteristic	Applications	Study	Highlight
Zero	Evenly distributed, completely blended, and devoid of flow	Assessment of lakes, ponds, and tanks with minimal or no flow	Becker et al. (2023)	• This finding implies that zero-dimension models are useful instruments for preliminary examination of the dynamics of water quality in reservoirs, especially in situations where data are poor • When computing resources and specific spatial data are scarce, these models can effectively assist in the planning and management of water quality by offering a dependable substitute. • Even though Zero D models have potential, the study points out a number of shortcomings and areas that might be strengthened, such as applying first-order decay coefficients and settling velocities consistently throughout all sectors.
			Boudreau et al. (2015)	• The goal of the study is to make the zero-dimensional model that is used to determine the water contamination capacity better. With this update, the Code of Practice's current model (GB/T25173-2010) will take into consideration water bodies' ability to purify themselves, which is something that is not fully taken into account. • The paper addresses how designed hydrological conditions are determined, offering methods and insights that can be consulted for determining the water pollution capacity. This makes the evaluation of water resources and pollution thresholds more precise. • The study pinpoints and resolves a few shortcomings in the way water pollution capacity is currently calculated. The suggested enhancements are meant to increase the models' reliability and applicability in real-world situations.

(*Continued*)

Table 15.1: (Continued)

Model dimensions	Characteristic	Applications	Study	Highlight
One	Well-mixed, shallow, narrow flow that ignores spatial variation in both transverse and vertical directions	Rivers, streams, long creeks, and small channels	Boyce (2011)	• This model turns governing differential equations into systems of linear or nonlinear equations, providing powerful mathematical tools for studying various water resource problems. • The model was calibrated using data from the Euphrates River during the early spring flood of 2015 and verified through application to both an ideal canal and a specific reach of the Euphrates River. • The comparison between model results and field data showed high accuracy ($R^2 = 0.997$), indicating the model's readiness for future applications when field observations are available.
			Bürgler et al. (2023)	• Using HEC-RAS 5.0, the study successfully created a one-dimensional hydrodynamic model for the Pahang River by combining surveyed data with cross-sections that were spatially extracted and stream flow measurements. The purpose of this model is to improve the region's flood risk management. • The Pahang River area's extreme flood occurrences in 2007 and 2014, which resulted in considerable property damage and fatalities, were particularly examined in this study. The hydrodynamic model was validated by using these incidents as case studies. • With a correlation coefficient (R^2) of 0.82 for the 2014 flood event and 0.96 for the 2007 flood event, the model showed good accuracy in predicting flood levels. These outcomes demonstrate how well the model simulates flood conditions.

Two (vertical)	Deeply constricted flow, disregarding spatial variation in the lateral direction	Lakes, reservoirs, and river sections that are deep and narrow (highly stratified)	Chapra (2008)	• Using CE-QUAL-W2, the study created a vertical 2D hydrodynamic model for the Yangtze and Xiangxi rivers. Accurate model performance depended on critical parameters such as wind sheltering coefficient, eddy diffusivity, eddy viscosity, and channel roughness. • The study showed that reservoir shape, outflow position, wind shear effect, bottom shear effect, and variations in water temperature all had a complete impact on the water flow field in the mainstream. • The reservoir's longitudinal velocity and flow patterns are determined by the combination of these parameters.
Two (horizontal)	Wide, shallow flow that is well mixed and ignores spatial changes in the vertical direction	Ponds, wide shallow river stretches, and open, poorly stratified lakes and estuaries	Childs (2010)	• To simulate flood events and analyze inundation parameters, hazard assessments, and delineation of flood extents for the Koshi River, the study uses a two-dimensional hydrodynamic model with Nays 2DH software. • The proposed modeling approach is an effective option for simulating exceptional flood events in regions with limited data and resources, offering potential applications to other rivers with similar characteristics. • The model is calibrated with measured water surface elevations, achieving a high correlation ($R^2 = 0.98$) and low root mean square errors (RMSE = 0.95 m) between measured values and simulation results, demonstrating the model's accuracy and feasibility for predicting flood scenarios.

(Continued)

Table 15.1: (Continued)

Model dimensions	Characteristic	Applications	Study	Highlight
Three	Consideration of all three directions, substantial vertical direction variation, and deep, stratified flow	Consideration of all three directions, substantial vertical direction variation, and deep, stratified flow	DHI (1998)	• For improved accuracy and lower computational costs, the study presents the Fantom-Refined three-dimensional hydrodynamic simulator, which was created especially to solve issues with horizontal grid discretization. • It analyzes multiresolution phenomena from coastal bay to river upstream by using an organized gridding technique with local mesh refinement. • Chikugo River estuary data from Japan were used to precisely calibrate and validate antom-refined. It displayed reasonable accuracy (R^2 of 0.98) in predicting the vertical salinity distribution along the channel during spring-neap tidal cycles, and it correctly recreated measured water levels with an exceptional R^2 value of 0.99 during calibration. • The findings demonstrated the importance of channel meanderings and width on the estuary's horizontal salinity distribution.

The momentum equation and continuity equation for two-dimensional (horizontal) models are given in Eqs. (15.6) and (15.7) below (Liu, 2018).

$$\text{Continuity equation}\frac{\partial(Bu)}{\partial x}+\frac{\partial(Bw)}{\partial z}=0 \tag{15.6}$$

$$\text{Momentum equation}\frac{\partial(Bu)}{\partial t}+\frac{\partial\left(Bu^2\right)}{\partial t}+\frac{\partial(uwB)}{\partial z}+gB\frac{\partial(z_s)}{\partial x}-\frac{\partial}{\partial t}\left(BA_H\frac{\partial u}{\partial x}\right)-\frac{\partial}{\partial x}\left(BA_v\frac{\partial u}{\partial z}\right)-\tau_x=0 \tag{15.7}$$

where A_H is the horizontal Eddy viscosity, A_v is the vertical turbulent mixing coefficient, and Z_s is the water surface elevation.

Three dimensions In a three-dimensional model, fluid behavior can be portrayed with more complexity and realism. The mass conservation equation and momentum equations along the x, y, and z axes are shown in Eqs. (15.8), (15.9), (15.10), and (15.11) (Zainalfikry et al., 2019).

$$\text{Mass conservation equation}\frac{\partial u}{\partial x}+\frac{\partial v}{\partial y}+\frac{\partial w}{\partial z}=0 \tag{15.8}$$

$$\text{Momentum conservation equations}\frac{\partial u}{\partial t}+u\frac{\partial u}{\partial x}+v\frac{\partial u}{\partial y}+w\frac{\partial u}{\partial z}=F_1-\frac{1}{\rho}\frac{\partial P}{\partial x}+v_t\left[\frac{\partial^2 u}{\partial x^2}+\frac{\partial^2 u}{\partial y^2}+\frac{\partial^2 u}{\partial y^2}\right] \tag{15.9}$$

$$\frac{\partial u}{\partial t}+u\frac{\partial u}{\partial x}+v\frac{\partial u}{\partial y}+w\frac{\partial u}{\partial z}=F_2-\frac{1}{\rho}\frac{\partial P}{\partial y}+v_t\left[\frac{\partial^2 u}{\partial x^2}+\frac{\partial^2 u}{\partial y^2}+\frac{\partial^2 u}{\partial y^2}\right] \tag{15.10}$$

$$\frac{\partial u}{\partial t}+u\frac{\partial u}{\partial x}+v\frac{\partial u}{\partial y}+w\frac{\partial u}{\partial z}=F_3-\frac{1}{\rho}\frac{\partial P}{\partial z}+v_t\left[\frac{\partial^2 u}{\partial x^2}+\frac{\partial^2 u}{\partial y^2}+\frac{\partial^2 u}{\partial y^2}\right] \tag{15.11}$$

15.7.3.3 Water quality models

Water quality modeling is a crucial tool for managers and policymakers to address multiscale integrated water resources and environmental challenges. Compliance with regulations is now the primary motivation for using water quality models, necessitating the identification of contamination hotspots, standardization of models, and assessment of their relevance to policy (Zhang & Fu, 2023; Zhang et al., 2023). Agriculture, particularly in developing countries, contributes significantly to water quality issues through the release of industrial wastewater effluents into downstream areas of urban centers. Mathematical water quality models, integrated with hydrodynamic modeling, are employed to study the fate, transport, and degradation of pollutants in water bodies globally.

Several factors influence the selection of water quality models, including model complexity, data availability, water body type, simulation capabilities, programming accessibility, validation, application area, cost, familiarity, and support (Zhang et al., 2023). Various water quality models have been developed to provide more scientifically rigorous analysis, tailored to address specific quality issues in diverse environmental and socioeconomic contexts.

The classification of water quality models and their categories consider factors such as computational complexity (1D, 2D, and 3D models), data requirements (large databases vs minimal data models), approach type (physically based, conceptual, empirical), pollutant type (nutrients, sediment, salts), application area (catchment, groundwater, rivers, lakes, coastal waters), nature (deterministic or stochastic), state analysis (steady-state or dynamic simulation), and spatial analysis (lumped, distributed) (Zhang et al., 2023). Key parameters such as BOD degradation, DO impact on BOD decay, BOD settling velocity, resuspension of BOD from sediment, nitrification oxygen demand, ammonium release rate, nitrification rate, concentration dependence of nitrification, and heat radiation from rivers are considered in comprehensive water quality modeling (Zhu et al., 2017). Table 15.2 summarizes studies employing 1D, 2D, and 3D water quality models and their applications. Below are brief descriptions of some developed water quality models:

DRAINMOD: The DRAINMOD software simulates water flow and nitrogen movement in streams. Originally developed to study drainage and subirrigation systems' impact on water use, crop response, and pollution control. Negm et al. (2017) used regression equations to quantify annual drainage flow and nitrate loss reductions due to drainage water management in eastern North Carolina. Golmohammadi et al. (2016) employed DRAINMOD to assess the effects of different tillage techniques on water and nitrate dynamics and predict the impacts of climate change.

Export Coefficient Model (ECM): The ECM predicts nutrient loading at water sampling locations based on nutrient export from all pollution sources in the catchment above each location. Rao et al. (2022) enhanced ECM with land-use data and a grid-based rainfall-runoff module, improving temporal and spatial variability management of nonpoint source (NPS) nutrient loads. Zhang et al. (2023) developed an agro-hydrological process-based ECM to estimate monthly NPS loads accurately in semiarid agricultural regions.

MIKE-11: MIKE-11 is a 1D hydrodynamic model used to simulate water dynamics in rivers, integrating modules for sediment transport, advection-dispersion, water quality, eutrophication, and rainfall runoff (Zhu et al., 2017). Thu Minh et al. (2022) applied MIKE-11 in the Long Xuyen Quadrangle of the Mekong Delta to analyze spatiotemporal dynamics of water quality parameters and recommended enhanced water quality monitoring for sustainable aquaculture development.

Table 15.2: Features of 1D, 2D, and 3D water quality models (Balcerzak, 2000; Ejigu, 2021).

Model dimensions	Advantages	Limitations	Study	Highlight
One	• Employed quickly, without prior calibration for lake and reservoir water, and with a small library of measurements. • The most fundamental and widely used techniques for evaluating the quality of river water.	It is not designed to compute how concentration changes over time, nor is it intended to describe the intricate chemical, physical, and biological processes that take place in water.	Ejigu (2021)	• The study focuses on the Tigris River in Baghdad. It applies the QUAL2K model to a 22-kmsection between Jadiriyah and the Tigris-Diyala confluence, and it analyzes water quality metrics. • The biochemical oxygen demand (BOD5), dissolved oxygen (DO), temperature, conductivity, alkalinity, and pH are important water quality metrics that have been researched. • 2019 saw both low flow and high flow times for the calibration and validation of the QUAL2K model.
Two	• Evaluating water quality at various depths and looking at different parameters on an hourly, daily, weekly, monthly, and annual basis	• The models have three requirements above 1D-dimensional models: (1) more data are needed; (2) strict calibration is needed; and (3) they are more susceptible to changes in a variety of water quality parameters.	El kadi Abderrezzak and Paquier (2009)	A mercury (Hg) simulation module was added to the CE-QUAL-W2 (W2) model, a two-dimensional, laterally averaged hydrodynamic and water quality model. The modified W2 model is capable of simulating the physical and biological processes involved in Hg transport and cycling, including volatilization, settling, resuspension, adsorption/desorption on multisolids, diffusive exchange, and biogeochemical transformations. The efficacy of the W2 model in simulating Hg dynamics is demonstrated by its ability to accurately forecast the reported quantities of methylmercury and total Hg in the Xiaxi River.

(*Continued*)

Table 15.2: (Continued)

Model dimensions	Advantages	Limitations	Study	Highlight
Three	• Essential for determining the geographical distribution of the concentrations of simulated water quality parameters	Models are rarely used because of the extreme complexity of the subjects being examined and the need for highly skilled analysts and massive amounts of data.	Fen et al. (2010)	• To support the total maximum daily load (TMDL) program, which aims to reduce watershed load, a three-dimensional hydrodynamic and water quality model was constructed for Lake Fuxian, China. • The water surface height, temperature fluctuations, and concentrations of total nitrogen (TN), total phosphorus (TP), and chemical oxygen demand (COD) were all precisely replicated by the model, which also successfully mimicked flow circulation and pollutant fate and transport.

15.8 Case studies and industrial applications

15.8.1 Municipal water treatment plants

The implementation of automated systems in municipal water treatment plants has significantly enhanced operational efficiency and water quality management. For instance, the city of Copenhagen, Denmark, implemented an advanced water treatment automation system that integrates real-time monitoring and predictive analytics. This system uses sensors to continuously monitor water quality parameters such as pH, turbidity, and chlorine levels, and it employs ML algorithms to predict future water quality issues. The result has been a marked reduction in chemical usage and operational costs, as well as improved compliance with stringent water quality standards. The automated system also provides early warnings of potential issues, allowing for proactive maintenance and reducing the risk of water supply disruptions.

15.8.2 Industrial effluent management

Industries such as manufacturing, chemical, and mining face stringent environmental regulations regarding the discharge of effluents into water bodies. Automated systems have proven crucial in ensuring compliance with these regulations. A critical example is the case of a large chemical manufacturing plant in Germany that implemented an automated effluent management system. The system continuously monitors the concentration of pollutants in the wastewater and automatically adjusts the treatment processes to ensure compliance with environmental regulations. This has led to a significant reduction in pollution levels in the discharged water, avoiding hefty fines and negative environmental impacts. Additionally, the system's real-time data collection and reporting capabilities have streamlined regulatory compliance reporting and auditing processes, further enhancing the plant's operational efficiency.

15.8.3 Agricultural water management

The agricultural sector has increasingly adopted automated systems for irrigation and drainage to improve water management practices. A notable example is the adoption of automated irrigation systems in the vineyards of California's Napa Valley. These systems use a combination of soil moisture sensors, weather data, and satellite imagery to optimize irrigation schedules. The automated system ensures that vines receive the precise amount of water needed at different growth stages, significantly reducing water usage compared to traditional irrigation methods. This precise water management has not only conserved water resources in the drought-prone region but also improved grape quality and yield. The system's ability to provide detailed data on soil moisture and plant water needs has enabled vineyard managers to make informed decisions, further enhancing the sustainability and profitability of their operations.

In each of these case studies, the implementation of automated systems has demonstrated significant benefits, including cost reduction, improved accuracy, and enhanced compliance with environmental regulations. These examples highlight the transformative potential of automation in various sectors, leading to more sustainable and efficient water management practices.

15.9 Challenges and limitations

The implementation of automated water management systems faces several technical challenges, primarily concerning sensor accuracy and reliability. Sensors, crucial for monitoring water quality and flow rates, often suffer from issues like drift and fouling, leading to inaccurate readings over time. These issues necessitate frequent calibration and maintenance, which can be labor-intensive and costly. Furthermore, integrating data from various sources, such as field sensors and external weather data, poses significant difficulties due to differences in data formats and protocols, complicating real-time data analysis and decision-making processes.

Economic and financial barriers also play a substantial role in hindering the adoption of automated water management systems. The initial investment required for advanced sensors, control systems, and software is often prohibitively high, especially for small municipalities and agricultural enterprises. Ongoing maintenance costs further add to this financial burden. Moreover, assessing the return on investment for these systems is complex and uncertain. While potential benefits include cost savings and improved efficiency, quantifying these against the high upfront costs can be challenging, particularly for industries operating under tight financial margins.

Regulatory and policy issues add another layer of complexity to the adoption of automated water management systems. Compliance with varying local and international standards regarding water quality and effluent discharge can be costly and complex. For example, a water treatment plant must adhere to national water quality standards and international guidelines, navigating differing regulations across jurisdictions. Additionally, privacy and data security concerns are significant, especially as these systems increasingly rely on digital technologies and the IoT. Protecting sensitive data from cyber threats and ensuring robust data encryption and secure communication protocols are essential to mitigate risks.

Addressing these technical, economic, and regulatory challenges requires a collaborative approach involving technology developers, regulatory bodies, and end-users. By working together, these stakeholders can develop solutions that enhance sensor reliability, ensure data integration and interoperability, and make the financial investment in automated systems more viable. Additionally, clear regulatory frameworks and robust data protection measures will be crucial in fostering the widespread adoption of automated water management systems, ultimately leading to more sustainable and efficient water management practices.

15.10 Future directions and innovations

Future directions in automated water management systems are poised to benefit significantly from advancements in sensor technologies and integration with cutting-edge technologies. The development of new sensors capable of detecting emerging contaminants such as microplastics, pharmaceuticals, and other pollutants will enhance the accuracy and comprehensiveness of water quality monitoring. Additionally, the introduction of smart sensors with self-calibration capabilities promises to address current issues related to sensor drift and fouling, thereby reducing maintenance requirements and ensuring consistent data accuracy. These innovations will enable more precise monitoring and management of water resources, crucial for both environmental protection and public health.

Integration with advanced technologies like blockchain and AI represents another transformative avenue. Blockchain can provide a secure and transparent method for managing and sharing water quality data, ensuring data integrity and reducing the risk of tampering. AI, particularly in the form of machine learning algorithms, can facilitate adaptive management systems that automatically adjust water treatment processes in response to real-time data, optimizing performance and efficiency. On a global scale, trends in computational automation and the adoption of these technologies are likely to accelerate, fostering greater international collaboration. As countries recognize the benefits of these innovations, there is potential for widespread global adoption, driving improvements in water quality management practices worldwide.

15.11 Conclusion

In conclusion, computational automation in water quality management represents a transformative approach to addressing the complex challenges associated with monitoring and maintaining water resources. By leveraging advanced technologies such as sensor networks, hydrodynamic modeling, and real-time data analytics, automated systems provide a high degree of accuracy, efficiency, and responsiveness that traditional methods cannot match. These systems are critical in various applications, including municipal water treatment, industrial effluent management, and agricultural water use, demonstrating substantial benefits in cost reduction, regulatory compliance, and resource conservation.

Despite the significant advantages, several technical, economic, and regulatory challenges must be addressed to fully realize the potential of computational automation. Issues related to sensor accuracy and data integration, high implementation costs, and compliance with evolving standards are significant barriers that require ongoing research and development. Looking forward, advancements in sensor technology, integration with blockchain and AI, and global collaborative efforts are promising pathways for overcoming these obstacles. As these innovations continue to evolve, the future of water quality management will likely be more efficient, sustainable, and capable of meeting the growing demands of a changing world.

Acknowledgments

The authors gratefully acknowledge the support and collaboration of the Center of Research and Innovation (CORI) at Universiti Kuala Lumpur (UniKL), Universiti Teknologi Malaysia, LOSC Langsat-OSC Sdn. Bhd., and Indah Water Konsortium in facilitating and supporting this research. Their contributions and expertize have been invaluable to the successful completion of this work.

References

Ahmed, U., Mumtaz, R., Anwar, H., Shah, A. A., Irfan, R., & García-Nieto, J. (2019). Efficient water quality prediction using supervised machine learning. *Water*, *11*(11), 2210. Available from https://doi.org/10.3390/w11112210.

Akomea-Frimpong, I., Dzagli, J. R. A. D., Eluerkeh, K., Bonsu, F., Opoku-Brafi, S., Gyimah, S., . . . Kukah, A. S. K. (2023). A systematic review of artificial intelligence in managing climate risks of PPP infrastructure projects . Engineering Construction & Architectural Management. Available from https://doi.org/10.1108/ecam-01-2023-0016.

Al Kindi, G. Y., Hussain, T. A., & Abed, A. N. (2021). Application of a one-dimensional steady state model for simulation the water quality in Tigirs River. *Journal of Engineering Science and Technology*, *16*(6), 4638–4649.

Al-Mansori, N. J. H., Al-Zubaidi, L. S. A., & Ashoor, A. Z. L. S. (2020). One-dimensional hydrodynamic modeling of the Euphrates River and prediction of hydraulic parameters. *Civil Engineering Journal*, *6*(6), 1074–1090.

Alshehri, F., & Rahman, A. (2023). Coupling machine and deep learning with explainable artificial intelligence for improving prediction of groundwater quality and decision-making in arid region, Saudi Arabia. *Water*, *15*(12), 2298. Available from https://doi.org/10.3390/w15122298.

Balasundaram, S., & Venkatagiri, S. (2020). A structured approach to implementing robotic process automation in HR. *Journal of Physics: Conference Series*, *1427*(1), 012008. Available from https://doi.org/10.1088/1742-6596/1427/1/012008.

Balcerzak, W. (2000). Application of selected mathematical models to evaluate changes in water quality. In *Presented at the International Conference on Water Supply, Water Quality, and Protection, Kraków*.

Basuli, D., & Roy, S. (2023). Beyond human limits: Harnessing artificial intelligence to optimize immunosuppression in kidney transplantation. *Journal of Clinical Medicine Research*, *15*(8-9), 391. Available from https://doi.org/10.14740/jocmr5012.

Becker, A. C. C., et al. (2023). Zero-dimensional modelling as tool for reservoir water quality planning and management. *RBRH*, *28*, e4.

Boudreau, P., et al. (2015). Environment Canada for the International Lake Champlain–Richelieu River Technical Working Group, Technical Report. Development of an experimental 2D hydrodynamic model of Lake Champlain using existing bathymetric data.

Boyce, M. P. (2011). *Gas turbine engineering handbook*. Elsevier.

Bürgler, M., Vetsch, D. F., Boes, R. M., & Vanzo, D. (2023). Systematic comparison of 1D and 2D hydrodynamic models for the assessment of hydropeaking alterations. *River Research and Applications*, *39* (3), 460–477.

Chapra, S. C. (2008). *Surface water-quality modeling*. Waveland Press.

Childs, P. (2010). *Rotating flow*. Elsevier.

Choi, D., R'bigui, H., & Cho, C. (2022). Enabling the gab between RPA and process mining: User interface interactions recorder. *IEEE Access*, *10*, 39604–39612. Available from https://doi.org/10.1109/access.2022.3165797, vol.

DHI. (1998). *MIKE 11: a microcomputer based modeling system for rivers and channels*. Hoersholm, Denmark: Reference manual of the Danish Hydraulic Institute.

Duraisamy, P., Khan, S., & Seetharaman, S. (2018). Integrating water quality index, GIS and multivariate statistical techniques towards a better understanding of drinking water quality. *Environmental Science and Pollution Research*, *25*(29), 29124–29132.

Ejigu, M. T. (2021). Overview of water quality modeling. *Cogent Engineering*, *8*(1), 1891711.

El kadi Abderrezzak, K., & Paquier, A. (2009). One-dimensional numerical modeling of sediment transport and bed deformation in open channels. *Water Resources Research*, *45*(5).

Eze, E., Kirby, S., Attridge, J., & Ajmal, T. (n.d.). Aquaculture 4.0: hybrid neural network multivariate water quality parameters forecasting model, Scientific.

Fen, L., Ran-hao, S., Li-rong, Y., & Li-ding, C. (2010). Assessment of freshwater ecosystem services in Beijing based on demand and supply. *Yingyong Shengtai Xuebao*, *21*(5).

Geographic information system (GIS), Wikipedia. (n.d.). https://en.wikipedia.org/wiki/Geographic_information_system Accessed 12.06.2023.

Glock, K., Tritthart, M., Habersack, H., & Hauer, C. (2019). Comparison of hydrodynamics simulated by 1D, 2D and 3D models focusing on bed shear stresses. *Water*, *11*(2), 226.

Golmohammadi, G., et al. (2016). Modeling the impacts of tillage practices on water table depth, drain outflow and nitrogen losses using DRAINMOD. *Computers and Electronics in Agriculture*, *124*, 73–83.

Hoover, W. G. (1991). *Studies in Modern Thermodynamics (Vol. 11)* (pp. 181–199). Elsevier.

Htun, A. M., Shamsuzzoha, M., & Ahamed, T. (2023). Rice yield prediction model using normalized vegetation and water indices from Sentinel-2A satellite imagery datasets. *Asia-Pacific Journal of Regional Science*, *7*, 491–519, https://doi.org/10.1007/s41685-023-00299-2.

Human-machine interface (HMI) SCADA, Gefran. (n.d.). https://www.gefran.com/products/panel-pc-cpu-and-hmi/hmi-terminal/ Accessed 12.04.2023.

Infrastructure for climate action. (2022). https://doi.org/10.1163/9789004322714_cclc_2021-0252-837.

Introduction to industrial automation. https://www.youtube.com/watch?v = 372DqUtJnrw Accessed 12.03.2000.

Ji, Z. G. (2007). *Hydrodynamics and water quality- modeling rivers, lakes, and estuaries* (1st ed., p. 676) Hoboken, New Jersey: John Wiley & Sons, Inc.

Johnes, P. J. (1996). Evaluation and management of the impact of land use on the nitrogen and phosphorus load delivered to surface waters: The export coefficient modelling approach. *Journal of Hydrology*, *183*, 323–349.

Joshi, J. (2024). Transition through Urban Barriers, Master's thesis, University of Cincinnati.

Kafle, M. R., & Shakya, N. M. (2018). Two-dimensional hydrodynamic modelling of Koshi River and prediction of inundation parameters. *Hydrology: Current Research*, *9*, 30.

Klompenburg, T. V., Kassahun, A., & Çatal, Ç. (2020). *Crop yield prediction using machine learning: A systematic literature review*. Available from https://doi.org/10.1016/j.compag.2020.105709.

LeakNavigator: Water conservation and leak detection, Ovarro. (n.d.). https://ovarro.com/en/global/case-studies/leaknavigator-saves-millions-of-litres-for-thames-water/2/.

Kumawat, S. K., Singh, R., Jain, R., & Sharma, V. (2021). A comprehensive review on recent advancements in sustainable bioplastics. *Journal of Cleaner Production*, *278*, 123218. Available from https://doi.org/10.1016/j.jclepro.2020.123218.

Lee, S., Kim, J., & Kim, S. (2016). A decision support system for predicting membrane fouling in seawater desalination plants. *Desalination, 391*, 124–133.

Lim, B., Spanger-Siegfried, E. (n.d.). Adaptation policy frameworks for climate change: Developing strategies, policies, and measures.

Liu, L. (2018). *Application of a hydrodynamic and water quality model for inland surface water systems. Applications in water systems management and modeling (Vol. 10)*. IntechOpen.

Masoom, A., et al. (2021). A GIS-based groundwater contamination assessment using modified drastic geospatial technique. *Water, 13*(20), 2868.

Mehta, V., & Reddy, K. (2015). SCADA systems for power system protection and control. *International Journal of Electrical Power & Energy Systems*, 69–82.

Negm, L. M., Youssef, M. A., & Jaynes, D. B. (2017). Evaluation of DRAINMOD-DSSAT simulated effects of controlled drainage on crop yield, water balance, and water quality for a corn-soybean cropping system in central Iowa. *Agricultural Water Management, 187*, 57–68.

Niazi, A., Prasher, S. O., Adamowski, J., & Gleeson, T. (2014). A system dynamics model to conserve arid region water resources through aquifer storage and recovery in conjunction with a dam. *Water, 6*(8), 2300–2321.

OCSAN Orange County Sanitation District. (n.d.). https://www.ocsan.gov/residents/construction-program-projects/current-construction/headworks-rehabilitation-at-plant-no-1.

Ottaviano, E. (2020). Prediction of water quality monitoring indicators based on random forest. *Water Pollution Prevention and Control Project, 1*(2), 11–19. Available from https://doi.org/10.38007/wppcp.2020.010202.

Papanicolaou, A. N., Elhakeem, M., Krallis, G., Prakash, S., & Edinger, J. (2008). Sediment transport modeling review-current and future developments. *Journal of Hydraulic Engineering, 134*, 1–14.

Power, D. (2008). *Decision support systems: Concepts and resources for managers*. Elsevier.

Rao, P., Wang, S., Wang, A., Yang, D., & Tang, L. (2022). Spatiotemporal characteristics of nonpoint source nutrient loads and their impact on river water quality in Yancheng city, China, simulated by an improved export coefficient model coupled with grid-based runoff calculations. *Ecological Indicators, 142*, 109188.

Rietveld, L. C., van der Helm, A. W. C., van Schagen, K. M., & van der Aa, L. T. J. (2010). Environmental modelling & software.

SCADA. (n.d.). https://en.wikipedia.org/wiki/SCADA Accessed 13.06.2023.

Skaggs, R.W. (1981). DRAINMOD Reference Report. Methods for design and evaluation of drainage-water management systems for soils with high water tables, USDA-SCS, p. 329.

Techet, A. H. (2005). Introduction to basic principles of fluid mechanics. *Hydrodynamics*.

The sustainable development goals report. (2023).

Thu Minh, H. V., et al. (2022). A model-based approach for improving surface water quality management in aquaculture using MIKE 11: A case of the Long Xuyen Quadangle, Mekong Delta, Vietnam. *Water, 14*(3), 412.

Tsakiris, G., & Alexakis, D. (2012). Water quality models: An overview. *European Water**, *37*, 33–46.

Tunji-Olayeni, P., Omuh, I. O., Afolabi, A., Ojelabi, R. A., & Eshofonie, E. E. (2019). *Climate change mitigation and adaptation strategies for construction activities within planetary boundaries: Limitations of developing countries*. Available from https://doi.org/10.1088/1742-6596/1299/1/012006.

Veerapaga, N., Azhikodan, G., Shintani, T., Iwamoto, N., & Yokoyama, K. (2019). A three-dimensional environmental hydrodynamic model, Fantom-Refined: Validation and application for saltwater intrusion in a meso-macrotidal estuary. *Ocean Modelling, 141*, 101425.

What are the main components of a SCADA system? Pacific Blue Engineering (n.d.). https://cb-pacific.com/automation-product-solutions/hmi-scada/.

Wright, E. (2008). *SCADA systems for industrial control and data acquisition*. Butterworth-Heinemann.

Yi, Y., et al. (2017). Evaluating the ecological influence of hydraulic projects: A review of aquatic habitat suitability models. *Renewable and Sustainable Energy Reviews, 68*, 748–762.

Zainalfikry, M. K., Ab Ghani, A., Zakaria, N. A., & Chan, N. W. (2019). *HEC-RAS one-dimensional hydrodynamic modelling for recent major flood events in Pahang River. AWAM international conference on civil engineering* (pp. 1099–1115). Cham: Springer International Publishing.

Zhang, C., & Fu, T. (2023). Recalibration of a three-dimensional water quality model with a newly developed autocalibration toolkit (EFDC-ACT v1.0.0): Does the model calibrated with a wider hydrological variability become more robust? *Geoscientific Model Development Discussions*, 1–24.

Zhang, S., Cheng, G., Tan, Q., Zhao, H., & Zhang, T. (2023). An agro-hydrological process-based export coefficient model for estimating monthly non-point source loads in a semiarid agricultural area. *Journal of Cleaner Production*, *385*, 135519.

Zhu, S., Zhang, Z., & Liu, X. (2017). Enhanced two-dimensional hydrodynamic and water quality model (CE-QUAL-W2) for simulating mercury transport and cycling in water bodies. *Water*, *9*(9), 643.

CHAPTER 16

Case studies: Successful implementation of automation for water quality assessment

Srinibas Pattanaik[1], Anshu Mehta[1], Sartajvir Singh Dhillon[1] and Alessandro Vinciarelli[2]

[1]*Computer Science Engineering, Chandigarh University, Mohali, Punjab, India,* [2]*Computer Science Intelligence Society, University of Glasgow, Glasgow, Scotland, United Kingdom*

16.1 Introduction

The necessity for accurate water quality evaluation has become increasingly crucial in the face of expanding environmental issues, such as pollution, global warming, and population increase. Traditional techniques of monitoring the quality of water, which generally rely on sampling by hand and analysis in laboratories, can be laborious, costly, and highly susceptible to human error. These constraints underscore the demand for more modern, efficient, and dependable procedures. In the past few decades, technology for automation has come about as a disruptive answer in the hands of water quality evaluation (Amato, de Lillo, & Mosca, 2017). Automation spans a broad spectrum of methods, such as networks of sensors, the Web of Information (IoT), artificial intelligence (AI), machine learning, satellite imaging, and unmanned airliners. These methods enable constant, immediate monitoring and evaluation of water's environmental parameters that provide timely and precise data required for sustainable water management. This study explores many case studies that highlight the successful adoption of robotics for water quality testing (An, Huang, & Yao, 2016). The case studies span numerous geographic regions and biological circumstances, from urban water systems to agricultural drainage areas and natural water bodies (Bartoli, Racchetti, & Soana, 2012). They highlight that technology for automation has been implemented to improve the precision, efficacy, and reliability of tracking the quality of water (Behera, Kim, & Park, 2011). The integration of sensors and IoT devices enables the continual tracking of several water quality indicators, like the pH level, heat, oxygen concentration, and turbidity (Benke & Hamilton, 2008). Real-time transmission of data and advanced computational models permit instant quantitative and qualitative evaluation of data, permitting quick responses to changes in water conditions (Brockett & Krzyżanowski, 2021). Machine learning techniques and AI further augment

Computational Automation for Water Security.
DOI: https://doi.org/10.1016/B978-0-443-33321-7.00007-X

these capabilities by delivering statistical analysis, which can foresee problems with pollution and enable proactive management measures (Cai, Wang, & Yu, 2015). In addition, autonomous underwater vehicles, or AUVs, enhance the reach of monitoring activities to cover broad or hard-to-access locations (Chhipi-Shrestha, Hewage, & Sadiq, 2017). These technologies give high-resolution geographical and temporal data, enabling a comprehensive perspective of water quality across huge and diverse ecosystems (Choi, Lee, & Kim, 2018). Despite the clear benefits, the introduction of automated technologies in water quality testing is not without hurdles. High initial expenses, continuous maintenance requirements, and the need for specific technical competence can pose substantial impediments. This study analyzes these problems and explores alternative solutions, such as staggered adoption, training programs, and rigorous cost–benefit evaluations (Fang, Li, & Shi, 2015). Overall, the successful adoption of automation in the quality of water evaluation constitutes a substantial development in environmental monitoring procedures. It enables more sustainable and effective administration of water resources, facilitates decision making based on data, and encourages long-term ecological health. By studying these individual instances, this article seeks to provide significant insights and practical assistance for future projects in the sector (Goonetilleke & Thomas, 2003; Hamann & O'Brien, 2019). Water is a tremendously vital demand of every living thing. With a growth in people worldwide, the demand for maintaining water assets is obtaining increased relevance (Li & Li, 2018). Several metrics are utilized for identifying the water's quality depending on the sample's level of water, pH level, viscosity, and salt content (Fisher & Elder, 2013). The machine learning algorithms are paired with geographic information systems (GIS) based on the availability of information and quality. Machine learning methods must account for geographical dispersion and time dynamics to predict the health of water (Ahmed & Ali, 2020). Incorporating spatial analytic tools within GIS, such as geographic statistics, can help capture these fluctuations and increase the accuracy of forecasts. By tackling these difficulties, effective machine instruction and GIS-based framework can be developed for the index of water quality calculation, assisting in efficient water administration and choice-making processes. The assessment of water quality plays a critical role in determining and tracking the overall condition and safety of drinking water sources. It entails examining numerous physical, chemical, and biological factors to evaluate whether the water quality fulfills the needed standards for its intended use, such as drinking reasons, tourism, and ecosystem preservation. Traditional water supply monitoring systems have depended on conventional laboratory-based approaches in which water samples have been taken from diverse resources and transferred to the laboratory for examination. Water samples are gathered manually from different locations of fascination, such as rivers, lakes, and wells, and extensive examinations are undertaken using specialist tools and methods in laboratory circumstances (Alam, Mehmood, & Katib, 2019; Al-Fuqaha, Guizani, & Mohammadi, 2019). The gathered data areevaluated and presented to determine the water quality. It is measured against the specified norms or standards to meet the requirements for its specific purpose. The disadvantages of standard laboratory-based drinking water monitoring devices include that they are labor-intensive and expensive, lack immediate assessment, and are not accessible to remote locations (Al-Fuqaha et al., 2019; Al-Saadi & Abdalla, 2021).

16.2 Literature review

The academic scholarship on robotics in water quality evaluation covers a wide range of research illustrating the improvements, benefits, and issues connected with these technologies. This study synthesizes major discoveries from recent research focused on sensor networks, the Internet of Things (IoT), computer vision (AI), learning by accident, satellite imagery, and drones (UAVs). The sensor networks are essential parts of computerized water quality monitoring networks. Numerous studies have proved their efficiency in giving continuous, real-time data regarding different water quality indicators (Barakat & Mollah, 2022). As an example, the implementation of networked sensors to track pH, contamination, and oxygen dissolved levels, highlighting the enhancement in accuracy of data and temporal precision in comparison with traditional methods. The networks in question consist of many sensors positioned at critical points, delivering information to a central database for real-time evaluation and decision making. The incorporation of IoT in monitoring the water's quality has been thoroughly studied. IoT connects systems of sensors to the World Wide Web, providing smooth data transmission for remote monitoring. According to Fang et al. (2015) (Boulos & Koti, 2022), IoT-enabled devices boost the ability to monitor water quality over wide areas, allowing for more comprehensive and coordinated water-related methods. These technologies enable users to access current information from home, hence enhancing the reaction on water quality issues. Machine learning and AI have been extensively used for assessments of water quality to evaluate big datasets and build predictive models. Castillo and Venegas (2022) emphasized the application of AI algorithms to predict changes in water quality and identify probable contamination occurrences. These methods enable the analysis of huge volumes of data from networks of sensors, yielding insights that might guide proactive management techniques. AI-driven models can detect trends and connections that could be overlooked by conventional statistical procedures, leading to much more precise and relevant actions. Satellite imagery technologies, including satellite images and aerial photography, provide essential data for water quality monitoring. Studies by Chang and Lee (2023) show that satellite imagery can effectively track large bodies of water by detecting shifts in variables such as chlorophyll content, temperature at the surface, and turbidity. These advancements supplement ground-based networks of sensors by delivering geographically vast information, which is vital for monitoring huge and difficult-to-access locations. UAVs, or unmanned aircraft, have become known as excellent tools for environmental monitoring. Research by Cho and Han (2020) showed the use of drones that had spectral and thermal imaging sensors to analyze water quality in remote and vast places. UAVs give high-resolution temporal and spatial data, enabling extensive investigation of aquatic bodies that are tough to track using traditional approaches. These can quickly survey huge regions, delivering real-time data that aid prompt decision making. Although the advantages of robotics in water quality monitoring are well documented, some difficulties continue. Large beginning costs, maintenance needs, and a requirement for technical skills are significant impediments to adoption. Research by Ding and Wang (2021) and others recommend techniques such as gradual deployment, building capacity via educational initiatives, and cost–benefit evaluations to overcome these problems. Such strategies can help stakeholders overcome financial and technological challenges, guaranteeing the effective adoption of automated surveillance systems (González and Estrada, 2023).

16.3 Case study 1

Urban water systems are crucial for delivering safe drinking water, regulating wastewater, and facilitating industrial processes. Yet they are often hindered by large population density, industrial wastes, and complex infrastructure. Traditional water quality surveillance gets closer, which utilizes periodic manual sampling but fails to record the rapid and unpredictable shifts in water quality that occur in urban areas. These approaches are also labor-intensive or vulnerable to error by humans, which can delay the detection and solution of water quality concerns.

16.3.1 Automation deployment and implementation

To solve these issues, a robotic monitoring of water quality system was established in a major metropolis. The basic technology contains sensor networks related to the Internet of Things (IoT). Sensors were deliberately deployed at important spots in the municipality's water distribution and collecting systems. These connectors included intake terminals and transmission nodes in wastewater discharge sites. The sensors continually detect a range of water quality indicators, comprising pH, turbidity, humidity, temperature, chemical compounds (such as heavy metals and fertilizers), and microbiological pathogens (such as *E. coli*). Information gathered through the sensors is transferred in actual time from a centralized management system via IoT. This system gathers the data and performs basic processing and analysis, identifying any unusual patterns or trends that suggest probable water quality issues. The automated approach dramatically boosted the capability to detect impurities quickly and accurately. Real-time monitoring enables fast detection of deviations from typical water supply levels of difficulty, facilitating swift action and mitigation.

The technology lowered labor costs linked to sample collection and analysis. In addition, a decrease in errors made by humans improved the total trustworthiness of the data. Continuous monitoring gave a broader and accurate image of the water's condition compared to the snapshot perspective supplied by periodic sampling by hand. Enhanced Decision Making: The provision of real-time, reliable information promoted better decision-making procedures for water management bodies. They might proactively handle possible water supply issues before they develop into health concerns or regulatory infractions. The initial expenditure necessary for the setting up of sensors, connectivity to the IoT, and info management tools was substantial. Technical Expertise: Monitoring and running the automated system needed specific technical abilities, which entailed further training for existing workers or employing new personnel. Phased Implementation: To limit expenses, the execution took place in phases. The initial phase concentrated on the most essential places with the highest danger of contamination. Successive stages extended the structure to cover the whole metropolitan water network. Comprehensive courses were designed to equip personnel with the essential skills to run and maintain the computerized system. These programs comprised both initial instruction and continued education to stay up with technology improvements. The use of automation in the municipal water system of this huge city has proven to be an effective technique for boosting water quality monitoring. The automated system offered perpetual, real-time data that increased the finding of contaminants, decreased expenses, and raised the dependability of water quality examinations. Despite constraints relating to cost plus technical competence, staggered implementation and focused training programs permitted an effortless adjustment to the new system. This case study highlights the possibilities of automation technology to improve city management of water quality, providing more secure and trustworthy water supplies for metropolitan populations.

16.4 Case study 2: future trends in automated water quality assessment

With advancements in microelectromechanical systems, smaller, more compact sensors have been developed, making them conveniently deployable in a variety of situations. There is a broader range of potential uses for these little sensors because they are less invasive and easier to deploy. Thanks to developments in materials science and widespread manufacturing, the price of sensors has dropped dramatically. Now that sensors are more reasonably priced, comprehensive surveillance networks that were before unaffordable are within reach. Improved sensitivities in newly developed sensor technologies are enabling the identification of pollutants at very low concentrations. To achieve demanding water quality requirements and for early warning systems, this is essential. Improvements in biochemicals and biosensing have enhanced the sensitivity of detectors, allowing them to differentiate between identical molecules and accurately recognize specific pollutants. This precision is critical for focused remediation operations and compliance with regulations, and by combining water quality monitoring with Internet of Things technology, data may be collected and transmitted in real time. With the help of the IoT, sensors can talk to one another and to central monitoring systems, painting a full picture of the water quality in cities. Monitoring the quality of water is becoming an integral part of smart city programs' larger municipal management systems.

The long-lasting quality of urban water supplies, efficient use of water resources, and wastewater management are all aided by this integration. Water quality data can be connected with additional city offerings, such as health services, monitoring the environment, or emergency response systems. This integrated strategy enables a better and more coordinated administration of urban resources. By merging information on water quality with various other information from urban streams, communities can utilize predictive analytics to detect and alleviate possible problems with water quality before they become significant. This proactive strategy strengthens the capacity for resilience of metropolitan water systems. As scientific awareness of water pollutants and their health implications advances, regulatory regulations are getting increasingly strict. Automatic systems for monitoring must continuously adjust to meet these increasing demands. Real-time monitoring capabilities supplied by automated systems ensure continual compliance with regulatory criteria. This decreases the probability of infractions and promotes confidence in water management authority. Governments are progressively offering rewards, such as loans, rebates, and assistance, to promote the adoption of computerized monitoring of water quality systems. These incentives help cover the initial expenditure costs and speed up the deployment of new technology. Industry associations and standard-setting agencies are establishing certifications and optimal standards for automated quality of water monitoring systems. These standards safeguard the stability and efficiency of the technology, enabling increased adoption across multiple sectors. The future of computerized water quality assessment has been influenced by rapid improvements in technology for sensors, unity with smart city infrastructure, and developing legislative and regulatory frameworks. As sensors grow more compact, inexpensive, sensitive, and particular, their deployment in full monitoring networks could grow increasingly possible. The combination of these technologies with IoT and urban systems of management would enable greater effectiveness and proactive water quality management. Concurrently, evolving standards and legislative incentives will encourage the implementation of these technologies, guaranteeing that water quality monitoring keeps pace with new obstacles and societal needs. These trends combined promise a future wherein water quality is regularly monitored, safeguarding the health and sustainability of the water that we use (Fig. 16.1).

A connection to the internet of smart monitoring of water quality systems uses smart water condition sensor technology to assess parameters related to water such as temperature, oxygen concentration, the amount of dispersed particulates, TDS level, pH, viscosity, and more. The major purpose of smart monitoring of water quality gadgets is to make certain these metrics are within the appropriate boundaries in the water. The IoT-based quality of water measuring system incorporates intelligent water quality sensors that keep track of the parameters of the water in actual time and deliver vital insights so that you can control and monitor your water purity.

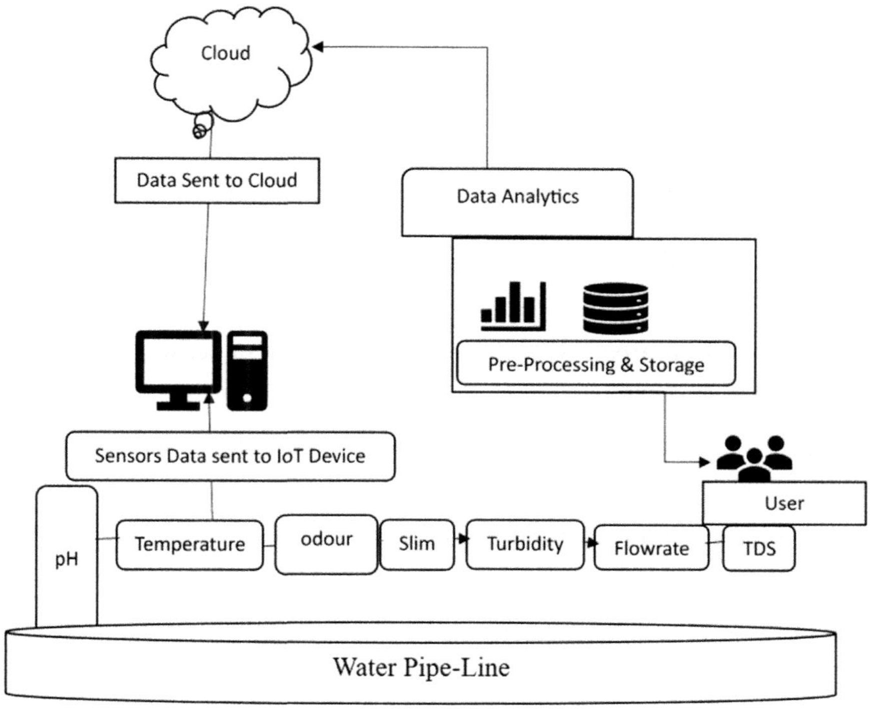

Figure 16.1
Intelligent deep water condition supervising system.

16.5 Case study 3: farm runoff and mitigation

Farm drainage, including pesticides, fertilizers, and various other compounds, is an important cause of water contamination. This case study explores an initiative aimed at regulating and reducing runoff from agriculture in a broad farming region. The key aims are to computerize the evaluation of the water's quality negatively impacted by agricultural runoff, apply mitigation techniques for minimizing pollutants, and increase the sustainability of agricultural operations. Specific goals included decreasing nutrient pollution, boosting soil and water health, and ensuring compliance with environmental requirements, all of which were closely linked to the development of best practices and technology. Ensure rules for water safety and monitor compliance. Supplied sensors, data collecting devices, and programming for automation. Knowledge on sustainable farming and runoff control strategies was provided. The impact of the project on regional water bodies and ecosystems was monitored. It was used to measure phosphorus, potassium, and nitrogen levels in runoff water. These sensors were put at important sites where discharge enters water bodies. Data on rainfall, temperatures, and other climatic conditions impacting runoff were provided, as well as soil conditions tracked to inform irrigation methods.

Information from each sensor was collected and transferred to a cloud platform for processing and reporting. Computerized control systems were employed to control mitigation measures like built wetland areas, buffering strips, and retention ponds.

16.5.1 Deployment techniques

The first stage (0–3 a while): Plan and analysis of requirements, including stakeholder engagements and site surveys.

The second stage (4–6 months): System layout, acquisition of sensing and machinery, and selection of mitigating strategies.

The third stage (7–9 months): Deployment of sensors, data collection systems, and mitigating infrastructure.

The fourth stage (10–12 months): Integrating the system, validating the detectors and instruction of farmers and workers.

Phase 5 (13–15 months): Complete implementation, final testing, and continual monitoring.

16.6 Regional weather prediction models

The automated system made sure that the water quality was monitored continuously, so the farming operations could constantly satisfy all the regulatory criteria. Any deviations from acceptable levels could be promptly identified and addressed thanks to this actual time monitoring and reporting. With its automated data logging, compliance paperwork became easier, with less room for human error and faster, more accurate report submission to regulatory agencies. There was a considerable decrease in nutrient discharge following the deployment of the automated surveillance and mitigation measures. This reduced the likelihood of nutrient enrichment and algal blooms by reducing the amounts of phosphorous and nitrogen that were entering adjacent bodies of water.

As a result of fewer contaminants, local aquatic ecosystems improved, and bodies of water became healthier. Recovery and maintenance of diversity in the afflicted areas were aided by the decrease in dangerous chemical runoff. Sustainable farming methods including buffer strips, cover crops, and precision agriculture were advocated for in the study. In addition to lessening negative effects on the environment, these methods enhanced soil health and increased crop yields. Connecting everyone involved, particularly farmers along with government organizations, through the project was vital. Regular conversations and updates assured that every individual was aligned on the project's aims and could contribute meaningful feedback.

Designing an apparatus that could tolerate shifting weather conditions ensured over time reliability and accuracy. This involved picking sturdy sensors and setting a regular servicing and calibration plan. Scalability planning enables the system to respond to future growth and technology improvements, ensuring its continuous relevance and efficacy. Implementing a strict maintenance program for monitors and other devices was critical to ensuring system efficiency and precision. Regular calibration and service prevented technical concerns and enhanced the service life of the equipment. Providing full training for farms and personnel on the operation of the machinery and sustainable farming methods was crucial. This covered practical education sessions, instructional materials, and continuing support. Training participants on the positive effects of a system and environmentally friendly methods helped in increasing their support and involvement, contributing to a more successful implementation. Keeping the system able to embrace new technologies and improvements assured that the endeavor could grow and become better over time. Being open to altering the system and processes based on feedback and new insights ensured continuous development. This adaptive management style enables the initiative to keep being strong in dealing with shifting conditions and new obstacles.

16.7 Financial and environmental effects of the use of automation

The costs of developing automated systems were weighed against potential benefits such as reduced labor, increased productivity, and reduced regulatory fines. Tthe return on investment (ROI) was calculated via indicators like the net present worth (The net present value), the internal amount of return (IRR), and the period of payback. The reduction in pollutants emitted was calculated throughout the surrounding environment brought about by computerized monitoring and mitigation technologies. The impact of technology on water conservation measures, such as reduced water consumption and better water use efficiency, was analyzed. The influence of increased water purity on public health was analyzed, particularly lower incidence of waterborne infections. How automation projects improve community comprehension and participation regarding water quality issues raised an awareness of sustainability.

16.8 Summary and improvements

The summary ofthe key results and advantages spotted in the particular case studies provided, highlighting the broader favorable influence of technology on the performance of water testing and executive leadership. A collection of recommendations was provided based on training gained from the case research studies. Policy changes and measures were proposed to facilitate the greater implementation of autonomous water integrity monitoring methods. New technologies and improvements that might considerably improve water quality monitoring were identified. Shortfalls in current understanding were addressed, indicating areas where further inquiry might be beneficial for improving the water cleanliness evaluation and control.

16.9 Conclusion (which includes shortcomings from what was investigated and further ideas)

The discussion chapter creates the data from the actual studies, stressing the accomplishments, problems, and general effects of computerized water quality monitoring systems. This presents an unbiased evaluation by recognizing the limits of this inquiry and proposing additional recommendations for strengthening the execution and efficacy of such systems. Computerized systems considerably increase the accuracy and responsiveness of the water's condition monitoring, enabling improved compliance with law enforcement criteria. Decreases in airborne contaminants alongside effective resource management contribute to beneficial environmental effects and, over an extended period, economic savings. Effective stakeholder participation and instructional materials are important to ensuring the effectiveness of automation programs.

16.9.1 Challenges belonging to the research

Notwithstanding advancements, gadgets with sensors are still subject to difficulties like slide, contamination, and susceptibility to outside factors, reducing the preciseness of data. Combining information from many devices can be challenging, possibly causing disparities and discrepancies. The initial expenditure for computerized systems, including sensors and infrastructures with software, can be unacceptably pricey for some firms. Continuing expenses such as upkeep, calibration, and system improvements can also be costly. Achieving that all essential staffs are sufficiently trained to utilize and service computerized equipment is a continuing challenge. Systems built for surroundings may not work optimally in diverse climatic or geographical circumstances without adaptation. Computerized systems are subject to cyber assaults, which somewhat could affect the confidentiality of information and reliability. The acquisition and use of historical information are requirement to be handled cautiously to satisfy confidentiality and moral demands.

16.9.2 Future prosperity

Participate in the creation of greater durability, precise, yet less expensive sensors that may function under various climate situations. Increase the application of machine learning and artificial intelligence in analytical forecasting, identifying anomalies, and machine learning to make decisions regarding water quality management. Develop financial methods and incentives that will promote the initial expenditures and recurring costs corresponding to automated systems. Support collaborations between public agencies and commercial enterprises to collaborate on costs and expertise in installing monitoring of water quality systems. Develop uniform instruction and certification for workers involved in managing and operating automated systems. Build modular and adaptable solutions that may be quickly adapted to diverse circumstances and local needs. Push for regulations that encourage the deployment of automated pollution monitoring equipment and offer clear guidelines for data usage along with privacy. Advocate the standardization of protocols and technologies to ensure interoperability and interoperablility across distinct networks and areas. Increase public knowledge about the significance of the condition of water and the function of technology in safeguarding it. Develop venues for stakeholders to communicate information, expertise, and best practices relevant to automated monitoring of water quality. Establish a rigorous structure for ethical gathering of information, administration, and use to promote openness and responsibility. Implement modern cybersecurity mechanisms to protect information regarding water quality from breaches and unwanted access.

16.10 Conclusion

The use of computerized mechanisms for quality of water evaluation is an effective method in monitoring and enhancing water quality across diverse sectors. The case examples given in this book provide persuasive proof of the considerable advantages associated with robotics, including increased monitoring transparency, adherence to regulations, protection of the environment, and operational efficiency. The aforementioned case studies—covering urban drinking water monitoring, wastewater from industry supervisors, and runoff from agriculture control—highlight the real-life uses and beneficial effects of adopting robotic quality of water monitoring systems. The successful deployment of automation in the quality of water evaluation constitutes a big step in controlling and protecting water resources. By addressing the limitations at present and taking into account future developments, stakeholders may enhance both the efficiency and long-term viability of these networks, ultimately leading to healthier environments and enhanced public health results. Continued investigation, creativity, and partnerships are necessary to drive advancement in this critical subject.

References

Ahmed, M., & Ali, R. (2020). IoT-based water quality monitoring system for smart cities. *Journal of Environmental Science and Technology*, *13*(2), 104–116.

Alam, F., Mehmood, R., & Katib, I. (2019). Data fusion and IoT for smart water management. *Future Generation Computer Systems*, *87*, 368–381.

Al-Fuqaha, A., Guizani, M., & Mohammadi, M. (2019). Enabling smart water networks through internet of things: A review. *IEEE Internet of Things Journal*, *6*(2), 2386–2415.

Al-Saadi, S., & Abdalla, A. (2021). Real-time water quality monitoring using cloud-based IoT solution. *Environmental Monitoring and Assessment*, *193*(2), 71.

Amato, F., de Lillo, E., & Mosca, A. (2017). Automated systems for continuous monitoring of water quality: A review of literature. *Environmental Monitoring and Assessment*, *189*(10), 503.

An, C., Huang, G., & Yao, Y. (2016). An integrated approach to assess the environmental impact of industrial effluent on urban river water quality. *Journal of Cleaner Production*, *112*, 4626–4636.

Barakat, A., & Mollah, M. (2022). Implementation of machine learning algorithms for water quality assessment. *Water Resources Management*, *36*, 987–1001.

Bartoli, G., Racchetti, E., & Soana, E. (2012). Nitrogen balance and fate in a heavily impacted watershed: A case study of the Oglio River (Northern Italy). *Environmental Monitoring and Assessment, 184*(4), 2281–2292.

Behera, S. K., Kim, H. W., & Park, H. S. (2011). Performance evaluation of a pilot-scale constructed wetland treating industrial wastewater. *Ecological Engineering, 37*(10), 1546–1553.

Benke, K. K., & Hamilton, A. J. (2008). The performance of water quality indices in assessing water quality and land use effects in rivers in Victoria, Australia. *Ecological Indicators, 8*(6), 1018–1028.

Boulos, P., & Koti, R. (2022). Advanced analytics and IoT for smart water quality monitoring. *Smart Water, 7*(1), 8.

Brockett, B. F. T., & Krzyżanowski, J. (2021). A systematic review of the impact of climate change on waterborne diseases. *Environmental Research Letters, 16*(6), 063004.

Cai, H., Wang, C., & Yu, L. (2015). Internet of Things-based water quality monitoring and management system for large-scale aquaculture. *Computers and Electronics in Agriculture, 114*, 181–191.

Castillo, A., & Venegas, P. (2022). Water quality monitoring using wireless sensor networks in aquaculture. *Aquaculture (Amsterdam, Netherlands), 546*, 737330.

Chang, H., & Lee, S. (2023). Development of a real-time water quality monitoring system using LoRa technology. *Sensors, 23*(3), 1198.

Chhipi-Shrestha, G., Hewage, K., & Sadiq, R. (2017). Water–energy–carbon nexus modeling for urban water systems: A case study of Fort McMurray, Canada. *Journal of Cleaner Production, 143*, 701–713.

Cho, K., & Han, S. (2020). Deep learning-based water quality monitoring for sustainable development. *Sustainability, 12*(7), 2944.

Choi, Y., Lee, S. H., & Kim, J. H. (2018). Real-time water quality monitoring and controlling system based on Internet of Things. *International Journal of Distributed Sensor Networks, 14*(4), 1550147718765003.

Ding, Y., & Wang, S. (2021). A real-time monitoring and early warning system for water quality based on IoT. *Water Science and Technology, 83*(5), 1212–1220.

Fang, F., Li, J., & Shi, P. (2015). A real-time water quality monitoring system for large freshwater bodies based on UAV and WSN. *Sensors, 15*(9), 22578–22591.

Fisher, J. C., & Elder, R. S. (2013). Water quality assessment using sensor networks and data mining techniques. *Journal of Water Resources Planning and Management, 139*(3), 220–228.

González, R., & Estrada, H. (2023). Advances in automated water quality monitoring for aquaculture applications. *Aquacultural Engineering, 97*, 102193.

Goonetilleke, A., & Thomas, E. (2003). Understanding the role of land use in urban stormwater quality management. *Journal of Environmental Management, 68*(4), 401–410.

Hamann, M., & O'Brien, K. (2019). A review of the application of social-ecological systems framework to water governance research. *Ecology and Society, 24*(1).

Li, Y., & Li, W. (2018). Remote sensing and GIS applications for urban water resource management and planning. *Environmental Monitoring and Assessment, 190*(9), 520.

CHAPTER 17

Enhancing water security through automation: case studies and technical advancements in water quality management

Inam Ul Haq[1], Akib Mohi ud Din Khanday[2], Hilal Ahmad Shah[1] and Syed Zoofa Rufai[1]

[1]*Faculty of Engineering and Technology, SGT University, Gurugram, Haryana, India,* [2]*Department of CS and Software Engineering, United Arab Emirates University, Al Ain, United Arab Emirates*

17.1 Introduction

The management of water quality is an intricate and diverse undertaking that involves a variety of obstacles arising from natural phenomena, human actions, and environmental fluctuations. A thorough grasp of these issues is necessary to guarantee the availability of clean and safe water for human consumption, agriculture, industry, and ecosystem health. We examine some of the major obstacles to managing water quality below in Fig. 17.1:

- Pollution Sources: Finding and reducing pollution sources is one of the biggest problems in managing water quality. Numerous point and nonpoint sources, such as sewage systems, incorrect waste disposal, urban stormwater runoff, agricultural runoff, and industrial discharges, can produce pollution (Kaserzon, Heffernan, Thompson, Mueller, & Gomez Ramos, 2017). The public's health and the purity of the water are seriously endangered by contaminants such as pesticides, heavy metals, infections, nutrients, and newly discovered pollutants.
- Waterborne Diseases: Water bodies contain pathogens that have the potential to spread waterborne illnesses and endanger human health. Water that has been contaminated may contain bacteria, viruses, protozoa, and other germs that can lead to diseases like cholera, hepatitis, typhoid fever, and diarrhea. A crucial component of managing water quality is guaranteeing the microbiological safety of drinking water sources, especially in areas with poor sanitary facilities.
- Eutrophication: The process of eutrophication, which is typified by the overgrowth of algae and aquatic plants, can be brought on by excessive nutrient inputs, particularly those of nitrogen and phosphorus (Tröger, Klöckner, Ahrens, & Wiberg, 2018). Oxygen

Computational Automation for Water Security.
DOI: https://doi.org/10.1016/B978-0-443-33321-7.00002-0

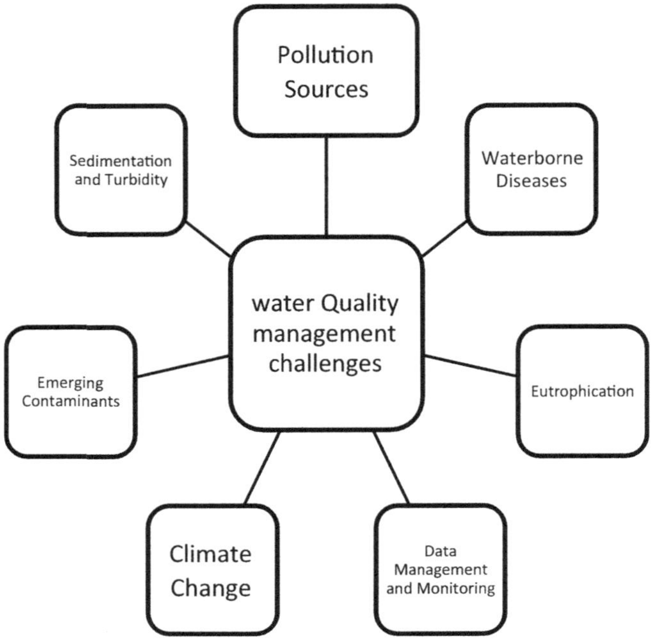

Figure 17.1
Overview of data management and monitoring.

depletion, toxic algal blooms, and ecological imbalances brought on by eutrophication can result in fish mortality, habitat degradation, and a decline in biodiversity. Using techniques to lessen nutrient loading from sources such as agricultural fertilizers, wastewater discharges, and atmospheric deposition is necessary to manage nutrient contamination.

- Sedimentation and Turbidity: Due to soil erosion and land disturbance, sedimentation and turbidity can reduce water clarity, clog streams, and suffocate aquatic habitats, all of which can worsen the quality of water. Pollutants, including pesticides, heavy metals, and fertilizers, can be carried by sediments, causing pollution in the water and harming aquatic environments. Mitigating sedimentation and maintaining water quality requires the implementation of effective sediment control measures, such as riparian buffer zones and erosion control techniques (Hollender et al., 2023).
- Emerging Contaminants: The management of water quality is facing new issues due to the emergence of contaminants such as industrial chemicals, microplastics, personal care products, and medications. These pollutants can enter water bodies through a variety of channels and harm the environment and human health since they are frequently not sufficiently controlled or monitored. To assess hazards and create mitigation methods, addressing new pollutants calls for continuous research, monitoring, and regulatory frameworks.

- Climate Change: By changing precipitation patterns, increasing the frequency and intensity of extreme weather events, and making droughts and floods worse, climate change exacerbates already-existing problems with water quality. Water quality and ecosystem dynamics may be impacted by these changes in the distribution of contaminants, nutrient cycling, temperature, and pH levels. To reduce environmental effects and adapt to climate change, plans for managing water quality must incorporate resilience measures. Sustainable practices must also be adopted (Place et al., 2021).
- Data Management and Monitoring: To evaluate trends in water quality, pinpoint the sources of pollution, and guide decision-making, efficient water quality management depends on reliable data gathering, analysis, and monitoring systems (González-Gaya et al., 2021). However, obstacles to spatial-temporal coverage, quality control, and data accessibility may make it difficult to apply evidence-based management techniques. To overcome these difficulties, it is imperative to improve the infrastructure for data management, encourage data-sharing activities, and make use of modern monitoring tools.

17.2 The rationale for adopting automation in water quality assessment

Several strong arguments support the use of automation in water quality evaluation, which is motivated by the need to manage water resources with scalability, precision, and efficiency. Fig. 17.2 shows the salient features that highlight the reasons to embrace automation in this particular context:

- Real-time Monitoring: Real-time monitoring of water quality measurements is made possible by automation, which offers immediate insights into patterns and fluctuations. The inability to acquire data quickly in traditional manual sampling and laboratory analysis procedures makes it more difficult to react quickly to new problems with water

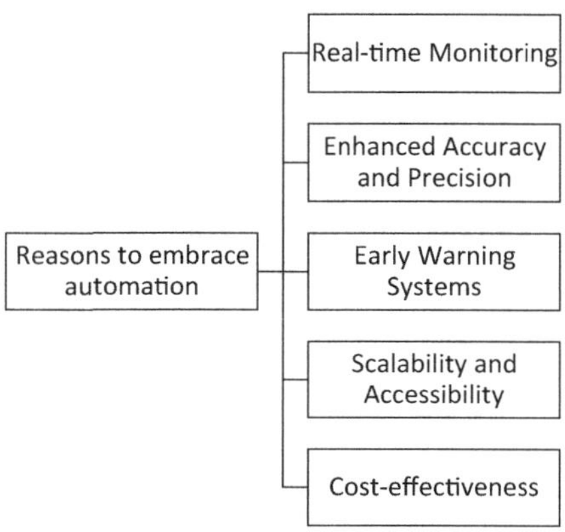

Figure 17.2
The reasons to embrace automation.

quality (Haq, Rather, & Kaur, 2023). In contrast, automated monitoring systems and sensor networks provide high-frequency, continuous data collection, enabling prompt identification of pollution occurrences and pre-emptive management actions.

- Enhanced Accuracy and Precision: By reducing human error and variability linked to manual sampling and analysis methods, automation raises the precision and accuracy of water quality assessments. Standardized protocols and calibration processes are followed by automated sensors and analytical devices to provide consistent and trustworthy data quality. This improved precision is crucial for monitoring pollutant concentrations, determining if pollution management strategies are working, and determining whether water quality standards are being met (Schymanski et al., 2020).
- Cost-effectiveness: Automation technologies can have a large upfront cost, but by simplifying monitoring processes and lowering labor and material costs, they frequently result in long-term cost benefits. Automated solutions reduce the need for sample transportation and laboratory processing, need fewer workers for data collection and analysis, and maximize resource allocation through focused monitoring initiatives. Additionally, automation makes data-driven decision-making and predictive modeling possible, which makes it possible to allocate resources and investments in water quality management more effectively (Krier et al., 2022).
- Scalability and Accessibility: Automation makes water quality monitoring programs more accessible and scalable, allowing for increased temporal resolution and broader regional coverage. Water bodies and watersheds can be completely covered by automated sensor networks, which can be installed over large geographic areas, even inaccessible and distant ones. Furthermore, real-time access to data on water quality is now possible from any location because of developments in data transmission networks and remote sensing technology. This gives stakeholders the power to decide wisely and efficiently plan their reactions.
- Early Warning Systems: Early warning systems for identifying and addressing hazards to water quality, such as chemical spills, algal blooms, and bacterial contamination, can be developed thanks to automation. Automated systems can provide alerts and notifications ahead of possible water quality crises, enabling prompt interventions and mitigating measures, by integrating sensor data with predictive models and decision support tools (Choudhary, Thakur, Bhardwaj, Haq, & Rather, 2023). To preserve ecosystems, protect public health, and guarantee that water supply systems remain resilient in the face of new dangers and uncertainties, early warning systems are very important.

17.3 Fundamentals of computational automation

Automating various tasks and processes involved in water quality evaluation through the use of computer methodologies and algorithms is known as computational automation or automated computational analysis. Fundamentally, computational automation uses ideas from statistics, computer science, and machine learning to create algorithms that can evaluate big datasets, draw conclusions from them, and help make decisions without the need for human interaction. Fig. 17.3 shows the principles that are underlying the computational automation.

a) **Data Acquisition**: To gather data on water quality from various sources and situations, computational automation depends on automated data acquisition systems, such as sensor networks, remote sensing platforms, and autonomous sampling devices. These technologies make it possible to collect data continuously and often, guaranteeing a thorough understanding of the temporal and spatial fluctuations in water quality indicators.
b) **Data Processing and Analysis**: Data on water quality are obtained, and then computational approaches and algorithms are used for preprocessing, quality control, and analysis. Cleansing, filtering, and modifying raw data to eliminate mistakes and noise is known as data preprocessing. Quality control processes, on the other hand, guarantee data accuracy and dependability. Following data processing, computational algorithms are used to analyze the results and derive useful information about trends, patterns, and anomalies in water quality. These algorithms include statistical methods, machine learning models, and data mining techniques.
c) **Modeling and Prediction**: The creation of prediction models and simulations to project future water quality conditions and evaluate the possible effects of environmental stressors and management measures is made easier by computational automation. These models predict water quality indicators, such as nutrient concentrations, dissolved oxygen levels, and pollutant loads, under various scenarios and conditions by utilizing historical data, environmental variables, and system dynamics.
d) **Decision Support Systems**: The process of computational automation facilitates the incorporation of modeling outputs and analytical results into decision support systems (DSS), which help stakeholders, policymakers, and water managers make well-informed decisions about water quality management tactics. To prioritize resource allocation, examine alternative management solutions' efficacy, and reduce potential dangers to human health and the environment, DSS offers visualization tools, scenario analysis capabilities, and risk assessment frameworks (Alygizakis et al., 2019).

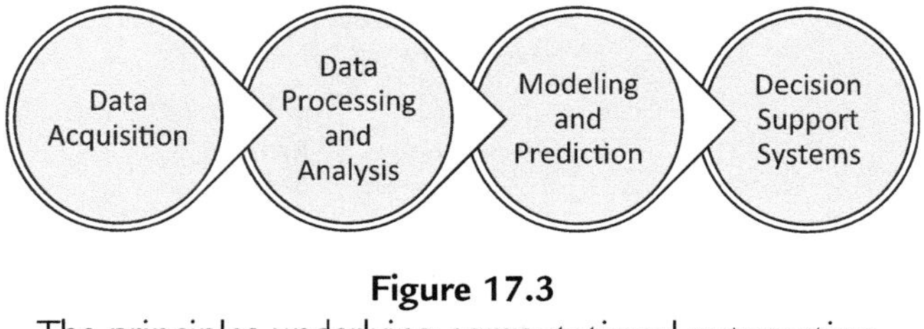

Figure 17.3
The principles underlying computational automation.

17.3.1 Methodologies and techniques

Computational automation, as used in practice, includes a variety of approaches and strategies adapted to particular goals and difficulties in water quality monitoring. Fig. 17.4 shows the typical techniques including the following:

a) **Statistical Analysis**: In computational automation, statistical techniques like regression analysis, time series analysis, and spatial analysis are frequently employed to find trends, correlations, and spatial patterns in data on water quality. By defining the connections between environmental factors and water quality metrics, these techniques facilitate trend analysis and predictive modeling.
b) **Machine Learning**: Supervised, unsupervised, and reinforcement learning are a few examples of machine learning approaches that provide strong tools for analyzing large, intricate datasets related to water quality. Based on past data and model training, machine learning algorithms can identify patterns, classify water quality samples, identify anomalies, and optimize decision-making processes.
c) **Geospatial Analysis**: Geographic information systems (GIS) and computational methods are combined in geospatial analysis to examine spatial linkages and patterns in data on water quality. Spatial decision-making and spatially explicit management strategies are made easier by GIS-based techniques, which allow the display, mapping, and spatial interpolation of water quality indices.
d) **Remote Sensing**: Remote sensing technologies, which provide temporally and geographically dense observations of water bodies, are essential to computational automation. Examples of these technologies include underwater sensors, aerial drones, and satellite imaging. Data from remote sensing are used to track indicators of water quality, identify changes in the environment, and evaluate the health of ecosystems across wide geographic areas.

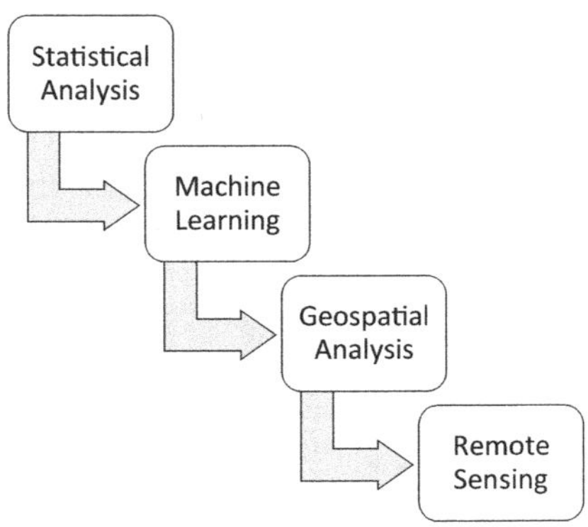

Figure 17.4
Methodologies and techniques.

17.4 Definition and principles of computational automation

The term "computational automation," which is frequently used interchangeably with "automated computational analysis," describes the use of computational techniques, algorithms, and technologies to automate procedures and activities related to the evaluation, management, and monitoring of water quality (Menger, Gago-Ferrero, Wiberg, & Ahrens, 2020). It includes a broad range of methods, such as modeling, data processing, statistical analysis, machine learning, and decision support, to maximize the computational capacity to improve the efficiency and promptness of water quality-related tasks.

The convergence of environmental engineering principles, automation technology, and computational science forms the foundation of computational automation in water quality control. Among the fundamental ideas are:

a) The objective of computational automation is to automate labor-intensive and repetitive procedures connected to assessing the quality of water, such as data collection, processing, analysis, and reporting. By automating these tasks, computational automation reduces human error, boosts operational effectiveness, and frees up resources for more strategic and value-added work.
b) Computational automation supports water quality management decision-making by employing data-driven techniques (Smith, Want, O'Maille, Abagyan, & Siuzdak, 2006). Computational automation helps to guide the development of policies, management plans, and resource allocation decisions by analyzing massive amounts of data collected from several sources, including sensors, satellites, and monitoring networks.
c) Principles and techniques from a variety of fields, such as computer science, statistics, environmental science, and hydrology, are integrated with computational automation. Computational automation promotes multidisciplinary collaboration and creativity by bringing together knowledge from other sectors, making it possible to find comprehensive and all-encompassing solutions to challenging problems related to water quality (Pluskal, Castillo, Villar-Briones, & Orešič, 2010).
d) The scalability and adaptability of computational automation solutions to various contexts, spatial scales, and data sources are built in. Computational automation frameworks can handle different levels of complexity and data availability, making them flexible tools for addressing a range of water quality management objectives, whether they are used in large-scale watersheds or small-scale freshwater systems.
e) Transparency and interpretability are prioritized in data analysis and modeling procedures through computational automation. Computational automation enables open communication of findings, assumptions, and uncertainties related to water quality assessments and management decisions by utilizing interpretable algorithms and visualization approaches. This promotes stakeholder participation and comprehension.

f) By utilizing feedback loops and adaptive algorithms, computational automation promotes a culture of ongoing learning and development. Computational automation systems can develop and improve over time, increasing their prediction accuracy, reliability, and robustness in capturing complex environmental dynamics through iterative model refinement, validation, and calibration (Böcker, Letzel, Lipták, & Pervukhin, 2009).

17.5 Role of computational methods in water quality assessment

Because computational methods offer strong tools for data analysis, modeling, prediction, and decision support, they are essential to the advancement of water quality assessment. The diverse role that computational approaches play in the assessment and management of water quality emphasizes how they improve understanding, guide decision-making, and help solve difficult environmental problems. In Fig. 17.5, the role of computational methods in water quality assessment is being explained.

a) **Data Analysis and Interpretation:** Large and complicated datasets collected from a variety of sources, such as in-situ sensors, remote sensing platforms, and laboratory experiments, can be analyzed and interpreted thanks to computational tools (Tsugawa et al., 2015). Through the use of statistical methodologies, machine learning algorithms, and data mining techniques, researchers and practitioners can find trends, patterns, and anomalies in data related to water quality. Computational methods offer important

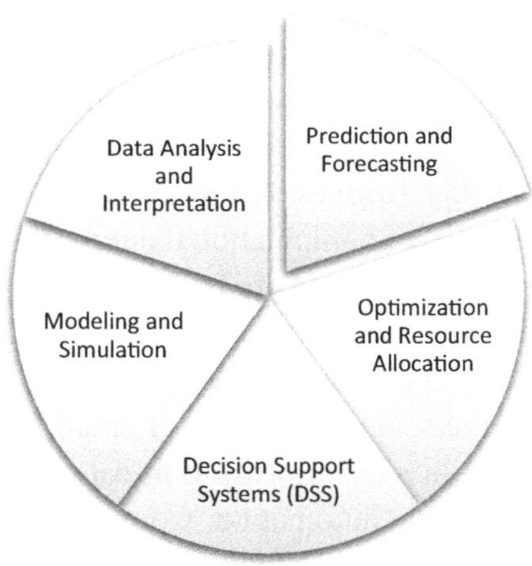

Figure 17.5
Role of computational methods in water quality assessment.

insights into the spatial and temporal variations in water quality parameters through the analysis of real-time observations and historical data. This makes it easier to identify sources of pollution, evaluate trends in water quality, and identify new concerns.

b) **Modeling and Simulation:** By simulating the behavior and dynamics of aquatic systems, computational models allow researchers to investigate the relationships between anthropogenic activities, environmental factors, and water quality parameters. The physical, chemical, biological, and hydrodynamic processes influencing the dynamics of water quality are captured by these models, which include hydrological, ecological, coupled systems, and water quality models (Haq, Hassan, & Shah, 2023). Computational models help scenario analysis, impact assessment, and management planning by predicting the distribution of pollutants, nutrient cycling, habitat suitability, and ecosystem responses to environmental changes by incorporating spatial and temporal variability.

c) **Prediction and Forecasting:** The creation of forecasting methods and prediction models to foresee future changes in the conditions of water quality is made easier by computational methods. These techniques forecast factors related to water quality, like dissolved oxygen levels, nutrient concentrations, and pollutant loads, by using historical data, environmental indicators, and predictive variables (Helmus, ter Laak, van Wezel, de Voogt, & Schymanski, 2021). Computational models produce both short- and long-term projections through the use of machine learning algorithms, time series analysis, and regression analysis. This allows for proactive management practices, early warning systems, and risk assessment techniques.

d) **Decision Support Systems:** The creation of decision support systems (DSS), which combine data, models, and tools for decision-making to help stakeholders, policymakers, and managers of water resources make well-informed choices, is supported by computational techniques (Feraud et al., 2023). To analyze management choices, prioritize interventions, and gauge the efficacy of mitigation measures, these systems offer interactive interfaces, visualization tools, and scenario analysis capabilities. In water quality assessment and management, computational DSS improves openness, accountability, and adaptive management by combining scientific knowledge, stakeholder preferences, and uncertainty analysis.

e) **Optimization and Resource Allocation:** Computational techniques balance conflicting goals, limitations, and uncertainties to maximize resource allocation and management tactics. These strategies include multicriteria decision analysis methodologies, mathematical programming techniques, and optimization algorithms. They aid in finding affordable solutions, setting investment priorities, and improving monitoring networks. Computational optimization techniques facilitate sustainable and effective water quality management practices by taking into account trade-offs between environmental, economic, and social objectives (Manz et al., 2023).

17.6 Case study 1: automated sensor networks for real-time monitoring

17.6.1 Background

Real-time monitoring in a variety of industries, such as industrial processes, infrastructure management, and environmental monitoring, has been transformed by automated sensor networks. These networks are made up of linked sensors that can gather and send data on their own, enabling ongoing analysis and monitoring (Hohrenk, Vosough, & Schmidt, 2019).

17.6.2 Objective

Increasing monitoring capabilities, optimizing decision-making procedures, and facilitating prompt responses to events or abnormalities picked up by the sensors are the goals of implementing automated sensor networks.

17.6.3 Implementation

a) *Sensor Selection*: The initial stage is choosing the right sensors, following the environmental factors and monitoring specifications. Temperature, humidity, pressure, motion, and other types of sensors are examples of sensors.
b) *Network Design*: When building a network, sensors must be positioned to provide the best possible coverage and connectivity. To guarantee dependable data transmission, variables including range, transmission power, and data aggregation techniques are taken into account.
c) *Data Transmission*: Using communication protocols like Wi-Fi, Bluetooth, Zigbee, or LoRaWAN, data gathered by the sensors are wirelessly communicated to a central data repository or cloud platform. This makes the gathered data accessible in real-time from any location with internet access.
d) *Alerting Mechanism*: When predetermined thresholds or conditions are satisfied, an alerting mechanism is put in place to instantly notify pertinent stakeholders. This guarantees a quick response to important occurrences or departures from standard operating procedures.
e) *Visualization and Reporting*: To display the tracked data in an understandable manner, such as graphs, charts, or dashboards, data visualization tools are employed. This makes it easier for stakeholders to understand the data and supports decision-making procedures. Furthermore, periodic reports summarizing the monitoring data may be generated by implementing automatic reporting elements.

17.6.4 Benefits

- *Better Monitoring*: Real-time, continuous parameter monitoring is made possible by automated sensor networks, which give a thorough picture of the process or environment under observation.
- *Early Anomaly Detection*: By enabling prompt intervention in the event of abnormalities or departures from typical operating circumstances, the danger of expensive damages or disruptions is reduced.
- *Enhanced Decision-Making*: Decision-makers may make more informed decisions faster and more productively when they have access to real-time data and actionable insights.
- *Cost Savings*: Automated sensor networks can assist decrease downtime, lower maintenance costs, and maximize resource use by automating the monitoring process and enabling early detection of anomalies.
- Automated sensor networks have great scalability, facilitating the installation of additional sensors or network growth to encompass broader regions or additional parameters as required.

In a variety of applications, automated sensor networks have proven to be an invaluable tool for real-time monitoring. These networks let firms monitor better, make better decisions, and react more quickly to events or changing situations by utilizing advancements in sensor technology, wireless connectivity, and data analytics (Rather & Haq, 2023). Automated sensor networks have enormous potential to stimulate efficiency and innovation across many industries as technology develops.

17.7 Analysis of the effectiveness of real-time data collection in identifying pollution sources and assessing water quality trends

Examining how well the system detects anomalies, records trends, and facilitates decision-making processes is a key component in evaluating how successful real-time data gathering is in locating pollution sources and evaluating water quality trends. This is an explanation:

1. **Early Detection of Pollution Events:**
   a. Water quality metrics are continuously monitored by real-time data collection, which enables the early detection of pollution occurrences. Pollutants entering the water system may be indicated by abrupt spikes or differences from baseline levels.
   b. Authorities can take swift action to lessen the effects of pollution by quickly identifying its sources and implementing source control procedures, cleanup initiatives, or legal action against polluters.
2. **Spatial Analysis and Source Identification:**
   a. Water quality parameters can be spatially mapped through the use of GIS analysis and real-time data gathering. The locations of pollution sources and the routes taken by pollutants within the river basin can both be determined with the aid of this spatial analysis.

   b. Authorities can effectively target interventions where they are most needed and prioritize monitoring efforts by establishing correlations between data on water quality and information on industrial activity, land use patterns, and other potential sources of pollution.
3. **Assessment of Water Quality Trends:**
   a. An extensive dataset for examining trends in water quality is produced by ongoing data gathering throughout time. Trends in water quality indicators might show long-term shifts, seasonal oscillations, or a slow decline.
   b. Trend analysis enables authorities to evaluate the success of pollution control strategies, monitor changes in water quality over time, and spot new problems before they become serious ones.
4. **Data-driven Decision Making:**
   a. Real-time data collection gives decision-makers accurate and timely information, which supports data-driven decision-making processes.
   b. Authorities can analyze prospective situations, estimate the likely consequences of various management techniques, and make well-informed decisions to safeguard and enhance water quality by combining real-time monitoring data with predictive analytics and decision support tools.
5. **Public Engagement and Transparency:**
   a. Real-time monitoring data can improve openness and public participation by making water quality conditions easily accessible.
   b. Public access to real-time data via web portals, mobile applications, or billboards raises awareness of problems with water quality, encourages community participation in monitoring programs, and strengthens relationships between stakeholders and authorities.
6. **Regulatory Compliance and Enforcement:**
   a. Real-time data collection supports regulatory compliance by providing the evidence needed to assess compliance with water quality standards and regulations.
   b. Authorities can use real-time monitoring data as a basis for regulatory enforcement actions, such as issuing permits, imposing penalties on polluters, or implementing pollution control measures to bring water bodies into compliance with regulatory requirements.

17.8 Case study 2: exploration of a case study employing machine learning algorithms to develop predictive models for water quality assessment

17.8.1 Background

An extensive urban area's water supply is managed by a municipal water utility corporation, which aims to enhance its water quality monitoring and management procedures. Conventional techniques for evaluating water quality depended on routine laboratory analysis and sampling, which took time and offered little understanding of fluctuations in water quality indices in real time. The utility business set out to create prediction models that could estimate water quality situations based on data from real-time monitoring by utilizing machine learning algorithms (Khatoonabadi, Vosough, Hohrenk, & Schmidt, 2021).

17.8.2 Implementation

1. **Data Collection and Preprocessing:**
   a. The utility company collected extensive datasets comprising real-time monitoring data from sensors deployed across the water distribution network. These datasets included measurements of various water quality parameters such as pH, turbidity, chlorine levels, temperature, and conductivity.
   b. Data preprocessing techniques were applied to clean, normalize, and transform the raw sensor data into a format suitable for machine learning analysis. This involved handling missing values, outlier detection, and feature engineering to extract relevant patterns and relationships from the data.
2. **Feature Selection and Model Development:**
   a. Machine learning algorithms were employed to develop predictive models for water quality assessment. Supervised learning techniques such as regression, classification, and time series forecasting were used to train models on historical monitoring data.
   b. Feature selection methods were applied to identify the most relevant variables influencing water quality and reduce dimensionality in the dataset. Features such as time of day, weather conditions, flow rates, and water source characteristics were considered in model development.
   c. Various machine learning algorithms were evaluated, including linear regression, decision trees, random forests, support vector machines, and neural networks, to determine the most suitable models for predicting water quality parameters.
3. **Model Training and Validation:**
   a. The dataset was divided into training and testing sets for model training and validation purposes. Cross-validation techniques were employed to assess the performance of different machine learning models and mitigate overfitting.
   b. Model performance metrics such as accuracy, precision, recall, and mean squared error were used to evaluate the predictive performance of the models against observed water quality data.
4. **Model Deployment and Integration:**
   a. Once trained and validated, the predictive models were deployed into operational systems for real-time water quality assessment. Integration with existing monitoring

infrastructure and data management systems allows for seamless data flow and model execution.
   b. The utility company developed user-friendly dashboards and visualization tools to display model predictions, alert operators to potential water quality issues, and support decision-making processes.
5. **Benefits:**
   a. *Improved Water Quality Monitoring*: Machine learning models enabled the utility company to forecast water quality conditions in real time, providing early warnings of potential contamination events or anomalies.
   b. *Enhanced Operational Efficiency*: Predictive models facilitated proactive management of the water distribution network, allowing operators to optimize treatment processes, adjust operational parameters, and allocate resources more effectively.
   c. *Cost Savings*: By reducing the need for manual sampling and laboratory analysis, predictive modeling helped the utility company save time and resources while maintaining water quality standards.
   d. *Risk Mitigation*: Early detection of water quality fluctuations allowed for timely intervention to mitigate risks to public health and prevent service disruptions.
   e. *Data-Driven Decision Making*: Predictive modeling supported data-driven decision-making processes by providing actionable insights into water quality trends, patterns, and potential sources of contamination.

The potential of sophisticated analytics approaches to improve water management practices and guarantee the security and dependability of municipal water supply was proved by the implementation of machine learning algorithms for predictive water quality evaluation. The utility firm was able to reduce risks, maximize operational efficiency, and proactively monitor and manage water quality by utilizing real-time monitoring data and predictive modeling capabilities (Cairoli et al., 2023). The use of machine learning in water quality evaluation shows potential for solving difficult problems in environmental sustainability and water resource management as it develops further.

17.9 Considerations for scalability and integration with existing monitoring frameworks

Scalability and integration with existing monitoring frameworks are critical considerations when developing predictive models for water quality forecasting. Ensuring that the system can accommodate growing data volumes, new monitoring stations, and evolving technology is essential for long-term effectiveness and sustainability. Here, we delve into various considerations to address these aspects comprehensively.

a) Scalability is vital to accommodate increasing data volumes, the addition of new monitoring stations, and the expansion of monitoring coverage. A scalable system can handle growing demands without sacrificing performance or reliability. When designing for scalability, modularity and flexibility are key principles. The system should be composed of modular components that can be easily scaled horizontally or vertically (Vosough, 2018). This modularity enables the addition of new functionalities or resources without disrupting the existing system.
b) Integration with existing monitoring frameworks is crucial for seamless data exchange and interoperability. Many water utilities and environmental agencies already have established monitoring systems in place, comprising sensors, data loggers, and data management platforms. Integrating predictive modeling capabilities into these frameworks requires compatibility with existing data formats, communication protocols, and interfaces. Standardization plays a crucial role in ensuring interoperability, as it allows different systems to communicate effectively and share data without compatibility issues (Dávila-Santiago et al., 2022).
c) Open application programming interfaces (APIs) facilitate integration by providing standardized interfaces for accessing and exchanging data. By implementing open APIs, predictive modeling systems can communicate with existing monitoring frameworks and data management platforms using common protocols and formats. This simplifies the integration process and enables seamless data exchange between different systems.
d) Data standardization is another important consideration for integration. Standardizing data formats, metadata, and vocabularies ensures consistency and interoperability across different systems (Purschke, Vosough, Leonhardt, Weber, & Schmidt, 2020). Adopting industry-standard formats such as CSV, JSON, or XML simplifies data exchange and reduces the need for data transformation or conversion.
e) Cloud-based infrastructure offers scalability and flexibility, making it an ideal platform for hosting predictive modeling systems. Cloud platforms provide elastic computing resources that can scale up or down based on demand. This scalability enables the system to handle varying workloads efficiently, from processing large datasets to running complex machine learning algorithms. Additionally, cloud services offer built-in features for data storage, processing, and analysis, reducing the need for extensive infrastructure management and maintenance (Bonnefille et al., 2023).
f) Real-time data streaming and processing capabilities are essential for handling continuous data streams from monitoring sensors. Technologies like Apache Kafka, Apache Flink, or AWS Kinesis enable the ingestion, processing, and analysis of streaming data in real time. By processing data as it arrives, predictive modeling systems can provide timely insights and responses to changes in water quality conditions, enhancing their effectiveness for decision-making and management.
g) Interoperability with Internet of Things (IoT) devices and sensors is critical for integrating predictive modeling systems with monitoring networks. IoT devices play a

key role in collecting real-time data from various sources, including water quality sensors, weather stations, and flow meters (Hohrenk-Danzouma, Vosough, Merkus, Drees, & Schmidt, 2022). By supporting interoperability standards such as MQTT or CoAP, predictive modeling systems can communicate with IoT devices and ingest data from distributed sensor networks seamlessly.

17.10 Identification of challenges encountered in implementing automation for water quality assessment

Implementing automation for water quality assessment presents several challenges that organizations may encounter during the deployment and operation of automated systems. Here in Fig. 17.6, there are some key challenges that organizations may encounter during the deployment and operation of automated systems:

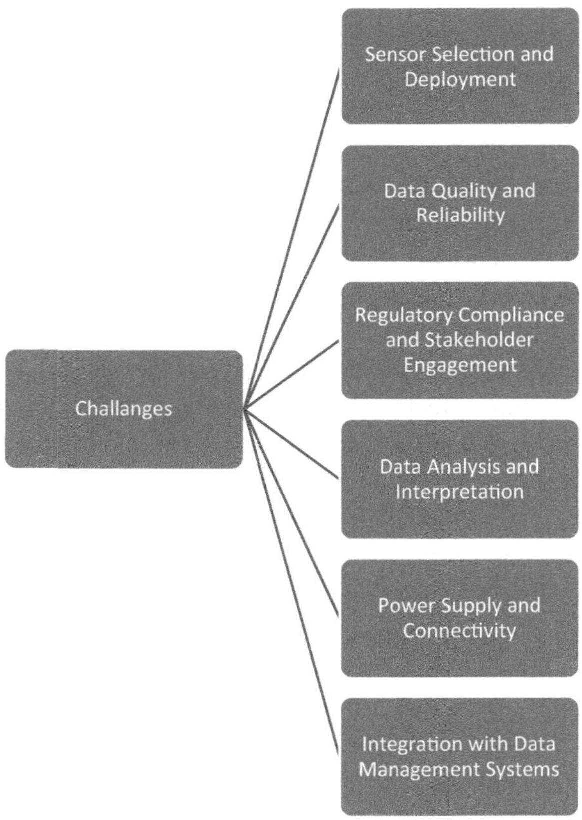

Figure 17.6
Challenges that organizations may encounter during the deployment and operation of automated systems.

- **Data Quality and Reliability:** Ensuring the accuracy, reliability, and consistency of data collected by automated sensors is essential for producing valid water quality assessments. Challenges may arise due to sensor calibration issues, measurement errors, drift over time, and environmental factors affecting sensor performance (e.g., fouling, biofouling). Regular maintenance, calibration checks, and quality assurance procedures are necessary to mitigate these challenges and maintain data integrity.
- **Sensor Selection and Deployment**: Selecting appropriate sensors for water quality parameters of interest requires consideration of factors such as sensor accuracy, reliability, compatibility with environmental conditions, and suitability for continuous monitoring. Challenges may arise in deploying sensors in diverse environmental settings (e.g., rivers, lakes, coastal waters) with varying conditions (e.g., temperature, salinity, turbidity). Ensuring proper sensor placement, protection from physical damage, and secure attachment to monitoring platforms or infrastructure is essential for reliable data collection.
- **Integration with Data Management Systems**: Integrating automated sensor networks with existing data management systems and IT infrastructure can be challenging due to differences in data formats, communication protocols, and system architectures. Ensuring compatibility, interoperability, and seamless data exchange between automated sensors, data loggers, communication networks, and data repositories requires careful planning and coordination. Challenges may arise in developing custom interfaces, APIs, or data pipelines to facilitate data integration and synchronization between different components of the monitoring system.
- **Power Supply and Connectivity**: Providing reliable power supply and connectivity to automated sensors in remote or inaccessible locations can be challenging. Deploying solar panels, battery systems, or alternative energy sources may be necessary to power autonomous monitoring stations in off-grid areas. Ensuring robust communication networks (e.g., cellular, satellite, radio) and backup systems to transmit data from remote sensors to central data repositories is essential for real-time monitoring and data access.
- **Data Analysis and Interpretation**: Processing, analyzing, and interpreting large volumes of data generated by automated sensor networks can be challenging. Developing algorithms, models, and data analytics tools to extract meaningful insights, detect anomalies, and identify trends from raw sensor data requires expertize in data science and machine learning. Ensuring that automated data analysis processes are transparent, reproducible, and validated against independent datasets is essential for generating accurate water quality assessments and actionable insights.
- **Regulatory Compliance and Stakeholder Engagement**: Ensuring compliance with regulatory standards and requirements for water quality monitoring and reporting is essential for organizations responsible for managing water resources. Challenges may arise in aligning automated monitoring systems with regulatory guidelines, data quality

standards, and reporting protocols. Engaging stakeholders, including regulatory agencies, policymakers, local communities, and environmental advocacy groups, in the design, implementation, and operation of automated monitoring systems is critical for fostering trust, transparency, and accountability.

17.11 Strategies for addressing challenges and maximizing the benefits of automation

To effectively address the challenges and maximize the benefits of automation in water quality assessment, organizations should adopt a multifaceted approach that encompasses technological innovation, procedural enhancements, and stakeholder collaboration. First, investing in state-of-the-art sensor technology is paramount, ensuring the selection of reliable, accurate sensors tailored to the specific environmental conditions of the monitoring site (Fisher, Peter, Newton, Schaub, & Sobus, 2022). Regular calibration, maintenance, and quality assurance protocols must be established to uphold sensor accuracy and reliability over time. Additionally, leveraging advanced sensor technologies, such as optical or multispectral sensors, can offer enhanced performance and expanded capabilities for comprehensive water quality monitoring. Second, optimizing sensor deployment and maintenance procedures is crucial (Boelrijk, van Herwerden, Ensing, Forré, & Samanipour, 2023). Conducting thorough site assessments to identify optimal sensor locations and deployment strategies is essential for maximizing data coverage and accuracy. Implementing robust sensor protection measures, including physical safeguards and biofouling prevention techniques, can help mitigate environmental factors that may compromise sensor performance. Moreover, establishing efficient maintenance schedules and protocols ensures timely sensor upkeep, minimizing downtime and data gaps (Yang, van Herwerden, Preud'homme, & Samanipour, 2022). Third, integrating automated sensor networks with existing data management systems and infrastructure requires careful planning and coordination. Standardizing data formats, communication protocols, and interfaces facilitates seamless data exchange and interoperability across different platforms. Developing custom interfaces, APIs, or data pipelines can streamline data integration processes, enabling efficient data transmission and synchronization. Moreover, leveraging cloud-based infrastructure and services offers scalability, flexibility, and accessibility, supporting the seamless operation of automated monitoring systems (Pourchet et al., 2020). Fourth, employing advanced data analysis techniques and predictive modeling algorithms enhances the value of automated sensor data. Leveraging machine learning algorithms and data analytics tools enables the extraction of actionable insights, trend identification, and anomaly detection from raw sensor data. By harnessing the power of predictive modeling, organizations can forecast water quality trends, anticipate potential issues, and optimize resource allocation and management strategies. Lastly, fostering stakeholder engagement and collaboration is essential for the successful implementation and operation of automated water quality monitoring systems (Chaurasia & Haq, 2023). Engaging regulatory agencies, policymakers, local communities, and environmental stakeholders throughout the process ensures alignment with regulatory requirements, promotes transparency, and builds trust and confidence in the monitoring outcomes. Furthermore, involving end-users in system design, implementation, and evaluation processes ensures that the automated monitoring system meets their needs and addresses relevant concerns (Schulze et al., 2020). By embracing these strategies, organizations can overcome challenges, unlock the full potential of automation, and realize the myriad benefits of advanced water quality assessment technologies in safeguarding precious water resources and promoting environmental sustainability (Jewell et al., 2021).

17.12 Future directions and opportunities

Looking ahead, the future of water quality assessment holds immense potential for leveraging advanced technologies such as artificial intelligence (AI), the IoT, and remote sensing to revolutionize monitoring practices and address complex environmental challenges. Here is a discussion of potential applications and opportunities for these technologies:

17.12.1 Artificial intelligence

- *Predictive Modeling:* AI algorithms can be used to develop predictive models for water quality assessment, forecasting trends, and identifying potential pollution sources. Machine learning techniques enable the analysis of large datasets to uncover patterns, correlations, and anomalies indicative of water quality changes.
- *Anomaly Detection*: AI-powered anomaly detection systems can automatically identify abnormal water quality conditions, such as contamination spikes or unusual patterns, in real-time sensor data. This enables timely responses and intervention measures to mitigate risks to public health and the environment.
- *Optimization of Treatment Processes*: AI algorithms can optimize water treatment processes by analyzing sensor data in real time and adjusting operational parameters accordingly. This ensures efficient resource utilization, minimizes energy consumption, and maintains water quality standards.

17.12.2 Internet of Things

- *Sensor Networks*: IoT-enabled sensor networks provide real-time monitoring of water quality parameters across large geographic areas. Wireless sensors deployed in rivers, lakes, and reservoirs continuously collect data on temperature, pH, dissolved oxygen, and pollutants, transmitting it to centralized data repositories for analysis.

- *Smart Water Management*: IoT technology facilitates smart water management systems that monitor water usage, detect leaks, and optimize distribution networks. By integrating IoT sensors with infrastructure components such as pumps, valves, and meters, utilities can improve operational efficiency, reduce water loss, and enhance service reliability.
- *Environmental Monitoring*: IoT devices equipped with environmental sensors can monitor water quality in remote or inaccessible areas, including coastal regions, wetlands, and marine ecosystems. These data aid in understanding ecosystem dynamics, detecting environmental changes, and informing conservation efforts.

17.12.3 Remote sensing

- *Satellite Imagery*: Remote sensing techniques, such as satellite imagery and aerial drones, provide valuable spatial data on water quality parameters, land use patterns, and environmental conditions. High-resolution satellite images enable the monitoring of large-scale water bodies, identifying pollution hotspots, and tracking changes over time.
- *Hyperspectral Imaging*: Hyperspectral imaging technology enables the detection and characterization of water quality parameters, including chlorophyll concentration, turbidity, and sedimentation, with high precision. These data enhance our understanding of aquatic ecosystems and support ecosystem management and restoration initiatives.
- *Integration with AI*: Remote sensing data can be integrated with AI algorithms to improve the accuracy and efficiency of water quality assessment. Machine learning models trained on remote sensing data can predict water quality parameters, classify water bodies, and identify environmental trends, complementing ground-based monitoring efforts.

17.13 Literature review

The development, application, and effects of smart water management systems in the United States and Africa are the main topics of this review study. It covers legislative and regulatory frameworks, technological advancements, theoretical frameworks, repercussions on the environment, society, and economy, as well as difficulties and future directions. The study highlights how crucial it is for smart water management initiatives to incorporate technology, legislation, and community involvement and urges more investigation to fully realize their potential (Olatunde, Michael, Adelani, & Sikhakhane, 2024).

This thorough analysis examines the advantages, difficulties, potential legal ramifications, and emerging trends of these technologies. Smart water management solves data security and infrastructure investment issues while improving operational effectiveness, predictive maintenance, and resource conservation. To guarantee data privacy and moral application, regulatory frameworks are essential in guiding the appropriate application of AI and IoT. Future developments in water data security include blockchain, decentralized systems, quantum computing, and sophisticated sensors. The promise of smart water management to achieve universal access to clean water, climate resilience, and inclusive, sustainable development is highlighted by its alignment with the sustainable development goals (SDGs). The development of intelligent and equitable water management systems will be guided by ethical considerations, public awareness, and collaboration as we embrace these technologies (Dada et al., 2024).

Climate change, pollution, and population increase are making safe water increasingly rare. In the world, 40% of deaths are related to contaminated water. Resources are being protected, and water management techniques are being enhanced by the deployment of smart water quality monitoring systems that include IoT sensors, data analytics, and machine learning algorithms. These solutions guarantee clean drinking water for both cities and villages, revolutionizing conventional techniques for assessing water quality (Vanshika, Ambati, Kuber, & Poojitha, 2024).

The need for clean drinking water is growing worldwide, and new technologies like cloud computing and smart sensors are required to meet this need. To enhance current irrigation systems, this study suggests a safe, intelligent framework that provides automated control depending on weather and seasonal variations. To provide effective and secure administration, it also has four special algorithms and water management guidelines (Bhardwaj et al., 2024).

This study highlights the importance of improving public communication of water quality data while examining the benefits and drawbacks of employing citizen science and the IoT as alternative monitoring methodologies. We identify and suggest ways to address five major obstacles that citizen science and IoT monitoring systems need to overcome to develop reliable sources of water quality data through a thorough evaluation of studies conducted in the field. Furthermore, we delineate three basic issues that impact the communication process on water quality and provide approaches for effectively communicating this subject to the general public (Amador-Castro et al., 2024).

To improve stakeholder collaboration, the study also includes publication statistics. Results demonstrate that researchers use a variety of technologies, including sensors, communication modules, embedded programming languages, microcontrollers, and protocols, to achieve their goal of creating IoT-based smart water solutions. In developing their smart water applications, none of the publications used 5G mobile networks as a communication module. Additional research demonstrates how innovative and perhaps more sustainable it is to incorporate solar power and 3D printing into IoT-based smart water applications. Suggestions for future directions are made to guarantee that produced smart water apps are generally implemented to aid in water resource management and conservation (Okoli, Joy, & Kabaso, 2024).

The article provides a list of fundamental guidelines for organizations that actively work to advance sustainability. It provides decision-makers with insightful information that makes it easier for them to comprehend the opportunities and difficulties involved in achieving sustainable performance (Antony et al., 2024).

The chapter addresses the drawbacks of traditional irrigation techniques, such as soil deterioration, water waste, and energy consumption, and offers creative solutions including automation, intelligent irrigation controllers, and IoT-based systems. It places a strong emphasis on water-saving methods and the advantages of effective practices, such as decreased environmental effects, energy savings, and water conservation (Ray & Majumder, 2024).

17.14 Key findings, interpretations, and implications

17.14.1 Key findings

1. **Integration of Technology, Legislation, and Community Involvement**: Effective smart water management initiatives must incorporate technology, regulatory frameworks, and community engagement to achieve optimal results.
2. **Benefits of Smart Water Management**: These systems enhance operational efficiency, predictive maintenance, and resource conservation, addressing key issues such as data security and infrastructure investment.
3. **Technological Advancements**: Emerging technologies like blockchain, decentralized systems, quantum computing, and advanced sensors offer promising advancements in water data security and management.
4. **Global Context**: The study highlights the significance of smart water management in achieving SDGs, particularly in providing universal access to clean water and promoting climate resilience.
5. **Challenges and Future Directions**: The review identifies several challenges, including the need for improved data privacy, ethical application of AI and IoT, and the incorporation of sustainable and innovative technologies like solar power and 3D printing.

17.14.2 Interpretations

1. **Legislative and Regulatory Frameworks**: Effective legislation and regulatory frameworks are essential for guiding the ethical and appropriate use of AI and IoT in water management.

2. **Public Engagement**: Improving public communication of water quality data and involving communities in monitoring efforts, such as through citizen science, can enhance the reliability and acceptance of smart water solutions.
3. **Technological Integration**: The integration of advanced technologies and innovative approaches into water management systems has the potential to significantly improve efficiency and sustainability.
4. **Research Gaps**: There is a noticeable gap in the use of 5G mobile networks in existing smart water applications, indicating a potential area for future research and development.

17.14.3 Implications

1. **Policy and Decision-Making**: Policymakers and decision-makers are provided with valuable insights to understand the opportunities and challenges associated with implementing smart water management systems.
2. **Sustainable Development**: The alignment of smart water management with SDGs underscores its potential in promoting sustainable and inclusive development, ensuring access to clean water, and enhancing climate resilience.
3. **Future Research Directions**: Future research should focus on addressing the identified challenges, such as enhancing data privacy, incorporating new technologies, and improving public engagement and communication.
4. **Application in Developing Regions**: The findings suggest significant potential for smart water management systems in both urban and rural areas, particularly in regions facing water scarcity and quality issues.

This comprehensive review study provides a roadmap for developing intelligent and equitable water management systems, emphasizing the importance of ethical considerations, public awareness, and collaborative efforts in harnessing the full potential of smart water technologies.

17.15 Conclusion

In terms of guaranteeing water security, the effective adoption of automation for water quality evaluation is a major advancement. This chapter has highlighted the revolutionary potential of automation in the management and protection of water resources through the examination of several case studies. Utilizing technical innovations such as computational automation, the assessment of water quality has improved in terms of speed, accuracy, and efficiency. Automation facilitates early detection of impurities or anomalies and timely reaction to possible hazards to water quality by enabling continuous monitoring, fast analysis, and proactive intervention. There are obstacles in the way of fully automated water quality control, though. To fully utilize automation in this field, concerns including cybersecurity, system interoperability, and data integration must be resolved. To overcome these obstacles and further optimize automated systems for water quality evaluation, academics, practitioners, and policymakers must continue their investigation, innovate, and work together. However, the case studies in this chapter provide insightful information about the benefits, challenges, and possible uses of automation in the management of water quality. This chapter offers a roadmap for maximizing automation's potential to improve water security by drawing on practical experiences and technological advancements.

Essentially, one of the most important tools in our combined efforts to guarantee that there will be safe and clean water for current and future generations is the efficient application of automation in water quality testing. Let us stay dedicated to the mission of protecting this invaluable resource for the benefit of both people and the environment as we develop and improve automated systems.

References

Alygizakis, N. A., Oswald, P., Thomaidis, N. S., Schymanski, E. L., Aalizadeh, R., Schulze, T., et al. (2019). NORMAN digital sample freezing platform: A European virtual platform to exchange liquid chromatography high resolution-mass spectrometry data and screen suspects in "digitally frozen" environmental samples. *Trends in Analytical Chemistry*, *115*, 129–137. Available from https://doi.org/10.1016/j.trac.2019.04.008.

Amador-Castro, F., González-López, M. E., Lopez-Gonzalez, G., Garcia-Gonzalez, A., Díaz-Torres, O., ... Gradilla-Hernández, M. S. (2024). Internet of Things and citizen science as alternative water quality monitoring approaches and the importance of effective water quality communication. *Journal of Environmental Management*, *352*, 119959.

Antony, J., Bhat, S., Sony, M., Fundin, A., Sorqvist, L., & Molteni, R. (2024). Sustainable development through quality management: a multiple-case study analysis of triumphs, trials and tribulations. *The TQM Journal*.

Bhardwaj, A., Kumar, M., Alshehri, M., Keshta, I., Abugabah, A., & Sharma, S. K. (2024). Smart water management framework for irrigation in agriculture. *Environmental Technology*, *45*(12), 2320–2334.

Boelrijk, J., van Herwerden, D., Ensing, B., Forré, P., & Samanipour, S. (2023). Predicting RP-LC retention indices of structurally unknown chemicals from mass spectrometry data. *Journal of Cheminformatics*, *15*, 28. Available from https://doi.org/10.1186/s13321-023-00699-8.

Bonnefille, B., Karlsson, O., Rian, M. B., Raqib, R., Parvez, F., Papazian, S., et al. (2023). Nontarget analysis of polluted surface waters in bangladesh using open science workflows. *Environmental Science & Technology*, *57*, 6808–6824. Available from https://doi.org/10.1021/acs.est.2c08200.

Böcker, S., Letzel, M. C., Lipták, Z., & Pervukhin, A. (2009). SIRIUS: Decomposing isotope patterns for metabolite identification†. *Bioinformatics (Oxford, England)*, *25*, 218–224. Available from https://doi.org/10.1093/bioinformatics/btn603.

Cairoli, M., van den Doel, A., Postma, B., Offermans, T., Zemmelink, H., Stroomberg, G., et al. (2023). Monitoring pollution pathways in river water by predictive path modelling using untargeted GC-MS measurements. *npj Clean Water*, *6*, 48. Available from https://doi.org/10.1038/s41545-023-00257-7.

Chaurasia, A., & Haq, I. U. (2023). Housing price prediction model using machine learning. In *2023 international conference on sustainable emerging innovations in engineering and technology (ICSEIET), Ghaziabad, India* (pp. 497–500). Available from https://doi.org/10.1109/ICSEIET58677.2023.10303359.

Choudhary, C., Thakur, J., Bhardwaj, H., Haq, I. U., & Rather, A. H. (2023). Revolutionizing wind energy: A enhanced machine learning perspective on wind turbines fault prediction. In *2023 second international conference on automation, computing and renewable systems (ICACRS)*. IEEE (pp. 1763–1769).

Dada., Ayorinde, M., Majemite, M. T., Obaigbena, A., Daraojimba, O. H., Oliha, J. S., & Nwokediegwu, Z. Q. S. (2024). Review of smart water management: IoT and AI in water and wastewater treatment. *World Journal of Advanced Research and Reviews*, *21*(1), 1373–1382.

Dávila-Santiago, E., Shi, C., Mahadwar, G., Medeghini, B., Insinga, L., Hutchinson, R., et al. (2022). Machine learning applications for chemical fingerprinting and environmental source tracking using non-target chemical data. *Environmental Science & Technology*, *56*, 4080–4090. Available from https://doi.org/10.1021/acs.est.1c06655.

Feraud, M., O'Brien, J. W., Samanipour, S., Dewapriya, P., van Herwerden, D., Kaserzon, S., et al. (2023). InSpectra - A platform for identifying emerging chemical threats. *Journal of Hazardous Materials*, *455*131486. Available from https://doi.org/10.1016/j.jhazmat.2023.131486.

Fisher, C. M., Peter, K. T., Newton, S. R., Schaub, A. J., & Sobus, J. R. (2022). Approaches for assessing performance of high-resolution mass spectrometry-based non-targeted analysis methods. *Analytical and Bioanalytical Chemistry*, *414*, 6455–6471. Available from https://doi.org/10.1007/s00216-022-04203-3.

González-Gaya, B., Lopez-Herguedas, N., Bilbao, D., Mijangos, L., Iker, A. M., Etxebarria, N., et al. (2021). Suspect and non-target screening: The last frontier in environmental analysis. *Analytical Methods*, *13*, 1876–1904. Available from https://doi.org/10.1039/D1AY00111F.

Haq, I. U., Hassan, I. U., & Shah, H. A. (2023). Machine learning techniques for result prediction of one day international (ODI) cricket match. In *IEEE eighth international conference for convergence in technology (I2CT)*. IEEE (pp. 1–5).

Haq, I. U., Rather, A. H., & Kaur, G. (2023). A comparative analysis of machine learning algorithms for the early prediction of cardiovascular disease. In *2023 second international conference on edge computing and applications (ICECAA)*. IEEE (pp. 987–993).

Helmus, R., ter Laak, T. L., van Wezel, A. P., de Voogt, P., & Schymanski, E. L. (2021). patRoon: open source software platform for environmental mass spectrometry based non-target screening. *Journal of Cheminformatics*, *13*(1). Available from https://doi.org/10.1186/s13321-020-00477-w.

Hohrenk, L. L., Vosough, M., & Schmidt, T. C. (2019). Implementation of chemometric tools to improve data mining and prioritization in LC-HRMS for nontarget screening of organic micropollutants in complex water matrixes. *Analytical Chemistry*, *91*, 9213–9220. Available from https://doi.org/10.1021/acs.analchem.9b01984.

Hohrenk-Danzouma, L. L., Vosough, M., Merkus, V. I., Drees, F., & Schmidt, T. C. (2022). Non-target analysis and chemometric evaluation of a passive sampler monitoring of small streams. *Environmental Science & Technology*, *56*, 5466–5477. Available from https://doi.org/10.1021/acs.est.1c08014.

Hollender, J., Schymanski, E. L., Ahrens, L., Alygizakis, N., Béen, F., Bijlsma, L., et al. (2023). NORMAN guidance on suspect and non-target screening in environmental monitoring. *35*. Available from https://doi.org/10.1186/s12302-023-00779-4, https://www.google.com/search?client = firefox-b-d&q = Environ + Sci + Eur.

Jewell K. S., Hermes N., Ehlig B., Thron F., Köppe T., Thorenz U., et al. (2021). Methodik zur Anwendung von Non-Target-Screening (NTS) mittels LC-MS/MS in der Gewässerüberwachung. Umweltbundesamt.

Kaserzon, S. L., Heffernan, A. L., Thompson, K., Mueller, J. F., & Gomez Ramos, M. J. (2017). Rapid screening and identification of chemical hazards in surface and drinking water using high resolution mass spectrometry and a case-control filter. *Chemosphere*, *182*, 656–664. Available from https://doi.org/10.1016/j.chemosphere.2017.05.071.

Khatoonabadi, R. L., Vosough, M., Hohrenk, L. L., & Schmidt, T. C. (2021). Employing complementary multivariate methods for a designed nontarget LC-HRMS screening of a wastewater-influenced river. *Microchemical Journal, Devoted to the Application of Microtechniques in all Branches of Science*, *160*105641. Available from https://doi.org/10.1016/j.microc.2020.105641.

Krier, J., Singh, R. R., Kondi, T., Lai, A., Diderich, P., Zhang, J., et al. (2022). Discovering pesticides and their TPs in Luxembourg waters using open cheminformatics approaches. *Environment International*, *158*106885. Available from https://doi.org/10.1016/j.envint.2021.106885.

Manz, K. E., Feerick, A., Braun, J. M., Feng, Y. L., Hall, A., Koelmel, J., et al. (2023). Non-targeted analysis (NTA) and suspect screening analysis (SSA): A review of examining the chemical exposome. *Journal of Exposure Science & Environmental Epidemiology*, *33*, 524–536. Available from https://doi.org/10.1038/s41370-023-00574-6.

Menger, F., Gago-Ferrero, P., Wiberg, K., & Ahrens, L. (2020). Wide-scope screening of polar contaminants of concern in water: A critical review of liquid chromatography-high resolution mass spectrometry-based strategies. *Trends in Environmental Analytical Chemistry*, *28*e00102. Available from https://doi.org/10.1016/j.teac.2020.e00102.

Okoli., Joy, N., & Kabaso, B. (2024). Building a smart water city: IoT smart water technologies, applications, and future directions. *Water*, *16*(4), 557.

Olatunde., Michael, T., Adelani, F. A., & Sikhakhane, Z. Q. (2024). A review of smart water management systems from Africa and the United States. *Engineering Science & Technology Journal*, *5*(4), 1231–1242.

Place, B. J., Ulrich, E. M., Challis, J. K., Chao, A., Du, B. L., Favela, K. A., et al. (2021). An introduction to the benchmarking and publications for non-targeted analysis working group. *Analytical Chemistry*, *93*(49), 16289–16296. Available from https://doi.org/10.1021/acs.analchem.1c02660.

Pluskal, T., Castillo, S., Villar-Briones, A., & Orešič, M. (2010). MZmine 2: Modular framework for processing, visualizing, and analyzing mass spectrometry-based molecular profile data. *BMC Bioinformatics*, *11*, 395. Available from https://doi.org/10.1186/1471-2105-11-395.

Pourchet, M., Debrauwer, L., Klanova, J., Price, E. J., Covaci, A., Caballero-Casero, N., et al. (2020). Suspect and non-targeted screening of chemicals of emerging concern for human biomonitoring, environmental health studies and support to risk assessment: From promises to challenges and harmonisation issues. *Environment International*, *139*105545. Available from https://doi.org/10.1016/j.envint.2020.105545.

Purschke, K., Vosough, M., Leonhardt, J., Weber, M., & Schmidt, T. C. (2020). Evaluation of nontarget long-term LC-HRMS time series data using multivariate statistical approaches. *Analytical Chemistry*, *92*, 12273–12281. Available from https://doi.org/10.1021/acs.analchem.0c01897.

Rather, A. H., & Haq, I. U. (2023). Intelligent framework for early prediction of diabetic retinopathy: A. In A. Yadav, S. J. Nanda, & M. H. Lim (Eds.), *Proceedings of international conference on paradigms of communication, computing and data analytics. PCCDA 2023. Algorithms for intelligent systems*. Singapore: Springer. deep learning approach. https://doi.org/10.1007/978-981-99-4626-6_32.

Ray, S., & S. Majumder. (2024). Water management in agriculture: innovations for efficient irrigation (pp. 169–185).

Schulze, B., Jeon, Y., Kaserzon, S., Heffernan, A. L., Dewapriya, P., O'Brien, J., et al. (2020). An assessment of quality assurance/quality control efforts in high resolution mass spectrometry non-target workflows for analysis of environmental samples. *Trends in Analytical Chemistry*, *133*116063. Available from https://doi.org/10.1016/j.trac.2020.116063.

Schymanski, E. L., Kondi, T., Neumann, S., Thiessen, P. A., Zhang, J., & Bolton, E. E. (2020). Empowering large chemical knowledge bases for exposomics: PubChemLite meets MetFrag. *Journal of Cheminformatics*, *13*, 19. Available from https://doi.org/10.1186/s13321-021-00489-0.

Smith, C. A., Want, E. J., O'Maille, G., Abagyan, R., & Siuzdak, G. (2006). XCMS: Processing mass spectrometry data for metabolite profiling using nonlinear peak alignment, matching, and identification. *ACS Publications*, *78*, 779–787. Available from https://doi.org/10.1021/ac051437y.

Tröger, R., Klöckner, P., Ahrens, L., & Wiberg, K. (2018). Micropollutants in drinking water from source to tap - method development and application of a multiresidue screening method. *The Science of the Total Environment*, *627*, 1404–1432. Available from https://doi.org/10.1016/j.scitotenv.2018.01.277.

Tsugawa, H., Cajka, T., Kind, T., Ma, Y., Higgins, B., Ikeda, K., et al. (2015). MS-DIAL: Data-independent MS/MS deconvolution for comprehensive metabolome analysis. *Nature Methods*, *12*, 523–526. Available from https://doi.org/10.1038/nmeth.3393.

Vanshika, A., Kuber, B. R., & Poojitha, N. (2024). *Revolutionizing water quality monitoring: The smart tech frontier. Innovations in machine learning and IoT for water management* (pp. 152–171). IGI Global.

Vosough, M. (2018). Current challenges in second-order calibration of hyphenated chromatographic data for analysis of highly complex samples. *Journal of Chemometrics*, *32*, 1–15. Available from https://doi.org/10.1002/cem.2976.

Yang, F., van Herwerden, D., Preud'homme, H., & Samanipour, S. (2022). Collision cross section prediction with molecular fingerprint using machine learning. *Molecules (Basel, Switzerland)*, *27*, 6424. Available from https://doi.org/10.3390/molecules27196424.

CHAPTER 18

Jal Jeevan Mission and role of smart technologies in achieving water security in rural India

Ankita Mukherjee[1], Tarakeshwar Senapati[2] and Arpita Ghosh[3]

[1]Indian Institute of Engineering Science and Technology, Shibpur, West Bengal, India, [2]Sidho-Kanho-Birsha University, Purulia, West Bengal, India, [3]Symbiosis Centre for Management Studies, Symbiosis International (Deemed University), Pune, Maharastra, India

18.1 Introduction

Water is of utmost importance for every living being on earth. It is the optimum temperature of the earth, based on its distance from the sun, that the planet has water present in various forms. The presence of water makes Earth a habitable planet. Oceans cover about 96.5% of global water resources; about 2.5% is freshwater, of which around 1.2% is available as surface water (USGS, Water Science School, 2019). Water is present as surface and groundwater. Surface water mainly comprises the lentic and lotic water bodies like ponds, rivers, seas, and oceans. On the contrary, groundwater refers to the available water in the subsurface. Additionally, approximately 97.2% of earth's water is found in the oceans, 0.009% in inland seas, 2.15% in glaciers and frozen water, 0.61% in groundwater, and 0.0001 % in rivers (National Ground Water Association, 2024).

It is usually the surface water that we consume. With rapid industrialization and urbanization, there has been an increase in population, which has led to unsustainable use of water. Per capita consumption globally has increased by 600% since the last 100 years (Boretti & Rosa, 2019) and is also expected to increase by at least 20% in the coming 20 years (Lahlou, Govindan, Hazrat, Mackey, & Al-Ansari, 2023; Wada et al., 2016). Additionally, the cumulative effects of the ongoing global warming and climate change are leading to increased scarcity of water around the globe (Datta, Chakraborty, & Roychoudhury, 2025; Franklin & MacDonald, 2024; Thorne, MacDonald, Chavez, Ambrose, & Barnard, 2024; Wang et al., 2024).

India is facing the worst water crisis in its history, and millions of lives and livelihoods are under serious threat (Subramanian, 2020). Several states like Rajasthan, Maharashtra, Gujarat,

Computational Automation for Water Security.
DOI: https://doi.org/10.1016/B978-0-443-33321-7.00015-9

Jharkhand, Andhra Pradesh, and Karnataka have already been facing this problem since 2017, which currently includes parts of Uttar Pradesh, Tamil Nadu, and West Bengal (Outlook Planet, 2021). Generally, rural parts of India face several issues with groundwater crises for drinking and irrigation (Ghosh et al., 2024; Ghosh, 2024; Patra, Roy, Ghosh, & Malik, 2023). Many a times, crop vegetables got persistent pollutant contamination during farming with polluted water. India is a country where about 55% of the country is dependent upon agriculture (DAC & FW, n.d., Annual Report, 2020–21). Agriculture is a water-intensive industry, followed by thermal power plants, fertilizers, wood and pulp, livestock, several other industries, and daily consumption. All these demands are met primarily by surface water, which has led to a decrease in water table in most areas. However, Indian agriculture in most parts is still based on the salvation of the monsoon. Climate change plays a key role in adding to the water scarcity. Perennial rivers are mainly the snow-fed ones, so they usually have water throughout the year; the nonperennial rivers are entirely dependent upon seasonal precipitation and thus may be unreliable. As temperature rise, ice and glaciers melt, increasing the flow rate downstream. However, eventually, due to sporadic distribution of rainfall and temperature extremities, the flow rate can go below the threshold levels (Kumar, 2022). The ocean water cannot be consumed directly despite being in plethora. It has to be desalinated, which costs extensive. The freshwater resources are replenishing faster (600 billion cubic meters per year) than the recharge rate (17 to 960 mm per year) (Bhanja & Mukherjee, 2019; Groundwater Scarcity, 2019). The National Commission on Integrated Water Resources Development (NCIWRD) has predicted an increase in water demand by 250 billion cubic meters by 2050 (Water Resources of India, ADRI, n.d.).

This article explores the Jal Jeevan Mission's role in improving water security in India, highlighting its impact on public health and socio-economic development.

18.2 Global water status

The infamous "Water War" in Latin America has been an eye opener to the consequence of water scarcity. Bolivia is a country that is progressing well; however, it is a prey to water crisis due to the conjugate effect of its geography (in Andean Highlands) and climate change (Resilient Bolivia, 2017).

Kuang et al. (2024) reported that to guarantee a sustainable supply of fresh groundwater resources for ecosystems and people, it is imperative to quantify these changes. Hekmatnia, Ardakani, Isanezhad, and Monibi (2024) revealed that wheat virtual water trade positively impacts global freshwater consumption and reduces irrigation demand due to the majority of it being green water.

Kinouchi, Nakajima, Mendoza, Fuchs, and Asaoka (2019) conducted a case study of water security status and prospective water management measures in Bolivia. They established population as a function of water demand and its dearth. Parts of Bolivia, especially El Alto and La Paz, are in severe demand, yet there is a scarcity of water, mainly due to retreat of glacial water (Soruco et al., 2015). Kinouchi et al. (2019) have shown that there is an ongoing significant increase in population in El Alto, but the population will decrease in La Paz in the coming future due to lack of water. Upon analyses, they have suggested diversion of the surface runoff to one of the main lakes, i.e., Tuni Lake, with the Tuni Lake catchment being the main source catchment.

As the Chinese economy is significantly influenced by agricultural production, it requires huge amounts of water. The Chinese Government has mapped the total water resources and the total water that has been withdrawn and has also established a goal of 0.6 as an effective irrigation water use coefficient. In a study, Guo et al. (2019) classified different Chinese provinces on the basis of hydro-economy. They have identified low pressure of the water in the subsurface as the key hydrological challenge behind dearth of water and thus leading to higher demand. Extreme hydrological environments lead to higher water use efficiency over the ones with less extreme hydrological scenarios. They have also emphasized extreme hydrological conditions as the impetus behind efficient use of water.

Three African countries—Algeria, Morocco, and Tunisia—are facing water stress due to inadequate precipitation, climate change-induced rising temperatures and droughts, and a lack of judicious use of water due to increasing population. Kadi (2004) has emphasized the importance of a holistic strategy. Morocco is giving importance to exploring more surface water to ensure water management. However, it could be challenging owing to the increasing population, which is caused by both innate and immigration-led increase. To cope with the existing problems, Morocco has decided to take up institutional transformation by developing a national "Water Sector Development Policy." They also developed a novel water law in 1995 according to which the public owns the water resources. This law is a comprehensive one and inculcates the provision of penalties for polluting water bodies and paying taxes while using water. It is a good way of achieving water security.

Water scarcity is quite a common problem, even in the developed nations. Countries like Israel and Jordan face acute water scarcity due to low availability of surface and groundwater resources and the higher immigration rate of refugees from Syria, leading to overpopulation and thus leading to water stress conditions (Influx of Syrian Refugees, 2021). Hence, the policy programs include transboundary nexus strategy to solve the water-based issues. Two ventures have started as a means of transboundary water management, viz., the Water Conveyance Project and the National Water Carrier Desalination Project. The former is a connection between Israel and Palestine, was implemented in 2013, whereas the latter entirely encompasses Jordan and will desalinate water from the Gulf of Aqaba and will commence in 2027 (Chenoweth & Al-Masri, 2021). Al-Addous, Bdour, Alnaief, Rabaiah, and Schweimanns (2023) have reported the incorporation of collaboration between different states like the Joint Water Commission (JWC) between Israel, Palestine, and Jordan for attaining water security.

Germany always had a rich water resource, but climate change and increasing population have been leading to a huge loss of water (Climate Crisis, n.d.). Deforestation, extreme heat waves, and increased concentration of carbon dioxide in the atmosphere have led to the water crisis. Prevalent forest fires due to extreme heat, drought, and insect infestations are also leading to death of trees (Climate & CO_2 Agriculture, 2021).

18.2.1 Water security

Allocation of water to every person is the most challenging part of attaining water security. This is crucial for all nations, since water security in the long run directly affects food security. It also impacts social, economic, and geopolitical dimensions of a region. Water security requires proper infrastructure, as well as periodic and proper management.

18.2.1.1 Importance of water security in rural India

India has about 4% of the global available water resources (The Water Crisis in India, 2023). Despite the fact that urban India uses more water than rural India, rural India has significant problems in terms of water security. Lack of availability of adequate resources, unhygienic living conditions, lack of proper sanitation facilities, and some traditional methods of water usage like bathing, washing clothes, and utensils in rivers that are used for consumption lead to water pollution and water-borne diseases (Ghosh & Manchanda, 2019; Ghosh, Dastidar, & Sreekrishnan, 2017; Ilame & Ghosh, 2021). Rural areas depend on the available perennial rivers, while nonperennial rivers depend on seasonal rainfall. In rural India, nonclimate-related factors are more significant in terms of water security than climate-associated factors (Basu, Hoshino, & Hashimoto, 2015). They have also pointed out that social and economic factors contribute to water security, and that strategies to mitigate water scarcity should take into consideration local areas as well as regional and national basis. Fig. 18.1 shows the earth's population based on water security levels. It can be envisaged that 5.60% of the population has insecure water distribution, while only 1% has secure water availability.

Water-borne diseases like cholera, diarrhea, amebiasis, giardiasis, leptospirosis, and hepatitis are the root cause of mortality and fatality in many people (Praveen et al., 2016). Srikanth (2009) has reported that about 21% of all the communicable diseases in India are caused by water-borne diseases. Kumar et al. (2022) and Kumar (2022) have reported that about 37.7 million people are affected by water-induced diseases in rural India annually. They have also stated that about 1.5 million children die of gastrointestinal disorders annually. It is mainly the summer months and rainy months when the outbreak of waterborne diseases is the most. This could be attributed to lack of adequate and potable water due to intense heat during the summer months and abundant rainfall leading

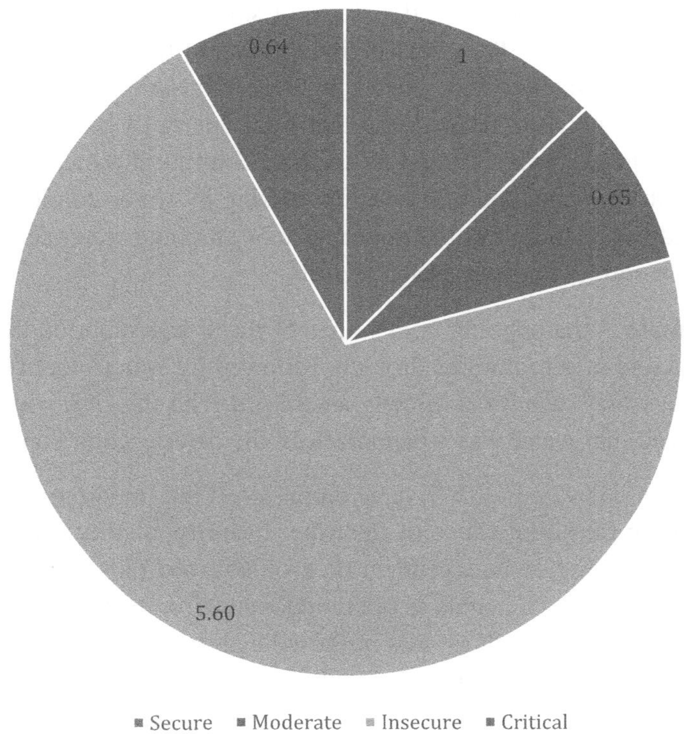

Figure 18.1
Global population (in billions) based on water security levels 2023. From *https://www.statista.com/statistics/1458133/global-population-water-security-levels/ (n.d.)*.

to near-flooding conditions in the rural regions. It has also been reported that the elderly people are affected significantly by water-borne diseases. Over time, reachability of water in households has increased, but about 18% of people in rural India consume locally filtered, untreated drinking water (Biswas & Mandal, 2010).

18.2.1.2 Dimensions of water security

Multiple factors are significantly responsible for attaining water security, which is known as dimensions of not only attaining water security but also other parameters like food availability. Some studies have taken into consideration different dimensions of water security, as described below.

Varis, Keskinen, and Kummu (2017) have envisaged a direct link between water security and water availability. They have reported four different dimensions, viz., peace—conflict, macro—micro, direct—indirect, and technical—political.

In their study, direct refers to the sole water use, whereas indirect refers to the cumulative effect of various interrelated features with respect to water. Macro and micro factors relate to the international or large scale, as well as local or small scales. They can have both short-term and long-term goals. Technical-political refers to the use of different advancing technologies in adopting water security, like rainwater harvesting procedures and water treatment facilities. The political factors envisage significance of governance in regulations governing water resource distribution and utilization. The fourth dimension, viz., peace-conflict pertains to the existing and past conflicts arising by waterscarcity, civil wars, and climate change, additively adding to the importance of attaining water security. All these factors cumulatively add to attaining water security.

A remarkable example is the ongoing civil war in Syria. It was reported that the war was due to multifold factors of a prolonged drought, followed by water scarcity. In fact, Beck (2014) reported that water scarcity is directly associated with the civil war in Syria. King (2015) also envisaged that water was weaponized in the deteriorating condition of Syria.

Janakarajan, Llorente, and Zérah (2006) have emphasized the importance of human-induced water scarcity, inciting the significance of the water quantity, quality, and accessibility. Chennai being a rain-fed region is dependent on groundwater. This has led to depletion in groundwater levels. Presently, Chennai is obtaining water from the periurban areas. The Chennai Metro Water Board has dug wells in the villages, and water is supplied throughout the state, resulting in some social conflicts. In 2019 Chennai was declared Day Zero due to severe water scarcity (Water Crisis: Day Zero in Chennai, n.d.). The same incident took place in South Africa in 2018 (Ghosh & Bandyopadhyay, 2009). According to reports, the transboundary dispute regarding the Cauvery River, which involves some states like Tamil Nadu and Karnataka, has been ongoing for quite some time and is based on water distribution and rights (Interstate Water Disputes, 2016).

18.3 Multifarious nexus of water security in India

India has been facing water security challenges due to the cumulative effects of multiple factors, viz., population growth, climate change, water pollution, and lack of execution of proper water use practices.

India surpassed China in being the most populous country in the globe in 2023, with a current population of 1441.7 million (World Population Dashboard, n.d.). It is expected to increase further unless the carrying capacity is reached. China's population decline took place due to rapid decline in fertility due to adoption of several policies (Jiang, Li, & Feldman, 2013). India's unsustainable increase in population is likely to pose a threat to food and water security (Mishra, Kumar, Saraswat, Chakraborty, & Gautam, 2021; Sahoo, Rout, & Jakovljevic, 2023; Zakar, Zakar, & Fischer, 2020). Scarcity of food and water is already visible. With population growth, per capita water consumption rises, increasing demand for water. Bassi, Schmidt, and De Stefano (2020) study forecast a significant increase in water demand between 2025 and 2030. It has been shown that nonconsumptive uses, such as power plants, will contribute significantly more to water demand in the near future. It is followed by domestic and industrial applications. Irrigation provides the least contribution.

India has a plethora of major resources, in which a certain amount of water is utilizable, as 100 percent of water cannot be accessed or put to potable use. This is because most are saline water, desalination of which would be quite expensive. Fig. 18.2 shows the proportion of utilizable water resources in the major rivers in India. It can be observed that different rivers have different water resources, but among the available water, the utilizability is less.

India is one of the water-stressed nations in South Asia, with substantial inequalities in water availability due to uncertainties in water supplies caused by increasing population and differences in water supply. Hence, there is a dire need for water security in a nation like India, where population and demand for resources are increasing at a very fast pace. Since rural areas are the most underdeveloped ones, they need more surveillance and attention.

Fig. 18.3 exhibits the increasing demand of water for people in India over the years. Due to unprecedented increase in population, there has been a subsequent increase in per capita water demand and thus consumption. However, the per capita availability of water to different people around the globe is decreasing.

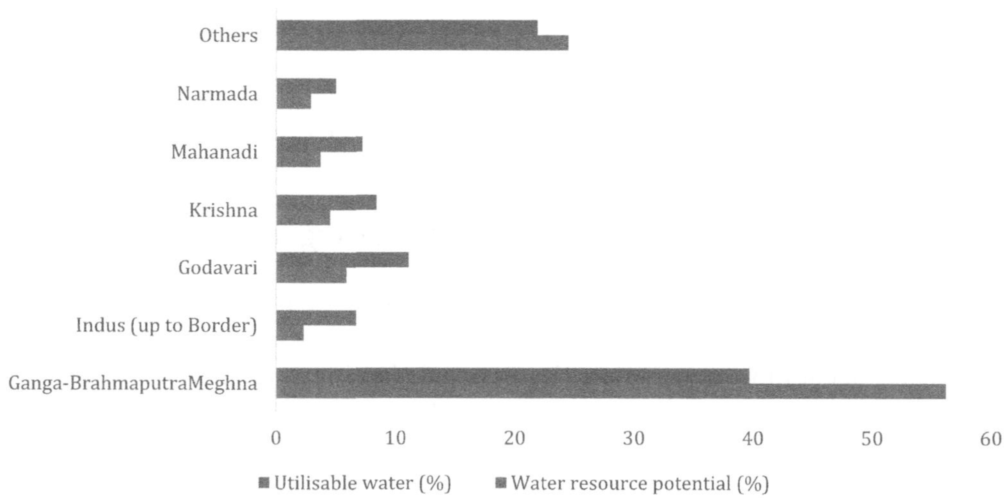

Figure 18.2
Variability of the total water resource potential and its usage. From *https://cwc.gov.in/sites/default/files/water-and-related-statistics-2021compressed-2.pdf (n.d.)*.

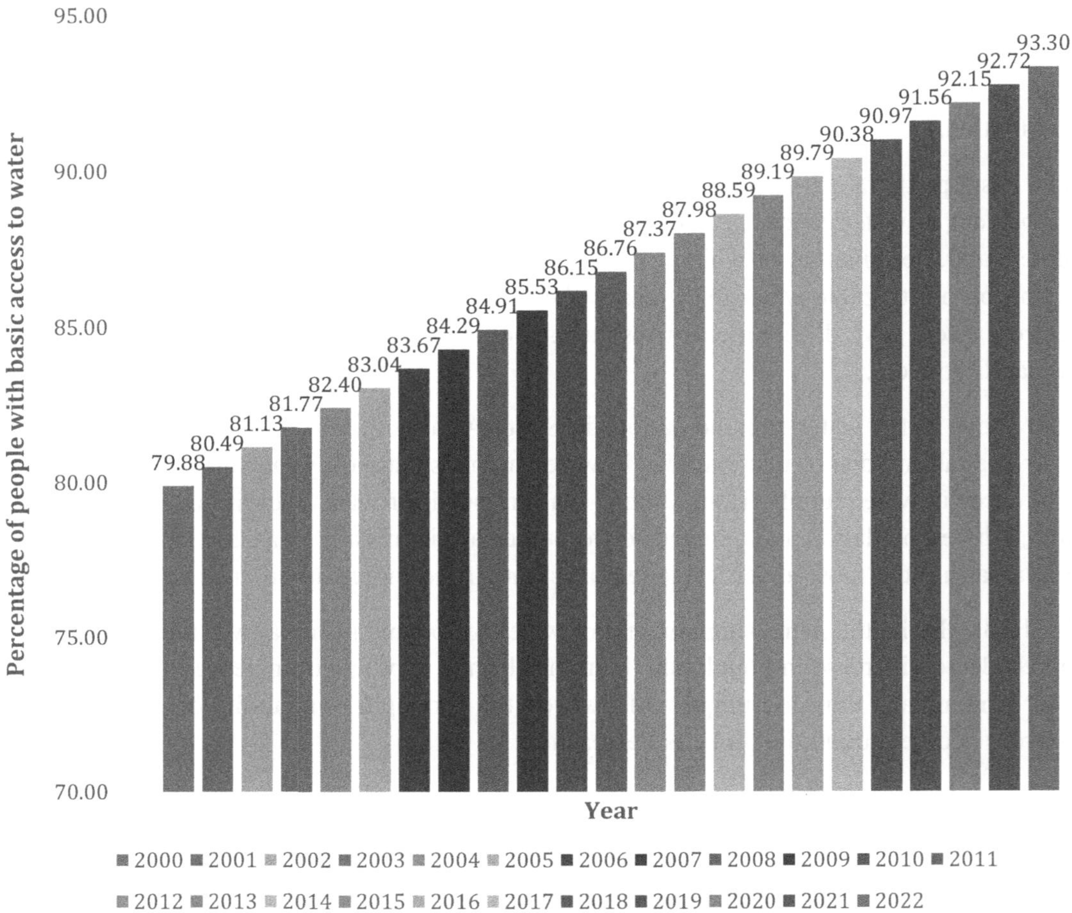

Figure 18.3

Share of people with basic access to water India 2000–22. From *https://www.statista.com/statistics/1266791/india-share-of-households-with-improved-water-sources/* (n.d.).

18.4 Institutionalization of water security in changing climate in India

India is clearly competing for water supply, ranking 130 out of 180 countries (World Sustainable Development Summit, 2022). India is clearly facing one of the worst water crises, which is about to worsen. Despite lesser population in the rural areas and lesser demand for water, rural people are the most vulnerable ones. Moreover, over time with increasing population, the per capita water availability decreases. Thus, to address and resolve the issues of water scarcity and availability, the Ministry of Jal Shakti, Government of India, proposed some short-term and long-term schemes, such as the Jal Shakti Abhiyan and Jal Jeevan Mission, respectively, both of which are administered by the Ministry of Jal Shakti, India.

18.4.1 Jal Shakti Abhiyan

Jal Shakti Abhiyan (JSA) was designed for management and conservation of water, especially rainwater. It came into force on July 1, 2019. It is a program targeting water security in India (Ministry of Jal Shakti, 2019). They identified 256 districts under water stress through surveys. During the survey, they identified about 756 urban areas as water stressed; this led to the inclusion of the Ministry of Housing and Urban Affairs (Jal Shakti Abhiyan, Water Aid, 2019).

The Jal Shakti Abhiyan report also has a detailed program that has three different arenas, viz., core, special intervention, and outcome areas, respectively.

Core intervention areas: It comprises strategies pertaining to conservation of water, including provisions for rainwater harvesting and development of watersheds. It also proposed refurnishing existing surface water bodies such as ponds, lakes, and streams to make them fit for consumption, hence increasing the availability of accessible water. In addition to this, extensive reforestation in the adjoining areas would not only balance the water cycle but also restore the offset in the carbon cycle.

Special intervention areas: These consist of the conduction of different fairs like Krishi Vigyan Kendra and exhibitions to raise awareness among people. These are also used to formulate and execute plans for conservation at local or district levels.

Outcome intervention areas: These consist of different result-oriented goals, such as conserving water resources, achieving water sustainability at both local and national levels, and forming an inventory. It also emphasizes mitigation measures based on the status of water security in respective situations.

18.4.2 Strategies of the Jal Shakti Abhiyan

The JSA was implemented in two phases to enhance its impact and distribution in different regions based on their monsoon patterns. The following are the detailed strategies and components of the JSA:

Phase 1:
Duration: July 1, 2019 to September 15, 2019
Target Regions: All states and union territories
Focus area: Immediate measures to conserve rainwater and prepare for the upcoming monsoon.
Phase 2:
Duration: November 30, 2019 to present day
Target Regions: States receiving retreating monsoon.
Focus area: Continued and intensified efforts on water conservation post-monsoon.

The JSA aims to adopt rainwater harvesting, watershed development, borewells reuse and water recharge structures, water body renovation, and afforestation. The strategies involved inculcate both rural and urban areas. Rainwater harvesting and watershed management help to store abundant water for later use. These technologies already exist; however, JSA provides enhancements of the adoption of these techniques.

18.4.2.1 Outcomes of JSA

As an outcome, approximately 2.96 lakhs of water bodies were renovated, 16.2 lakhs of water harvesting structures were built, 36.75 crores of plantations around the country, and 19 lakhs of watershed development have been carried out till 2022 (Press Release, Ministry of Jal Shakti, 2022). In April 2021, the Union Government launched "Jal Shakti Abhiyan 2." As of 2024, the JSA is still playing a pivotal role. This year's scheme aims to leverage women's participation in promoting water sustainability under the theme Catch the Rain—2024. To increase green cover, more than 43,000trees were planted (Status of JJM and JSA, n.d.).

Digitalization of the JSA is carried out with the help of its amalgamated portal and JSA Android mobile application. The portal receives data from all throughout the country and often updates it every 30 seconds. It has the advantage of notifying the customers about every update pertaining to the campaign via short message service and e-mail operations. It also inculcates feedback from the users, authoritative visits, and public participation. Moreover, it also exhibits the top 10 active districts in the campaign. The Android application of the campaign, the Jal Shakti application, was aimed at higher mass involvement. It is a potable version of the desktop module. The biggest advantage of this application is that upon installation and registration, it works in offline mode also. It also provides the user to geographically tag the regions where the various activities are carried out, thus adding to the database (Jal Shakti Application). Table 18.1 shows the different methodologies adopted and their respective outcomes by different scientists around the globe.

18.4.3 Jal Jeevan Mission

Jal Jeevan Mission is a long-term community-based initiative taken up by the Central Government and launched in 2019. It aims to provide potable drinking water supply to the rural parts of the country. It also envisions adapting sustainability measures while providing the facility to the people across. It is also a community-based strategy like the JSA. This could be achieved by opting for obligatory measures like gray water management, rainwater harvesting measures, and conservation of water, which will also aid in recharge of groundwater (JJM objectives). It is also aimed at creating mass awareness.

18.4.3.1 Strategies of Jal Jeevan Mission

The primary objective of the JJM is to provide water security for rural communities. This involves providing functional household tap connections to various commercial buildings,

Table 18.1: Methodologies and outcomes of Indian water security status.

Aim	Place of study	Methods	Findings	References
Providing water facilities at home to all those families living in rural areas who have to go through problems like water.	India	Under Jal Jeevan Mission	Under this mission of the central government, so far 50% of the families have benefited from the JJM mission.	Status of JJM and JSA (n.d.)
Financial implementation and funding pattern under JJM	Uttarakhand	Under Jal Jeevan Mission	90% of the fund sharing pattern will be payable by the central government and 10% by the state government.	Ministry of Water Resources, Government of India, Central Ground Authority (n.d.)
Financial implementation and funding pattern under JJM	Himachal Pradesh	Under Jal Jeevan Mission	For Himachal Pradesh and North Eastern states, 100% fund sharing will be payable by the Central Government.	Central Water Commission, Government of India. National Register of Large Dams-2002 (2024)
Water conservation, agriculture, animal care	Zabo Nagaland	Rainwater harvesting	It reduces wastage of water in the form of a reduction in annual rainwater runoff, helps in soil conservation, controls unexpected flooding, and it also improves water quality	National Water Mission, Ministry of Jal Shakti, Department of Water Resources, RD and GR, Government of India (n.d.)
Adopted rainwater harvesting and raised the groundwater level by 1.7 m	Garden Estate, Gurgaon	Guidelines for Urban Water Conservation Jal Shakti Abhiyan, 2019	Best practices in rainwater harvesting are discussed, which succeeded due to public support.	Agrawal, Pandey, and Bechan Sharma (2010)
Effectiveness of implementation of Jal Shakti Abhiyan in the five main intervention areas	Gurakonda, Devarkadra, Mahabubnagar, Telangana	Survey with villagers such as farmers, youth, students, and old people	Increase in catchment area from 200 acres to 500 acres and production of crops and ground level water have been increased.	Ramappa, Reddy, and Patil (2014)
Study on groundwater problems, artificial recharge techniques	Musunuru, Andhra Pradesh	Collects the required data like number of households, number of population, number of borewells, geographical area, crop details, etc.	Exploited the groundwater, and there this study considers the groundwater recharge technique to increase the groundwater level in the Musunuru.	Reddy, Kumar, and Venkatesh (2020)

(*Continued*)

Table 18.1: (Continued)

Aim	Place of study	Methods	Findings	References
Spatiotemporal analysis of water bodies	Ranchi	Advanced technologies such as remote sensing and geographic information system accentuate alterations in land use patterns over time and the extent to which the changes affect the human population and the natural habitat	Pre and post-monsoon water spread, aquatic vegetation, and turbidity of major water bodies were mapped using multitemporal, moderate resolution satellite data	Sharma, Kullu, and Anshula (2020)
Strategies of implementing drinking water project	Bihar	Modern technology in planning and implementation of water supply schemes, development of water sources, treatment and supply of water, and development of participatory models of local communities	Implementing the drinking water supply scheme in rural areas was given to three-tier Panchayati Raj systems	Seetharam and Rao (2010)
Renovation of traditional and other water bodies/tanks, reuse and recharge structures, watershed development	India	Block and District Water Conservation Plan, Krishi Vigyan Kendra Mela, Scientists and IITs	Water resource management by focusing on accelerated implementation of five target interventions. The JSA aims at making water conservation a Jan Andolan through asset creation and extensive communication.	Chatterjee, Kim, Swaminathan, Kumar, and Subramanian (2019)

Note: *JJM*, Jal Jeevan Mission; *JSA*, Jal Shakti Abhiyan.

households, educational institutions, Anganwadi centers, and medical facilities. Water-stressed regions and arid or semiarid regions like deserts will be prioritized. It also inculcates the villages under the Sansad Adarsh Gram Yojana. Under the Sansad Adarsh Gram Yojana, a single village is taken care of under Panchayat (Saansad Adarsh Gram Yojana, 2018). The scheme also seeks to engage more stakeholders to secure additional funding and improve the program and its duration. In addition to this, monitoring provisions will be established to ensure the effective implementation and maintenance of these water connections.

There is a dearth of sanitation and hygiene awareness in villages. Thus, people in rural India still bathe, wash clothes and utensils, and defecate in the nearby water resources, which also serve to be their only source of drinking water. To solve this problem, provision of safe water will be carried out by pollutant and contaminant removal. For this, periodic water quality analysis and its monitoring will be conducted (JJM Objectives). Proper and sustainable distribution of water will also take place depending upon the sectors concerned.

Fig. 18.4 shows the most common water contaminants in water bodies in India. Due to high concentration of saline water in the nation, people depend on freshwater sources, mainly groundwater. Groundwater is easily contaminated due to excessive exploitation of water, which is when the contamination by different pollutants like fluoride and arsenic takes place.

Jal Jeevan Mission assigns about 135 liters per capita per day for urban water supply. As per the scheme, about 55 liters of water will be supplied per capita every day for rural areas (Per Capita Water Availability, 2020). There has been a significant increase in provision of tap water across the nation, as given in Fig. 18.5. The Jal Jeevan Mission has been a successful venture in providing tap water connections around the nation.

Eleven states and union territories have achieved 100% tap water connections, viz., Goa, Andaman and Nicobar Islands, Punjab, Haryana, Himachal Pradesh, Daman and Diu, and Dadra and Nagar Haveli, Telangana, Arunachal Pradesh, Mizoram, Gujarat, and Puducherry. Kerala, West Bengal, and Rajasthan have not achieved even 50% of the tap water provisions, but there has been significant progress in the last 5 years.

Outcomes of JJM: It has been a successful venture, with every state, district, and block involved. The program has resulted in significant health benefits by reducing the number of affected and the number of deaths caused by water-borne diseases.

> Goa has turned out to be the first state to accomplish the "Har Ghar Jal" mission. Moreover, Dadra and Nagar Haveli and Daman Diu are also the pioneer Union Territories accomplishing the "Har Ghar Jal " mission. All the 96 villages of the afore-mentioned Union Territories and 378 villages of Goa have access to tap water at homes and commercial sectors, along with provisions for mending and maintenance.

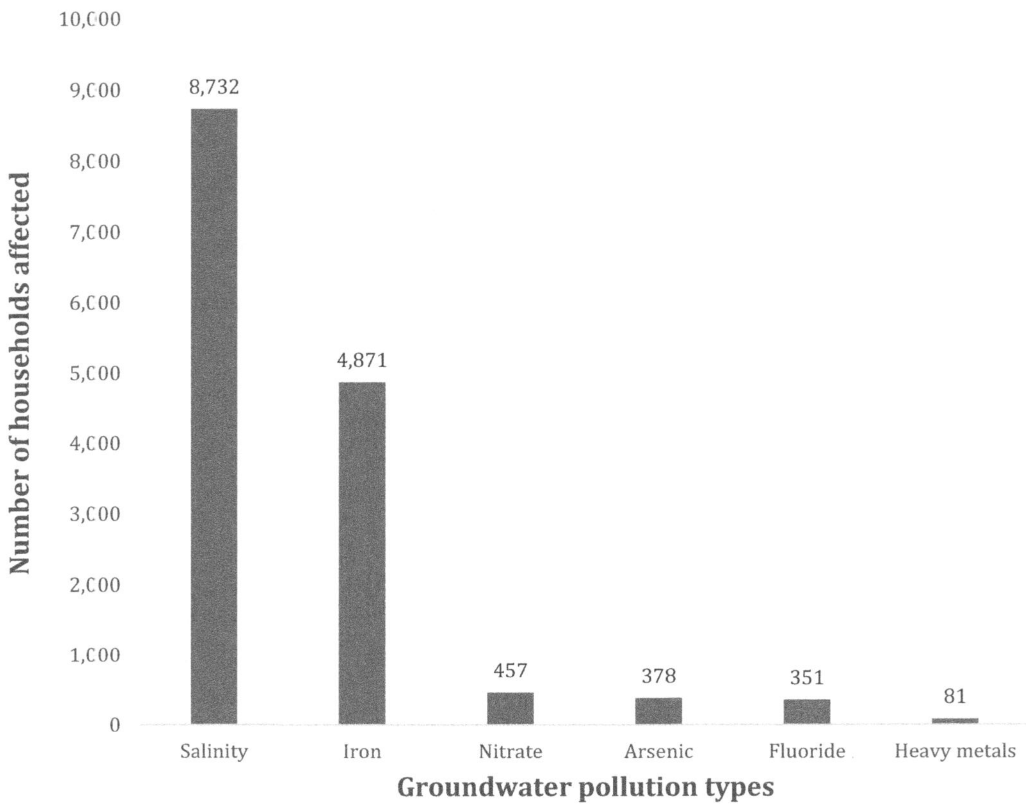

Figure 18.4
Households affected by groundwater pollution in India 2024, by contaminant. From *https://www.statista.com/statistics/1454605/households-affected-by-groundwater-contamination-by-contaminant/ (n.d.).*

There is a Jal Jeevan Mission app that provides monitoring of provided water quality. It also gives quantitative information on the households that have proper access to water. By 2024, all the states will have achieved success (Ministry of Jal Shakti, 2019).

There are about 13.91 crores of household tap water systems currently in India.

It has led to improvements in the health of people, as access to more potable water has decreased water-borne diseases. In 2023 WHO published a report citing the same and its contribution being cost-effective (Jal Jeevan Mission, 2019).

The mission has also led to employment generation of skilled people, labors, engineers, operators, operations, and management personnels. This has led to improvements in the lives of several people.

India has been successfully adding to the sixth Sustainable Development Goal of providing clean drinking water and hygiene (Biswas, Dandapat, Alam, & Satpati, 2022).

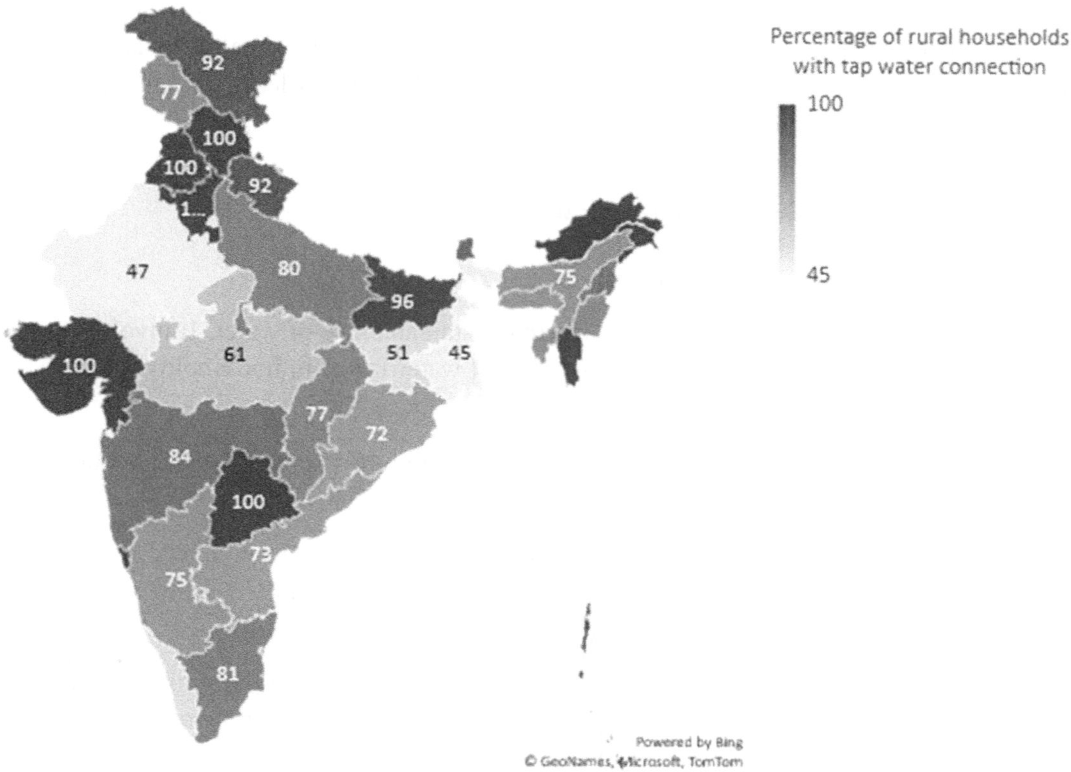

Figure 18.5

Share of rural households with tap water connection in India 2024, by state. *Note*: Map lines delineate study areas and do not necessarily depict accepted national boundaries. From *https://www.statista.com/statistics/1454729/share-households-tap-water-connection-india-by-state/ (n.d.)*.

18.5 Importance of smart technology in attaining water security through water management

Ahmed, Sayed, Abdoulhalik, Moutari, and Oyedele (2024) explored the use of advanced machine learning techniques in the water industry to ensure sustainable water supply, quality, and flood and drought mitigation. Water is a precious environmental resource and it is very vital for the survival of every creature. Our every economic operation needs water. As the human population grows the water demand is also increasing. In this scenario, the water security is very important for our survival and keeps our economic growth steady. The increasing water demand is putting severe strain on the water supply system. So, in water management system, the use of smart technologies such as application of sensors and telemetry can be a viable solution to achieve water security through smart water management (Mezzera et al., 2018; Yasin et al., 2021).

Smart water technology is an inevitable tool to attain water security. It involves using innovation and technology in managing water utilities. By providing real-time, automated data, these systems enable utilities to monitor water quality, detect leaks, and manage distribution with greater efficiency.

18.5.1 Different smart technologies for water management

18.5.1.1 Strategies of smart water system

Smart water technologies are a key to promoting sustainable use of water. It involves recognizing leakage in the water system and overall water usage in households and commercial places.

Additionally, there is an alert system that will identify and provide rapid alerts about any indiscretions, hence reducing water loss.

Water usage patterns and trends can be studied and evaluated to help in identifying water demand in a given location. Thus it will ensure that there is uniform water distribution, additionally leading to energy efficiency.

However, only some states in the country have access to smart water systems. It is, however, anticipated that the demand is increasing with time. By 2035, there will be significant usage of smart water technologies.

18.5.1.2 Some examples of smart water management with special reference to rural India

It has been reported that in 2018, Shirpur, in Maharashtra, became the first place in the nation to obtain 100% smart water usage (Smart Water Metering, n.d.).

Smart water systems join in modern and latest technology and create a resourceful and effective water distribution. It can make use of remote sensing and GIS, computer architecture methodologies like supervisory control and data acquisition systems (Dwarkanath et al., 2023), and smart meters to achieve water security. The combination of these technologies is used to monitor, regulate, and optimize the water distribution, as well as to improve resource management.

GIS is a reliable method of identifying regions of water scarcity and changes, whether significant or not. GIS aids in determining land use and land utilization, which helps in prediction and also in evaluating any change in patterns and their respective evolutions (Chaudhuri, Singh, & Rai, 2018). Puri and Sahay (2003) have reported the significance of using GIS applications to envisage and alleviate land degradation. They also mentioned that it will involve massive participation of local people to obtain water security.

Several software programs, such as LabVIEW, WiFi, and data acquisition systems, are used with the respective software, which can easily detect even minute leakages (Palermo et al., 2022).

18.6 Critical discussion

The present study investigates the use of smart technologies in water conservation, monitoring, and management, specifically in India's Jal Jeevan Mission, aiming to achieve universal affordability. Access to potable drinking water is the right of every person, and it provides impetus to the economy of a nation as it forms a nexus with other parameters like food production, social life, and geopolitical ones (Namany, Govindan, & Al-Ansari, 2023). Water security is the ability of a population to ensure sustainable availability and access to water. The "clean water security and sanitation" are important goals of UN-SDG 6. Allocation of water to every person is the most challenging part of attaining water security. This is crucial for all nations since water security in the long run directly affects food security. After the successful implementation of the project (Jal Jeevan Mission, 2019), a sustainable, clean, and safe drinking water supply to every household in rural India will be a reality.

18.6.1 Social implications

Water security is crucial for social and economic development, but its absence can have significant social implications. It can lead to inequality, conflict, and migration, as marginalized communities are more affected. Water scarcity can also lead to waterborne diseases, malnutrition, and mental health issues. Economic development can be affected by water scarcity in agriculture, industry, and tourism. Political instability can also arise due to shared water resources (Gabrielyan, 2024). Effective water governance and management are essential for addressing water security challenges. Additionally, overexploitation can lead to ecosystem degradation and climate change, further exacerbated by water scarcity.

Water authorities can make creative use of technologies like satellite remote sensing, semantic sensor web, and geographic information systems to track and forecast river levels, find new freshwater sources, and get real-time information about water use. ICT uses web-enabled sensors and communication networks to provide real-time information about physical and environmental variables, enabling water stakeholders to make accurate forecasts and decisions. Smart metering technology helps people become more conscious, identify leaks, and manage water demand. In recent smart city concept, smart water metering is being used to reduce and monitor water losses. However, many parts of the globe face a potable water crisis. There is a requirement for more stringent policies to achieve water security in this manner.

18.6.2 Challenges and probable mitigation strategies for JJM

18.6.2.1 Institutional challenges

In India, the majority of GPs across the country do not demonstrate the institutional capacity to take up such comprehensive planning and management roles.

18.6.2.1.1 Action points 1

Engage the private sector to promote infrastructure expansion and manage operations of drinking water services on a pilot basis. For locations where private sector is not present or willing, build capacities of local institutions to undertake the O&M of the scheme. Include participatory development and enforcement of water safety plans. Adequately resource and provide SWSM with operational autonomy (other than line departments) and improve their capacities to implement contextual state solutions.

Resource Planning Challenges: There is a continued reliance on groundwater to meet drinking water demand in water-stressed parts of the country. Groundwater in such areas (identified as priority areas under JJM) runs the risk of competing demands for groundwater with agriculture and industrial purposes and poses a long-term threat to the sustainability of rural water supply systems. The availability of groundwater resources is also subject to wide environmental and climate variability in the current state of climate change and its eventual fate. While the policy recognizes the need for state governments to implement a regulatory policy to abate the risk of groundwater depletion, community engagement should be sought in supporting planning and implementation of environmental assessments. Indian experience suggests that taps, tube wells, and hand pumps were most problematic in terms of quality of water (per NSS 54th round) (Ravichandran & Muthalagu, 2002). Factors such as groundwater contamination, poor maintenance of rural water supply systems, and the lack of awareness on the process of contracting diarrheal diseases add to the problem and are often difficult to address completely. Research has shown (Wright, Gundry, & Conroy, 2004) that the reduction in the incidence of waterborne diseases is contingent on the consumption of safe water at the point of use.

18.6.2.1.2 Action points 2

Seek community engagement to support planning and implementation of environmental assessments via public consultations at the village level. Strengthen social audit mechanisms through community engagement. Introduce a pilot insurance program to cover hazard risks in completed schemes. Promote demonstrated household water treatment solution like JalKalp water filter, a low-cost sustainable solution for safe water consumption at the point of use. This technology offers integration with other suitable technologies to address specific contaminants.

18.6.2.2 Issues of slippage

The water supply schemes in rural India are facing a tremendous challenge due to slippage. It is an issue of not having safe and adequate amounts of supplied tap water. The quality of water is inevitable for human health and well-being, but due to poor operation and maintenance, the adequate quality of supplied water is not maintained in our public water supply schemes, such as JJM. Due to improper maintenance of pipeline pathogenic contamination may happen. In the JJM scheme, mostly the source of water supply is groundwater but in some areas of our country, groundwater is contaminated with harmful chemicals like arsenic and fluoride. According to the report of the National Report by Department of Drinking Water and Sanitation (2022), nearly 4.76 lakh habitations, which were previously covered by the scheme of supplied tap water (JJM), are now partially covered or under the category of "no safe source."

18.6.2.2.1 Action plan 3

Technological intervention is needed to detect the contamination and decontaminate the contaminated water before supply. Setting up water quality monitoring and testing facilities at the development block level can be a possible solution.

18.6.3 Recommendations

India, the most populous country, is facing a water crisis due to its limited access to clean water. The government has prioritized water governance, acknowledging its limitations. However, citizens should also take responsibility for conserving and managing water. The government is involving rural communities, particularly women, in a people-centric water governance system. Cities and urban residents should also participate in this effort to preserve this limited resource, as global temperatures are rising.

18.7 Conclusion

This present study explored smart technologies' role in water conservation, monitoring, and management, particularly in drinking water, focusing on India's Jal Jeevan Mission's efforts to achieve universal affordability. Water authorities can make creative use of technologies like satellite remote sensing. The JJM has changed the life and environmental health conditions of millions of people in rural India with the availability of safe drinking water. However, due to several constraints, the target of connecting every household by piped water supply within 2024 is yet to be achievable. These challenges and constraints can be neutralized by using different smart technologies and positive approaches of administration. Ultimately, it will be the prime influencing factor in achieving water security for Indian people. Water is the fundamental need of human life. So, the successful implementation of the JJM project will meet the fundamental needs of the rural people of India. The significance of JJM is not only supplying clean and safe drinking water but also ensuring the dignity of humans, especially of women. The successful implementation of the Jal Jeevan Mission (JJM) will help to achieve UN-SDG 3 (Good health and well-being) and UN-SDG 6 (Clean water and sanitation) sustainable development goals by 2030.

References

Agrawal, A. A., Pandey, R. S., & Bechan Sharma, B. S. (2010) Water pollution with special reference to pesticide contamination in India.

Ahmed, A. A., Sayed, S., Abdoulhalik, A., Moutari, S., & Oyedele, L. (2024). Applications of machine learning to water resources management: A review of present status and future opportunities. *Journal of Cleaner Production*, 140715.

Al-Addous, M., Bdour, M., Alnaief, M., Rabaiah, S., & Schweimanns, N. (2023). Water resources in Jordan: A review of current challenges and future opportunities. *Water*, *15*(21), 3729.

Bassi, N., Schmidt, G., & De Stefano, L. (2020). Water accounting for water management at the river basin scale in India: Approaches and gaps. *Water Policy*, *22*(5), 768–788.

Basu, M., Hoshino, S., & Hashimoto, S. (2015). Many issues, limited responses: Coping with water insecurity in rural India. *Water Resources and Rural Development*, *5*, 47–63.

Beck, A. (2014). Drought, dams, and survival: Linking water to conflict and cooperation in Syria's civil war. *International Affairs Forum*, *5*(1), 11–22, Routledge.

Bhanja, S. N., & Mukherjee, A. (2019). In situ and satellite-based estimates of usable groundwater storage across India: Implications for drinking water supply and food security. *Advances in Water Resources*, *126*, 15–23.

Biswas, P. K., & Mandal, K. (2010). Drinking water in rural India: A study of deficiency, quality and some social implications. *Water Policy*, *12*(6), 885–897.

Biswas, S., Dandapat, B., Alam, A., & Satpati, L. (2022). India's achievement towards sustainable Development Goal 6 (Ensure availability and sustainable management of water and sanitation for all) in the 2030 Agenda. *BMC Public Health*, *22*(1), 2142.

Boretti, A., & Rosa, L. (2019). Reassessing the projections of the World Water Development Report. *npj Clean Water*, *2*, 15. Available from https://doi.org/10.1038/s41545-019-0039-9.

Central Water Commission, Government of India. National Register of Large Dams-2002. (2024). < https://www.indiawaterportal.org/data/national-register-large-dams-2002 > Accessed 04.01.2024.

Chatterjee, P., Kim, R., Swaminathan, A., Kumar, R., & Subramanian, S. V. (2019). From administrative to political evaluation: Estimating water, sanitation, and hygiene indicators for parliamentary constituencies in India. *Journal of Development Policy and Practice*, *4*(2), 188–212.

Chaudhuri, A. S., Singh, P., & Rai, S. C. (2018). Modelling LULC change dynamics and its impact on environment and water security: Geospatial technology-based assessment. *Ecology, Environment and Conservation*, *24*, S292–S298.

Chenoweth, J., & Al-Masri, R. A. (2021). The impact of adopting a water-energy nexus approach in Jordan on transboundary management. *Environmental Science & Policy*, *118*, 49–55.

Climate & CO_2 Agriculture. (2021). *Consequences of the climate crisis in Germany are becoming more severe.* < https://www.cleanenergywire.org/news/german-forests-massively-damaged-due-droughts-heat-waves-2020-report > Accessed 23.05.2024.

Climate Crisis Consequences of the climate crisis in Germany are becoming more severe. (n.d.). < https://www.umweltbundesamt.de/en/press/pressinformation/consequences-of-the-climate-crisis-in-germany-are#: ~ : text = Germany%20is%20one%20of%20the,led%20to%20significant%20crop%20losses > Accessed 20.05.2024.

DAC & FW (Department of Agriculture, Cooperation & Farmers' Welfare): Annual Report 2020-21. (n.d.). < https://agriwelfare.gov.in/Documents/annual-report-2021-22.pdf > Accessed 6.06.2024.

Datta, K., Chakraborty, S., & Roychoudhury, A. (2025). Management of soil, waste and water in the context of global climate change. *Environmental nexus for resource management.* (pp. 1–26). CRC Press.

Dwarkanath, B., Devi, P. K., Kumar, A. R., Metwally, A. S. M., Ashrafe, G. A., & Thamineni, G. M. (2023). Smart IoT-based water treatment with a Supervisory Control and Data Acquisition (SCADA) system process. *Water Reuse*, *13*(3), 411. Available from https://doi.org/10.2166/wrd.2023.052.

Franklin, J., & MacDonald, G. M. (2024). Climate change and California sustainability—Challenges and solutions. *Proceedings of the National Academy of Sciences*, *121*(32), e2405458121.

Gabrielyan, H. (2024). *Turkey's water policy as part of the political strategy*. Springer Nature.

Ghosh, A. (2024). Challenges and mitigation techniques for clean rural water supply in Himachal Pradesh, India. In *Water resources management for rural development.* (pp. 237–248). Elsevier.

Ghosh, A., & Manchanda, N. (2019). Phytoremediation of heavy metals from water of Yamuna River by Tagetes patula, Bassica scoparia, Portulaca grandiflora. *Mercury*, *1*, 3.

Ghosh, A., Dastidar, M. G., & Sreekrishnan, T. R. (2017). Response surface modeling of bioremediation of acid black 52 dye using Aspergillus flavus. *Water Science and Technology*, *75*(12), 2864–2874.

Ghosh, A., Sharma, P., Vashisht, D., Malik, P., Mondal, A., & Mondal, S. (2024). Socio-economic-environmental challenges at himachal villages: Findings from five Unnat Bharat Abhiyan adopted villages. *GeoJournal*, *89*(1), 3.

Ghosh, N., & Bandyopadhyay, J. (2009). A scarcity value based explanation of trans-boundary water disputes: The case of the Cauvery River Basin in India. *Water Policy*, *11*(2), 141–167.

Government of India, Central Ground Authority. (n.d.). <https://cgwb.gov.in/ground-water-regulation> Accessed 02.06.2023.

Groundwater Scarcity. (2019). *Groundwater scarcity, pollution set India on perilous course*. <https://www.newsecuritybeat.org/2019/01/groundwater-scarcity-pollution-set-india-perilous/> Accessed 01.2024.

Guo, A., Jiang, D., Zhong, F., Ding, X., Song, X., Cheng, Q., & Huang, C. (2019). Prediction of technological change under shared socioeconomic pathways and regional differences: A case study of irrigation water use efficiency changes in Chinese provinces. *Sustainability*, *11*(24), 7103.

Hekmatnia, M., Ardakani, A. F., Isanezhad, A., & Monibi, H. (2024). A novel classification of virtual water trade for the sustainability of global freshwater resources. *Environment, Development and Sustainability*, *26*(3), 7377–7408.

https://cwc.gov.in/sites/default/files/water-and-related-statistics-2021compressed-2.pdf (n.d.).

https://www.statista.com/statistics/1266791/india-share-of-households-with-improved-water-sources/ (n.d.).

https://www.statista.com/statistics/1454605/households-affected-by-groundwater-contamination-by-contaminant/ (n.d.).

https://www.statista.com/statistics/1454729/share-households-tap-water-connection-india-by-state/ (n.d.).

https://www.statista.com/statistics/1458133/global-population-water-security-levels/ (n.d.).

Ilame, T., & Ghosh, A. (2021). The promising applications of nanoparticles for synthetic dyes removal from wastewater: Recent review. *Management of Environmental Quality: An International Journal*, *33*(2), 451–477.

Influx of Syrian Refugees. (2021). *Influx of Syrian refugees in Jordan effects on the water sector*. <https://reliefweb.int/report/jordan/influx-syrian-refugees-jordan-effects-water-sector> Accessed 20.02.2024.

Interstate Water Disputes. (2016). Why interstate water disputes are difficult to manage or control. The Economic Times. <https://economictimes.indiatimes.com/news/politics-and-nation/why-interstate-water-disputes-are-difficult-to-manage-or-control/articleshow/54632801.cms?from = mdr> Accessed 02.05.2024. Accessed.

Jal Jeevan Mission. (2019). <https://pib.gov.in/PressReleaseIframePage.aspx?PRID = 1930998> Accessed 04.06.2023.

Jal Shakti Abhiyan, Water Aid. (2019). *Jal Shakti Abhiyan, analysis and learnings*. <https://www.wateraid.org/in/sites/g/files/jkxoof336/files/jal-shakti-abhiyan-analysis-and-learning.pdf> Accessed 20.12.2022.

Janakarajan, S., Llorente, M., & Zérah, M. H. (2006). Urban water conflicts in Indian cities: Man-made scarcity as a critical factor. In *Urban water conflicts: An analysis of the origins and nature of water-related unrest and conflicts in the urban context* (pp. 91–111). Paris: UNESCO.

Jiang, Q., Li, S., & Feldman, M. W. (2013). China's population policy at the crossroads: Social impacts and prospects. *Asian Journal of Social Science*, *41*(2), 193–218.

Kadi A., & Marquina, A. (Eds.), (2004). *Environmental Challenges in the Mediterranean* 2000–2050, (pp.175–185). Kluwer Academic Publishers.

King, M. D. (2015). The weaponization of water in Syria and Iraq. *The Washington Quarterly*, *38*(4), 153–169.

Kinouchi, T., Nakajima, T., Mendoza, J., Fuchs, P., & Asaoka, Y. (2019). Water security in high mountain cities of the Andes under a growing population and climate change: A case study of La Paz and El Alto, Bolivia. *Water Security*, *6*, 100025.

Kuang, X., Liu, J., Scanlon, B. R., Jiao, J. J., Jasechko, S., Lancia, M., & Zheng, C. (2024). The changing nature of groundwater in the global water cycle. *Science (New York, N.Y.)*, *383*(6686), eadf0630.

Kumar, L. (2022). Assessing the impact of water resources vulnerability under climate change in India.

Kumar, P., Srivastava, S., Banerjee, A., & Banerjee, S. (2022). Prevalence and predictors of water-borne diseases among elderly people in India: Evidence from Longitudinal Ageing Study in India, 2017–18. *BMC Public Health*, *22*(1), 993.

Lahlou, F. Z., Govindan, R., Hazrat, B., Mackey, H. R., & Al-Ansari, T. (2023). Water resource portfolio design for optimum water productivity in agriculture. *Computer aided chemical engineering* (Vol. 52, pp. 569–574). Elsevier.

Mezzera, L., Carminati, M., Di Mauro, M., Turolla, A., Tizzoni, M., Antonelli, M. (2018). A-Parameter Platform for Smart and Wireless Networks Monitoring On-Line Water Quality. *In Proceedings of the 2018 25th IEEE international conference on electronics, circuits and systems (ICECS)* (pp. 709–712). Bordeaux, France, 9–12 December 2018. Bordeaux, France: IEEE.

Ministry of Jal Shakti. (2019). *Concept note Jal Shakti Abhiyan: Catch the rain-2022 campaign.* <https://jsactr.mowr.gov.in/website/help-documents/Concept-Note-JSA-2022.pdf> Accessed 06.06.2024.

Mishra, B. K., Kumar, P., Saraswat, C., Chakraborty, S., & Gautam, A. (2021). Water security in a changing environment: Concept, challenges and solutions. *Water*, *13*(4), 490.

Namany, S., Govindan, R., & Al-Ansari, T. (2023). Operationalising transboundary cooperation through game theory: An energy water food nexus approach for the Middle East and North Africa. *Futures*, *152*, 103198.

National Ground Water Association. (2024). *Information on earth's water*, <https://www.ngwa.org/what-is-groundwater/About-groundwater/information-on-earths-water> Accessed 20.05.2024.

National Report by Department of Drinking Water and Sanitation. Ministry of Jal Shakti. Functionality Assessment of Household Tap Connections 2022, Jal Jeevan. (n.d.). <https://jaljeevanmission.gov.in/sites/default/files/2022-10/national_report_of_functionality_assessment_2022.pdf> Accessed 08.062024.

National Water Mission, Ministry of Jal Shakti, Department of Water Resources, RD & GR, Government of India (n.d.).

Outlook Planet. (2021). *It takes a village: how two states dealt with a severe water crisis.* <https://poshan.outlookindia.com/story/poshan-news-water-scarcity-scares-the-states-of-gujarat-and-maharashtra/377897#:~:text=The%20states%20of%20Maharashtra%2C%20Gujarat,fallen%20alarmingly%20over%20the%20years> Accessed 17.03.2024.

Palermo, S. A., Maiolo, M., Brusco, A. C., Turco, M., Pirouz, B., Greco, E., & Piro, P. (2022). Smart technologies for water resource management: An overview. *Sensors*, *22*(16), 6225.

Patra, P., Roy, A., Ghosh, A., & Malik, P. (2023). Socio-economic impact on the availability of basic amenities: A comparative analysis of villages of hilly states, India. *Management of Environmental Quality*, *34*(1), 37–58.

Per capita water availability. (2020). <https://pib.gov.in/pressreleasepage.aspx?prid=1604871#:~:text=Per%20Capita%20Availability%20of%20Water&text=As%20per%20Ministry%20of%20Housing,to%20higher%20level%20by%20states>. Accessed 01.04.2024.

Praveen, P. K., Ganguly, S., Wakchaure, R., Para, P. A., Mahajan, T., Qadri, K., & Dalai, N. (2016). Water-borne diseases and its effect on domestic animals and human health: A review. *International Journal of Emerging Technology and Advanced Engineering*, *6*(1), 242–245.

Press Release, Ministry of Jal Shakti. (2022). <https://pib.gov.in/PressReleasePage.aspx?PRID=1805805> Accessed 14.122023.

Puri, S. K., & Sahay, S. (2003). Participation through communicative action: A case study of GIS for addressing land/water development in India. *Information Technology for Development*, *10*(3), 179–199. Available from https://doi.org/10.1002/itdj.1590100305.

Ramappa, K. B., Reddy, B. S., & Patil, S. K. (2014). Water conservation in India: An institutional perspective. *Ecology Environment and Conservation*, *20*(1), 303–311.

Ravichandran, M. (2002). Economic and environmental status of drinking water provision in Rural India.

Reddy, M. S. S., Kumar, N. S., & Venkatesh, K. (2020). A study on ground water problems, artificial recharge techniques in Musunuru. *Materials Today: Proceedings*, *33*, 353–359.

Resilient Bolivia. (2017). *Resilient Bolivia, an example to overcome the water crisis.* <https://www.wearewater.org/en-in/insights/resilient-bolivia-an-example-to-overcome-the-water-crisis/#:~:text=Beyond%20the%20Water%20War%20and,have%20access%20to%20adequate%20sanitation> Accessed 20.05.2024.

Saansad Adarsh Gram Yojana. (2018). https://www.india.gov.in/spotlight/saansad-adarsh-gram-yojana Accessed 01.01.2024.

Sahoo, P. M., Rout, H. S., & Jakovljevic, M. (2023). Consequences of India's population aging to its healthcare financing and provision. *Journal of Medical Economics*, *26*(1), 308–315.

Seetharam, K. E., & Rao, B. (2010). Index of drinking water adequacy (IDWA) for the states of India. *Journal of Infrastructure Development*, *1*(2), 179–192.

Sharma, N. K., Kullu, N. J., & Anshula, K. (2020) Spatiotemporal analysis of waterbodies in Ranchi District, Jharkhand using satellite data: An input for Jal Shakti Abhiyan.

Smart Water Metering. (n.d.) *Building a sustainable water supply with smart metering.* https://www.kamstrup.com/en-en/customer-references/water/shirpur-warwade Accessed 04.05.2023.

Soruco, A., Vincent, C., Rabatel, A., Francou, B., Thibert, E., Sicart, J. E., & Condom, T. (2015). Contribution of glacier runoff to water resources of La Paz city, Bolivia (16° S). *Annals of Glaciology*, *56*, 147–154.

Srikanth, R. (2009). Challenges of sustainable water quality management in rural India. *Current Science*, 317–325.

Status of JJM and JSA. (n.d.). *Status of Jal Jeevan Mission and Jal Shakti Abhiyan.* https://pib.gov.in/PressReleasePage.aspx?PRID=1805805 Accessed 14.10.2023.

Subramanian, S. (2020). The fast lane to India's water management. *World Water Policy*, *6*(2), 279–285.

The Water Crisis in India. (2023). *The water crisis in India: Harnessingtechnology to tackle it.* <https://parati-in.medium.com/the-water-crisis-in-india-harnessing-technology-to-tackle-it-2f75df7f6e41> Accessed 01.06.2024.

Thorne, K. M., MacDonald, G. M., Chavez, F. P., Ambrose, R. F., & Barnard, P. L. (2024). Significant challenges to the sustainability of the California coast considering climate change. *Proceedings of the National Academy of Sciences*, *121*(32), e2310077121.

USGS, Water Science School. (2019). *The distribution of water on, in, and above the Earth.* <https://www.usgs.gov/media/images/distribution-water-and-above-earth#:~:text=About%2071%20percent%20of%20the,in%20you%20and%20your%20dog> Accessed 20.05.2024.

Varis, O., Keskinen, M., & Kummu, M. (2017). Four dimensions of water security with a case of the indirect role of water in global food security. *Water Security*, *1*, 36–45.

Wada, Y., et al. (2016). Modeling global water use for the 21st century: The water futures and solutions (WFaS) initiative and its approaches. *Geoscientific Model Development*, *9*(1), 175–222.

Wang, M., Bodirsky, B. L., Rijneveld, R., Beier, F., Bak, M. P., Batool, M., & Strokal, M. (2024). A triple increase in global river basins with water scarcity due to future pollution. *Nature Communications*, *15*(1), 880.

Water Crisis: Day Zero in Chennai. (n.d.). <https://healthissuesindia.com/2019/06/19/water-crisis-day-zero-arrives-in-chennai/> Accessed 01.05.2024.

Water Resources of India, ADRI. (n.d.). *India water facts.* <https://www.adriindia.org/adri/india_water_facts#:~:text=According%20to%20the%20projections%20by,(Press%20Information%20Bureau%202013)> Accessed 14.05.2024.

World Population Dashboard. (n.d.). India <https://www.unfpa.org/data/world-population/IN> Accessed 12.05.2024.

World Sustainable Development Summit. (2022). *Water – key facts for its sustainable management in India.* < https://www.teriin.org/sites/default/files/2021-06/WSDS_India-Water-Factsheet_Report_2021.pdf > Accessed 01.04.2024.

Wright, J., Gundry, S., & Conroy, R. (2004). Household drinking water in developing countries: A systematic review of microbiological contamination between source and point-of-use. *Tropical Medicine & International Health*, *9*(1), 106–117.

Yasin, H. M., Zeebaree, S. R. M., Sadeeq, M. A. M., Ameen, S. Y., Ibrahim, I. M., Zebari, R. R., ... Sallow. (2021). IoT and ICT based smart water management, monitoring and controlling system: A review. *Asian Journal of Research in Computer Science*, *8*, 42–56.

Zakar, M. Z., Zakar, D. R., & Fischer, F. (2020). Climate change-induced water scarcity: A threat to human health. *South Asian Studies*, *27*(2).

CHAPTER 19

Computational automation for enhancing water security and quality management

Shaheen Efrah Ali[1], A.K. Sarkar[2] and Mohammad Afzal[3]

[1]Department of Management, Invertis University, Bareilly, Uttar Pradesh, India, [2]Business & Administration Department, M. J. P. Rohilkhand University, Bareilly, Uttar Pradesh, India, [3]Department of Management, Rajshree Institute of Management & Technology, Bareilly, Uttar Pradesh, India

19.1 Introduction

Water scarcity is a global issue that turned worse due to massive population, adverse climate change, and high pollution rates. Indeed, due to the increasing demand for water management, the use of automation through computational means has become one of the possible approaches to improve water quality control and management. This new approach implies the use of advanced technologies like the IoT, AI, and machine learning to generate enhanced adaptive and proactive systems of water management (Gleick, 2018). However, there are some difficulties in adoption of these technologies, and some of them are ethical in nature. There is an issue of incorporating smart solutions into ageing water systems and maintaining the reliability of the obtained data along with preventing data breaches (Smith and Brown, 2021a, 2021b, 2021c; Harris, 2020a, 2020b). Budget limitations include large initial capital and development, as well as costs the search for feasible operational modalities that will sustain the activity for a long time (World Bank, 2022a, 2022b, 2022c; Lee and Xu, 2021a, 2021b, 2021c). It includes questions such as how to handle the issues of automation that are already replacing many workers and how to close the gap concerning the availability of technology to everyone (Doe and Green, 2020a, 2020b, 2020c; UNESCO, 2019a, 2019b, 2019c). In addition, there are ethical requirements that may include issues to do with data privacy, the principles of governance, and community welfare protection (Fig. 19.1). There must be rules on how data are to be used, standardized measures as to who should be given access to the newer technologies, and how those who stand to benefit from the applications of automation should share these benefits (Kumar and Patel, 2021a, 2021b, 2021c, 2021d; Miller and Smith, 2020a, 2020b, 2020c, 2020d; Taylor, 2021a, 2021b, 2021c, 2021d). It is with this background that an in-depth analysis of these

Computational Automation for Water Security.
DOI: https://doi.org/10.1016/B978-0-443-33321-7.00017-2

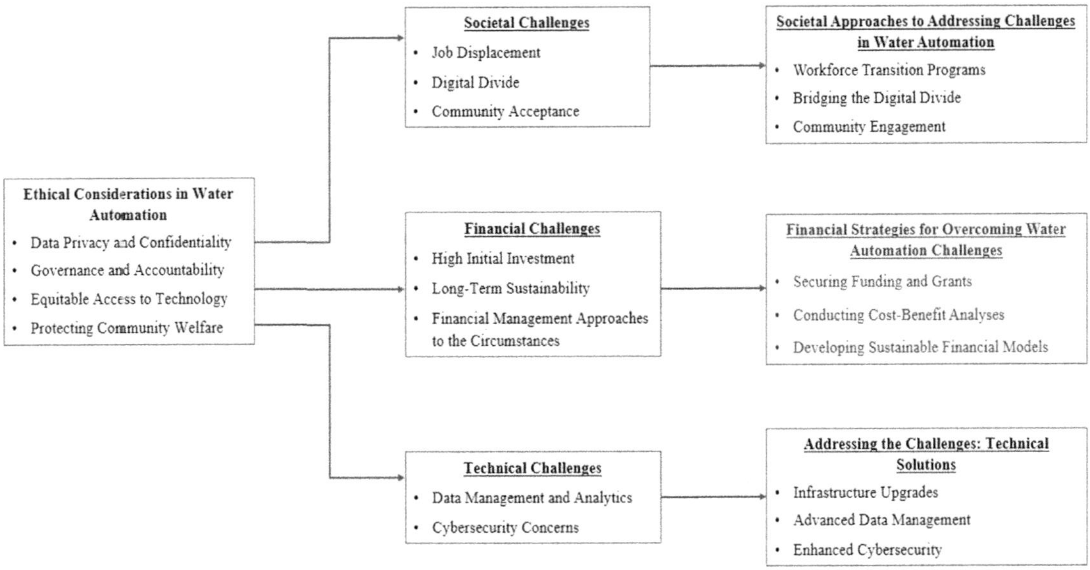

Figure 19.1
Challenges and ethical considerations in water automation.

challenges and issues is done in the subsequent chapters of this paper in what is approximated to offer a fair setting of the premises for employing computational automation in water management.

Computational automation is one of the essential elements of water security that improves the quality of methodologies used in water management. But this new expansion also brings certain problems and questions of ethics, which need to be discussed. The first of these is related to the technical factors that are inherent in the process of integrating automatic systems into an organization's operations. For example, implementation of latest sensors and data triangulation tools is capital intensive—an aspect that eludes many developing states (Zhao et al., 2023a, 2023b). There are also ethical concerns regarding the automation of water issues, especially on issues to do with data privacy and fairness in sharing of water. The deployment of automated systems entails the gross collection of data; it creates doubts regarding the use of these data and who has a right to it (Smith and Lee, 2024a, 2024b). This paper finds that there has been an emerging conflict on the ownership of water data, particularly in situations where private companies operate, manage, and control water resources that are perceived to be the property of the public. Other ethical issues raised by the mishandling of data either in unauthorized form or in the form of algorithms are gripping (Liang et al., 2024a, 2024b). Also, the employment relation issue arises in relation to the automation of water management systems. Automation technology is a threat in the water management sector since there are possibilities of employing few people due to the reduction in employment opportunities due to automation (Ramirez and Thompson, 2023a, 2023b). This displacement, not only on an individual worker basis but in areas where water management jobs are an important source of employment, has a social impact. Environmental ethics also feature in the discourses of water automation. In the case of decision-making supported by automated systems, decisions made may not consider environmental implications in the long run since the decision-making system is focused solely on efficiency as well as cost (Anderson, 2020a, 2020b, 2020c). For instance, the applications' emphasis on enhancing quick water quality through automation might avoid important sustainable water system characteristics required to protect water ecosystems in the foreseeable future. Equity nature is another aspect of ethical consideration that is important when it comes to the provision of automated water management systems. One reason that could be potentially problematic is that automation has the potential to deepen the inequalities in water supply, particularly in supporting communities that do not have resources to apply technologies such as automation systems to support them. This inequality is made worse by the fact that world regions with ample money will invest in sophisticated automation systems, leaving the less fortunate regions open to such attacks.

19.2 Technical challenges and integration with existing infrastructure

It is important to stress that the implementation of computational automation in water systems is accompanied by numerous technical issues that must be resolved for successful application of innovative technologies. From this, one can deduce that these challenges include matters to do with the adoption of new technologies in relation to the existing infrastructure, issues to do with the storage of massive data as well as matters to do with security. It is equally important regarding technological and infrastructural frameworks of water management to implement changes with reference to water automation. As of now, there are many water systems, particularly in the developing countries, that are old and the structures that were established did not install complicated automation subsystems. Adapting these systems to include the newer brands of technology is, however, usually a technical and expensive endeavor. Examples of these include pipelines, treatment plants, and distribution networks that are as old as several decades, and most of these are not compatible with high-end sensors and control systems and such will need a lot of retrofitting (Smith and Brown, 2021a, 2021b, 2021c). This process requires technical knowledge so that it does not interfere with its functioning and, in addition to that, does not alter the characteristics and dependability of water. Also, problem areas may be present related to a number of attributes of the original structures, physical configuration, and organizational layouts. In fact, a great number of systems were designed and built from materials and designs that are now considered obsolete; this transforms into a problem when introducing new technologies. For instance, embedding IoT gadgets into a conventional system may involve designing new interfaces or redesigning the existing units of the system (Smith and Brown, 2021a, 2021b, 2021c). Other concerns are the financial repercussions of such retrofits, which are massive because a formidable restoration, as demonstrated, maybe both fiscally chargeable and time-consuming. The identification of such costs and their manner of being manageable with less interruption to water services organizes the integration of automation technologies.

19.2.1 Data management and analytics

Water automation is clearly built on the conception of data collected from a large number of sensors and interconnected monitoring tools used in water systems. These devices generate voluminous data and hold the details of numerous factors, such as the quality of water, the rate at which it flows, and the efficiency of the system's operation. However, the key issue arises when it comes to managing these data: Big Data must hence be addressed to satisfy the 3V's of volume, velocity, and variety to competently process Big Data in real-time (Jones et al., 2019a, 2019b). The correctness, coherence, as well as credibility of the data collected, are invaluable since even minor discrepancies may lead to making wrong conclusions and reducing the possibility of providing inhabitants and communities with clean water. For instance, wrong sensor information leads to wrong actions that include the addition of more treatment steps or the failure to act when there is a contamination incident. Thus data validation and verification techniques must be vigorous to ensure that the collected data are accurate (Jones et al., 2019a, 2019b). Moreover, other issues include data integration problems from various data sources throughout the organization, even new or old sensors embedded in the equipment. Structuring the format of the data sources and attempting to have integrated systems is imperative in facilitating the integration of the data set for analyses. Sophisticated analysis techniques as well as the intelligent application of such mechanical computations as machine learning are also important in deriving insights out of the acquired data. These technologies can help analyze patterns, determine systems' behavior, and make correct operation decisions. Nevertheless, the creation and implementation of these business analytics solutions are computationally demanding and demand data scientists and engineers who are well-trained and experienced. The key challenge of the implementation of efficient automation processes is the availability of the skills and technologies necessary to support the water management organizations (Jones et al., 2019a, 2019b).

19.2.2 Cybersecurity concerns

By making water management systems smart as well as integrating them, they end up being exposed to cyber-criminal activities. Security is a major matter of discussion in water automation because any sort of violation can lead to an uncontrolled impact on quality, safety, and reliability of water supplies. The safeguarding of these systems from hostile activities is crucial to ensure the availability of water services without any interruptions or interferences. Employment of strong security measures such as data encryption, user authentication, and security check-ups is crucial for the protection of crucial information and business structures (Harris, 2020a, 2020b). There are serious perceptions of what cyber-attacks can do to water systems. An attack on such targets can interrupt water treatment and purification actions and consequently produce contamination or interruption of supply. Additionally, threats such as the ability to view the control systems without authorization mean that the operations within systems can be manipulated in a variety of ways, which leads to physical destruction or system failures (Harris, 2020a, 2020b). To address these concerns, it is necessary to incorporate proper cybersecurity measures into every facet of water automation systems. Encryption can be regarded as one of the most basic concepts of cybersecurity because it prevents the interception and subsequent reading of the data by third parties. The encryption of data at the times of communication between the sensors, control systems, and data management platforms might help in preventing the cases of data breaches. The practices like two-factor authentication confirm that only the authorized persons should be able to access critical systems and data. It is also important to conduct periodic reviews of security requirements and considerations, standardly known as security space scans but usually referred to as vulnerability scans, to determine and fix any flaws in the system (Harris, 2020a, 2020b). Besides the factual prevention measures, there is a need to focus on the staff of water management regarding the understanding of cybersecurity. Promoting awareness among the employees is a crucial measure that can significantly improve the security of water automation systems with the help of training dedicated to existing cyber threats, good and bad practices, and ways to act in the case of an attack. Another vital component is the cooperation with IT specialists as well as regular monitoring of new threats. Thus the management of technical issues that are linked with water automation will help to achieve the goals of increasing the quality of water monitoring. This is a primary element of this process that involves incorporating new technologies in existing structures, handling and processing large quantities of information, and protecting the networks from cyber-attacks. If water management organizations effectively acquire these essential aspects, such problems can be addressed and solved by using automation technology, which would enhance water security and sustainability.

19.3 Financial challenges

Automating water management systems is expected to provide numerous advantages, such as optimization of water supply and demand and high and accurate responses to events affecting water resources. Nonetheless, as with every decision that is made, there are always certain costs that must be incorporated to make the intended plans work most effectively, as is the case with such systems. This section focuses on the acquisition-related issues; difficulty in financing such a high initial investment for automation technologies and the problem of long-term financial viability.

19.3.1 High initial investment

When it comes to financial capital, one of the biggest hurdles in implementing automatic water management systems is major capital outlay. This comprises the expenses of procuring modern sensors, networks, information systems, and other essential automation equipment. To many municipalities, especially the developing ones and the smaller ones, these costs are quite expensive (World Bank, 2022a, 2022b, 2022c). Some of the areas that compete for capital needed to move from the older, more manual systems include infrastructure, healthcare, and education. Sensors, which are used primarily in the assessment of different qualities and characteristics of water as well as the efficiency of the system, can be very costly. Most of these sensors must be durable and adequately stable for real-time operations, which can be costly during acquisition (Smith and Brown, 2021a, 2021b, 2021c). Also, the deployment of these sensors over vast and sometimes far-flung water distribution systems is the second factor. Bearing in mind that data management systems are central to the processing and analysis of large volumes of the data collected, substantial investments in terms of hardware, software, and human capital are usually inevitable. Fund sourcing for such costs is a big question mark. Thus municipalities and water management authorities need to look for external funds, which may again be in the form of government grants, international donors, or private investments. Sometimes, it is challenging to explain why investment in such technologies is required since the gains are only realized in the future and not always tangible, as in the automation case (Lee and Xu, 2021a, 2021b, 2021c). The gains that are expected to happen in the future through healthcare reforms and implemented policies must be weighed against the costs reimbursed in the present, and this has mostly to happen within the given budget limits, and the objective is to explain these expenses to the public.

19.3.2 Long-term sustainability

As for the automation, a number of crucial positive effects, like the cost savings due to enhanced productivity and decreased prices for labor, cannot be refuted, but the question of how to make these systems financially effective in longer terms remains an issue. While automation technologies may be initially costly, it is important to consider operation costs, regular injections, and technological losses (Lee and Xu, 2021a, 2021b, 2021c). Some of the costs associated with using automated systems, such as maintenance expenses, may, on occasion, be as high as the initial investment. The maintenance expenses may, on occasions, be as big as the initial investment you made to acquire an automated system. Sensors and all other related circuits or other structures that consist of a part of the hardware are susceptible to wear and tear, environmental conditions, and possible technical problems. By so doing, it becomes possible to recommend that the people from the business should ensure that they provide regular maintenance and repair services of the systems to enhance their continuous and reliable services. This calls for financial capital as well as technical manpower that would be able to undertake the expected maintenance tasks. Moreover, it is highly likely that due to changing technology, there may be a gradual requirement for the use of existing technology, hence implying that more funds will be required for updating or acquiring new technology (Jones et al., 2019a, 2019b). It is for these ongoing costs that the sustainability of the automated water management systems is planned for. This entails the use of financial models that take into consideration the cost of initiation and subsequent costs for recurrent maintenance and replacement, among others. The costs associated with these balances prove to respect a forward calendaring of such costs based on the expected lifespan of the technologies used across the electronic commerce platforms and the expected improvements within the field of study. The municipalities and the water management authorities need to set up bodies or ensure constant revenue sources for these future costs (World Bank, 2022a, 2022b, 2022c). Another factor that defines the realm of long-term sustainability is the training and retention of qualified human resources. The functioning of automatic systems of water management requires the professional level of technicians, engineers, analysts, and other specialists capable of working with these systems, identifying existing problems, and making appropriate corrections. Thus the general investment into workforce development is critical to foster, grow, and sustain a team that will deliver these sophisticated solutions. This encompasses professional development within the first year but also the growth of career professionals continuing personal education to incorporate new technology and research advancements (Doe and Green, 2020a, 2020b, 2020c). It must also be remembered that financial planning cannot ignore the economic consequences of robotic automation. However, automation, which aims to minimize the cost of labor using machines, has negative implications for employees since machines take over most tasks. It is therefore important to formulate sound policies and more so programs that would enhance the social and economic reintegration of displaced workers, for instance, retraining programs and placement services (Doe and Green, 2020a, 2020b, 2020c). All things considered, it is necessary to guarantee that automation is implemented in a way that does not have a negative effect on some segments of the population and does not deepen existing inequalities in financial planning.

19.3.3 Financial management approaches to the circumstances

These are some of the financial problems that municipalities and water management authorities can deal with: Another method of action is to implement PPPs, which will facilitate private capital and experience for the improvement of the carried out automated systems. The competencies of PPPs can arrange sufficient funds and efficient expertize along with fair dissemination of risks and profits over the contractors of the public and private sectors (Miller and Smith, 2020a, 2020b, 2020c, 2020d). Another intervention is to seek funds from grant-making bodies, international organizations, and government-sponsored programs and agencies with a mandate for providing funds toward supporting development of sustainable water management programs. Some of the above funding options can be used to subsidize the initial costs that are involved and provide funding for sustaining and rehabilitating the facilities. Clarifying how long-term gains and environmental effects of systems that are automated could add to funding applications is a way of boosting investment (UNESCO 2019a, 2019b, 2019c). Municipalities are also able to implement new financing structures like green bonds or impact investment funds that target the sustainability aspect. These financial instruments can attract investors who are eager to invest in such projects that will socially and environmentally benefit society and thus are suitable for providing capital for water automation (World Bank, 2022a, 2022b, 2022c).

Altogether, a range of different measures can be suggested to influence the finance-related aspects of the fully automated water management systems. Thus municipalities and water management authorities can eliminate these obstacles and enjoy the advantages of automation with the help of initial funding, long-term funding, and applying innovative financing models. It is therefore crucial that these systems are sustainably financially based to boost water security and even wider goals of sustainable development.

19.4 Societal challenges

It is found that several challenges, which are technical, financial, and at the human resources end, are vertical in nature that reached a certain benchmark of automation in water management systems. The challenges can have significant implications on communities, especially those that are already vulnerable. This section explores three main societal challenges: job displacement, the digital divide, and community acceptance. It is crucial to address these issues to guarantee that the advantages of water automation are equitably and sustainably realized.

19.4.1 Job displacement

One of the primary societal challenges presented by water automation is the potential displacement of workers engaged in manual tasks within water management. The transition to automated systems can diminish the requirement for human labor in various functions, such as overseeing water quality, supervising distribution networks, and preserving infrastructure (Doe and Green, 2020a, 2020b, 2020c). In underdeveloped nations, where the job market might not be adaptable enough to absorb displaced workers, the effects can be even more prominent (Doe and Green, 2020a, 2020b, 2020c). Dealing with job displacement necessitates proactive and considerate workforce planning. This encompasses the implementation of retraining and upskilling programs to aid affected workers in acquiring new skills that are sought-after in the changing job landscape. For instance, training initiatives in digital literacy, data analysis, and maintenance of automated systems can equip workers with the necessary skills to transition to new roles within the water management sector or other industries (Smith, 2021). These support mechanisms should be reinforced to offer temporary assistance to displaced workers. Unemployment benefits, job placement services, and career guidance can help alleviate the immediate consequences of job loss and aid workers in transitioning to new work opportunities (Doe and Green, 2020a, 2020b, 2020c).

19.4.2 Digital divide

The digital divide denotes the disparity between individuals and communities with access to digital technologies and those without. In the realm of water automation, this gap can exacerbate existing inequities, leaving marginalized communities devoid of the advantages of enhanced water management (UNESCO, 2019a, 2019b, 2019c). The absence of access to digital technologies can impede these communities from reaping the benefits of the efficiencies and enhancements brought about by water automation. For example, real-time water quality monitoring and automated leak detection systems may be out of reach due to lack of essential infrastructure and resources to implement such technologies. This can lead to persistent water management issues and perpetuate current disparities in water availability and quality (UNESCO, 2019a, 2019b, 2019c). Ensuring fair access to technology is imperative to avert perpetuating or exacerbating disparities. Policymakers and stakeholders must prioritize endeavors that bridge the digital divide and foster inclusive access to water automation technologies. This required investing in infrastructure development, such as expanding broadband internet access to rural and underserved areas and providing financial support to enable communities to acquire necessary technologies (Anderson, 2020a, 2020b, 2020c). The enhancement of digital literacy and technical skills among marginalized populations should be the core focus of capacity-building endeavors. Empowering individuals to effectively utilize and benefit from automated water management systems can be achieved through training programs and community outreach initiatives. By promoting digital inclusion, these initiatives can contribute to ensuring universal access and advantages of technological advancements in water management irrespective of socioeconomic status (Smith, 2021).

19.4.3 Community acceptance

Securing community acceptance for automated water management systems is imperative for their successful deployment. Addressing public concerns regarding privacy, data security, and potential employment implications necessitates transparent communication and active community involvement. Establishing trust and illustrating the tangible merits of automation can facilitate community endorsement (Anderson, 2020a, 2020b, 2020c). Privacy apprehensions hold particular significance within the realm of automated systems that amass and analyze extensive data. Communities may express apprehension regarding data utilization, accessibility, and safeguarding against misuse or breaches. Enforcing stringent data security protocols and transparent data governance practices is indispensable in mitigating these concerns and cultivating public trust (Lee and Xu, 2021a, 2021b, 2021c). Initiating engagement with communities early in the planning and execution phases can aid in addressing these concerns and ensuring their perspectives are acknowledged. Platforms such as public consultations, community assemblies, and informational campaigns can provide avenues for stakeholders to articulate their opinions, seek clarifications, and acquire precise and comprehensive information concerning the advantages and risks associated with water automation. By fostering candid and open dialog, water management authorities can instill trust and secure community backing (Anderson, 2020a, 2020b, 2020c). Moreover, illustrating the concrete benefits of water automation, such as enhanced water quality, diminished service disruptions, and cost-effectiveness, can engender public endorsement. Demonstrating successful implementations through pilot projects and case studies can serve as compelling illustrations of how automation can positively influence communities. Showcasing such success stories can assuage apprehensions and underscore the potential of technology in augmenting water management (Smith, 2021). Mitigating potential job displacements via retraining and upskilling programs, as previously discussed, can also contribute to garnering community acceptance. Facilitating opportunities for affected workers to transition to new roles and minimizing the overall economic repercussions can fortify broader support for automation initiatives.

The societal obstacles linked with water automation are substantial and multifaceted. Job displacement, the digital gap, and community endorsement are pivotal challenges that must be tackled to ensure equitable and sustainable realization of automation benefits. Through prudent workforce planning, bridging the digital disparity, and transparent engagement with communities, stakeholders can navigate these challenges and foster an inclusive stance toward water automation.

19.5 Ethical considerations in water automation

The incorporation of computational automation in water management represents an innovative approach aimed at enhancing water quality and security. The integration of various ethical considerations is essential to ensure the responsible and equitable deployment of automation, such as data privacy, confidentiality, governance, accountability, and providing fair access to technology. It is crucial to address these ethical concerns meticulously to build trust and ensure sustainable distribution of automation benefits.

19.5.1 Data privacy and confidentiality

The data utilization in automated water management systems introduces ethical dilemmas concerning privacy and confidentiality. These systems are dependent on data acquired from sensors and surveillance tools, prompting worries about privacy violations and unauthorized entry to confidential data. Stringent procedures for data gathering, retention, and utilization are imperative to uphold data privacy. The formulation of explicit guidelines, securing informed consent, and executing data anonymization techniques are crucial measures to uphold privacy and deter unauthorized data retrieval.

19.5.2 Governance and accountability

Effective governance structures play a crucial role in overseeing automated water management systems. Governance frameworks ensure responsible and ethical use of these systems through clear accountability mechanisms and compliance with regulations. Transparency in governance is key to preventing technology misuse and promoting ethical decision-making. Regulatory frameworks should evolve to address the unique challenges posed by automation, including the implementation of robust cybersecurity measures to safeguard water management systems.

19.5.3 Equitable access to technology

Ensuring equitable access to automated water management technologies is a significant ethical consideration. It is essential to fill this gap to promote fairness and inclusivity in the deployment of automation technologies. Efforts and programs directed toward fostering all-inclusive technology access can aid in narrowing the digital gap. This encompasses investment in infrastructure to extend internet reach in underserved regions, provision of financial aid or grants to assist low-income communities in acquiring essential technologies, and execution of educational schemes to enhance digital literacy (Miller and Smith, 2020a, 2020b, 2020c, 2020d). Through the promotion of digital inclusivity, these actions can guarantee a broad distribution of the advantages of water automation, eliminating any bias toward already privileged factions. Moreover, fair access also entails the development of user-friendly technologies that cater to individuals with diverse technical competencies. Streamlined interfaces, support for multiple languages, and user education can ensure that a wider spectrum of people can proficiently use and gain from automated water control systems (Miller and Smith, 2020a, 2020b, 2020c, 2020d).

19.5.4 Protecting community welfare

The principal objective of water automation should be to advance community well-being. While enhancing water quality and security stands as a pivotal factor, it is imperative to also consider the broader societal and economic implications of automation. This encompasses the consideration of potential adverse outcomes like job displacement and heightened inequality. The introduction of automation may result in the displacement of laborers engaged in manual tasks within water resource management. To alleviate this repercussion, measures such as training and enhancement initiatives can be instituted to assist affected workers in transitioning to new positions within the field or across other sectors (Taylor, 2021a, 2021b, 2021c, 2021d). These initiatives should be customized to suit the workforce requirements and harmonized with emerging job prospects stemming from the automation itself. Social welfare mechanisms, such as unemployment benefits and job placement services, offer provisional support to displaced employees during their quest for alternative employment opportunities (Taylor, 2021a, 2021b, 2021c, 2021d). Addressing these societal repercussions is indispensable in guaranteeing that the shift to automated water management systems is all-encompassing and equitable. Community engagement emerges as another crucial element in safeguarding community well-being. Involving communities in the planning and execution phases can help assuage public apprehensions and foster trust. This entails transparent communication regarding the advantages and drawbacks of automation, alongside active involvement of community members in decision-making procedures (Anderson, 2020a, 2020b, 2020c). By nurturing a sense of ownership and involvement, water management authorities can bolster public backing and ensure that the systems are attuned to the necessities and priorities of the communities they cater to.

19.6 Addressing the challenges: technical solutions

19.6.1 Infrastructure upgrades

To confront the technical obstacles of water automation, a substantial investment in the enhancement of current water management infrastructure is imperative. Numerous existing systems, particularly in developing areas, are antiquated and lack the capacity to accommodate sophisticated automation technologies. Although the retrofitting of these systems may be intricate and expensive, it is indispensable for the effective assimilation of modern technologies (Smith and Brown, 2021a, 2021b, 2021c). The enhancement of infrastructure may encompass the substitution of outdated pipes and pumps with intelligent counterparts capable of real-time data transmission, the installation of advanced sensors for monitoring water quality and flow, and the upgrading of control systems to enable remote and automated operations. These enhancements not only facilitate the integration of automation technologies but also enhance the general efficiency and dependability of water management systems (i.e., assimilation of modern technologies).

19.6.2 Advanced data management

Water automation generates extensive data from diverse sources, such as sensors, meters, and monitoring devices. Proficient data management is pivotal for managing this magnitude and intricacy of data. The implementation of robust data management and analytics systems can aid in processing and scrutinizing data instantly, thereby empowering more well-informed decision-making (Jones et al., 2019a, 2019b). Advanced data management systems should guarantee data precision, uniformity, and dependability. This entails the use of sophisticated algorithms and machine learning methodologies to detect patterns and irregularities in the data, anticipate potential issues, and optimize water management procedures. Furthermore, these systems should be scalable to accommodate the escalating volume of data as more automated devices are deployed (Jones et al., 2019a, 2019b).

19.6.3 Enhanced cybersecurity

With the increasing digitization of water management systems, they are becoming more susceptible to cyber threats. Safeguarding these systems from malicious activities is crucial to ensure the uninterrupted and secure provision of water services. The implementation of comprehensive cybersecurity strategies is vital to protect sensitive data and infrastructure (Harris, 2020a, 2020b). Enhanced cybersecurity measures may involve data encryption to thwart unauthorized access, robust authentication protocols to validate user identities, and routine security assessments to pinpoint and rectify vulnerabilities. Moreover, devising a prompt response plan to alleviate the repercussions of potential cyber-attacks can uphold the resilience of water management systems (Harris, 2020a, 2020b). The ethical considerations associated with the implementation of automated water management systems are intricate and demand meticulous attention to guarantee responsible and fair implementation. Addressing concerns regarding data privacy and confidentiality, establishing resilient governance and accountability frameworks, ensuring fair access to technology, and safeguarding community well-being are imperative for navigating these ethical dilemmas. Through the execution of comprehensive strategies and the cultivation of inclusive and transparent procedures, stakeholders can ensure that the advantages of water automation are realized in a manner that fosters justice, inclusivity, and sustainability.

19.7 Financial strategies for overcoming water automation challenges

The transition toward automated water management systems introduces considerable financial obstacles that demand thorough planning and strategic allocation of resources. It is imperative to employ effective financial tactics to oversee the substantial initial investment outlays, guarantee enduring viability, and optimize the advantages of water automation. Critical strategies encompass the acquisition of funding and grants, execution of cost–benefit analyses, and formulation of sustainable financial frameworks. This section delves into these strategies, emphasizing their significance and offering insights into their execution.

19.7.1 Securing funding and grants

Securing sufficient funding presents a primary challenge in the implementation of water automation systems. The significant initial capital needed for acquiring and installing advanced sensors, data management systems, and other automation technologies must not be underestimated. To tackle this issue, authorities responsible for water management should explore various funding channels, including governmental and nongovernmental grants, loans, and investments from the private sector. Governmental funding opportunities are frequently accessible through national and international initiatives aimed at enhancing infrastructure and fostering sustainable development. For example, financial support offered by the World Bank for water management projects in developing nations, aiding them in upgrading their infrastructure and incorporating modern technologies (World Bank, 2022a, 2022b, 2022c). Moreover, regional development banks and environmental agencies extend grants and low-interest loans to bolster water automation endeavors. Non-governmental organizations and private foundations also influence financing of water management ventures. These entities often possess distinct mandates to advance environmental sustainability and social equity, rendering them valuable collaborators in water automation endeavors. For instance, grants from organizations like the Bill & Melinda Gates Foundation can furnish crucial financial backing for innovative water projects that tackle both technological and social hurdles.

19.7.2 Conducting cost–benefit analyses

Thorough cost–benefit analyses are imperative for validating the expenditures linked to water automation. Such analyses serve to elucidate the enduring financial advantages of automation, thereby facilitating funding acquisition and garnering stakeholder endorsement. A cost–benefit analysis examines the economic, environmental, and social ramifications of a project, juxtaposing implementation costs against projected benefits. Key considerations in a cost–benefit analysis for water automation revolve around the initial capital outlays, encompassing expenses associated with the procurement of equipment, installation of software, and improvements to the infrastructure. Operational cost reductions constitute another pivotal aspect, as automation has the potential to result in lower labor costs, decreased energy consumption, and reduced maintenance expenditures owing to increased efficiency. Improved monitoring and management capabilities contribute to enhanced water quality and security, consequently leading to better water quality and minimizing risks of contamination or disruptions in the water supply. Furthermore, environmental and societal benefits emanate from the bolstered water governance, yielding positive impacts on public health, conservation of ecosystems, and overall community welfare. The execution of a thorough cost–benefit analysis serves to quantify these benefits, offering a transparent financial rationale for engaging in water automation. Additionally, it supports the recognition of hazards and uncertainties, thus enabling more well-informed decision-making (Lee and Xu, 2021a, 2021b, 2021c).

19.7.3 Developing sustainable financial models

Ensuring the enduring financial viability of automated water management systems necessitates the formulation of robust financial frameworks that encompass continual maintenance and operational expenses. Sustainable financial models assist in strategizing for the entire lifecycle of automation systems, guaranteeing their continued functionality and efficacy over time. Key components of sustainable financial models encompass various elements such as lifecycle costing, revenue generation, financial reserves, performance-based funding, etc. Lifecycle costing is a methodology that considers all expenses linked to a project throughout its complete lifecycle, encompassing initial capital outlays, continuous upkeep, operational charges, and eventual decommissioning or replacement. This comprehensive perspective aids in ensuring adequate resource allocation for the enduring viability of the system (World Bank, 2022a, 2022b, 2022c). Revenue generation is the identification of potential revenue streams to sustain ongoing expenditures. This could involve user charges, service fees, or collaborations between the public and private sectors. For instance, water management entities can introduce tiered pricing systems that levy higher fees for greater water consumption levels, encouraging conservation while also generating extra income. Financial reserves is the establishment of financial reserves or emergency funds to manage unforeseen expenses or crises. These reserves offer a financial cushion that can uphold system stability and ensure continuous operation during financially challenging periods. Performance-based funding is the connection of funding to performance results to promote efficiency and efficacy. For example, financial rewards can be granted for achieving specific targets related to water quality, energy efficiency, or customer contentment. Performance-based funding fosters ongoing enhancement and answerability in water management approaches. In addition to these aspects, sustainable financial models must also encompass risk management tactics to tackle potential financial uncertainties. This involves recognizing and alleviating risks associated with technological obsolescence, regulatory alterations, and market fluctuations. By proactively addressing these risks, water management authorities can bolster the resilience and sustainability of their financial frameworks (World Bank, 2022a, 2022b, 2022c). The financial hurdles linked to water automation are substantial but can be effectively tackled through strategic planning and resource allocation. Acquiring funding and grants from various sources, conducting thorough cost–benefit evaluations, and formulating sustainable financial models are pivotal tactics for surmounting these challenges. By implementing these approaches, water management entities can guarantee that the shift to automated systems is financially feasible and enduring eventually. Securing funding from governmental and nongovernmental sources furnishes the indispensable financial backing for initial investments, while cost–benefit analyses substantiate these disbursements by showcasing the long-term advantages of automation. Formulating sustainable financial models ensures that ongoing maintenance and operational expenses are suitably accounted for, fostering the enduring success and steadiness of automated water management systems.

Therefore effective financial strategies are imperative for conquering the financial obstacles of water automation. By embracing a comprehensive and strategic approach to financial planning, water management authorities can facilitate the successful deployment of automated systems, enhancing water quality and security for communities globally.

19.8 Societal approaches to addressing challenges in water automation

Water automation presents a multitude of advantages, such as heightened efficacy, enhanced management of water quality, and improved allocation of resources. Nevertheless, the societal obstacles linked to this technological transition must not be disregarded. These obstacles encompass job displacement, the digital gap, and community approval. Effectively dealing with these challenges necessitates proactive steps that back affected communities and guarantee comprehensive access to the perks of water automation. Essential strategies encompass the implementation of workforce transition programs, bridging the digital disparity, and involving communities via transparent communication and participatory decision-making procedures.

19.8.1 Workforce transition programs

One of the foremost societal hurdles of water automation is the potential displacement of workers engaged in manual duties within the water management field. Tackling this predicament demands deliberate workforce planning, which includes retraining and upskilling initiatives to support displaced workers in transitioning to novel roles within or beyond the water sector. These transition programs should be tailored to equip workers with the necessary skills to conform to the evolving job market. This involves technical instruction in domains like data analysis, systems management, and cybersecurity, which are progressively pertinent in automated water management setups.

Collaboration among governmental bodies, educational establishments, and private sector entities is vital in formulating and executing effective workforce transition schemes. Governmental regulations can endorse these endeavors through financial support and rewards for employers investing in retraining their workforce. Educational institutions can provide specialized training programs. Successful workforce transition initiatives not only aid displaced workers in securing novel employment but also bolster the overall resilience and adaptability of the labor market. By investing in the continuous advancement of the workforce, societies can adeptly manage the repercussions of technological progressions and ensure that workers are not left behind.

19.8.2 Bridging the digital divide

In the realm of water automation, this division can worsen preexisting disparities, leaving marginalized groups devoid of the advantages of enhanced water management. It is imperative to bridge the digital divide to ensure universal access and utilization of automated water management technologies. The investment in digital infrastructure stands as a crucial measure in tackling the digital gap. This encompasses the extension of broadband internet availability to underserved regions, provision of cost-effective and dependable internet services, and assurance of the resilience and capacity of digital infrastructure to support cutting-edge technologies. Collaborations between the public and private sectors can significantly contribute to financing and executing digital infrastructure endeavors. In conjunction with infrastructure investments, initiatives promoting digital literacy and skill enhancement are indispensable. Offering training on digital technology usage, cybersecurity awareness, and data management can empower these communities to actively engage in and reap the benefits of water automation endeavors (UNESCO, 2019a, 2019b, 2019c). Policies and programs advocating for inclusive technology access can also aid in narrowing the digital gap. For instance, governments can introduce subsidies or grants for low-income households to obtain digital devices and services. Educational institutions and community groups can deliver complimentary or affordable training programs, while technology firms can devise economical and user-friendly solutions tailored to the requirements of underserved communities.

19.8.3 Community engagement

Securing community approval for automated water management systems is pivotal for their successful establishment. Public apprehensions regarding privacy, data security, and potential job displacement must be alleviated through transparent communication and community involvement. Establishing trust and highlighting the concrete benefits of automation can facilitate community endorsement (Anderson, 2020a, 2020b, 2020c). Community engagement entails active involvement of stakeholders in decision-making processes and ensuring their viewpoints are acknowledged. This can be achieved through public meetings, training sessions, and platforms where community members can voice their apprehensions, offer input, and contribute to the design and execution of water automation schemes. Engaging with communities in the initial stages aids in building trust and ensuring that the solutions devised are attuned to their necessities and preferences.

Transparent communication plays a pivotal role in addressing public concerns and fostering trust. This encompasses furnishing clear and precise details about the advantages and risks of water automation, along with the steps taken to mitigate potential adverse effects. Regular updates and open communication channels can aid in keeping communities informed and engaged throughout the implementation phase. Involving community members in decision-making processes further enhances community acceptance. By engaging residents in the planning and execution of water automation projects, authorities can guarantee that the solutions formulated are inclusive and fair. Addressing the societal challenges linked to water automation necessitates proactive and inclusive methodologies that uphold impacted communities and ensure fair access to technology. Implementation of workforce transition initiatives, investments in digital infrastructure, and community involvement emerge as pivotal tactics in surmounting these obstacles. Through the execution of these actions, stakeholders can contribute to ensuring widespread sharing of the advantages of water automation and the engagement of all communities in and benefit from the technological progressions in water management.

Efficient workforce transition programs equip displaced workers with the essential skills to adjust to fresh roles, fostering resilience and adaptability within the labor market. Closing the digital gap via investments in infrastructure and digital literacy initiatives guarantees that marginalized communities are not excluded. Open communication and participatory decision-making mechanisms establish trust and community backing, thus facilitating the successful execution of water automation schemes.

Therefore addressing societal challenges related to water automation demands a comprehensive and cooperative approach that gives precedence to the necessities and apprehensions of impacted communities. Through the adoption of these methodologies, stakeholders can advocate for inclusive and fair entry to water automation technologies, consequently amplifying water quality and security for all.

19.9 Discussion

The use of computational automation in water handling matters is also a paradigm shift that will facilitate better security and usefulness of water. However, it is not without its challenges and ethical considerations that, for a good reason, need proper assessment for a fair implementation of strategies. The technical issues are the most important, starting from integration of new technologies into an active network. Existing water networks in many places, especially in the developing world, are old and nonintelligent; hence, they cannot support automation tools like IoT and AI. This means a significant amount of retrofitting is required, and such works are both expensive and challenging in terms of technological possibilities (Smith and Brown, 2021a, 2021b, 2021c). Additionally, other issues such as data management and cybersecurity arise since substantial amounts of data generated through these systems have to be managed, secured, and analyzed to counter hack attacks (Zhao et al., 2023a, 2023b). Another challenge relating to the finance of automation of water provision is another factor that acts as a barrier. The capital cost involved in implementing these technologies is high, which is a big turn-off for many municipal bodies and water management authorities, especially in the developing countries where financial constraints might be a big constraint (World Bank, 2022a, 2022b, 2022c). However, achieving this seeks long-term sustainability of the funding of these systems. If these systems are not supported sustainably within areas of financial management, they can deteriorate in functionality or fail to be utilized as intended, thus reversing the purpose of their formation (Lee and Xu, 2021a, 2021b, 2021c). From a societal perspective, the problems are not much less complicated. Different segments of automation may lead to job losses as it does through skill substitution for workers like manual workers in water management. This displacement can affect social aspects, especially where such jobs are vital for employment purposes (Doe and Green, 2020a, 2020b, 2020c). Third, improved water management may not be evenly distributed across world populations due to the existence of the digital divide that may result in the worsening of the social inequalities, whereby groups such as low-income groups will be further locked out from enjoying the benefits of efficient water management (UNESCO, 2019a, 2019b, 2019c). Another aspect is public acceptance, as the public is concerned with the issues relating to privacy and security, data, and job automation (Anderson, 2020a, 2020b, 2020c). Ethically, therefore, the water automation causes some important questions to be answered. The issues of data rights and data security are extreme since these systems may gather a massive amount of data that, if misused or leaked, may compromise the privacy and trust of individuals (Smith and Lee, 2024a, 2024b). Another consideration is governance and accountability because there must be regulation of the proper use of these technologies to comply with the laws set down (Miller and Smith, 2020a, 2020b, 2020c, 2020d). It should be noted that an attempt must be made to ensure that the different communities are able to benefit from the automation of the water resource (Taylor, 2021a, 2021b, 2021c, 2021d). These challenges can therefore only be tackled by different and new strategies. In that regard, only substantial amounts of investment can be required to retrofit the existing infrastructure to support automation technologies. This infrastructure includes not only the physical framework of the system but also the data handling and security, not only for use by these systems but also for the protection of these structures against threats (Harris, 2020a, 2020b). In a financial aspect, new approaches to funding and financial planning are crucial to support these systems and find the means for their constant function and progress (Ramirez and Thompson, 2023a, 2023b). In the social domain, the preventive strategies require intervention in the effects of mass job loss, closing the digital divide, and ensuring the inclusion of stakeholders in the decision-making process (Anderson and Patel, 2023a, 2023b, 2020c). All in all, computational automation has the potential to revolutionize the approaches to water management; however, it is crucial to consider the technical, financial, social, and ethical implications of the use of such technologies. Overall, one has to understand that with the correct planning, investments, and participation of the communities, it is possible to reap the advantages of the automation in a way that is beneficial for all, equal, sustainable, and able to withstand the future's challenges.

19.10 Future directions and recommendations

Over time, the role of computation automation in water management has been introduced, and it is vital to examine the complex issues and the moralities found in them. More specifically, the following issues deserve increased attention in the future to facilitate water automation technologies that are sustainable and equitable in the future.

- Several issues, including a review of the status of IoT, AI, and machine learning applications in the water sector, will persist. Future studies should also aim at finding solutions that are more flexible and can easily work within the modern complex and older systems.
- As for research, they need to develop technologies that can be integrated easily into the existing structures to cause minimal intrusions but maximize performance improvements.
- Technology developers and water management authorities should work hand in hand toward setting standard platforms and communicating interfaces, therefore improving the interoperability and interconnection of the systems.
- Since water automation systems produce enormous data, future efforts should focus on enhancing the Big Data management as the three "V"s of Big Data, namely, volume, velocity, and variety of the data.
- Big data techniques such as analytics and machine learning can provide more insights and pattern recognition; however, it has to be coupled with strong data validation and verification measures.
- Information security or cybersecurity will remain as the key issue. Such directions of development include advanced protective measures, including the use of encryption, access control, and security audits.
- Further work should be focused on searching for new methods of financing, including public–private partnerships, bond issues for environmentally friendly projects, and grants from international funds needed to replenish initial capital and cover costs of maintenance of automations.
- Direct misapplication of the automation on societal water utility consists of employment loss alongside the era's division into the digital haves and the have-nots, which means that there is a need to proactively tweet societal utilities of water automation.
- The precedence of the future policies should be given to skill demand and supply considering automation; these are training and redeployment of workers.
- It is also important that corporate governance structures consider issues of accountability and oversights of the automation practices and these in relation to the legal and ethical frameworks. The sad reality that present-day worldwide adjustments to water require advanced computing implies that the automated computing will be the approach to the future in water management and even lift the quality and security of

water compared to the past. However, the solution of the problems and issues arising at the technical, financial, societal, and ethical levels requires a systematic and long-term approach. All these challenges can be aptly dealt with by applying innovation, varied planning, and integration of end policies to achieve the best outcome for AM while implementing the same that is fair, efficient, and sustainable. Further innovations in water automation must be done under consideration of these difficulties, encouragement of policymakers, technology creators, and community, and indication that the fresh technologies will be useful for all the parties interested and the world.

19.11 Conclusion

The integration of computational automation in water management offers a promising strategy to enhance water quality and security by leveraging advanced technologies such as IoT, AI, and machine learning to improve water quality control and management efficiency. Nevertheless, this transition presents a multitude of obstacles and moral dilemmas that necessitate careful attention to ensure an equitable and accountable implementation of these technologies. The technical hurdles primarily revolve around adapting current infrastructure to incorporate modern technologies, effective data management and analytics, and robust cybersecurity measures. Numerous water systems, particularly in developing nations, are antiquated and demand substantial modifications to integrate modern sensors and control systems. This procedure is not only technically challenging but also financially burdensome. Efficient data management is vital, as water automation heavily relies on data gathered from various sensors. Securing the precision, uniformity, and trustworthiness of this data is paramount, as any irregularities could lead to inaccurate decisions that impact water quality and safety. Moreover, the significance of cybersecurity cannot be exaggerated. Water management systems must be protected from cyber-attacks that have the potential to disrupt water supply and pollute water sources. Implementing stringent security measures, conducting regular vulnerability assessments, and promoting cybersecurity awareness among personnel are crucial measures to safeguard these systems. Financial challenges involve substantial initial investment costs, long-term sustainability, and the necessity for innovative financing mechanisms. The initial capital needed to acquire modern sensors, networks, and information systems is significant, often surpassing the capacity of smaller and developing municipalities. Ensuring long-term sustainability entails not only maintenance expenses but also the requirement for periodic updates and replacements due to advancing technology. Financial management strategies such as public–private partnerships, soliciting funds from international benefactors, and utilizing financial tools like green bonds can offer the essential capital and expertize for sustainable automation. Societal challenges encompass job displacement, the digital divide, and community acceptance. The shift to automated systems can diminish the requirement for human labor, resulting in job displacement. Proactive workforce planning, retraining, and upskilling initiatives are crucial to aid displaced workers in transitioning to new roles. The digital divide, which signifies the gap between individuals with access to digital technologies and those without, can exacerbate existing disparities. Addressing this gap necessitates investments in infrastructure, financial aid for acquiring technologies, and enhancing digital literacy among marginalized communities. Community acceptance is pivotal for the successful implementation of automated water management systems. Transparent communication, active community participation, and demonstration of the tangible benefits of automation can foster trust and secure public approval. Moral considerations are pivotal in the responsible deployment of water automation technologies. Data privacy and confidentiality are paramount, necessitating stringent protocols for data collection, storage, and utilization. Governance and accountability frameworks must adapt to tackle the distinct challenges presented by automation, ensuring the ethical and responsible utilization of these systems. Equitable access to technology is another crucial ethical consideration. Endeavors to bridge the digital divide and guarantee that all communities, irrespective of their socioeconomic status, can reap the advantages of water automation are essential. Securing the welfare of the community involves not only enhancing water quality and security but also considering the broader societal and economic consequences of automation. This entails tackling issues such as job displacement through the provision of retraining initiatives and ensuring active community involvement in the stages of planning and execution. While computational automation has the capacity to transform water management by improving water quality and security, its implementation demands a meticulous approach that considers technical, financial, societal, and ethical hurdles. Through the proactive deployment of strategies and the establishment of inclusive policies to address these challenges, stakeholders can guarantee that the advantages of water automation are experienced in a manner that fosters justice, inclusivity, and sustainability. This holistic approach is imperative for policymakers, technologists, and professionals in water management who seek to navigate the intricacies of incorporating advanced automation technologies in the water sector.

References

Anderson, J. (2020a). Community engagement in the digital age: Building trust in automated water management. *Journal of Public Administration and Policy*, *12*(3), 145–162.

Anderson, P. (2020b). Community engagement in water management: Building trust and acceptance. *Journal of Environmental Communication*, *12*(4), 234–245.

Anderson, P., & Patel, S. (2023a). The role of automation in water management and its environmental implications. *Journal of Environmental Studies*, *28*(3), 112–129.

Anderson, R. (2020c). Building trust in water automation systems: A community perspective. *Journal of Water Resources Management*, *34*(2), 145–158.

Anderson, T., & Patel, R. (2023b). Environmental ethics in water management automation: Balancing efficiency and sustainability. *Water Resources Management Journal*, *18*(2), 135–149.

Doe, J., & Green, M. (2020a). The impact of automation on employment in water management sectors. *Journal of Water Resources*, *15*(2), 89–101.

Doe, J., & Green, R. (2020b). Workforce transition in the age of automation: Strategies for upskilling and reskilling. *Journal of Labor Economics*, *18*(2), 89–110.

Doe, J., & Green, S. (2020c). Workforce transition in the age of automation: A case study in the water sector. *Journal of Employment Studies*, *28*(4), 233–247.

Gleick, P. H. (2018). The world's water volume 8: The biennial report on freshwater resources. *Island Press*.

Harris, M. (2020a). Cybersecurity challenges in the automation of water management systems. *Journal of Cybersecurity*, *7*(3), 102–119.

Harris, M. (2020b). Cybersecurity in water management: Protecting critical infrastructure. *Journal of Water Security*, *9*(2), 56–72.

Jones, A., Smith, B., & Thompson, C. (2019a). Advanced data management in water systems: Challenges and solutions. *International Journal of Water Resources Development*, *35*(4), 567–582.

Jones, T., Smith, K., & Brown, M. (2019b). Data management in automated water quality monitoring: Challenges and solutions. *Environmental Science & Technology*, *53*(12), 6942–6950.

Kumar, N., & Patel, R. (2021a). Governance frameworks for ethical water management automation. *Environmental Policy and Governance*, *34*(4), 275–290.

Kumar, R., & Patel, S. (2021b). Data privacy and security in automated water management systems. *Journal of Environmental Management*, *30*(2), 115–130.

Kumar, R., & Patel, S. (2021c). Data privacy in automated water management systems: Ethical considerations and best practices. *Journal of Information Ethics*, *30*(2), 85–98.

Kumar, R., & Patel, S. (2021d). Ethical considerations in data privacy for water automation systems. *Journal of Information Ethics*, *30*(1), 45–60.

Lee, J., & Xu, H. (2021a). Financial implications of implementing automated water management technologies. *Journal of Environmental Economics*, *12*(4), 217–230.

Lee, J., & Xu, Y. (2021b). Cost-benefit analysis in water automation: Assessing the long-term financial benefits. *Journal of Environmental Economics*, *45*(3), 233–248.

Lee, T., & Xu, Y. (2021c). Financial strategies for sustainable water management automation. *Water Economics and Policy*, *9*(2), 210–225.

Liang, S., et al. (2024a). Data privacy concerns in the age of water automation: Ethical implications and regulatory frameworks. *Journal of Environmental Informatics*, *12*(1), 45–62.

Liang, X., Zhang, Y., & Wang, L. (2024b). Ethical dilemmas in data utilization within automated water management. *Ethics in Water Management*, *12*(1), 34–48.

Miller, A., & Smith, L. (2020a). Bridging the digital divide in water automation: Policy recommendations. *Global Water Journal*, *14*(1), 78–91.

Miller, D., & Smith, A. (2020b). Transparency and accountability in the deployment of automated water systems. *Governance and Technology*, *8*(2), 133–150.

Miller, D., & Smith, J. (2020c). Bridging the digital divide: Ensuring equitable access to water automation technologies. *Journal of Technological Equity*, *18*(3), 210–225.

Miller, T., & Smith, J. (2020d). Promoting equity and inclusion in digital technology access. *International Journal of Digital Society*, *22*(4), 333–350.

Ramirez, F., & Thompson, H. (2023a). Long-term financial viability of automated water management systems. *Water Resource Management*, *13*(4), 219–238.

Ramirez, L., & Thompson, D. (2023b). Labor displacement and social implications of automation in water management. *Journal of Human Resource Management in Utilities*, *9*(3), 215–229.

Smith, A., & Brown, T. (2021a). Modernizing water infrastructure for the digital age. *Journal of Infrastructure Development*, *22*(1), 44–59.

Smith, B., & Brown, C. (2021b). Technical challenges in integrating automation into existing water infrastructure. *International Journal of Water Management*, *14*(3), 172–188.

Smith, J., & Lee, A. (2024a). The ethics of data ownership in automated water management systems. *Journal of Digital Ethics*, *15*(1), 78–92.
Smith, J., & Lee, K. (2024b). Data privacy concerns automated water management. *Journal of Information Security*, *10*(2), 67–84.
Smith, K. (2021). Digital inclusion and the future of water management. *International Journal of Water Resources Development*, *37*(2), 310–328.
Smith, K., & Brown, M. (2021c). Integrating automation technologies into existing water infrastructure: Technical challenges and solutions. *Journal of Engineering for Sustainable Development*, *26*(3), 311–329.
Taylor, B. (2021a). Community-centric approaches to technology implementation. *Community Development Journal*, *19*(1), 87–102.
Taylor, J. (2021b). Equitable access to technology in water management: Challenges and solutions. *Technology and Society*, *5*(1), 59–74.
Taylor, P. (2021c). Protecting community welfare in the age of water automation. *Social Impact Journal*, *15*(2), 133–148.
Taylor, R. (2021d). Social impacts of automation in water management: Strategies for mitigating job displacement. *Journal of Socio-Economic Studies*, *40*(2), 133–150.
UNESCO. (2019a). Addressing the digital divide in developing regions: A focus on water management. *UNESCO Policy Briefs*, *22*(1), 1–15.
UNESCO. (2019b). Addressing the digital divide in water management. *UNESCO Water Report*, *22*, 34–48.
UNESCO. (2019c). *Bridging the digital divide: Ensuring equitable access to technology. UNESCO Report.* https://www.unesco.org/digital-divide.
World Bank. (2022a). Financing models for water management automation in developing countries. *World Bank Report*, *48*, 112–135.
World Bank. (2022b). Financing water automation: Challenges and opportunities. *World Bank Working Papers*, *39*(4), 98–114.
World Bank. (2022). Financing water infrastructure: Strategies for sustainable investment. *World Bank Report*. https://www.worldbank.org/water-infrastructure.
Zhao, H., et al. (2023a). Financial and technical barriers to water automation: A comparative study of developed and developing regions. *Water Science and Technology*, *88*(4), 965–980.
Zhao, L., Wang, Y., & Li, J. (2023b). Technological advancements in water automation and their challenges. *Journal of Automation in Environmental Science*, *19*(2), 89–103.

CHAPTER 20

Synergizing sustainability: integrating computational automation for water security and sustainable tourism

Shikha Dhakad

School of Studies Tourism and Travel Management, Jiwaji University, Gwalior, Madhya Pradesh, India

20.1 Introduction

Water quality regulation is a crucial component of sustainable tourism (Lehmann, 2011; Prakash, Sharma, Singh, & Vijayvargy, 2023); effective control and measurement tools are needed to reduce water use and promote recycling, and tourist accommodations must incorporate sustainable water management techniques These initiatives can be strengthened by implementing the water quality index as an indication of tourist effect management (Fonseca, Rangel, Lima, & Oliveira, 2017). Polycarp (2014) emphasized the responsibility to invest in sufficient water management throughout the value chain to ensure sustainable access to water resources.

Artificial intelligence (AI) and machine learning, in particular, are increasingly being used in computational automation to improve water treatment processes. Al Aani, Bonny, Hasan, and Hilal (2019) and Bongards, Gaida, Trauer, and Wolf (2014) highlighted the use of AI and computational intelligence techniques to optimize desalination and water treatment process. Their research highlighted five distinct applications that are rising at the convergence of environmental engineering and computational intelligence. Furthermore, Icke et al. (2020) demonstrated the practical application of machine learning in enhancing wastewater treatment processes, namely in anomaly detection and predictive control. The automation in water treatment facilities signifies the necessity of automation in water treatment plants to enhance safety and operating conditions (Al-Sum, Sattar, & Aziz, 1993). Hence, computational automation is becoming increasingly important for improving water treatment processes.

Technological innovations, sustainable tourism, and water security are intricately and dynamically interconnected. The evolving network structure of tourism highlights the need

Computational Automation for Water Security.
DOI: https://doi.org/10.1016/B978-0-443-33321-7.00013-5

for efficient business and destination environmental management systems and hotel industry water demand management strategies, stressing the importance of cross-sectoral stakeholder engagement (Gössling et al., 2012; Hu, Ying, Lovelock, & Mager, 2023). Water is a vital resource, attraction, and risk in sustainable tourism, with direct links to a particular UN Sustainable Development Goals (SDGs) (Moyle, Weaver, Gössling, McLennan, & Hadinejad, 2022). There is an interdependent relationship between water and wastewater infrastructure and tourism, which Frone and Frone (2013) highlighted in their study in the context of Romania's goals for developing sustainable tourism and water supply and sanitation by considering the intrinsic connection between environmental sustainability and tourism growth, which draws not only visitors but also preserves natural resources for future generations. Highlighting the involvement of stakeholder participation, policy changes and infrastructure are necessary for integrating sustainable approaches to water management in the tourism industry.

Computational automation has the potential to enhance water quality control in the tourism sector. Numerous companies have looked at the incorporation of computational automation into water quality control in the tourist sector. Likely, Yasruddin, Ismail, Husin, and Tan (2022) and Lin and Zhai (2013) created automated real-time water quality monitoring and regulating systems, with Yasruddin focusing on the aquaculture sector and Lin on reservoirs. The economic benefits and production efficiency were significantly enhanced by these systems. Meanwhile, in the problems and solutions related to automated networks for monitoring water quality, Copp et al. (2010) emphasized the need for fully documented long-term databases of usable data. However, Blomenhofer, Groß, Procelewska, Delgado, and Becher (2013) introduced a hybrid automation approach for managing water quality in the food and beverage business, which could potentially be used in the tourism industry.

20.2 Literature review

The future green transformation is cloud computing and web technology, with an emphasis on the functions of AI, the Internet of Things (IoT), and secure enterprise corporate systems to foster sustainable practices. Organizations are under pressure to be sustainable, and incorporating advanced digital technologies presents unprecedented opportunities to enhance and increase resource efficiency, cut waste, and increase carbon footprints.

The use of computational automation in water quality monitoring is supported by a range of theoretical frameworks and models and highlights the importance of robust, dynamic, and integrated models in computation automation for water quality control. Kartbayev (2023) highlighted the necessity of dynamic verification of control algorithms in cyber-physical systems, particularly in hyperprocesses. Hof and Schuurmans (2000) discussed the application of linear descriptions and control theory to develop dependable control systems for water level and discharge management, while Galelli, Castelletti, and Goedbloed (2015) introduced a high-performance integrated control system that included model predictive control and a 3-D simulation model. Nonlinear programming is used to design waste treatment systems; Marsden, Pingry, and Whinston (1973) demonstrated the use of optimization theory and mathematical models in this field.

Technologies that can revolutionize water treatment processes like machine learning, real-time monitoring, and automated data sampling are revolutionizing water treatment processes. Lowe, Qin, and Mao (2022) emphasize the potential of AI and machine learning in enhancing water treatment and monitoring.

Computational automation has the potential to enhance water sustainability and efficiency in water supply networks connected to tourism. Numerous studies have explored the application of computational automation in tourism-related water supply chains. According to Fernández-Pacheco, Ferrández-Villena, Molina-Martínez, and Ruiz-Canales (2015), irrigation system automation can result in notable reductions in energy and water use, along with brief payback periods. Soboll and Schmude (2011) used agent-based modeling to simulate tourism water consumption under climate change circumstances, offering insights into how automation can affect water demand. Automation has the potential to result in significant water and energy savings. Styles, Schoenberger, and Galvez-Martos (2015) identify best practices and performance benchmarks for water management in the European hospitality industry. Finally, Petnga and Austin (2016) proposed a model-based design in their study and a formal verification procedure for automated water system operations, which could be used to improve the effectiveness of water supply chains in the tourism industry.

While there are many advantages of the IoT and artificial integration in water management, including operational efficiency and resource conservation, there are also ethical and data security issues (Dada et al., 2024). The use of these technologies is impacted by ethical issues and legal frameworks, but they are crucial in detecting and reducing cybersecurity risks in the water industry (Adelani, Okafor, Jacks, & Ajala, 2024). However, information technology can help with these issues, especially when it comes to predicting and preventing water-related disasters with the use of AI and machine learning (Christias & Mocanu, 2019). Nonetheless, the utilization of digital technology in water management necessitates a meticulous assessment of possible hazards, such as cybersecurity vulnerabilities and the requirement for human supervision (Savić, 2022).

20.3 Integrating frameworks

20.3.1 Integrated sustainability frameworks

The study emphasizes the comprehensive approach to sustainability by integrating several sustainability dimensions, including economic, social, and environmental.

Dao, Langella, and Carbo (2011) developed the Integrated Sustainability Framework, which is divided into four quadrants: internal today, external today, internal tomorrow, and external tomorrow. Munda (1997) highlighted numerous aspects of ecological and environmental economics, such as how the trash disposal puts pressure on the ecosystem, addresses environmental challenges such as the ineffectiveness of substituting man-made capital for natural capital, and highlights democratic decision-making as well as multidisciplinary and transdisciplinary approaches for ecological economics. Meanwhile, Sala, Ciuffo, and Nijkamp (2015) gave a methodological framework for sustainability assessment. It consists of assessment techniques and sustainability assessment principles. The assessment process includes defining the approach to sustainability, decision-making, and selecting methodological choices for the assessment, including handling ambiguity and stakeholder involvement. Principles include a clear goal, essential factors, adequate scope, framework and indicators, transparency, effective communications, continuity and capacity, and widespread involvement. Camarillo and Randhir (2023) gave the TCS (tears, consequences, solutions) framework, which was derived from the Multiscale Ecosystem Framework. It was created to evaluate watershed-based assessments. TCS framework seeks to identify the factors influencing stakeholder–resources interactions in local commons and to assess the social and ecological characteristics that influence this common resilience. Wescoat, Murty, Singh, and Verma (2022) gave a Sustainability Planning Framework for rural drinking water systems that included five major criteria of sustainability: source water sustainability, water service sustainability, operations and maintenance (O&M) financial sustainability, village institutional capacity, and asset management. Kimmel (2007) gave a new framework for assessing sustainability in small island developing states, addressing dimensions of sustainability and focused mainly on how tourism affects energy, solid waste, water resources, and the environment in local communities. Starkl, Brunner, Lopez, and Martínez-Ruiz (2013) introduced the Planning-Oriented Sustainability Assessment Framework (POSAF) for peri-urban water management. It highlights stakeholder participation and an integrative approach to evaluate sustainability. The four main components of the POSAF are participatory scenario analysis and planning, technical feasibility study, participatory assessment and analysis of trade-offs, and social network analysis. Chung, Pan, Zou, and Liu (2018) emphasize a conceptual model based on the telecoupling framework for understanding interrelationships in nature-based tourism, which is divided into five major components: systems, agents, flows, causes, and consequences. It focuses on the interaction between socioeconomic and environmental sectors.

The key components of these integrated sustainability frameworks are

1. Integrated sustainability framework: Four quadrants (internal today, external today, internal tomorrow, and external tomorrow).
2. Methodological Framework: Principles and assessment procedure.

3. TCS Framework: threats, consequences, and solutions for managing common watersheds.
4. Sustainability Planning Framework: source water sustainability, water service sustainability, operations and maintenance (O&M) financial sustainability, village institutional capacity, and asset management.
5. Framework for Assessing Sustainability: General indicator categories, application to specific locations, and issues/recommendations.
6. POSAF: Participatory scenario analysis, technical feasibility study, participatory assessment, and social network analysis.
7. Conceptual Model of Integrated Telecoupling Frameworks: Systems, agents, flows, causes, and effects in nature-based tourism.

20.3.2 Emerging frameworks in water management and computational technologies

Harshadeep and Young (2020) introduced a framework for Disruptive Technologies for Water Management, highlighted the need for adopting conventional methods to focus on water security outcomes, discussed how technology affects operations, decision-making, and stakeholder engagement, and evaluated the applications, implications, and institutional responsibilities of disruptive technologies. Govindan and Al-Ansari (2019) discussed numerous frameworks on computational decisions to enhance resilience in the energy-water-food (EWF) nexus, addressing the challenges and interactions focusing on sustainability development, risk management, and climate adaptation. Allen-Dumas, Xu, Kurte, and Rastogi (2021) address machine learning techniques for urban water security, focusing on EWF, water quality, soil erosion, and sediment transport, remotely sensed data for water hazard prediction and mitigation, multi-hazard risk assessment, and best management practice selection. Botta, De Donato, Persico, & Pescapé (2016) incorporated cloud computing, radio-frequency identification (RFID) systems, wireless sensor networks (WSNs) addressing schemes, middleware, and IoT for sustainable critical infrastructures in their study.

The key components of frameworks in Water Management and Computational Technologies are:

1. Disruptive Technologies for Water Management: Impact assessment, applications, implications, and institutional roles.
2. Computational Decision Framework: Enhancing resilience in the EWF nexus.
3. Machine Learning Techniques: Applications in Urban Water Security.
4. Integration of Cloud Computing and IoT: Leveraging RFID systems, WSNs, addressing schemes, middleware, cloud computing, and IoT for critical infrastructure sustainability.

20.3.3 Frameworks for sustainable tourism with emerging technologies

Majid, Tussyadiah, Kim, and Pal (2023) introduced the AI4GoodTourism framework, including four quadrants based on sustainability inclusions and the use of AI-based innovations to provide direction for future technological advancements in the tourist industry. Alsahafi, Alzahrani, and Mehmood (2023) gave a data-driven framework for sustainable tourism to improve sustainability and improving the quality of tourism experiences via the incorporation of data-driven insights and emerging technologies. Alsahafi et al. (2023) highlight objectives such as improving service quality, cultural sustainability, economic sustainability, environmental sustainability, and tourist experiences while also emphasizing tourism challenges such as climate change, air pollution, water quality, security, lack of awareness, and food safety. Kashem, Shamsuddoha, Nasir, and Chowdhury (2022) emphasize the role of AI and blockchain technologies in sustainable tourism in the Middle East and employ a methodological framework that combines the IMRAD framework and elements from Hall's book. Tsaih and Hsu (2018) gave a conceptual framework for implementing AI/ML technology in smart tourism, with an emphasis on creating smart destinations, connecting business ecosystems, and improving visitor experiences. It highlights the importance of breadth, scale, speed, and value generation when putting digital business strategies for smart tourism into practice and addresses important challenges like data privacy and technology adoption.

The key components of frameworks for sustainable tourism with emerging technologies are:

1. AI4GoodTourism Framework: Quadrants based on sustainability inclusion levels and tourist involvement of AI-based innovations.
2. Data-Driven Framework: Objectives, enablers, challenges, and goals for sustainable tourism.
3. Role of AI and Blockchain: Implications for sustainable tourism in the Middle East.
4. AI/ML Implementation Framework: Enhancing tourist experiences, integrating business ecosystems, and developing smart destinations.

20.3.4 Water security framework

Srinivasan, Konar, and Sivapalan (2017) gave a dynamic framework for water security, identifying culture/economy, resource availability, wealth, and governance as the key factors influencing water security in the framework. Also, Srinivasan et al. (2017) highlight the conflict between simplicity and complexity in defining and projecting water security across various periods. Aboelnga, Ribbe, Frechen, and Saghir (2019) introduced the Urban Water Security Assessment Framework, which includes urban water systems, urban water security, assessment criteria, results, and urban water index measurement. Aboelnga et al. (2019)

stressed the holistic approach to address several aspects such as the quality of drinking water, preservation of ecosystems, climate change, and socioeconomic aspects. Penn, Loring, and Schnabel (2017) gave an Environmental Security framework for rural areas. It has four dimensions, which are availability, access, utility, and stability, to assess water security physically and socio-economically to identify areas for improvement. Babel and Shinde (2018) emphasize the Driving forces–Pressure–State–Impact–Response (DPSIR) approach for basin-scale water security assessment given by the Organization for Economic Co-operation and Development (OECD). It involves components such as water availability, water productivity, water-related disasters, watershed health, and water governance. Cook and Bakker (2012) conducted a critical evaluation of water security concepts in which they assessed the variations and similarities of approaches to water security, concentrating on the frameworks, scales, and methodologies employed in various fields.

The key components of water security frameworks are:

1. Dynamic Framework: Factors influencing water security, pathways, and UN SDGs 6 targets.
2. Urban Water Security Assessment: Criteria for urban water systems, assessment methodology, and normalization.
3. Environmental Security Framework: Assessment of availability, access, utility, and stability in rural areas.
4. DPSIR Approach: Cause–effect relationships between environmental and human systems for basin-scale assessment.
5. Critical Review: Analysis of different approaches to water security across academic and policy literature.

20.4 New framework developed: integrated computational sustainability framework for water security and tourism

Integrated Computational Sustainability Framework for Water Security and Tourism (ICSF-WST). This framework gives a comprehensive strategy to enhance water security and encourage sustainable tourism. It addresses the intricate integration of water resources and tourism activities by combining data-driven insights to advance computational technology to maximize sustainability efforts. This proposed framework consists of six components: sustainability dimensions matrix, disruptive technologies integration, methodological and assessment frameworks, holistic water security approach, stakeholder engagement and governance, and environmental and socioeconomic integration.

Each component has been taken from the previously published frameworks by integrating them; likely the first component, which is sustainability dimensions matrix, is made by integrating AI4GoodTourism and integrated sustainability frameworks (Dao et al., 2011;

Majid et al., 2023), including the internal today, external today, internal tomorrow, and external tomorrow quadrants of integrated sustainability framework (Dao et al., 2011). The second component, i.e., disruptive technologies integration, is derived from disruptive technologies for water management (Harshadeep & Young, 2020) and AI/ML implementation frameworks (Tsaih & Hsu, 2018), including AI machine learning, blockchain IoT, and cloud computing. Methodological and Assessment Frameworks is the third component; it incorporates Methodological Framework (Sala et al., 2015) and POSAF (Starkl et al., 2013), which includes three steps: assessment principles, assessment techniques, and sustainability indicators. The water Security Approach component is based on the Urban Water Security Assessment (Aboelnga et al., 2019) and the DPSIR Approach (Babel & Shinde, 2018); it includes dynamic assessments, the DPSIR model, and urban and rural focus. Stakeholder Engagement and Governance, a critical component, is inspired by POSAF (Starkl et al., 2013) and the Sustainability Planning Framework (Wescoat et al., 2022), which includes participatory governance, institutional capacity building, and transparent communication. Environmental and Socioeconomic Integration is the last component; it is adapted from the Environmental Security Framework (Penn et al., 2017) and Data-Driven Frameworks (Alsahafi et al., 2023), including environmental sustainability, socioeconomic benefits, and data-driven insights (Fig. 20.1).

Six Components of ICSF-WST:

1. **Sustainability Dimension Matrix:**

   The Sustainability Dimension Matrix is adapted from principles found in AI4GoodTourism (Majid et al., 2023) and Integrated Sustainability frameworks (Dao et al., 2011) to help in organizing and prioritizing sustainability initiatives across various dimensions and periods. It can give a holistic, comprehensive, and progressive approach to sustainable development, balancing immediate needs with long-term goals. This component is divided into four quadrants, as given in the Integrated Sustainability Framework: internal today, external today, internal tomorrow, and external tomorrow. Each quadrant emphasizes different aspects of sustainability, such as enabling a thorough evaluation and action plan to address present and future challenges and opportunities.

   1.1. **Internal Today:**

   It is the first quadrant, which focuses on current internal processes, policies, and practices for assessing and improving the current water management and tourism operations, implementing sustainable practices within the organization, increasing productivity, and reducing the environmental impact. This includes auditing water usage in tourist facilities, training staff on sustainable tourism practices, and implementing recycling and waste reduction programs.

   1.2. **External Today:**

   External today is the second quadrant, which focuses on current external factors impacting sustainability to address compliance and mandatory regulations,

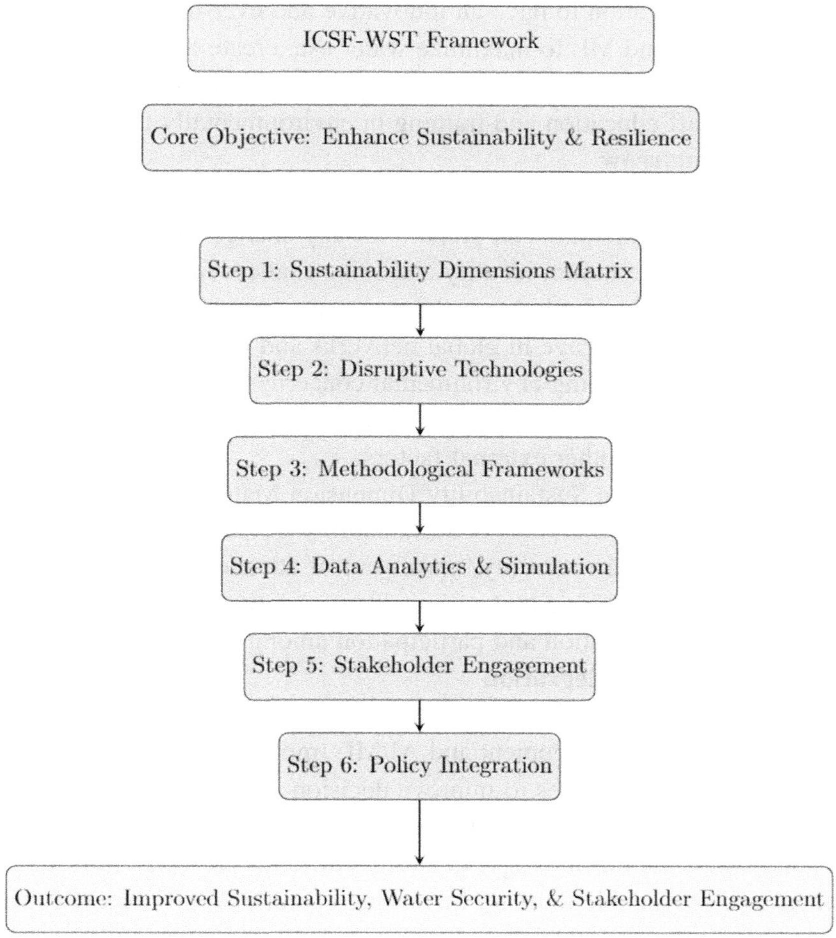

Figure 20.1
Diagram of ICSF-WST framework. *ICSF-WST*, Integrated Computational Sustainability Framework for Water Security and Tourism.

stakeholder engagement such as local communities and government and private agencies, and minimizing the effects of tourism and water activities by collaborating with regional authorities, engaging with local communities, and implementing measures to ensure adherence to water consumption guidelines, understand its needs and concerns related to tourism for reducing pollution, and protect local ecosystems.

1.3. **Internal Tomorrow:**

This third quadrant focuses on the future internal strategies and innovations to plan for the incorporation of advanced technology to improve sustainability, develop long-term strategies for optimizing resources and enhancing operations, foster the organization to have an innovative and ever-improving culture for investing in AI and ML to maximize water use, create a long-term sustainability strategy that includes benchmarks and objectives, and establish programs for continuing staff education and training in environmentally friendly methods.

1.4. **External Tomorrow**

It is the last quadrant, which focuses on future external challenges and opportunities to anticipate and prepare for any changes in regulations and environment in the future, identify and seize chances for collaboration with external stakeholders, and ensure the water management and tourism resilience and adaptability to engage in global networks and projects related to sustainability, developing environmental concerns by collaborating with research institutes, and finally, creating plans for adaptive management plans to respond to climate change and other external factors.

The benefits of the Sustainability Dimension Matrix are that it gives a holistic plan by addressing all aspects of sustainability. It balances short-term needs with long-term goals with versatility, makes use of advanced technology to support effective decision-making and optimized operations, and is stakeholder-centric, encouraging collaboration and participation among stakeholders.

2. **Disruptive technologies integration**

Disruptive technologies integration is derived from principles found in Disruptive Technologies for Water Management and AI/ML Implementation Frameworks. It leverages advanced technologies to improve decision-making and sustainable practices. This component focuses on incorporating advanced technical solutions, including AI, ML, IoT, and blockchain technology, to address difficult problems in water management and sustainable tourism. Increasing operational efficiency, enhancing the ability to make decisions, fostering innovation and adaptability, and ensuring sustainable and resilient practice are the main objectives of this component.

2.1. **Artificial Intelligence and Machine Learning:**

AI and ML are focused on predictive analytics of water demand, tourist influx, and environmental impacts; AI-driven optimization of water distribution systems, energy use, and resource allocation in tourism facilities; automation to predict water demand; and automation monitoring for controlling water quality and infrastructure maintenance.

2.2. **Internet of Things:**

IoT focuses on real-time monitoring, data collection, and integration via sensors and devices to monitor water quality, weather conditions, and tourist movements; environmental parameters, infrastructure performance, and user behavior to data gathering; and finally, incorporating IoT data with AI/ML systems for in-depth analysis and decision-making. Results include installing IoT sensors to continuously monitor water quality in tourist areas, using IoT devices to track energy consumption and optimize HVAC systems in real-time, and implementing smart irrigation systems in parks and resorts to conserve water.

2.3. **Blockchain Technology:**

Blockchain technology focuses on transparency enhancement, security, and efficiency in data management to secure water management and tourist activities data, ensure transparency and safe records of resource usage and sustainability practices, and automate and impose compliance with sustainability standards. This leads to the use of blockchain to securely store and share water usage data among stakeholders, as well as the development of smart contracts to ensure that tourism operations comply with environmental regulations.

2.4. **Cloud Computing:**

Cloud computing focuses on data storage for analytics by availing cloud platforms to collect and analyze data from multiple sources for comprehensive insights, processing via cloud-based platforms, and accessibility to water management systems for real-time data.

2.5. **Remote Sensing and Geographic Information Systems (GIS):**

Remote sensing provides satellite imagery that can be used to monitor land use, water bodies, and environmental changes. Meanwhile, employ GIS for spatial analysis, mapping, and visualization of sustainability metrics to develop interactive maps that visualize sustainability initiatives and their outcomes, as well as to analyze patterns of tourist activities and their environmental impact using GIS.

3. **Methodological and Assessment:**

Methodological assessment is another component that signifies a defined process for evaluating and enhancing sustainability practices, including rigorous methodological principles with practical assessment techniques to make certain thorough and transparent evaluations. The objectives of this component are to create assessment principles, implement effective assessment techniques, and screen sustainability indicators.

3.1. **Assessment Principles:**

This includes enacting clear goals and crucial factors to signify environmental, economic, and social variables by setting targets to reduce water consumption in tourism activities, considering factors such as water availability, tourist behavior, and local economic conditions.

3.2. **Assessment Techniques:**

This includes participatory scenario analysis to organize workshops to create scenarios for water use under various climate changes, as well as technical feasibility to install advanced water recycling systems.

3.3. **Screening Sustainability Indicators:**

To evolve and screen indicators for water resource sustainability, including daily water consumption per tourist, recycled water percentage, water quality index, and tourism impact benefits by establishing indicators, such as carbon footprint of tourism, local employment rates, and tourist satisfaction levels.

4. **Water Security**

   This component incorporates urban water security (Allen-Dumas et al., 2021) and DPSIR (Babel and Shinde, 2018). This component thoroughly addresses water security by incorporating the DPSIR model with dynamic assessments in the context of rural and urban water security influencing factors. This component directs to conduct dynamic assessments for evaluating water security, economy, resource availability, wealth, and governance using the DPSIR (Allen-Dumas et al., 2021) model to inspect the cause–effect relationships in water security and its challenges in urban and rural areas.

5. **Stakeholder Engagement and Governance:**

   This component contains participatory governance and institutional capacity building. Whereas participatory governance includes stakeholders from local government bodies and the private sector in decision-making processes, institutional capacity building focuses on improving and enhancing by giving training and education, resource allocation, and policy development (Starkl et al., 2013).

6. **Environmental and Socioeconomic Integration:**

   This component transformed from an environmental security framework (Penn et al., 2017) to a data-driven framework (Alsahafi et al., 2023). Includes environmental sustainability, socioeconomic benefits, and data-driven insights for conserving ecosystems, sustaining water quality, encouraging economic sustainability, and, finally, tourist water usage patterns.

20.5 Discussion and conclusion

This chapter stresses the integration of sustainability frameworks to address the complicated and mixed nature of sustainable development, especially when it comes to water management and tourism. Amalgamating frameworks such as the Integrated Sustainability Framework (Dao et al., 2011), the Methodological Framework for Sustainability Assessment (Sala et al., 2015), and the Planning-Oriented Sustainability Assessment Framework (Starkl et al., 2013) emphasize the value of comprehensive mythology in assessing sustainability initiatives to efficiently meet present and future demands by taking into account social, economic, and environmental factors. For instance, the TCS framework (Camarillo & Randhir, 2023) and the Sustainability Planning Framework (Wescoat et al., 2022) give vigorous ways for defining rural and watershed-based water systems with an emphasis on stakeholder engagement.

In the realm of water management and computational technologies, this chapter highlights the important roles of disruptive technologies; likewise, frameworks developed by Harshadeep and Young (2020) and Govindan and Al-Ansari (2019) stress the involvement of stakeholder participation can be improved through AI, ML, IoT, and blockchain technology. The use of ML techniques in urban water security highlights the potential of technologies for complex challenges in water management in variety of contexts (Allen-Dumas et al., 2021), as well as the integration of cloud computing and IoT for demanding infrastructure resilience (Botta et al., 2016).

Furthermore, in sustainable tourism, AI4GoodTourim (Majid et al., 2023) and the data-driven framework (Alsahafi et al., 2023) highlight the implication of AI and data-driven insights in improving service quality, cultural and economic sustainability, and environmental protection by integrating advanced technologies. Kashem et al. (2022) pointed out the challenges such as data privacy and technology adoption to demonstrate the role of AI and blockchain in ST, and the conceptual framework for implementing AI/ML technology in smart tourism indicates the potential for advancing destinations as smart destinations to enhance the visitors' experiences.

The vigorous framework for water security (Srinivasan et al., 2017) and the Urban Water Security Assessment Framework (Aboelnga et al., 2019) demonstrated dynamic features of water security as well as holistic and adaptive approaches, taking into account characteristics such as resource availability, governance, and socioeconomic aspects. Babel and Shinde (2018), the DPSIR approach, and The Environmental Security Framework (Penn et al., 2017) illustrate the importance of interconnection between environment and human systems and give a systematic method for evaluating water security.

On top of that, the chapter gave a new framework known as the ICSF-WST. This framework emphasizes a comprehensive approach to address the complex connection among technology, water resources, stakeholders, and tourists during tourism activities. This framework contains six key components: Sustainability Dimensions Matrix, Disruptive Technologies Integration, Methodological and Assessment Frameworks, Water Security Approach, Stakeholder Engagement and Governance, and Environmental and Socioeconomic Integration. Similarly, these components are further subdivided into a few categories. Sustainability Dimensions Matrix has four quadrants that are internal today, external today, internal tomorrow, and external tomorrow (Dao et al., 2011); Disruptive Technologies Integration includes advanced technical solutions including AI, ML, IoT, and blockchain technology; Methodological and Assessment Frameworks include assessment principles, techniques, and screening sustainability indicators; Water Security approach consists of dynamic assessments for evaluating water security, economy, resource availability, wealth and governance, DPSIR (Allen-Dumas et al., 2021) model; Stakeholder Engagement and Governance contains participatory governance and institutional capacity building; and finally, Environmental and Socioeconomic Integration includes environmental sustainability, socioeconomic benefits, and data. This whole framework allows for better and enhanced decision-making, operational efficiency, and stakeholder involvement.

Finally, there are numerous limitations in this chapter; selective and limited literature review was conducted to emphasize the integration of the framework. Moreover, ICSF-WST framework has several limitations; it is likely to be highly dependent on the availability of data and quality of data, as well as the required technological tools for accurate results, and it requires substantial technical expertize and infrastructure for developing AI, ML, IoT, and blockchain.

This study acknowledges several limitations of the ICSF-WST framework, including the availability and quality of data, the development and enhancement of technological integration as the advanced technologies such as AI, ML, IoT, and blockchain need substantial technical expertize and infrastructure, difficulty in finding skilled personnel and a lack of technological resources, and cost and resource constraints for adapting advanced computational technologies, adaptability to local contexts because of limitation of diverse regions and contexts, ethical and privacy concerns emphasizing the concerns of ensuring compliance with legal and ethical standards is important and finally the uncertainty and complexities due to complex nature of water management and tourism systems involve.

References

Aboelnga, H. T., Ribbe, L., Frechen, F. B., & Saghir, J. (2019). Urban water security: Definition and assessment framework. *Resources*, *8*(4), 178.

Adelani, F. A., Okafor, E. S., Jacks, B. S., & Ajala, O. A. (2024). Theoretical frameworks for the role of ai and machine learning in water cybersecurity: Insights from African and US applications. *Computer Science & IT Research Journal*, *5*(3), 681–692.

Al Aani, S., Bonny, T., Hasan, S. W., & Hilal, N. (2019). Can machine language and artificial intelligence revolutionize process automation for water treatment and desalination? *Desalination*, *458*, 84–96.

Allen-Dumas, M. R., Xu, H., Kurte, K. R., & Rastogi, D. (2021). Toward urban water security: Broadening the use of machine learning methods for mitigating urban water hazards. *Frontiers in Water*, *2*, 562304.

Alsahafi, R., Alzahrani, A., & Mehmood, R. (2023). Smarter sustainable tourism: Data-driven multi-perspective parameter discovery for autonomous design and operations. *Sustainability*, *15*(5), 4166.

Al-Sum, E. A., Sattar, A., & Aziz, M. A. (1993). Automation of water treatment plants and its application in power and desalination plants. *Desalination*, *92*(1-3), 309–321.

Babel, M., & Shinde, V. R. (2018). A framework for water security assessment at basin scale. *APN Science Bulletin*, *8*(1), 27–32.

Blomenhofer, V., Groß, F., Procelewska, J., Delgado, A., & Becher, T. (2013). Water quality management in the food and beverage industry by hybrid automation using the example of breweries. *Water Science and Technology: Water Supply*, *13*(2), 427–434.

Bongards, M., Gaida, D., Trauer, O., & Wolf, C. (2014). Intelligent automation and IT for the optimization of renewable energy and wastewater treatment processes. *Energy, Sustainability and Society*, *4*, 1–12.

Botta, A., De Donato, W., Persico, V., & Pescapé, A. (2016). Integration of cloud computing and internet of things: a survey. *Future Generation Computer Systems*, *56*, 684–700.

Camarillo, A. L. Q., & Randhir, T. O. (2023). A sustainability framework based on threats, consequences, and solutions (TCS) for managing watershed commons. *PLoS One*, *18*(12), e0295228.

Christias, P., & Mocanu, M. (2019). Information technology for ethical use of water. In Business Information systems workshops: BIS 2019 international workshops, Seville, Spain, June 26–28, 2019, Revised Papers 22 (pp. 597–607). Springer International Publishing.

Chung, M. G., Pan, T., Zou, X., & Liu, J. (2018). Complex interrelationships between ecosystem services supply and tourism demand: General framework and evidence from the origin of three Asian rivers. *Sustainability*, *10*(12), 4576.

Cook, C., & Bakker, K. (2012). Water security: Debating an emerging paradigm. *Global Environmental Change*, *22*(1), 94–102.

Copp, J.B., Belia, E., Hübner, C., Thron, M., Vanrolleghem, P., & Rieger, L. (2010). Towards the automation of water quality monitoring networks. In *2010 IEEE international conference on automation science and engineering* (pp. 491–496). IEEE.

Dada, M. A., Majemite, M. T., Obaigbena, A., Daraojimba, O. H., Oliha, J. S., & Nwokediegwu, Z. Q. S. (2024). Review of smart water management: IoT and AI in water and wastewater treatment. *World Journal of Advanced Research and Reviews*, *21*(1), 1373–1382.

Dao, V., Langella, I., & Carbo, J. (2011). From green to sustainability: Information Technology and an integrated sustainability framework. *The Journal of Strategic Information Systems*, *20*(1), 63–79.

Fernández-Pacheco, D. G., Ferrández-Villena, M., Molina-Martínez, J. M., & Ruiz-Canales, A. (2015). Performance indicators to assess the implementation of automation in water user associations: A case study in southeast Spain. *Agricultural Water Management*, *151*, 87–92.

Fonseca, E. M., Rangel, C. M., Lima, L. S., & Oliveira, P. S. (2017). *Use of the water quality index as indicator of tourism impact management in paraty*. RIO DE JANEIRO/BRAZIL.

Frone, S. M., & Frone, D. F. (2013). Sustainable tourism and water supply and sanitation development in Romania. *Journal of Tourism and Hospitality Management*, *1*(3), 140–153.

Galelli, S., Castelletti, A., & Goedbloed, A. (2015). High-performance integrated control of water quality and quantity in urban water reservoirs. *Water Resources Research*, *51*(11), 9053–9072.

Gössling, S., Peeters, P., Hall, C. M., Ceron, J. P., Dubois, G., & Scott, D. (2012). Tourism and water use: Supply, demand, and security. An international review. *Tourism Management*, *33*(1), 1–15.

Govindan, R., & Al-Ansari, T. (2019). Computational decision framework for enhancing resilience of the energy, water and food nexus in risky environments. *Renewable and Sustainable Energy Reviews*, *112*, 653–668.

Harshadeep, N. R., & Young, W. (2020). Disruptive technologies for improving water security in large river basins. *Water*, *12*(10), 2783.

Hof, A., & Schuurmans, W. (2000). Water quality control in open channels. *Water Science and Technology*, *42* (1-2), 153–159.

Hu, X., Ying, T., Lovelock, B., & Mager, S. (2023). Sustainable water policymaking for the hotel industry: A longitudinal network analysis of policy documents. *Sustainability*, *15*(22), 15890.

Icke, O., Van Es, D. M., de Koning, M. F., Wuister, J. J. G., Ng, J., Phua, K. M., & Tao, G. (2020). Performance improvement of wastewater treatment processes by application of machine learning. *Water Science and Technology*, *82*(12), 2671–2680.

Kartbayev, A. (2023). An initial study of quality assurance techniques for automated water level control systems. In *E3S web of conferences* (Vol. 402, p. 03039). EDP Sciences.

Kashem, M. A., Shamsuddoha, M., Nasir, T., & Chowdhury, A. A. (2022). The role of artificial intelligence and blockchain technologies in sustainable tourism in the Middle East. Worldwide Hospitality and Tourism Themes (ahead-of-print).

Kimmel, K. (2007). Tourism and the environment in small island developing states: Development of a new framework for assessing sustainability.

Lehmann, L. V. (2011). Tourism and fresh water, an essential blend!. In *CAUTHE* (21st: 2011: Adelaide, S. Aust.) (pp. 532–542). Adelaide, SA: University of South Australia, School of Management.

Lin, S. W., & Zhai, Y. W. (2013). Study on automation control with water quality monitoring system based on GPRS technology. *Advanced Materials Research*, *738*, 239–242.

Lowe, M., Qin, R., & Mao, X. (2022). A review on machine learning, artificial intelligence, and smart technology in water treatment and monitoring. *Water*, *14*(9), 1384.

Majid, G. M., Tussyadiah, I., Kim, Y. R., & Pal, A. (2023). Intelligent automation for sustainable tourism: A systematic review. *Journal of Sustainable Tourism*, *31*(11), 2421–2440.

Marsden, J. R., Pingry, D. E., & Whinston, A. (1973). Application of nonlinear programming to water quality control. *Water, Air, and Soil Pollution*, *2*, 155–169.

Moyle, B. D., Weaver, D. B., Gössling, S., McLennan, C. L., & Hadinejad, A. (2022). Are water-centric themes in sustainable tourism research congruent with the UN Sustainable Development Goals? *Journal of Sustainable Tourism*, *30*(8), 1821–1836.

Munda, G. (1997). Environmental economics, ecological economics, and the concept of sustainable development. *Environmental Values*, *6*(2), 213–233.

Penn, H. J., Loring, P. A., & Schnabel, W. E. (2017). Diagnosing water security in the rural North with an environmental security framework. *Journal of Environmental Management*, *199*, 91–98.

Petnga, L., & Austin, M. (2016). Model-based design and formal verification processes for automated waterway system operations. *Systems*, *4*(2), 23.

Polycarp, A. (2014). Tourism sector responds to water challenges to protect our common future.

Prakash, S., Sharma, V. P., Singh, R., & Vijayvargy, L. (2023). Adopting green and sustainable practices in the hotel industry operations-an analysis of critical performance indicators for improved environmental quality. *Management of Environmental Quality: An International Journal*, *34*(4), 1057–1076.

Sala, S., Ciuffo, B., & Nijkamp, P. (2015). A systemic framework for sustainability assessment. *Ecological Economics*, *119*, 314–325.

Savić, D. (2022). Digital water developments and lessons learned from automation in the car and aircraft industries. *Engineering*, *9*, 35–41.

Soboll, A., & Schmude, J. (2011). Simulating tourism water consumption under climate change conditions using agent-based modeling: The example of ski areas. *Annals of the Association of American Geographers*, *101* (5), 1049–1066.

Srinivasan, V., Konar, M., & Sivapalan, M. (2017). A dynamic framework for water security. *Water Security*, *1*, 12–20.

Starkl, M., Brunner, N., Lopez, E., & Martínez-Ruiz, J. L. (2013). A planning-oriented sustainability assessment framework for peri-urban water management in developing countries. *Water Research*, *47*(20), 7175–7183.

Styles, D., Schoenberger, H., & Galvez-Martos, J. L. (2015). Water management in the European hospitality sector: Best practice, performance benchmarks and improvement potential. *Tourism Management*, *46*, 187–202.

Tsaih, R.H., & Hsu, C.C. (2018). Artificial intelligence in smart tourism: A conceptual framework.

Wescoat, J. L., Jr, Murty, J. V., Singh, R., & Verma, P. (2022). A sustainability planning framework and methods for rural drinking water in Satara District, Maharashtra, India. *Frontiers in Water*, *4*, 804845.

Yasruddin, M. L., Ismail, M. A. H., Husin, Z., & Tan, W. K. (2022). Development of automated real-time water quality monitoring and controlling system in aquarium. In *2022 IEEE 12th symposium on computer applications & industrial electronics (ISCAIE)* (pp. 241–245). IEEE.

CHAPTER 21

The moral current: ethical dilemmas in water automation

Rakhi Chauhan[1] and B. Sundaravadivazhagan[2]

[1]Geeta University, Panipat, Haryana, India, [2]College of Computing and Information Sciences, University of Technology and Applied Sciences-Al Mussanah, Oman

21.1 Introduction

The advent of digital technology, in particular artificial intelligence (AI), has brought about tremendous changes to both our lives and the environment. These changes have been extremely significant. The development of digital technology was the driving force behind this transition. It is generally agreed that the management of the urban water cycle (UWC) is one of the most significant parts of this shift. All of the services that are related with the extraction, delivery, and cleanliness of water inside a city are included in the UWC definition. The management of the underground water cycle in metropolitan regions involves taking into account all of the water resources that are present in these locations. There are several different kinds of water resources that fall under this category.

These include natural surface water, groundwater, drinking water, sewage, stormwater, flood overflow water, and recycled water from managed aquifer recharge and stormwater harvesting (Oral et al., 2020). A significant amount of change has taken place in the administration of the University of Washington during the course of the past several years. The implementation of these modifications includes the utilization of sophisticated automation and monitoring methods, as well as the utilization of AI data sources and knowledge management systems.

During the past few years, the phenomenon commonly referred to as the "digital revolution" has garnered a significant amount of attention, according to experts and professionals. The investigation that is now being carried out to better understand this transition can be broken down into three primary categories.

The following is a list of the categories: In the beginning, technology companies that are developing digital solutions for the water industry give a primarily positive viewpoint, putting an emphasis on the advantages of digitalization (Willi et al., 2012). Experts in water, environmental

Computational Automation for Water Security.
DOI: https://doi.org/10.1016/B978-0-443-33321-7.00005-6

engineering, and data science, who mostly concentrate on developing and updating existing technologies (Corominas et al., 2018; Knüsel et al., 2019; Sun & Scanlon, 2019), believe that digitization is a trend that has inherent benefits. Another location that is of interest is this one.

As a result of the findings of the study, the incorporation of AI, Big Data science, and water management has emerged as one of the most significant developments in the industry. This approach places a strong emphasis on the junction of ethical and legal issues, despite the fact that there is relatively little debate on these subjects being discussed. There is a new option that involves the capability of analyzing, organizing, and extracting knowledge from vast amounts of heterogeneous data (Garrido-Baserba, Corominas, Cortés, Rosso, & Poch, 2020; Poch et al., 2020).

An increasing corpus of research is tackling the most problematic aspects of the process of digitalizing water (Doorn, 2020). Within the confines of this research, the aspects of the procedure that can be considered the most problematic are being investigated. A computerized water environment is being developed, and its development is being carried out. In spite of the fact that this is the case, there is still a lack of clarity about research on the larger social implications that this process has. It is remarkable that there is a dearth of significant study on this area, considering the extensive interest in digitization and the abundance of scholarly articles focused on larger aspects of AI (Fig. 21.1).

There are considerable concerns involved with the unrestricted use of AI applications, as highlighted by specialists in the disciplines of ethics, politics, and technology. The whole literary oeuvres of renowned academics such as Shoshana Zuboff, Noah Yuval Harari, and Byung-Chul Han have been devoted to the examination of the potential adverse impacts that AI may have, frequently evoking scenarios of catastrophic proportions through their writing. As indicated by sources, there has been a discernible increase in the number of references to the topic, which is another sign of the growing emphasis on ethics in AI. The insufficient scrutiny of AI's application in water management is particularly perplexing given the significance and extensive prevalence of water. Water is not only a vital source of sustenance but also a medium for transmitting information and, due to its growing scarcity, a subject of political contention. This study seeks to provide a thorough and all-encompassing analysis of the effects of digitalization on the water management sector, considering both the advantageous and disadvantageous aspects. While there are predominantly positive reports regarding the use of AI in this field, it is important to acknowledge the possible drawbacks and challenges that come with such a significant change.

21.2 Literature review

Precision agriculture is at the forefront of a revolutionary era in modern farming, driven by the integration of advanced artificial intelligence technology. Environmentally friendly

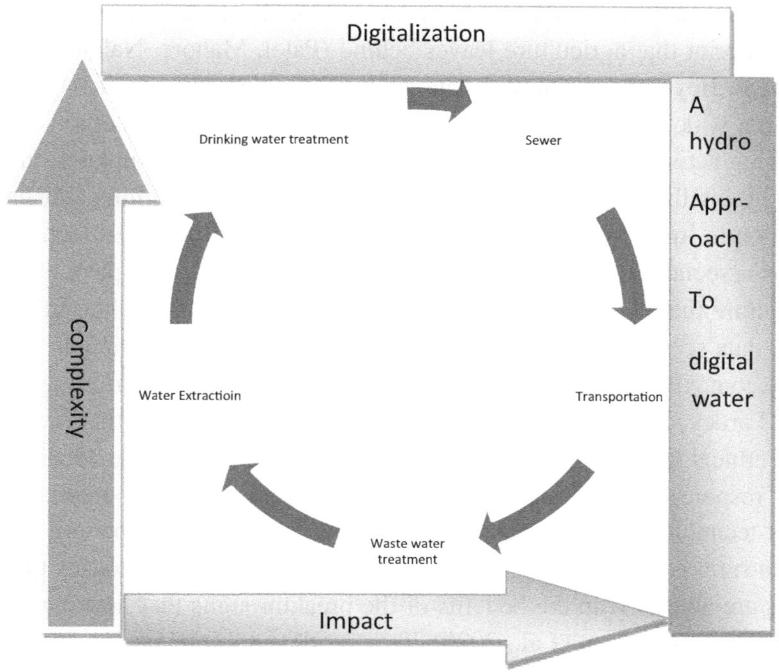

Figure 21.1
The impact of digitization on the UWC. *UWC*, Urban water cycle.

farming practices are becoming more and more necessary as the world's population and environmental concerns continue to rise. One potential solution to these challenges is the integration of AI with precision agriculture (Karunathilake, Le, Heo, Chung, & Mansoor, 2023). The application of AI in precision agriculture is centered on the utilization of advanced computing techniques, machine learning algorithms, computer vision, and sensor technologies to facilitate decision-making processes that are driven by data. The purpose of this analysis is to give a complete exploration of the multidimensional role that AI plays in transforming conventional agricultural practices. The review will focus on the technologies that are important and their aggregate impact on reaching sustainability goals. According to Misra and Ghosh (2024), the paradigm change from conventional, homogenous agricultural practices to a more individualized and adaptable approach is one of the most important aspects of the incorporation of AI in precision agriculture. This change is made possible by the real-time data collection capabilities of AI-driven technologies, such as drones, satellites, and ground-based sensors. These technologies provide farmers with extensive insights about the health of their crops, the conditions of the soil, and the elements that affect the environment. Farmers are able to make educated decisions regarding irrigation, fertilizer, and pest control when they have access to this plethora of information.

This allows them to minimize waste, maximize resource allocation, and reduce the environmental imprint that agriculture leaves behind (Patel, Mahore, Nalawade, Upadhyay, & Choudhary, 2023). Beyond the realm of data analysis, the applications of AI in precision agriculture span a wide range of activities, including robotics and autonomous systems. All aspects of farming activities, including planting, harvesting, and crop upkeep, are being revolutionized by intelligent machines that are loaded with AI algorithms (Mishra & Mishra, 2023). In addition to improving operational efficiency, these autonomous technologies also solve the issues associated with manpower shortages. As a result, they pave the way for a future in agriculture that is both more sustainable and economically viable. Ethical considerations, data security, and equal access to technology are emerging as significant concerns in the context of precision agriculture, which is becoming increasingly data-centric (Wilgenbusch, Pardey, Hospodarsky, & Lynch, 2022). The purpose of this review is to investigate the ethical implications of AI in agriculture, with a particular focus on the significance of responsible data management and the consideration of potential inequities in the adoption of technology. In addition, it investigates the difficulties that are related with the digital divide in rural regions, highlighting the necessity of inclusive policies that guarantee that all farmers are able to reap the benefits of the breakthroughs that have been made in precision agriculture (Robinson et al., 2020). To provide a comprehensive analysis of AI in precision agriculture, the purpose of this review is to provide insights into the technologies that are shaping sustainable agricultural methods. The purpose of this investigation is to contribute to the ongoing discussion on the utilization of advanced technologies to meet the growing demands of global food production while simultaneously ensuring environmental stewardship and long-term agricultural sustainability. This is accomplished by providing an explanation of the current state of AI integration, the challenges that are being faced, and the future prospects. According to researchers (Rodríguez-Maese, Rodríguez-Saldaña, Ferrer, & Leal, 2024), all sensors were calibrated and automated using Arduino Mega and C++. Descriptive statistical analysis was done to understand soil moisture data. The gravimeter approach matched equipment data. The equipment was validated using R^2, r, and root mean square error. According to a research (Zolotov, Ualiev, Ospanov, & Richart, 2024), it is possible to save resources by automating the water supply system for a community, which also improves safety and practicality. In addition to this, the paper discusses the successful installation of automated systems in residential complexes, as well as the development of these systems in the future.

21.3 Ethical considerations in AI-driven agriculture

In the field of agriculture, the use of AI leads to the introduction of breakthroughs that were previously imagined. These innovations, in turn, lead to the transformation of traditional farming practices. On the other hand, issues of ethics are becoming more and more significant as the agricultural environment continues to undergo transformations as a consequence of the introduction of new technology. The goal of this inquiry is to investigate the ethical challenges that are associated with agriculture that is driven by AI, with a specific emphasis on the requirement of responsible deployment and the consideration of potential social ramifications. AI-driven agricultural activities generate a massive amount of data, which includes information about crops, weather patterns, and farm management processes. These data may be used to make informed decisions about how to manage the farm.

The confidentiality of the data is called into question as a result of this. It is of the utmost importance that farmers and other stakeholders take the necessary steps to ensure that sensitive information is stored in a secure way and that individuals have control over how their data are utilized. Both the ownership of agricultural data and the distribution of such data create issues from an ethical standpoint. The establishment of clear standards about the rights to ownership of data, as well as the establishment of protocols for fair data sharing, is necessary to foster collaboration without compromising individual interests. Farmers, technology providers, and researchers are all responsible for establishing these norms. Inadvertently leading to farm surveillance is a possibility that can occur from the employment of modern monitoring technology such as satellite imagery and drones. This is a possibility that can arise. It is vital to establish a medium ground between monitoring the health of crops and respecting the privacy of farmers to prevent unwarranted intrusion. This is because the intent is to prevent unwarranted intrusion. When it comes to the consequences that monitoring has on individuals and groups at the local level, ethical considerations should also be taken into account. When it comes to the process of deploying AI technology, it is essential to be cognizant of cultural norms, to seek community consensus, and to take into consideration the potential implications that data collection could have on the social fabric of agricultural communities.

The use of AI should answer concerns over equitable access to technology when it comes to the agricultural sector. That both large-scale operations and small-scale farming operations have access to and are able to profit from innovations led by AI is important to prevent current disparities from becoming even more pronounced. This is necessary to prevent existing inequities from becoming even more pronounced. It is possible that ethical difficulties could arise as a result of the huge computer literacy gap that exists among farmers. To ensure that farmers, regardless of their level of technological competence, are able to effectively navigate the AI-driven agricultural landscape and make decisions that are based on accurate information, efforts should be made to provide training and support. This will ensure that farmers are able to make decisions that involve accurate information. When farming activities are automated through the use of technology powered by AI, there is the possibility for the displacement of agricultural labor, which raises ethical concerns. These concerns are related to the possible displacement of agricultural labor. The minimization of the impact on employment and the provision of a good transition for workers who are affected by the technological improvement should be elements that are integral to the ethical issues that are associated with agriculture that is driven by AI. AI should be implemented in a manner that respects and fits with the cultural and ethical norms of the community in which it is being applied.

To foster acceptance and prevent cultural disruptions, the adoption of agricultural technology should be carried out in a manner that is coherent with the values that inform the communities that they serve. This should be done with the intention of fostering acceptance. The fact that AI systems are shrouded in mystery has given rise to concerns over responsibility. Because of the establishment of openness in the algorithms that are utilized in decision support systems, it is guaranteed that farmers will be able to comprehend the foundation upon which recommendations are founded and will be able to have faith in the technology.

The duty of establishing comprehensive regulatory frameworks to manage the ethical use of AI in agriculture is one that falls on the shoulders of international institutions and national governments. These frameworks should address concerns surrounding the privacy of data, the transparency of algorithms, and the responsible deployment of AI technologies. It is crucial that these frameworks address these concerns. Even though this sort of agriculture holds immense promise for enhancing both productivity and sustainability, it is vital to successfully negotiate the deployment of AI-driven agriculture with ethical considerations at the forefront. This is the case despite the fact that this style of agriculture holds both of these promises. It is vital to establish a balance between innovation and responsibility to guarantee that technologies based on AI will continue to make a positive contribution to agriculture while also safeguarding the interests of farmers, communities, and society as a whole. Ethical concerns should be incorporated as an essential component into the ongoing discourse that surrounds the possible applications of AI in agriculture. This is necessary to achieve the goal of establishing an agricultural environment that is both resilient and equitable.

21.4 Ethical conundrums in water automation with artificial intelligence

There has been an increase in the number of techniques that are based on AI in recent years. In addition to this, there have been programs that have been launched to provide direction on how to develop "responsible AI" that is in line with human and ethical principles. The employment of AI-based approaches in the water domain is quite minimal when compared to other industries such as the energy sector, the healthcare industry, or the transportation industry.

Among the many serious moral questions raised by AI are the following: The ideas of impartiality and prejudice: It is possible for AI systems to pick up and may be magnify biases in the data used to train them. This has the potential to produce biased or unfair outcomes, especially in the areas of hiring, financing, and police officer applications. Water utilities can become more efficient and provide higher-quality services by incorporating AI into their operations.

This has several policy consequences. Among these ramifications is the fact that ethics and governance play a significant role in ensuring the security of personal and financial data held by water companies' clients, in addition to technical and financial data. Creating a risk-impact matrix for ethical dilemmas in water automation using AI involves identifying potential risks, assessing their likelihood, and evaluating the potential impact. Here is an example of how such a matrix might be structured, and that is described using Table 21.1. Two graphs have been created based on this table and are shown below.

The ethical challenges that arise in water automation require finding a balance between efficiency and fairness, which is influenced by the reliability of the technology and the equal allocation of resources. This is an obligatory condition that must be fulfilled (Fig. 21.1).

In terms of the accuracy and fairness of risk rankings, which have an impact on equitable access and resource allocation, the water automation business offers a dilemma from an ethical standpoint.

21.5 Challenges and future prospects

There are a number of fundamental ethical concerns that are related with automation, one of which is the phenomenon of job displacement because of automation. The ability of machines and algorithms to carry out tasks that were previously carried out by people is growing at an alarming rate. The reason behind this is that the capabilities of machines and algorithms are continuously expanding. This has the potential to result in significant employment losses, particularly in a variety of businesses that are dependent on tasks that are routine and repetitive in nature. It is possible to establish a set of ethical principles that can normatively guide water management and public health policy by introducing the ethical values of efficiency, equality, and stewardship. These values can serve as the foundation for their establishment. This is because AI is continuing to greatly revolutionize precision agriculture. This is the reason why this is happening. As the new era of sustainable farming techniques is getting launched, there are a number of challenges and exciting future prospects that are on the horizon. In addition, all of these things are happening simultaneously. All of the issues that AI is now confronting in the field of precision agriculture are thoroughly studied within the scope of this research. Additionally, the possible pathways for future advances are also thoroughly investigated. Many data sources, including satellite images, sensor data, and historical records, present a problem when it comes to the development of a reliable and consistent data set for AI algorithms. This challenge is posed by the quality of the data and the integration of these data sources. It is becoming increasingly vital to address concerns surrounding the privacy and security

Table 21.1: Ethical conundrums in water automation with artificial intelligence: a risk-impact matrix.

Ethical dilemma	Risk description	Likelihood (1–5)	Impact (1–5)	Risk score ($L \times I$)	Mitigation strategy
Data privacy	Unauthorized access to sensitive water usage data	4	5	20	Implement strong encryption and access controls
Bias in AI algorithms	Discriminatory practices in water distribution	3	4	12	Regular audits and bias testing in algorithms
System failure	AI system errors causing water supply disruptions	2	5	10	Establish robust monitoring and backup systems
Accountability and transparency	Lack of clarity on decision-making processes by AI	4	4	16	Ensure transparent AI decision documentation
Resource allocation inequity	Inequitable water distribution favoring certain areas	3	4	12	Implement fairness checks and community feedback mechanisms
Environmental impact	AI decisions leading to environmental degradation	2	5	10	Integrate environmental impact assessments into AI decision processes
Consent and autonomy	Use of AI without informed consent of affected communities	3	4	12	Ensure community engagement and consent processes
Security threats	Vulnerability to cyber-attacks on water infrastructure	5	5	25	Enhance cyber security measures and regular penetration testing
Reliability of AI predictions	Inaccurate predictions affecting water resource planning.	3	4	12	Validate AI models with real-world data and expert review

Note: *AI*, Artificial intelligence.

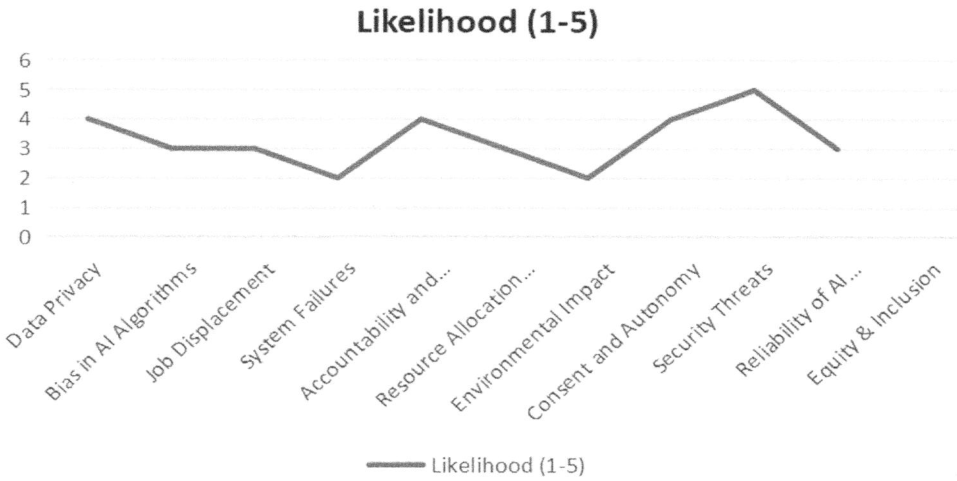

Figure 21.2
The ethical dilemma and the link between likelihood.

of data, particularly in light of the fact that the quantity of sensitive agricultural data that is being collected is continuing to expand even further. There are a number of challenges that must be overcome to guarantee equitable access to AI technology. This is especially true for small-scale farmers, who may not have the resources or the level of digital literacy that is necessary for successful adoption. To guarantee that the algorithms used in AI are both objective and comprehensible, there is a complex obstacle to overcome. This is because it is possible for biases to be introduced accidentally during the training of models, which can lead to inaccurate conclusions being drawn. Because of the challenges connected with network connectivity and access to current technology, the widespread implementation of AI-driven precision agriculture may be restricted in regions that have a limited technological infrastructure. Edge computing technologies, which enable data processing to take place closer to the source, can assist overcome limitations in infrastructure, which can lead to a reduction in the amount of dependence on centralized computing resources. This can be accomplished through the development of various edge computing technologies. To ensure that farmers are able to comprehend and have faith in the recommendations that are provided by AI systems, the development of Explainable AI methodology has the possibility to overcome the issue of algorithmic interpretability. This is done to ensure that farmers are able to have faith in the recommendations that are provided by AI systems. The usage of blockchain technology has the potential to increase data security and privacy by establishing a system that is both decentralized and resistant to tampering. This system would be employed for the management of agricultural data (Fig. 21.3).

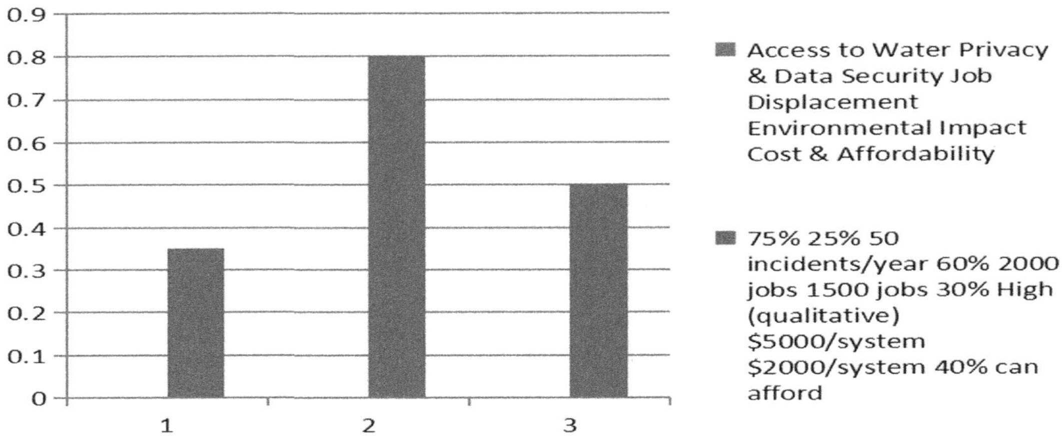

Figure 21.3
The ethical quandary and the connection between risk scores.

It is possible to acquire an understanding of the most effective methods in AI-driven precision agriculture by establishing collaborative platforms for the purpose of conducting research and exchanging knowledge. When it comes to resolving challenges that are associated with the quality of the data, this might be of assistance. In an effort to ensure that AI is used in a manner that is both ethical and responsible, it is possible to address concerns surrounding data privacy, security, and fairness by developing thorough legislative and regulatory frameworks. These frameworks can serve as a road map for the adoption of AI in agriculture. The digital divide has the potential to be overcome through the promotion of inclusive technology adoption programs that place an emphasis on digital literacy and offer help to small-scale farmers. This could open up opportunities for the digital divide to be closed. The developing solutions that are holistic and sustainable to handle the multiplicity of difficulties that AI presents in precision agriculture involves the combining of expertize from a range of fields, such as agriculture, computer science, ethics, and policymaking. This is necessary to handle AI's impact on precision agriculture. It is essential to involve farmers, technology developers, policymakers, and researchers in ongoing talks concerning AI technologies. This will ensure that the development and deployment of these technologies are in accordance with the requirements and ideals of the agricultural community. When it comes to precision agriculture, the challenges that AI must overcome present opportunities for innovation and advancement. The development of responsible frameworks, the advancement of technology, and the encouragement of collaboration among many stakeholders are the three fundamental factors that will determine the future possibilities. For the purpose of playing a significant part in the development of a future for agricultural methods around the world that is both environmentally sustainable and technologically advanced, it is vital that these challenges be solved and that the potential of AI in precision agriculture be embraced.

21.6 Discussion

In modern world, we have many digital technologies like Big Data and AI, and they have the ability to manage water resources. Many automated systems with any feedback loops are accurate and good, which also reduce the chances of error. The main implications of water systems in modern cities are that they are safe if they are used effectively. If it comes to managing water supplies, then the automated systems can be proactive. It can easily cut down the waste and make itdealing faster with problem. The limitation of water management has not been fully explored for its social consequences, according to the report. The ethical, legal, and community rights are ignored. For future work, we have to look into these issues. The collaboration across many disciplines can make a fair and secure digital transition in water management. The digital water management is more efficient and precise. This study stresses the need of understanding automation and balancing the benefits of technology. With the help of digital technologies, we can get more efficient and accurate results. This study emphasizes the automation of social and technological impacts of water automations. It is difficult to manage the ambiguity and complexity of water management. The researchers should address these social and technological impacts of water automation in a comprehensive framework to avoid operational approaches.

21.7 Conclusion

The objective of this study is to provide a comprehensive examination of the ethical and regulatory factors associated with the integration of digital technology in the water industry, specifically focusing on the urban water cycle. Applying the theoretical framework of urban political ecology and the study of hydrosocial cycles, we analyzed how digitization has affected the technical, operational, and sociopolitical components of the urban water cycle. Following discussions with specialists in the field, the applications were ultimately selected. The purpose of this essay was not to give a detailed study of the "digital water" problem but rather to emphasize the proper technique we should use to address this issue. The findings suggest that digitization has the potential to positively affect the urban water cycle. However, it may also raise concerns over individual and communal rights. Throughout the conversation, we suggest exploring other research prospects that examine the digitization of water from a critical, theory-driven, and multidisciplinary perspective. Undoubtedly, the integration of various disciplines is essential to ensure a secure and fair digital transition. This essay presents evidence to support its core argument by merging the opinions of specialists who are professionals in the fields of political science, water engineering, and AI. This initial attempt to integrate diverse viewpoints has the capacity to establish the foundation for a relevant discourse within the water business. This conversation would help prevent the adoption of wasteful operating habits, which might be extremely challenging to correct.

References

Corominas, L., Garrido-Baserba, M., Villez, K., Olsson, G., Cortés, U., & Poch, M. (2018). Transforming data into knowledge for improved wastewater treatment operation: A critical review of techniques. *Environmental Modelling & Software*, *106*, 89–103, [CrossRef].

Doorn, N. (2020). Artificial intelligence in the water domain: Opportunities for responsible use. *Science of the Total Environment*, *755*(142561), [CrossRef].

Garrido-Baserba, M., Corominas, L., Cortés, U., Rosso, D., & Poch, M. (2020). The fourth revolution in the water sector encounters the digital revolution. *Environmental Science & Technology*, *54*, 4698–4705, [CrossRef] [PubMed].

Karunathilake, E. M. B. M., Le, A. T., Heo, S., Chung, Y. S., & Mansoor, S. (2023). The path to smart farming: Innovations and opportunities in precision agriculture. *Agriculture*, *13*(8), 1593.

Knüsel, B., Zumwald, M., Baumberger, C., Hadorn, G. H., Fischer, E. M., Bresch, D. N., & Knutti, R. (2019). Applying big data beyond small problems in climate research. *Nature Climate Change*, *9*, 196–202, [CrossRef].

Mishra, H., & Mishra, D. (2023). Artificial intelligence and machine learning in agriculture: Transforming farming systems. *Research Trends in Agricultural Science*, *1*, 1–16.

Misra, S., & Ghosh, A. (2024). Agriculture paradigm shift: A journey from traditional to modern agriculture. *Biodiversity and bioeconomy* (pp. 113–141). Elsevier.

Oral, H. V., Carvalho, P., Gajewska, M., Ursino, N., Masi, F., van Hullebusch, E. D., Kazak, J. K., Exposito, A., Cipolletta, G., Andersen, T. R., et al. (2020). A review of nature-based solutions for urban water management in European circular cities: A critical assessment based on case studies and literature. *Blue-Green Systems*, *2*, 112–136, [CrossRef].

Patel, A., Mahore, A., Nalawade, R. D., Upadhyay, A., & Choudhary, V. (2023). Advancements in precision agriculture: Harnessing the power of artificial intelligence and drones in indian agriculture. *World Environment Day*, 43.

Poch, M., Garrido-Baserba, M., Corominas, L., Perelló-Moragues, A., Monclús, H., Cermerón-Romero, M., . . . Rosso, D. (2020). When the fourth water and digital revolution encountered COVID-19. *Science of the Total Environment*, *744*(140980), [CrossRef].

Robinson, L., Schulz, J., Dodel, M., Correa, T., Villanueva-Mansilla, E., Leal, S., . . . McMahon, R. (2020). Digital inclusion across the Americas and Caribbean. *Social Inclusion*, *8*(2), 244–259.

Rodríguez-Maese, R., Rodríguez-Saldaña, V., Ferrer, L., & Leal, L. O. (2024). The role of automation in the analysis of manganese in environmental water samples. *Water*.

Sun, A. Y., & Scanlon, B. R. (2019). How can Big Data and machine learning benefit environment and water management: A survey of methods, applications, and future directions. *Environmental Research Letters*, *14* (073001), [CrossRef].

Wilgenbusch, J. C., Pardey, P. G., Hospodarsky, N., & Lynch, B. J. (2022). Addressing new data privacy realities affecting agricultural research and development: A tiered-risk, standards-based approach,". *Agronomy Journal*, *114*(5), 2653–2668.

Zolotov, A. D., Ualiev, A. E., Ospanov, E. A., & Richart, A. (2024). Automation and control of the water supply system in a residential complex. *Bulletin of Shakarim University Technical Sciences*.

CHAPTER 22

Applications of machine learning and artificial intelligence in automating water quality monitoring, analysis, and management

Mamta[1] and B. Sundaravadivazhagan[2]

[1]Department of CSE, Chandigarh University, Mohali, Punjab, India, [2]College of Computing and Information Sciences, University of Technology and Applied Sciences-Al Mussanah, Oman

22.1 Introduction

Water is the most precious resource on our planet. It is not just something we drink; it is essential for cooking, cleaning, growing food, and supporting all life. But here iss the catch: not all water is safe to use. Some water sources might look clean but could be hiding harmful substances that can make us sick. This is why monitoring water quality is so crucial (Moretti, Ivan, & Skvaril, 2024).

In the past, checking water quality was a slow and challenging process. Scientists had to collect water samples, take them to a lab, and run many tests. This took a lot of time and effort, making it hard to keep up with the constant need for clean water. But now, we have some powerful helpers: machine learning (ML) and artificial intelligence (AI). Think of ML and AI as super-smart assistants (Nations, 2015). They can quickly analyze large amounts of data, spot patterns that humans might miss, and even predict future problems. In the world of water quality, these technologies are game-changers. They are helping us automate the process of monitoring, analyzing, and managing water quality, making it faster, more accurate, and more efficient than ever before.

22.1.1 Importance of water quality monitoring

(Fig. 22.1)

Why is water quality monitoring so important? Here are some key reasons:

- **Health:** Contaminated water can carry diseases like cholera, dysentery, and typhoid. These can lead to severe illnesses and even death, especially in children and the elderly. Regular monitoring helps detect these threats early.

Computational Automation for Water Security.
DOI: https://doi.org/10.1016/B978-0-443-33321-7.00014-7

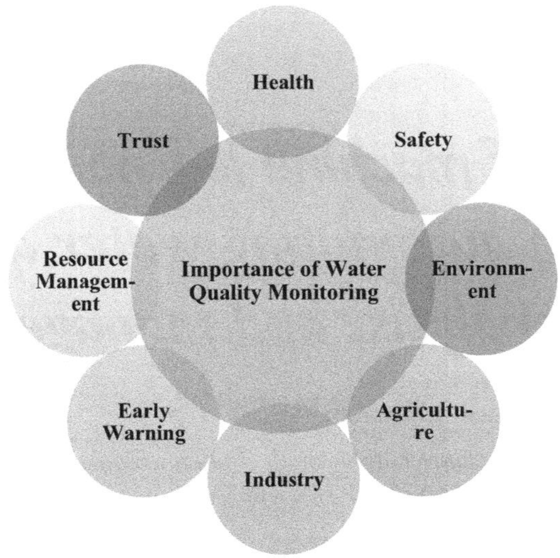

Figure 22.1
Key aspects related to the importance of water quality monitoring.

- **Safety:** Some water pollutants, like heavy metals (lead, mercury) or chemicals, do not always change how water looks or tastes. Without testing, we might not know they are there until people get sick.
- **Environment:** Water is not just for humans. Fish, plants, and other wildlife depend on clean water. Poor water quality can harm entire ecosystems.
- **Agriculture:** Farmers need good water to grow crops and raise animals. Polluted water can stunt plant growth or make livestock ill.
- **Industry:** Many industries, from food processing to electronics, require high-quality water. Impurities can damage equipment or ruin products.
- **Early Warning:** Regular monitoring can alert us to new pollution sources. For example, if we suddenly detect chemicals in a river, we can trace them back to the source and stop further contamination.
- **Resource Management:** Knowing water quality helps us decide how to use each water source best. A lake might be great for fishing but needs treatment before it is safe for drinking.
- **Trust:** When people see that their water is regularly tested and safe, they trust their water supply more. This trust is vital for public health and community well-being (Water, 2021).

22.1.2 Role of ML and AI in water quality automation

(Fig. 22.2)

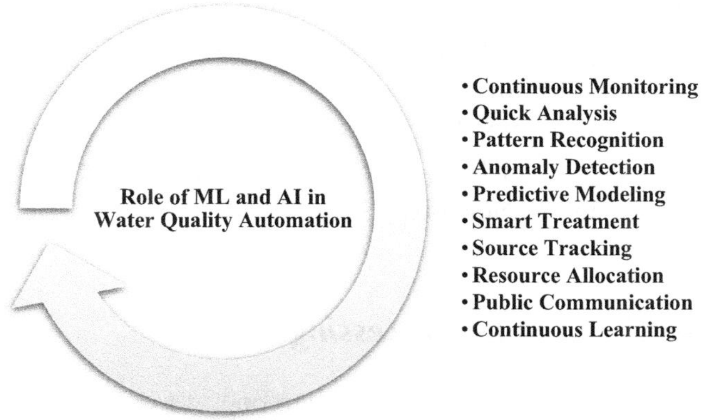

Figure 22.2
Role of ML and AI in water quality automation. *ML*, Machine learning; *AI*, artificial intelligence.

Now, let us see how ML and AI are revolutionizing water quality management:

- **Continuous Monitoring:** Traditional methods test water occasionally. We can monitor water quality 24/7 with AI-powered sensors, instantly detecting any changes.
- **Quick Analysis:** ML algorithms can analyze water samples much faster than humans. What once took days in a lab now takes minutes, allowing for rapid response to issues.
- **Pattern Recognition:** Water quality is not random. It follows patterns based on weather, seasons, and human activities. ML excels at spotting these patterns, helping predict problems before they occur.
- **Anomaly Detection:** AI can quickly spot anything unusual in water data. If a river suddenly has higher acidity, the system flags it immediately, even if human analysts might overlook the change.
- **Predictive Modeling:** Using past data, ML can forecast future water quality. For example, it might predict algae blooms in a lake next month, allowing preventive actions.
- **Smart Treatment:** AI can optimize water treatment plants. Analyzing incoming water quality, it adjusts real-time treatment processes, ensuring safety while saving energy and chemicals.
- **Source Tracking:** When contaminants are found, AI analyzes their "fingerprint" against a database. This helps pinpoint the pollution source—a factory, farm, or natural occurrence.
- **Resource Allocation:** ML helps decide where to focus efforts. It might show that a certain river needs more frequent testing or that a particular pollutant is the biggest threat in an area.

- **Public Communication:** AI can translate complex water data into easy-to-understand alerts. Instead of confusing charts, people get clear messages like "Boil water before drinking" or "Water safe for all uses."
- **Continuous Learning:** The more data these systems process, the smarter they become. An AI that monitors city water for a year will be even better at detecting issues in its second year (Shoushtarian & Negahban-Azar, 2020).

22.2 Data acquisition and preprocessing

In our quest to ensure safe and clean water for everyone, we need to understand what is in our water. This requires gathering a lot of information about water quality and making sure that information is accurate and useful. This process is called "Data Acquisition and Preprocessing" (Yang et al., 2019).

22.2.1 Sensor networks for water quality monitoring

Imagine having a super-smart water tester that never gets tired and can be in many places at once. That is what sensor networks are in water quality monitoring.

- **What They Are**
  - Small, smart devices placed in water sources (rivers, lakes, and taps).
  - Each sensor tests different water properties (cleanliness, safety).
- **Always Working**
  - Unlike humans, these sensors work day and night.
  - They check the water quality every few minutes.
- **Team Effort**
  - Many sensors work together, like a team.
  - If one sensor near a factory sees a problem, it warns others downstream.
- **What They Check**
  - Some look for chemicals (chlorine, fluoride).
  - Others check temperature, cloudiness, or acidity.
  - Advanced ones spot bacteria or metals.
- **Quick Alerts**
  - If the water becomes unsafe, sensors send instant messages.
  - Water plant workers can fix issues right away.
- **Low Upkeep**
  - These sensors need little care.
  - Some even clean themselves or use solar power.

22.2.2 Data cleaning and preprocessing

Our sensors collect tons of information, but not all of it is perfect. Some data might be wrong or messy (Shoushtarian & Negahban-Azar, 2020). "Data Cleaning and Preprocessing" is about fixing these issues.

- **Finding Mistakes**
  - Sometimes sensors report impossible things (like water being hotter than boiling).
  - We use tools to spot and remove these errors.
- **Filling Gaps**
  - A sensor might stop working briefly, leaving data gaps.
  - We use smart guesses based on past data to fill these gaps.
- **Smoothing Out Jitters**
  - Sensors can be jumpy, showing tiny, unimportant changes.
  - We use methods to smooth these out, showing the real trend.
- **Making Data Equal**
  - One sensor counts from 0 to 100, another from 0 to 14.
  - We convert all data to a standard scale (like 0–1) for fair comparison.
- **Time Matching**
  - Sensor A measures every minute, and B every 5 minutes.
  - We adjust data so they match in time.
- **Double-Checking**
  - We compare sensor data with lab tests.
  - This ensures our automated data stays accurate.
- **Same Format**
  - Different sensors use different data styles.
  - We convert all to one standard format so all our tools can use any data.

22.2.3 Feature engineering and selection

Now our data are clean, but there is so much of it! Not every piece of information is equally important (Bourgeois, Burgess, & Stuetz, 2001). "Feature Engineering and Selection" is about highlighting what matters most and even creating new, insightful information.

- **What's a Feature?**
  - In data talk, a "feature" is one type of measurement.
  - For water, features are things like pH, temperature, and chlorine level.
  - Each feature tells us something about water quality.
- **Making Super-Features**
  - Sometimes, combining features reveals more.
  - Example: pH ÷ Temperature = "Acidity Stability."

  - This new feature might better predict water safety.
- **Time Matters**
  - Water behaves differently at different times.
  - We create features like "Hour of Day" or "Is Weekend?"
  - This helps spot patterns, like pH changes every Friday.
- **Location Counts**
  - Where water affects its quality
  - We add features like "Distance to Nearest Farm."
  - This links water quality to nearby activities.
- **Speed of Change**
  - How fast something changes can be important.
  - We calculate things like "How Fast Chlorine Drops."
  - Quick changes often mean trouble.
- **Picking What's Important**
  - With so many features, we need to focus.
  - We use smart tools to pick the most telling ones.
  - Some tools even let the computer rank features (Yaroshenko et al., 2020).
- **Expert Knowledge**
  - Water experts guide us too.
  - They might say, "For algae, watch nitrogen and phosphorus."
  - Their wisdom helps us choose the right.
- **Goals Guide Us**
  - What matters changes with our goal.
  - Checking for factory pollution? Metal features are key.
  - Ensuring safe tap water? Chlorine and bacteria matter most.

22.3 Machine learning for water quality analysis

In the world of water quality, that is exactly what ML does. ML is a type of AI that learns from data to make smart decisions or predictions (Zhang, Tooker, & Mueller, 2020). In this section, we will explore how ML helps us analyze water quality, making it easier to ensure our water is safe and clean (Fig. 22.3).

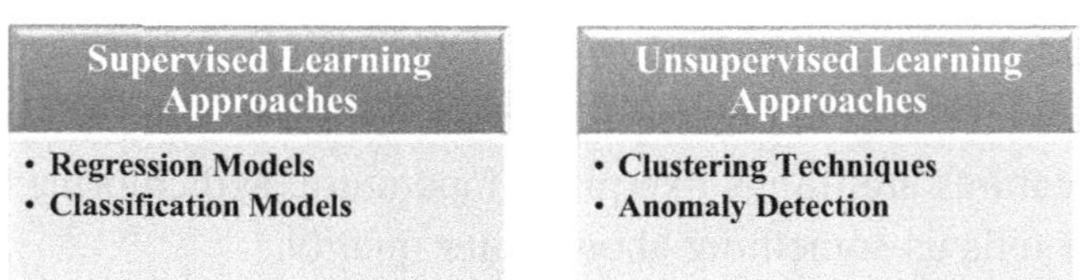

Figure 22.3
Categories of supervised and unsupervised learning approaches.

22.3.1 Supervised learning approaches

Think of supervised learning as teaching a child. You show them many pictures, saying "This is a cat" or "This is a dog." Soon, they can identify cats and dogs on their own (Harrou, Cheng, Sun, Leiknes, & Ghaffour, 2020). Similarly, in supervised learning, we show ML models lots of water quality data, telling them what each data point means. They learn patterns and can then make sense of new data.

22.3.1.1 Regression models

Regression is about predicting numbers. Imagine you are a weather forecaster (Kruse, 2018). You look at today's temperature, humidity, and wind, and then predict tomorrow's exact temperature. That is regression.

- **In Water Quality**
  - **What We Do:** We feed a model year of data about a river—daily rainfall, factory outputs, and temperature.
  - **What It Learns:** The model learns how these factors affect water quality measures like pH or chlorine levels.
  - **What It Predicts:** Given today's data, it predicts tomorrow's exact pH or chlorine level.
- **Why It is Great**
  - **Early Warning:** If it predicts unsafe pH next week, we can act now.
  - **Resource Planning:** Knowing next month's water quality helps plan treatment needs.
  - **Popular Models**
  - **Linear Regression:** Simple, fast, good for clear trends.
  - **Random Forest:** Combines many simple models for better accuracy.
  - **Neural Networks:** Best for complex relationships, like how 10 factors together affect pH.

22.3.1.2 Classification models

Classification is about sorting things into categories. Think of a fruit sorter machine that looks at each fruit and decides "Apple," "Orange," or "Banana." In water quality, we use this to categorize water safety.

- **In Water Quality**
  - **What We Do:** We show the model thousands of water samples labeled "Safe to drink," "Need Treatment," or "Unsafe."
  - **What It Learns:** It learns what makes waterfall into each category.
  - **What It Predicts:** For any new water sample, it says which safety class it belongs to.

- **Why It is Great**
  - **Quick Decisions:** Instantly tells if tap water is drinkable.
  - **Treatment Choices:** Suggest the right treatment based on water type.
  - **Risk Mapping:** Can classify every water source in a region, showing risk areas.
- **Popular Models**
  - **Decision Trees:** Work like a flowchart (if pH > 7 and chlorine <0.5, then "Need Treatment").
  - **Support Vector Machines:** Great for clear-cut decisions (Safe vs Unsafe).
  - **Convolutional Neural Networks:** Can even classify water by looking at its pictures!

22.3.2 Unsupervised learning approaches

Now imagine you are exploring a new planet. You find many rocks but have no idea what is normal. You group similar rocks, figuring out "types" on your own. That is unsupervised learning (Thakur & Devi, 2022). The ML model discovers patterns in water data without being told what to look for.

22.3.2.1 Clustering techniques

Clustering is about finding natural groups. Think of a school lunchroom where kids naturally sit in groups—athletes, musicians, bookworms—without being assigned. In water quality, clustering finds natural water "types."

- **In Water Quality**
  - **What We Do:** We give the model water data from hundreds of sources, without any labels.
  - **What It Learns:** It finds similarities and groups water samples into clusters.
- **What We Discover**
  - **Water Types:** Maybe it finds three clusters—"Mountain Spring," "Urban Runoff," and "Industrial Adjacent."
  - **Common Factors:** We see what makes each cluster unique (high minerals in mountain water).
  - **Source Tracing:** If tap water matches the "Industrial" cluster, we check nearby factories.
- **Popular Techniques**
  - **K-Means:** Simple, fast, and good when you guess how many clusters exist.
  - **DBSCAN:** Better when clusters have odd shapes or you are unsure how many.
  - **Hierarchical Clustering:** Shows relationships (Urban water is closer to Industrial than to Mountain).

22.3.2.2 Anomaly detection

Anomaly detection spots things that do not fit the norm. It is like a bank's fraud system. Most transactions are normal, but a sudden $10,000 purchase in a new country? That is flagged as unusual. In water, we use this to spot quality changes (Ul Haq, Lee, Rizzo, & Perdrial, 2024).

- **In Water Quality**
  - **What We Do:** The model learns what's "normal" for each water source.
  - **What It Learns:** It understands typical patterns—a river's usual pH range or how chlorine changes daily.
- **What It Flags**
  - **Sudden Changes:** If pH spikes beyond normal, it alerts immediately.
  - **Gradual Shifts:** It notices slow changes, like rising metal levels over months.
  - **Odd Combinations:** Normal pH but unusual mineral mix? Flagged!
- **Why It is Great**
  - **Instant Alerts:** Spots issues faster than any human could.
  - **Catches Subtle Changes:** Finds problems we might overlook.
  - **Adapts Over Time:** What is "normal" in summer differs from winter; it adjusts.
- **Popular Techniques**
  - **Isolation Forests:** Quick, good for high-dimensional data (lots of water properties).
  - **Local Outlier Factor:** Great for when "normal" varies across a water system.
  - **Auto encoders:** Advanced neural nets that learn to spot anomalies in complex data.

22.4 AI-driven automation systems

Imagine your home could take care of itself—adjusting the temperature, ordering groceries, and even cooking meals without you having to think about it. That is the idea behind automation. Now, imagine this for something even more important: ensuring our water is always safe and clean. That is what AI-Driven Automation Systems do for water quality management (Palmer & Hatley, 2018). AI is like having a super-smart assistant that can think, learn, and make decisions. In the world of water quality, AI does not just analyze data; it takes action. It can adjust how we treat water, predict issues, and even make complex decisions, all to keep our water safe. Let us explore the different types of AI that make this automation possible (Fig. 22.4).

Figure 22.4
Different analytical techniques.

22.4.1 Expert systems and rule-based approaches

Think about how you decide what to wear each day. You might have simple rules: If it is scorching, wear shorts. If it is raining, take an umbrella. If you have an important meeting, wear nice clothes. You do not overthink; you follow these "If-Then" rules. That is exactly how Expert Systems and Rule-based Approaches work in water quality. Water experts create a list of rules based on their years of experience (Gupta et al., 2018). For example: If the water is too acidic, add a chemical to increase its pH. If chlorine is too low in summer, add more chlorine. If the water gets cloudy after rain, switch to a different filter. The AI system follows these rules automatically (Di Lecce et al., 2017). It checks water data every minute and applies the right rule instantly. This means fast decisions without waiting for human approval. It also means small towns get the same expertize as big cities. And if asked why it added a chemical, it can point to the exact rule. Think of it as capturing a water expert's brain in a computer, making that expertize available 24×7 across many water systems.

22.4.2 Fuzzy logic and fuzzy inference systems

In real life, things are not always black or white. Water is not just "clean" or "dirty"; there are shades in between, like "somewhat clean" or "very dirty." Regular computers struggle with this, but Fuzzy Logic handles it beautifully. We describe water qualities in human terms like "Very Acidic," "Slightly Alkaline," and "Too Low Chlorine." Then we set rules using these terms: "If pH is Slightly Alkaline and Chlorine is Too Low, then Water Quality is good but Needs Attention." Fuzzy Logic translates these human terms into numbers the computer understands. It can handle mixed signals, like pH being great but chlorine being a bit off. This handles complexity since water quality is not just numbers; it is a balance (Meyer, Klein, Fünfrocken, Kautenburger, & Beck, 2019). It is more natural, talking in terms operators understand. And it works better in uncertainties when sensors give unclear readings. Imagine a very experienced water technician who understands the subtle interplay between different qualities. That is Fuzzy Logic in action (Shingare et al., 2024).

22.4.3 Artificial neural networks

Our brains are incredible. A child can learn to recognize safe water just by seeing and tasting many samples. They do not follow strict rules; their brain forms their understanding. Artificial neural networks (ANNs) work similarly, inspired by our brains. We create a computer version of brain cells and connections (Alam, Ahmed, Seefat, & Nahin, 2021). We show this "digital brain" thousands of water samples with all the properties as input and whether it is safe as output. The ANN adjusts its internal connections; learning patterns like high lead often mean low pH or certain bacteria suggesting farm runoff. Once trained, it can instantly assess any new sample. ANNs are great at finding hidden patterns humans miss. They learn and adapt, getting smarter with more data. And they can handle considering 50 + water properties at once. Think of training a water quality savant who develops an intuitive grasp after enough samples (Pan, Leng, & Liu, 2013).

22.4.4 Deep learning models

If an ANN is like a digital brain, then Deep Learning is like a huge, specialized brain. It has many neuron layers, each focusing on different aspects of understanding water in nuanced ways. A Deep Learning model might have layers spotting basics like pH and clarity, then combining those insights with deep layers grasping complex concepts like emerging contaminants. We feed it massive data: years of sensor readings, satellite images, and even social media posts about taste (Oladipo, Akinwumiju, Aboyeji, & Adelodun, 2021). It learns to predict quality days, suggest treatment recipes, and detect unknown pollutants. Special Deep Learning types include Convolutional Neural Networks which assess quality from images; Recurrent Networks that understand sequences and quality changes over time; and Generative Adversarial Networks where one party tries to generate fake data while another learns to spot it—mastering what "normal" water looks like. Deep Learning grasps water like a lifelong expert, with predictive power to foresee issues early (Lowe, Qin, & Mao, 2022).

22.5 Applications and case studies

- **Real-Time Water Quality Monitoring**
  - AI-powered systems with sensor networks for continuous monitoring.
  - Use neural networks, clustering, and anomaly detection.
  - Instant identification of deviations from normal water quality.
  - Enable rapid response to contaminants, spills, etc.
  - **Example:** Singapore's Smart Water Grid.
  - Extensive IoT sensor network integrated with AI analytics.
  - Monitors over 50 water quality parameters.
  - Identifies anomalies, locates sources, and recommends actions.
  - Reduced water losses and boosted public confidence.
- **Predictive Modeling and Forecasting**
  - ML techniques like regression, time-series analysis, and simulations.
  - Ingest meteorological, land use, and historical water quality data.
  - Forecast water quality dynamics under varying scenarios.
  - **Example:** Lake Erie algal bloom prediction.
  - Models predict bloom timing and severity weeks in advance.
  - Enables preemptive strategies—treatment, advisories, nutrient management.
- **AI-Augmented Decision Support Systems**

- Integrated platforms combining multiple AI models.
- Real-time monitoring, predictive simulations, optimization.
- Evaluate "what-if" scenarios: treatment changes, new sources, and land use.
- Use evolutionary algorithms and multiobjective optimization.
- **Example:** Cape Town drought management.
- Modeled emergency strategies: rationing, desalination, and incentives.
- Identified optimal solutions balancing cost, quality, and sustainability.
- Averted "Day Zero," minimized socioeconomic disruption.

- **Advantages**
  - Unprecedented spatiotemporal monitoring resolution.
  - Proactive management via prognostic modeling.
  - Rational decision frameworks for complex trade-offs.
  - Safeguarding water security under escalating stresses.

22.6 Challenges and future directions

While the applications of ML and AI in water quality automation are tremendously promising, several challenges must be addressed to fully realize their potential (Wang et al., 2023). Additionally, ongoing research and development are crucial to push the boundaries of these technologies and ensure their sustainable and responsible deployment (Yang et al., 2022).

22.6.1 Data quality and availability

The performance of ML and AI models is heavily dependent on the quality and quantity of data used for training and validation. In the context of water quality monitoring and management, acquiring high-quality, comprehensive data can be a significant challenge. Sensor networks may suffer from issues such as calibration drift, fouling, or temporary outages, leading to gaps or inconsistencies in the data (Li, Rong, Wang, & Yu, 2021). Additionally, data collection can be hindered by logistical and resource constraints, particularly in remote or underdeveloped regions.

To address these challenges, robust data quality assurance protocols must be implemented, including regular sensor maintenance, cross-validation with laboratory analyses, and advanced data imputation techniques. Furthermore, collaborative efforts among water utilities, research institutions, and government agencies can facilitate the sharing and integration of diverse data sources, contributing to the development of more comprehensive and representative datasets (Prasad et al., 2021).

22.6.2 Model interpretability and explain ability

While ML and AI models can achieve remarkable predictive accuracy, many advanced techniques, such as deep neural networks, are often criticized for their "black box" nature, making it difficult to interpret their decision-making processes (Huang, Ma, Ma, Huangfu, & He, 2021). In water quality management, where decisions can have significant public health and environmental implications, it is essential to ensure transparency and accountability.

Ongoing research efforts are focused on developing interpretable and explainable AI techniques that can provide insights into the reasoning behind model predictions. These methods aim to elucidate the relationships between input features and outputs, identify key contributing factors, and quantify the relative importance of different variables. By demystifying the inner workings of these complex models, water authorities and stakeholders can gain a deeper understanding and build trust in the automated decision-making processes (Zhao et al., 2021).

22.6.3 Integration with traditional methods

ML and AI techniques should not be viewed as a replacement for traditional water quality monitoring and management methods but rather as complementary tools that can enhance existing practices. The challenge lies in seamlessly integrating these advanced computational approaches with well-established analytical techniques, operational protocols, and human expertize (Liakos, Busato, Moshou, Pearson, & Bochtis, 2018). Hybrid models that combine the strengths of traditional mechanistic models with data-driven ML approaches show promise in bridging this gap. Additionally, interactive decision support systems that incorporate both AI-generated recommendations and human expert knowledge can facilitate effective collaboration and leverage the unique strengths of each component.

22.6.4 Ethical and regulatory considerations

The widespread adoption of ML and AI in water quality automation raises ethical and regulatory concerns that must be carefully addressed. Issues such as data privacy, algorithmic bias, and the potential for unintended consequences pose significant challenges that could undermine public trust and hinder the responsible deployment of these technologies. Robust governance frameworks and ethical guidelines must be established to ensure the fair and transparent use of ML and AI in water management (Al-Adhaileh & Alsaade, 2021). These frameworks should address data governance, model auditing, and accountability measures, as well as promote diversity and inclusivity in the development and deployment of these technologies. Furthermore, regulatory bodies must adapt to the rapidly evolving landscape of computational automation, updating existing policies and creating new guidelines to ensure the safe and responsible use of ML and AI in critical infrastructure sectors like water resource management.

22.7 Conclusion

Water is essential for life, but its quality and availability are under threat from population growth, industrialization, and climate change. Ensuring access to safe, clean water for all requires innovative solutions. This chapter explores how ML and AI can revolutionize water quality monitoring, analysis, and management through automation. The chapter discusses the importance of water quality monitoring and the pivotal role of ML and AI in enhancing this process. It covers various ML approaches for water quality analysis, including supervised learning techniques like regression and classification models, as well as unsupervised learning methods such as clustering and anomaly detection. Additionally, the chapter explores AI-driven automation systems that are transforming water quality management, including expert systems, rule-based approaches, fuzzy logic, fuzzy inference systems, ANNs, and deep learning models. Real-world applications are showcased, highlighting the tangible impact of these technologies in continuous monitoring, predictive modeling, and decision support systems. While the potential is immense, challenges in data quality, model interpretability, integration with traditional methods, and ethical and regulatory considerations must be addressed to ensure responsible deployment. Ongoing research and collaboration are crucial to pushing the boundaries of computational automation in water resource management. By harnessing ML and AI, we can unlock new frontiers in safeguarding water, ensuring its availability and quality for generations to come, and fostering a more proactive, resilient, and sustainable approach to water security.

References

Al-Adhaileh, M., & Alsaade, F. (2021). Modelling and prediction of water quality by using artificial intelligence.. *Sustainability*, *13*, 4259. Available from https://doi.org/10.3390/SU13084259.

Alam, R., Ahmed, Z., Seefat, S. M., & Nahin, K. T. K. (2021). Assessment of surface water quality around a landfill using multivariate statistical method, Sylhet, Bangladesh. *Environmental Nanotechnology, Monitoring & Management*, *15*, 100422. Available from https://doi.org/10.1016/j.enmm.2020.100422, ISSN 2215-1532.

Bourgeois, W., Burgess, J. E., & Stuetz, R. M. (2001). On-line monitoring of wastewater quality: A review. *Journal of Chemical Technology and Biotechnology*, *76*(4) 337–348.

Gupta, M., Giaccherini, F., Sridhar, G. R. D., Batstone, D., Santoro, D., & Nakhla, G. (2018). Application of respirometric techniques to determine COD fractionation and biokinetic parameters of sieved wastewater. *Proceedings of the Water Environment Federation*, *2018*, 106–121.

Harrou, F., Cheng, T., Sun, Y., Leiknes, T., & Ghaffour, N. (2020). A data-driven soft sensor to forecast energy consumption in wastewater treatment plants: A case study. *IEEE Sensors Journal*, *21*(4), 4908–4917.

Huang, R., Ma, C., Ma, J., Huangfu, X., & He, Q. (2021). Machine learning in natural and engineered water systems. *Water Research*, *205*117666. Available from https://doi.org/10.1016/j.watres.2021.117666.

Kruse, P. (2018). Review on water quality sensors. *Journal of Physics D: Applied Physics*, *51*(20), 203002.

Di Lecce, V., Petruzzelli, D., Guaragnella, C., Cardellicchio, A., Dentamaro, G., Quarto, A., & Dario, R. (2017). Real-time monitoring system for urban wastewater. In *2017 IEEE workshop on environmental, energy, and structural monitoring systems (EESMS)* (pp. 1–5). IEEE.

Li, L., Rong, S., Wang, R., & Yu, S. (2021). Recent advances in artificial intelligence and machine learning for nonlinear relationship analysis and process control in drinking water treatment: A review. *Chemical Engineering Journal*. Available from https://doi.org/10.1016/j.cej.2020.126673.

Liakos, K., Busato, P., Moshou, D., Pearson, S., & Bochtis, D. (2018). Machine learning in agriculture: A review. *Sensors (Basel, Switzerland), 18*. Available from https://doi.org/10.3390/s18082674.

Lowe, M., Qin, R., & Mao, X. (2022). A review on machine learning, artificial intelligence, and smart technology in water treatment and monitoring. *Water*. Available from https://doi.org/10.3390/w14091384.

Meyer, A. M., Klein, C., Fünfrocken, E., Kautenburger, R., & Beck, H. P. (2019). Real-time monitoring of water quality to identify pollution pathways in small and middle-scale rivers. *Science of the Total Environment, 651*, 2323–2333.

Moretti, A., Ivan, H. L., & Skvaril, J. (2024). A review of the state-of-the-art wastewater quality characterization and measurement technologies. Is the shift to real-time monitoring nowadays feasible? *Journal of Water Process Engineering, 60*105061. Available from https://doi.org/10.1016/j.jwpe.2024.105061.

Oladipo, J., Akinwumiju, A., Aboyeji, O., & Adelodun, A. (2021). Comparison between fuzzy logic and water quality index methods: A case of water quality assessment in Ikare community, Southwestern Nigeria. *Environmental Challenges, 3*, 100038. Available from https://doi.org/10.1016/j.envc.2021.100038.

Palmer, M., & Hatley, H. (2018). The role of surfactants in wastewater treatment: Impact, removal, and future techniques: A critical review. *Water Research, 147*, 60–72.

Pan, J., Leng, T., & Liu, Y. (2013). Shifosi reservoir water environmental assessment based on grey clustering. *Advanced Materials Research, 610*, 857–861.

Prasad, D., Kumar, P., Venkataramana, L., Prasannamedha, G., Harshana, S., Srividya, S., ... Indraganti, S. (2021). Automating water quality analysis using ML and auto ML techniques. *Environmental Research*111720. Available from https://doi.org/10.1016/j.envres.2021.111720.

Shingare, S. P., Mandake, M. B., Patil, R. N., Thokal, A. G., Qureshi, Y., & Thakar, C. M. (2024). Applicability of machine learning in waste water quality detection. In *2024 IEEE international conference on computing, power and communication technologies (IC2PCT), Greater Noida, India* (pp. 525–530). Available from https://doi.org/10.1109/IC2PCT60090.2024.10486814.

Shoushtarian, F., & Negahban-Azar, M. (2020). Worldwide regulations and guidelines for agricultural water reuse: A critical review. *Water, 12*(4), 971.

Thakur, A., & Devi, P. (2022). A comprehensive review on water quality monitoring devices: Materials advances, current status, and future perspective. *Critical Reviews in Analytical Chemistry, 54*(2), 193–218. Available from https://doi.org/10.1080/10408347.2022.2070838.

Ul Haq, I., Lee, B. S., Rizzo, D. M., & Perdrial, J. N. (2024). An automated machine learning approach for detecting anomalous peak patterns in time series data from a research watershed in the northeastern United States critical zone. *Machine Learning with Applications, 16*, 100543. Available from https://doi.org/10.1016/j.mlwa.2024.100543, ISSN 2666-8270.

United Nations. (2015). Sustainable development goals. United Nations, S-1018, New York, NY, 10017.

UN-Water. (2021). Summary progress update 2021: SDG 6—water and sanitation for all. Geneva, Switzerland.

Wang, Y., Cheng, Y., Liu, H., Guo, Q., Dai, C., Zhao, M., & Liu, D. (2023). A review on applications of artificial intelligence in wastewater treatment. *Sustainability*. Available from https://doi.org/10.3390/su151813557.

Yang, C., Barrott, W., Busch, A., Mehrotra, A., Madden, J., & Daigger, G. T. (2019). How much data is required for a robust and reliable wastewater characterization? *Water Science and Technology, 79*(12), 2298–2309.

Yang, L., Driscol, J., Sarigai, S., Wu, Q., Lippitt, C., & Morgan, M. (2022). Towards synoptic water monitoring systems: A review of AI methods for automating water body detection and water quality monitoring using remote sensing. *Sensors (Basel, Switzerland), 22*. Available from https://doi.org/10.3390/s22062416.

Yaroshenko, I., Kirsanov, D., Marjanovic, M., Lieberzeit, P. A., Korostynska, O., Mason, A., & Legin, A. (2020). Real-time water quality monitoring with chemical sensors. *Sensors, 20*(12), 3432.

Zhang, W., Tooker, N. B., & Mueller, A. V. (2020). Enabling wastewater treatment process automation: Leveraging innovations in real-time sensing, data analysis, and online controls. *Environmental Science: Water Research & Technology*, *6*(11), 2973–2992.

Zhao, S., Zhang, S., Liu, J., Wang, H., Zhu, J., Li, D., & Zhao, R. (2021). Application of machine learning in intelligent fish aquaculture: A review. *Aquaculture (Amsterdam, Netherlands)*. Available from https://doi.org/10.1016/J.AQUACULTURE.2021.736724.

Index

Note: Page numbers followed by "*f*" and "*t*" refer to figures and tables, respectively.

D

T

X

Y

Z